Rational Quadratic Forms

J. W. S. Cassels, F.R.S.
Department of Pure Mathematics
and Mathematical Statistics
University of Cambridge

Dover Publications, Inc.
Mineola, New York

Bibliographical Note

This Dover edition, first published in 2008, is an unabridged republication of the work originally published by Academic Press, London and New York, in 1968.

Library of Congress Cataloging-in-Publication Data

Cassels, J. W. S. (John William Scott)
Rational quadratic forms / J.W.S. Cassels.
p. cm.
Originally published: New York, N.Y.: Academic Press
ISBN-13: 978-0-486-46670-5
ISBN-10: 0-486-46670-1
1. Forms, Quadratic. I. Title.

QA243.C25 2008
512.7'4—dc22

2008015460

www.doverpublications.com

Preface

Obschon die rationalen quadratischen Formen zu den ältesten Gebieten der Zahlentheorie überhaupt gehören, haben sie doch bisher nur im binären Fall eine einigermassen abschliessende Behandlung erfahren. Was darüber hinaus angeboten wird, befindet sich in einem chaotischen unbefriedigenden Zustand, wie die enzyklopädische Darstellung von Bachmann deutlich erkennen lässt. Nirgends bemerkt man, dass leitende Gesichtspunkte in den Vordergrund gestellt und nebensächliche ihnen untergeordnet werden.

H. Brandt. Über Stammfaktoren bei ternären quadratischen Formen. *Ber. Verh. sächs. Akad. Leipzig* (Math.-Nat. Kl.) **100** (1952), Heft 1 (24 pp.)

. . . im Ganzen ist die Theorie nach wie vor, wie Brandt ganz richtig sagt, in einem chaotischen Zustand.

B. L. van der Waerden. Die Reduktionstheorie der positiven quadratischen Formen. *Acta Mathematica* **96** (1956), 265–309.

The above quotations may give a false impression. This is not a treatise in the high tradition of Bachmann, Eichler, Watson and O'Meara. Its aims are much more modest: to indicate some of the major themes of the classical arithmetical theory of quadratic forms in the light of our present knowledge but from a totally elementary point of view.

The discussion in this preface is intended for the reader who is acquainted with some of the theory already and wants to learn how the book fits in with his knowledge and prejudices. The débutant should proceed to the Introduction, which gives a more motivated discussion of the contents. When he has read the book he might return to this preface with profit.

The material of the book is largely nineteenth century but the treatment is structured by two twentieth century insights. The first, which seems to have come to its full recognition in the work of Hasse and Witt, is that the theory of forms over fields is logically simpler and more complete than that over rings. It is therefore appropriate, contrary to what seemed natural earlier, to study forms with rational coefficients and under rational equivalence before attacking integral forms and integral equivalence. The second major insight, due to Hensel and Hasse, is the perspective introduced by the p-adic view-point. This unveils the majestic simplicity of the logical structure: in particular it banishes for ever the need for the plethora of multifarious "characters" and "invariants" which earlier (and some later) authors use to distinguish forms which are p-adically inequivalent. The p-adic numbers are as natural as the reals—indeed it can plausibly be argued that they are logically simpler and that it is only by indoctrination that we feel that the reals are more familiar. However, as the p-adic numbers are not yet as well-known as they should be to the broad audience to which this book is addressed, no knowledge of them has been presupposed.

In our treatment attention is firmly focused on the forms themselves and not on structures associated with them. It could be claimed that the third major insight achieved in the twentieth century is that quadratic spaces are homogeneous spaces over their orthogonal groups and that much of the theory is an aspect of the general theory of linear algebraic groups. The view from that Pisgah is not, however, on the itinerary of this package tour. Nor do we use the theory of algebras apart from the very little which is needed at a comparatively late stage to discuss the spin group and spinor genera: and what is needed is developed from scratch.

The philosophy of this book, as of its author, is that new concepts are best assimilated in the simplest contexts. The treatment is confined to forms over the rational numbers or the rational integers (and their p-adic completions). Much of what is done extends simply and naturally to algebraic number fields: if the reader is familiar with the theory of such fields he will have little difficulty in seeing when this is so and in making the necessary modifications, if not, unnecessary obstacles would be put in his path by the more general framework.

There is, however, one place where the theory of forms over the rationals can be developed more simply than over an algebraic number-field, namely the Strong Hasse Theorem for forms in three variables. This is a special case of a general theorem about norms in cyclic extensions of algebraic number fields (a theorem in class field theory). In the rational field case, however, there is an easy proof on different lines (which appears to fail by a provocatively narrow margin for algebraic number fields).

Some of the proofs, in particular the one mentioned, use ideas from the geometry of numbers. As is explained in the text, for the major applications one could almost as easily have used Hermite's theorem about the minimum of a definite form. The author feels that the geometry of numbers gives a more perspicuous treatment, though this is doubtless a matter of taste. The basic ideas of the geometry of numbers are so natural that they should be part of the repertoire of every working mathematician. The author being a realist, however, they are developed briefly from scratch in the slightly non-standard shape best fitted for the applications. This has the advantage that the techniques are available in some peripheral passages where recourse to Hermite's theorem would be much less natural.

The one non-elementary theorem which is presupposed is Dirichlet's theorem about the existence of primes in arithmetic progressions. This is a clear-cut and simple proposition whose truth may be accepted on trust, even if one does not know the (comparatively elementary) analytic proof. Its use makes the proofs of the key theorems substantially more straightforward and, incidentally, as Dirichlet's theorem generalizes to algebraic number fields, so do the proofs of this text. The use of Dirichlet's Theorem may however seem paradoxical since it is utilized for theorems (existence of rational and integral forms with prescribed local behaviour) which had been proved by Gauss in the *Disquisitiones* before Dirichlet was born. In fact Atle Selberg used precisely these results from the theory of quadratic forms for his celebrated "elementary" proof of Dirichlet's Theorem (which was originally proved using analytic methods).

Binary forms enjoy special properties: in particular there is "composition". We reserve a discussion of these properties until late in the book, after the general theory is completed. This is the theory which was used by Gauss to prove the existence theorems mentioned in the last paragraph and so we retrospectively eliminate the use of Dirichlet's theorem from them. There is one further point where we use Dirichlet's theorem: the Strong Hasse Theorem for forms in four variables. It is shown that here, too, Dirichlet's theorem can be eliminated. It might be remarked that if one were developing the theory over a general algebraic number field the existence theorems for binary forms can be regarded as results about norms of quadratic extensions and, as said, are special cases of more general theorems. The method given for the elimination of Dirichlet's theorem from the Strong Hasse for quaternary forms does not, apparently, generalize because it invokes the elementary proof of the Weak Hasse discovered recently by Milnor and later, independently, by Conway. In the context of forms over algebraic number fields, however, there is a neat proof [O'Meara (1959) p. 187] which works by making an algebraic extension of the ground field: this appears to be the only place where we have been deprived of such a tactic by the decision to work over the rationals alone.

Perhaps something should be said about the definition used of an integral quadratic form. As is well known, Gauss in effect defined a form to be integral if it is of the shape $f(\mathbf{x}) = \mathbf{x}'\mathbf{A}\mathbf{x}$, where $\mathbf{A}$ is a symmetric matrix with integral coefficients; and in a well-known review of a book of the little-known mathematician Seeber, commented on the latter's simple-mindedness in adopting the definition that $f(\mathbf{x})$ is integral if it is a function with integral coefficients. Thus $x_1^2 + x_1x_2 + x_2^2$ is integral in the sense of Seeber but not in that of Gauss. The definition of Gauss (the "classical definition") was followed by subsequent writers; and Brandt in the article already quoted goes on to remark that Gauss thereby set back progress by 100 years. Brandt exaggerates, of course, but there is a point of view from which Seeber and Brandt are right. In particular the "non-classical" definition of Seeber makes better sense over fields of characteristic 2. For the topics treated in this book, however, the choice of definition of integrality is of little importance. What matters is the kind of equivalence under consideration and instead of referring to integral forms (of either kind) we could have formulated most of the theorems in terms of the behaviour of forms with rational coefficients under integral equivalence. Mainly because of the convenience of the matrix notation we have generally chosen to spite Brandt and use the "classical" definition. This means that in some places the 2-adic case looks rather more anomalous than it might do otherwise, but 2 is anomalous anyway (all primes are odd—and 2 is the oddest prime of all!). Perhaps the choice of the definition for integrality might have required more thought if we had been treating algebraic number-fields in general. There is, however, one chapter where the non-classical definition appears to offer definite advantages, namely that on composition of binary forms and so we use it there.

The spirit of the book should now be reasonably clear from the "Leitfaden" and from what is said above but perhaps a further brief discussion is in order. After the introduction there is a section on forms over fields in general (characteristic $\neq 2$) including the vital "Witt's Lemma". There follow forms over the p-adic field $\mathbf{Q}_p$ and then forms over the rational field $\mathbf{Q}$. Here the relation with the $\mathbf{Q}_p$ is very tight. Things happen over $\mathbf{Q}$ if and only if they happen over all $\mathbf{Q}_p$: and there are forms with any prescribed local ($=p$-adic) behaviour subject to some obviously necessary conditions. Then the same sequence is repeated for integral forms. First the theory over a general principal ideal domain, then over the p-adic integers, and finally over the rational integers. For integral forms the relation between the local and the global is not quite so tight as for rational forms and the notion of "genus" measures the discrepancy. Two forms are in the same genus if it cannot be shown by purely local methods that they are not integrally equivalent.

The notion of genus does not however exhaust the possibilities of the paradigm "local $\rightarrow$ global". Two forms in the same genus are rationally equivalent even though they need not be integrally equivalent and we can examine the new situation (i.e. including a given rational equivalence) locally as well as globally. Here the spin group (the 2-sheeted simply connected covering group of the orthogonal group) plays a rôle and must be introduced. Local considerations now lead to a new classification the "spinor genus" intermediate between the integral equivalence class and the genus. For indefinite forms in $\geqslant 3$ variables it turns out that the spinor genus and the equivalence class coincide, so again the global situation is entirely determined by the local ones. In this discussion we have, of course, by "indefinite" meant indefinite in the usual sense, that is with respect to the real embedding of the rationals or, in the language appropriate to this book "at the infinite prime". The theory can be developed in such a way that an ordinary p-adic field $\mathbf{Q}_p$ plays the rôle of the reals and this theory applies to forms which are definite in the ordinary sense. In accordance with our minimalist pedagogical philosophy we do not initially develop the theory in this generality, though we have occasion to mention it later.

We now turn to another classical theme, namely the reduction of definite forms. This is the theory introduced by Hermite and in a better form by Minkowski. It gives a means of selecting one (in general) canonical representative out of an integral equivalence class. More generally, the theory of reduction works with forms with real coefficients and determines a set of forms, defined (in the case of Minkowski reduction) by a finite set of linear inequalities. The investigation of this situation is closely bound up with the arithmetical structure of the integral unimodular group.

There is also a theory of reduction for indefinite forms. It is substantially more sophisticated, relies on the corresponding theory for definite forms and is much more arithmetical. An indefinite form can be equivalent to several "reduced forms": indeed a general real form can be integrally equivalent to infinitely many reduced ones and it is quite a delicate theorem (due to Siegel) that for an integral form there are only finitely many. The reduction of indefinite integral forms is closely bound up with the group of integral automorphs of an individual form and the way that this group is embedded in the group of real automorphs (the corresponding real orthogonal group). In low dimensions this situation was extensively investigated in the nineteenth century in connection with non-Euclidean geometry and discontinuous groups [see for example the massive treatises of Klein–Fricke (1890) and Fricke–Klein (1897)]. This book attempts to explain these connections in a unified manner taking advantage of the new perspectives and techniques which have subsequently emerged.

The next chapter deals with the composition of binary forms and its contents have already been sufficiently described. It could well have been placed earlier and its position is largely a matter of convenience. Much of the material on binary forms could come naturally in a variety of places. As it happens, reduction and automorphs of binary forms, which might well have come in this chapter, had already been described in the chapters on reduction and automorphs, which were actually the first two chapters of the book to be written.

The two appendices are of a different nature from the rest of the book. They do not attempt a full exposition but are, rather, surveys of areas which could not be covered, partly for reasons of space and partly because of the author's incompetence. Methods of analysis were quite early applied to quadratic forms, for example the class-number formula of Dirichlet and the weight-formulae (Massformeln) of Eisenstein and Minkowski. These and much more are subsumed in the grandiose formula of Siegel which summarizes quantitatively the relationship between local and global which is described qualitatively in this book. The formula of Siegel has itself been put in the framework of the arithmetic of algebraic groups and received a succinct formulation ("the Tamagawa number of an orthogonal group is 2"). Another analytical theory with implications for quadratic forms is that of modular forms. This gives, in particular, relations between the numbers of integral representations of integers by different quadratic forms which are, apparently, beyond the reach of elementary methods. (Something can, however, be done elementarily. It is discussed in Eichler's book, but not here). All this is surveyed briefly in Appendix B.

There are certain results on definite forms which make use of a variety of tools and which do not fall naturally into the Procrustean structure of this book. They are discussed in Appendix A.

This book is concerned only with that part of the theory of quadratic forms which is relevant to forms over the rational numbers. Recently there has been a renaissance of the study of quadratic forms over general fields largely stemming from Pfister's discovery of some quite unexpected results with simple and elegant proofs. There are accounts in Lam (1973), Lorenz (1970) and Scharlau (1969). It should also be mentioned that there is a theory of quadratic forms over more general rings which has applications in topology and which is contained in the framework of algebraic K-theory.

This is a book of mathematics, not of the history of mathematics. I have attached some names to theorems when this seems appropriate, especially when the theorems are habitually referred to by those names. Otherwise I have referred to original papers etc. only in an unsystematic and haphazard way. In general there is much that I have learned in the more than 30 years I have been interested in quadratic forms without my recollecting the source. If it might not have caused confusion in the trade, I could have chosen for this book the title of Field-Marshal Wavell's anthology "Other men's flowers". It has given me great pleasure to arrange those flowers in a bouquet to display their beauties to the best advantage. I hope that some of this pleasure will communicate itself to the reader.

Cambridge J.W.S.C.
October 1978

Acknowledgements

Professor G. L. Watson, Professor P. M. Cohn, Dr J. H. Conway and Dr S. J. Patterson made a number of helpful comments on the first draft, and Professor M. Kneser suggested some substantial improvements. Mrs B. Sharples and Mrs D. McClelland prepared and updated a clear and accurate typescript from my habitually illegible manuscript, Dr M. R. O'Donohoe programmed the Cambridge University computer to plot Figures 12.1 and 13.1, and Miss P. Hughes drew the others. My wife helped valiantly with the checking of the proofs. The staff of Academic Press and their printers have been helpful and efficient. I should like to express my gratitude to them all, and also to Professor P. M. Cohn for accepting the book of the London Mathematical Society's distinguished Monographs series.

Leitfaden

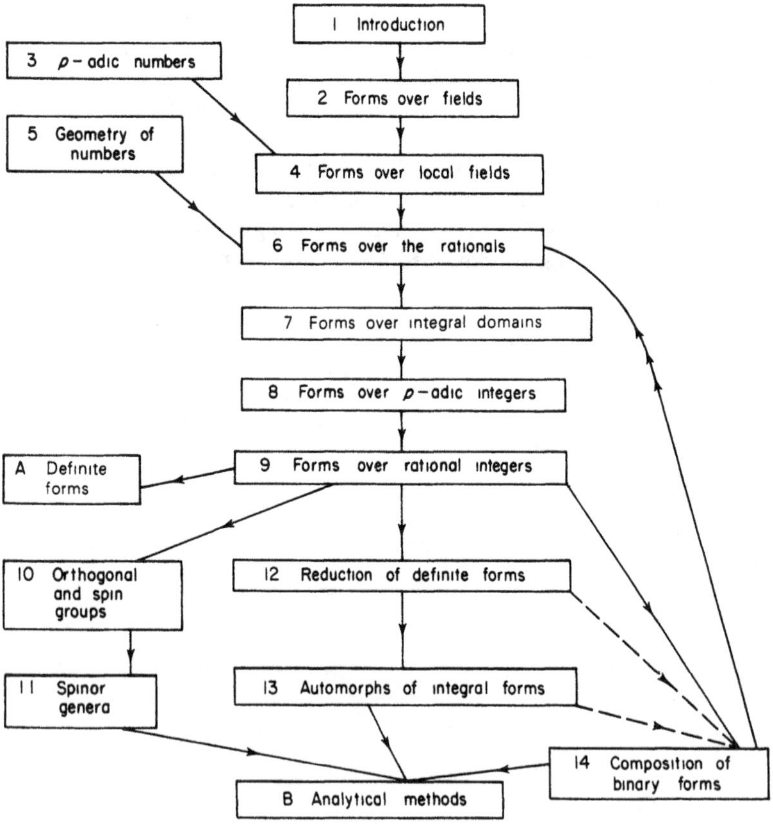

Note Dotted lines indicate that only the material about binary forms at the beginning of Chapters 12 and 13 is relevant to Chapter 14

Contents

CHAPTER 1

Introduction

1. INTRODUCTION

The arithmetical† theory of quadratic forms is almost as old as mathematics itself [e.g. Dickson (1919), vol. 2]. The solutions in integers of the equation

$$x^2 + y^2 - z^2 = 0 \tag{1.1}$$

give the right-angled triangles with integral sides. The equation

$$x^2 + y^2 - 2z^2 = 0, \tag{1.2}$$

which expresses that the squares x^2, z^2, y^2 are in arithmetic progression, has also a very respectable pedigree.

The equations (1.1) and (1.2) are homogeneous, so there is no essential difference between integral and rational solutions. If $(x, y, z) = (a, b, c)$ is a rational solution, say of (1.1), then there is an integer $l \neq 0$ such that la, lb, lc are integral and so give an integral solution. For inhomogeneous problems the difference between integral and merely rational solutions is important. For example if C is a positive integer it is trivial to find rational solutions of

$$x^2 - Cy^2 = 1. \tag{1.3}$$

The Hindu mathematicians Brahmegupta in the 7th and Bhascara in the 12th century made non-trivial contributions to the solution of (1.3) and more general equations in integers. Independently (1.3) was much considered by Fermat and his contemporaries. Fermat said in 1657 that he could prove that (1.3) always has an integral solution with $y \neq 0$ whenever $C > 0$ is not a

† We prefer "arithmetic" and "arithmetical" to the clumsy periphrasis "theory of numbers" and the barbarism "number-theoretical".

perfect square†: and a formal proof was published by Lagrange in 1769. [The equation is known as "Pell's equation" but this is due to a historical confusion by Euler: see Dickson (1919) or Whitford (1912).]

Again, the problem of representing a positive integer n as the sum of 2, 3 or 4 integral squares has a long and tangled history. Euler showed in 1749 that a necessary and sufficient condition that the integer $n > 0$ be representable in the shape

$$x^2 + y^2 = n \tag{1.4}$$

with integral x, y is that no prime $q \equiv 3 \pmod 4$ divide n to an odd power. In 1798 Legendre‡ showed that there is a representation

$$x^2 + y^2 + z^2 = n \tag{1.5}$$

except when $n = 4^a(8b + 7)$ for some integers a, b. Lagrange showed in 1772 that every positive integer is the sum of four integral squares:

$$x^2 + y^2 + z^2 + t^2 = n. \tag{1.6}$$

In every case the result had long been conjectured [again, see Dickson (1919) for the history].

Let us now look more generally at the problem of whether or not an integer is represented integrally by an integral quadratic form§. An obvious necessary condition is given by considerations of reality: it is hopeless to try to find an integral solution (x, y) of (say)

$$x^2 + y^2 = -3 \tag{1.7}$$

since there is not even a real solution. Other necessary conditions are given by congruence considerations. Any integer x satisfies

$$x^2 \equiv 0 \text{ or } 1 \pmod 4 \tag{1.8}$$

and so if the integer a can be represented as the sum of two squares we must have $a \not\equiv 3 \pmod 4$. A further type of congruence condition¶ is exemplified by the equation

$$x^2 + y^2 = 21. \tag{1.9}$$

† This condition is clearly necessary. For if $C = c^2$ then $x^2 - Cy^2 = (x + cy)(x - cy)$, so (1.3) implies $x + cy = x - cy = \pm 1$, so $y = 0$.

‡ *Essai d'une theorie des nombres* (Paris, anno VI). His proof is somewhat elaborate. Gauss independently found what is essentially the proof given in this book and gives it in his *Disquisitiones Arithmeticae* (Lipsiae, 1801).

§ The meaning of these terms should be clear from the context. There are formal definitions in Section 2.

¶ This type of condition occurs only for binary forms as we shall see.

Here $21 \equiv 1 \pmod 4$ but nevertheless there is no solution since (1.9) implies

$$x^2 + y^2 \equiv 0 \pmod 3,$$

and hence $x \equiv y \equiv 0 \pmod 3$, a contradiction.

The reality and congruence conditions are not, however, enough to ensure representability. There is clearly no integral representation

$$x^2 + 82y^2 = 2 \tag{1.10}$$

although there is a real representation and it can be shown that the congruence

$$x^2 + 82y^2 \equiv 2 \pmod M$$

has an integral solution (x, y) for every positive integer M. Again, the equation

$$x^2 - 82y^2 = 2$$

has a real solution and is soluble as a congruence to every modulus, but, rather less trivially, there are no integral solutions (cf. Chapter 13, Section 3).

Let us now turn to the homogeneous equation

$$f(x_1, \ldots, x_n) = 0 \tag{1.11}$$

where f is an integral quadratic form (e.g. (1.1) or (1.2)). There is always the trivial solution $x_1 = \ldots = x_n = 0$ and when we say that (1.11) is soluble we shall always mean that a non-trivial solution exists. If (1.11) has a non-trivial solution in integers then we can always divide out by the greatest common divisor of $x_1, \ldots, x_n$ and so ensure that

$$\text{g.c.d.}(x_1, \ldots, x_n) = 1. \tag{1.12}$$

Such a solution is called *primitive*. Again there are necessary reality and congruence conditions but the formulation of the latter requires a little care. The congruence

$$f(x_1, \ldots, x_n) \equiv 0 \pmod M \tag{1.13}$$

has always the trivial solution $x_1 \equiv \ldots \equiv x_n \equiv 0 \pmod M$. We shall say that $(x_1, \ldots, x_n)$ is a primitive solution of (1.13) if

$$\text{g.c.d.}(x_1, \ldots, x_n, M) = 1 \tag{1.14}$$

For example

$$x^2 + y^2 \equiv 0 \pmod 9$$

has non-trivial solutions but no primitive solution. It is now a fundamental theorem that if (1.11) has a non-trivial real solution and if (1.13) has a primitive solution for every positive integer M then (1.11) has a non-trivial solution. [Theorem 1.1 of Chapter 6.]

One of the major themes of this book, perhaps *the* major theme, will be the information that can be derived from reality and congruence considerations. However, the manipulation of congruences is exceedingly tiresome, essentially because the ring† $\mathbf{Z}$ (mod M) may have divisors of zero. There is an alternative formulation which is much more suitable, namely in terms of p-adic numbers. The properties of p-adic numbers will be developed from scratch in Chapter 3, so we give only a brief preliminary explanation here. Let p be some fixed prime number. Then any rational number $a \neq 0$ can be written in the shape

$$a = p^{\alpha}u/v \tag{1.15}$$

where α is an integer (positive, negative or zero) and u, v are integers not divisible by p: the representation is unique if we require that $v > 0$ and that u, v are coprime, but in any case α is uniquely determined by a. We define a function $| \; |_p$ (the *p-adic valuation*) on the field $\mathbf{Q}$ of rational numbers by putting

$$|a|_p = p^{-\alpha}$$

for $a \neq 0$ given by (1.15) and, further

$$|0|_p = 0.$$

Then

(i) $|a|_p \geqslant 0$, with equality only for $a = 0$.

(ii) $|ab|_p = |a|_p |b|_p$.

(iii) $|a + b|_p \leqslant \max\{|a|_p, |b|_p\}$,

as may be readily verified (Chapter 3, Section 1).

The property (iii) trivially implies the "triangle inequality"

(iii)* $|a + b|_p \leqslant |a|_p + |b|_p$.

We now recall that the ordinary absolute value $| \; |$ satisfies (i), (ii) and (iii)*. One way of obtaining the real numbers $\mathbf{R}$ is by "completing" the rationals $\mathbf{Q}$ with respect to the absolute value $| \; |$. Entirely analogously we can complete $\mathbf{Q}$ with respect to $| \; |_p$: the field so obtained is called the field of p-adic numbers and denoted by $\mathbf{Q}_p$. To emphasize the analogy we shall denote the ordinary absolute value by $| \; |_\infty$ and write $\mathbf{R} = \mathbf{Q}_\infty$. The use of the symbol ∞ is conventional for historical reasons which need not detain us here. The analogy between $\mathbf{R} = \mathbf{Q}_\infty$ and the $\mathbf{Q}_p$ is not complete since $| \; |_\infty$ does not satisfy (iii): but we shall find it suggestive.

In terms of p-adic numbers the theorem about the solubility of (1.11) which we enunciated above in terms of congruences may be reformulated succinctly as follows. Suppose that (1.11) has a non-trivial solution in every $\mathbf{Q}_p$ ($p =$ prime) and in $\mathbf{Q}_\infty$. Then there is a non-trivial solution in $\mathbf{Q}$.

† Here $\mathbf{Z}$ denotes the ring of rational integers. There is a list of commonly employed notations on p. 409.

We close this introduction by noting that Pell's equation (1.3) raises other questions. There is always the integral solution $(x, y) = (1, 0)$. Hence what is needed is a quantitative theory about the number of solutions. As a matter of fact, Pell's equation plays a very special role in the theory because it is connected with the automorphs of binary forms [Chapter 13, Section 3].

2. BASIC NOTIONS

Let I be a ring† in a field k of characteristic $\neq 2$. We allow the possibility $I = k$. By a *quadratic form* $f = f(\mathbf{x})$ in the n variables $\mathbf{x} = (x_1, \ldots, x_n)$ we shall mean a function

$$f(\mathbf{x}) = \sum_{i,j} f_{ij} x_i x_j = f_{11}x_1^2 + 2f_{12}x_1x_2 + \ldots \tag{2.1}$$

where

$$f_{ij} = f_{ji} \in k \qquad (1 \leqslant i, j \leqslant n). \tag{2.2}$$

We shall sometimes say that the form f has *dimension* n. The form f *represents* an element c of k *over* I if there is some $\mathbf{b} = (b_1, \ldots, b_n) \in I^n$ [that is, with $b_j \in I$ $(1 \leqslant j \leqslant n)$] such that

$$f(\mathbf{b}) = c. \tag{2.3}$$

In the particular case when k is the rational field $\mathbf{Q}$ we shall speak of *rational representations* if $I = \mathbf{Q}$ and of *integral representations* if $I = \mathbf{Z}$ (the ring of integers).

More generally, a form $f(\mathbf{x})$ in n variables *represents the form* $g(\mathbf{y})$ in m variables *over* I if there are $\mathbf{b}_1, \ldots, \mathbf{b}_m \in I^n$ such that

$$f(y_1\mathbf{b}_1 + \ldots + y_m\mathbf{b}_m) = g(y_1, \ldots, y_m) = g(\mathbf{y}) \tag{2.4}$$

for variables $\mathbf{y}$. Here we have used vector notation, so

$$y_1\mathbf{b}_1 + \ldots + y_m\mathbf{b}_m = (y_1b_{11} + y_2b_{12} + \ldots + y_mb_{1m}, \ldots, y_1b_{n1} + \ldots + y_mb_{nm}), \tag{2.5}$$

where

$$\mathbf{b}_j = (b_{1j}, b_{2j}, \ldots, b_{nj}). \tag{2.6}$$

Clearly f represents over I every $c \in k$ which is so represented by g.

We say that two forms f, g in the same number of variables are *equivalent over* I (or *I-equivalent*) if each represents the other. Clearly I-equivalence is an equivalence relation in the technical sense (reflexive, symmetric and transitive). We may therefore speak of an I-equivalence class of quadratic

† We insist that $1 \in I$. (All rings are "unitary").

forms. The remarks in the preceding paragraph imply that I-equivalent forms I-represent precisely the same elements of k. It thus makes sense to talk of I-representability by an I-equivalence class.

We now introduce matrix notation. The transpose of a matrix $\mathbf{T}$ will be written $\mathbf{T}'$. It is conventional to consider $\mathbf{x}$ as a column vector (though out of consideration to the printer we shall continue to write the components in a row). We can thus write (2.1) in the shape

$$f(\mathbf{x}) = \mathbf{x}'\mathbf{F}\mathbf{x}, \tag{2.7}$$

where $\mathbf{F}$ is the square symmetric matrix

$$\mathbf{F} = (f_{ij}). \tag{2.8}$$

The *determinant* $d(f)$ of the form f is by definition the determinant $\det(\mathbf{F})$ of $\mathbf{F}$. We say that f is *singular* if $d(f) = 0$: otherwise it is *non-singular* or *regular*. We shall concentrate attention on the regular forms since, at least in the cases of interest to us, the consideration of singular forms can be reduced to that of regular forms in a smaller number of variables [Chapter 2, Section 6, and Chapter 7, Section 4].

In matrix notation (2.4) becomes

$$\mathbf{G} = \mathbf{B}'\mathbf{F}\mathbf{B}, \tag{2.9}$$

where $\mathbf{F}, \mathbf{G}$ are the symmetric matrices corresponding to f, g and

$$\mathbf{B} = (b_{ij})_{1 \leqslant i \leqslant n,\, 1 \leqslant j \leqslant m} \tag{2.10}$$

is a matrix with $b_{ij} \in I$. Suppose, now, that f, g are equivalent forms, so that $m = n$ and in addition to (2.9) we have

$$\mathbf{F} = \mathbf{C}'\mathbf{G}\mathbf{C} \tag{2.11}$$

for some matrix $\mathbf{C}$ with entries in I. Then (2.9), (2.11) give respectively

$$d(g) = (\det \mathbf{B})^2\, d(f) \tag{2.12}$$

$$d(f) = (\det \mathbf{C})^2\, d(g) \tag{2.13}$$

It follows that if one of the forms f, g is singular so is the other, and so it makes sense to speak of a singular equivalence class or a regular equivalence class. Further, if the forms f, g are regular we have

$$(\det \mathbf{B})^2\ (\det \mathbf{C})^2 = 1.$$

Hence $\det \mathbf{B} \in U$, where U is the group of *units* (= invertible elements) of I:

$$U = \{u : u \in I, u^{-1} \in I\}. \tag{2.14}$$

There is a converse.

LEMMA 2.1. *A necessary and sufficient condition that the regular forms f, g in n variables be I-equivalent is that*

$$\mathbf{G} = \mathbf{B}'\mathbf{F}\mathbf{B} \tag{2.15}$$

for some matrix $\mathbf{B}$ *with entries in I and with*

$$\det \mathbf{B} \in U \tag{2.16}$$

Proof. We have already seen that the condition is necessary. To prove sufficiency it is enough to note that (2.16) implies that the matrix $\mathbf{C} = \mathbf{B}^{-1}$ has entries in I and that $\mathbf{F} = \mathbf{C}'\mathbf{G}\mathbf{C}$.

For the rest of this section we consider the case when I is strictly smaller than k. A typical case is $k = \mathbf{Q}$, $I = \mathbf{Z}$ and so we shall speak of the elements of I as the integers. We shall say that the quadratic form $f(\mathbf{x})$ is *integer-valued* if $f(\mathbf{b}) \in I$ for all $\mathbf{b} \in I^n$. Clearly being integer-valued is preserved under I-equivalence. In terms of the coefficients f_{ij} it is readily verified that a necessary and sufficient condition for f to be integer-valued is

$$f_{ii} \in I, \qquad 2f_{ij} \in I \qquad (1 \leqslant i \leqslant j \leqslant n). \tag{2.17}$$

We shall say that f is *classically integral* if

$$f_{ij} \in I \qquad (1 \leqslant i \leqslant n, 1 \leqslant j \leqslant n),$$

that is, if the elements of the corresponding matrix $\mathbf{F}$ are in I. By (2.9) and (2.11) all the forms of an I-equivalence class are classically integral if one of them is, and so we may speak of the equivalence class as being classically integral. We note that classically integral forms are integer-valued and that if $f(\mathbf{x})$ is integer-valued then $2f(\mathbf{x})$ is classically integral. Hence the two concepts coincide whenever 2 is a unit of I.

When $I = \mathbf{Z}$ there has been a long and often acrimonious debate as to the correct definition of an integral quadratic form, some preferring integer-valuedness and some (notably Gauss) preferring what we have called classical integrality. In fact each concept has its advantages in certain contexts: but often both definitions are equally convenient. Indeed it is often better not to make integrality conditions on the form at all, even though one is working with I-equivalence. In brief, the quarrel between the proponents of the two notions of integrality is reminiscent of that between the Big-endians and the Little-endians in Lilliput. When we require the notion of integrality for forms we shall generally use the classical definition except when (as in Chapter 14) the other definition is clearly more convenient.

3. PROSPECT

The object of this book is to study quadratic forms over the field $\mathbf{Q}$ of rational numbers and over the ring $\mathbf{Z}$ of rational integers. We give here a very brief outline of the contents.

As indicated in Section 1 the strategy will be to work via the fields $\mathbf{Q}_p$ of p-adic numbers and the rings $\mathbf{Z}_p$ of p-adic integers which will be introduced in Chapter 3. It is convenient to introduce the modern jargon and to speak of $\mathbf{Q}$ as the *global* field and of the $\mathbf{Q}_p$ (including $\mathbf{R} = \mathbf{Q}_\infty$) as the *local* fields.

Although historically quadratic forms were first considered over $\mathbf{Z}$ and only later over $\mathbf{Q}$ we now know that the theory of forms over fields is simpler and more satisfactory than that over integral domains in general. We therefore reverse the historical order and start with quadratic forms over fields.

Chapter 2 gives the general theory of quadratic forms over fields of characteristic $\neq 2$. As Witt was probably the first to remark, the appropriate language is that of vector spaces with a quadratic functional which generalizes the ordinary euclidean metric in $\mathbf{R}^n$. We must then break the orderly narrative so as to introduce the p-adic numbers and develop their properties from scratch: this is done in Chapter 3. In Chapter 4 we find a very satisfactory theory of forms over $\mathbf{Q}_p$: there is a convenient description of the equivalence classes of forms in terms of three invariants and a specification of the sets of invariants which actually belong to equivalence classes of forms. We must now break the narrative again in order to introduce some fundamental notions from the Geometry of Numbers: and this is done from scratch in Chapter 5. In Chapter 6 we consider quadratic forms over $\mathbf{Q}$. We use the Geometry of Numbers more because it provides the most suitable language than because it is absolutely necessary. We also use Dirichlet's theorem about the existence of primes in arithmetic progressions: this enables us to give a very transparent discussion of the relationship between global and local properties. We do not prove Dirichlet's theorem but treat it as a *deus ex machina*: in Chapter 14 we shall show how its use can, in fact, be obviated. In Chapter 6 we find that questions about quadratic forms over $\mathbf{Q}$ can be very satisfactorily reduced to those over the local fields. A quadratic form $f(\mathbf{x})$ with rational coefficients represents some $a \in \mathbf{Q}$ if it represents a over all $\mathbf{Q}_p$ including $\mathbf{Q}_\infty = \mathbf{R}$ or in the language we shall use "f represents a globally if it does so everywhere locally". Further, two forms f, g are equivalent globally if they are equivalent everywhere locally. There is also an elegant criterion for the existence of a global form with prescribed local properties.

The next three chapters 7, 8, 9 follow the same paradigm for integral equivalence. Chapter 7 gives the theory for a general integral domain, sometimes making the additional assumption that it is a principal ideal domain. Chapter 8 deals with forms over the p-adic integers $\mathbf{Z}_p$: there is a very neat description of the equivalence classes for $p \neq 2$ and a rather more messy one for $p = 2$. Next, Chapter 9 deals with forms over $\mathbf{Z}$. Here the local theory sheds considerable light but does not give a complete description. It is possible for two forms, for example $x^2 + 82y^2$ and $2x^2 + 41y^2$, to be integrally equivalent everywhere locally but not globally. Forms which are integrally equivalent everywhere locally are said to be in the same *genus*. It turns out that a genus contains only a finite number of integral equivalence classes. In a sense, the global situation is determined by the situation everywhere locally only up to a finite number of possibilities. The "local to global" theorem now merely asserts that if an integer a is representable integrally everywhere locally by a form f then it is represented globally by some form f^* (say) in the same genus as f. This tells us something about the representations by f only if we happen to know that the genus of f contains only one integral equivalence class; for example, for the forms $x^2 + y^2$, $x^2 + y^2 + z^2$ and $x^2 + y^2 + z^2 + t^2$. The chapter also contains a straightforward criterion for determining when there is a global integral form with prescribed local properties.

Chapters 10 and 11 deal with a "local to global" argument of a different kind which would require too much preliminary discussion for it to be described here. A consequence is that a genus of indefinite integral forms in three or more variables usually consists of only one integral equivalence class: there are very simple conditions which ensure this, and an explicit description of the exceptional situation when there is more than one class in the genus. This means that we have, at least usually, a good "local to global" theorem about integral representability by an indefinite form in three or more variables. The situation for definite forms is quite different: the normal situation is that a genus of definite forms contains many equivalence classes. However, the methods of Chapters 10 and 11 show that a definite integral form in five or more variables represents integrally all except finitely many of the integers which it represents everywhere locally; and the same is true for forms in four variables with suitable precautions.

The results mentioned in the last paragraph give fuller information about indefinite forms than definite ones. This is somewhat paradoxical since from the crudest point of view definite forms are easier to handle than indefinite ones. For example, if we wish to find whether a definite integral form $f(\mathbf{x})$ integrally represents an integer c, then the equation $f(\mathbf{x}) = c$ gives bounds on the size of the integers $x_1, \ldots, x_n$ and we can decide whether or not there is a representation by means of a finite search. Similarly we can always decide by a finite search whether or not two definite integral forms are equivalent. Chapter 12 studies the so-called reduction of definite quadratic forms. This is a purely real-variable theory and describes how one can select a single form (the *reduced* form) from an integral equivalence class of definite forms (exceptionally, an equivalence class may contain a finite number of reduced forms). There are several different theories of reduction: the one we use was invented by Minkowski and defines reduction by a collection of linear inequalities on the coefficients. It is shown that for given n the set of reduced forms can be defined by a finite number of such inequalities and that there is an interesting geometrical configuration.

There is also a theory of the reduction of indefinite quadratic forms but this is much more delicate and distinctly less useful. As will be explained in Chapter 13 the reduction of indefinite forms is made to depend on that of definite ones. The argument, which is due to Hermite, also gives a good description of the structure of the group $O_Z(f)$ of integral automorphs of a given indefinite integral form f: that is the group of integral linear substitutions $\mathbf{T}$ such that identically $f(\mathbf{Tx}) = f(\mathbf{x})$. The group $O_Z(f)$ is shown to be finitely generated. In a suitably vague sense $O_Z(f)$ is as large as is permitted by some obvious restraints placed by the geometry of the situation. The discussion of $O_Z(f)$ reveals interesting connections with non-euclidean geometry.

Chapter 14 is independent of much in the preceding chapters and considers integral binary forms. It was shown by Gauss that the classes under proper† integral equivalence of primitive† integral binary forms have a natural structure as an abelian group, the group law being a process called "composition". The genera play a special role. Gauss used composition to prove the existence of forms with given local properties. We give the Gauss theory in a version due to Dirichlet and show how it can be used to make the proofs in the earlier chapters independent of Dirichlet's theorem about primes in arithmetic progressions.

This ends the account of the material which is expounded in detail in the book. The appendices survey rather briefly certain related areas which do not fit into the general plan.

† The meaning of these terms will be explained in later chapters.

CHAPTER 2

Quadratic Forms over a Field

1. INTRODUCTION

Suppose that k is a field of characteristic $\neq 2$. By a *quadratic space* over k we shall mean a finite-dimensional k-vector space† U together with a symmetric bilinear form ϕ defined on U and taking values in k. We shall denote the dimension of U by n and will put

$$\phi(\mathbf{u}) = \phi(\mathbf{u}, \mathbf{u}) \qquad (\mathbf{u} \in U). \tag{1.1}$$

The function $\phi(\mathbf{u})$ of the single variable $\mathbf{u}$ determines the symmetric bilinear form $\phi(\mathbf{u}_1, \mathbf{u}_2)$ by the formula

$$\phi(\mathbf{u}_1, \mathbf{u}_2) = \tfrac{1}{4}\{\phi(\mathbf{u}_1 + \mathbf{u}_2) - \phi(\mathbf{u}_1 - \mathbf{u}_2)\}. \tag{1.2}$$

Thus instead of starting with the bilinear form $\phi(\mathbf{u}_1, \mathbf{u}_2)$ we could have started with a function $\phi(\mathbf{u})$ of a single variable subject to the conditions

(i) $\phi(\lambda\mathbf{u}) = \lambda^2\phi(\mathbf{u})$ $(\lambda \in k)$, and

(ii) that the right-hand side of (1.2) is a bilinear form in $\mathbf{u}_1$ and $\mathbf{u}_2$.

A degenerate case of the definition is the 0-dimensional vector space consisting only of the zero-vector $\mathbf{0}$ and with $\phi(\mathbf{0}) = 0$. There will be occasions when it is convenient to include this (in particular the definition of the Witt group in Section 5) but in general we shall exclude this case from the definition of a quadratic space.

For later reference we enunciate

LEMMA 1.1. *Suppose that the bilinear form* $\phi(\mathbf{u}_1, \mathbf{u}_2)$ *is not identically* 0. *Then there is a* $\mathbf{u} \in U$ *such that* $\phi(\mathbf{u}) \neq 0$.

Proof. Trivial.

† In this Chapter, the typical quadratic space is called U, ϕ. In the rest of the book it is usually V, ϕ. The discrepancy was noted too late to make a change.

If $\mathbf{u}_1, \ldots, \mathbf{u}_n$ is any basis for U, then clearly

$$\begin{aligned} f(x_1, \ldots, x_n) &= \phi(\sum_j x_j \mathbf{u}_j) \\ &= \sum_{i,j} x_i x_j \phi(\mathbf{u}_i, \mathbf{u}_j) \end{aligned} \tag{1.3}$$

is a quadratic form over k. If $\mathbf{u}'_1, \ldots, \mathbf{u}'_n$ is any other basis for U, then clearly

$$f'(x_1, \ldots, x_n) = \phi(\sum x_j \mathbf{u}'_j) \tag{1.4}$$

is equivalent to f, and every form equivalent to f arises in this way. Further, every quadratic form f over k arises from some quadratic space U, ϕ as one sees by taking any space U of the appropriate dimension with basis $\mathbf{u}_1, \ldots, \mathbf{u}_n$ and using (1.3) to define ϕ in terms of f. We shall see that the language of quadratic spaces is a very natural one for the discussion of quadratic forms over fields.

We denote the set of k-linear maps $U \to k$ by $\mathrm{Hom}(U, k)$ (the *dual* of U). The bilinear form ϕ determines a k-linear map

$$U \to \mathrm{Hom}(U, k) \tag{1.5}$$

where $\mathbf{w} \in U$ corresponds to the $\phi_{\mathbf{w}} \in \mathrm{Hom}(U, k)$ defined by

$$\phi_{\mathbf{w}} : \mathbf{u} \to \phi(\mathbf{u}, \mathbf{w}) \tag{1.6}$$

We shall say that the quadratic space U, ϕ is *regular* (or *non-singular*) if (1.5) is an isomorphism. Otherwise we say that U, ϕ is *singular*. We enunciate the trivial

LEMMA 1.2. *The following four statements are equivalent*

(i) U, ϕ *is regular.*

(ii) *If* $\mathbf{w} \in U$ *and* $\phi(\mathbf{u}, \mathbf{w}) = 0$ *for all* $\mathbf{u} \in U$, *then* $\mathbf{w} = \mathbf{0}$.

(iii) $\det_{1 \leq i \leq n,\, 1 \leq j \leq n} \phi(\mathbf{u}_i, \mathbf{u}_j) \neq 0$, *where* $\mathbf{u}_1, \ldots, \mathbf{u}_n$ *is any basis of* U.

(iv) *The form* f *given by* (1.3) *is regular.*

Proof. Clear.

We are now in a position to introduce an important invariant of a regular quadratic space. Let $\mathbf{u}_1, \ldots, \mathbf{u}_n$ and $\mathbf{v}_1, \ldots, \mathbf{v}_n$ be bases for U, so

$$\mathbf{v}_i = \sum_j s_{ij} \mathbf{u}_j \qquad (1 \leq i \leq n)$$

where $s_{ij} \in k$ and $\det(s_{ij}) \neq 0$. It is easily verified that

$$\det_{i,j} \phi(\mathbf{v}_i, \mathbf{v}_j) = (\det_{i,j}(s_{ij}))^2 \det_{i,j} \phi(\mathbf{u}_i, \mathbf{u}_j). \tag{1.7}$$

Hence if we denote, as usual, by k^* the multiplicative group of non-zero elements of k we see that $\det \phi(\mathbf{u}_i, \mathbf{u}_j)$ lies in a class of k^* modulo $(k^*)^2$ which is independent of the choice of basis $\mathbf{u}_1, \ldots, \mathbf{u}_n$. We call the element of $k^*/(k^*)^2$ thus defined the *determinant* $d(\phi)$ of U, ϕ. When U, ϕ is singular we define the determinant to be 0.

We have in a quadratic space U, ϕ a generalization of ordinary euclidean geometry. We shall say, for example, that $\mathbf{u}_1$ is *normal* (or *orthogonal* or *perpendicular*) to $\mathbf{u}_2$ if $\phi(\mathbf{u}_1, \mathbf{u}_2) = 0$. Note however that we cannot exclude the possibility that $\phi(\mathbf{u}) = \phi(\mathbf{u}, \mathbf{u}) = 0$, i.e. that $\mathbf{u}$ is normal to itself. (see Section 2).

Let U, ϕ be a quadratic space and let V be a linear subspace of U. Then ϕ gives a quadratic space structure on V. When we speak of linear subspaces of U we shall regard them as naturally endowed with their quadratic space structure. Further we shall denote by $V^\perp$ the set of all elements of U which are perpendicular to all the elements of V (the *orthogonal complement* of V):

$$V^\perp = \{\mathbf{u} : \phi(\mathbf{u}, \mathbf{v}) = 0 \text{ for all } \mathbf{v} \in V\}.$$

Clearly $V^\perp$ is also a linear subspace of U.

LEMMA 1.3. *Let U, ϕ be a quadratic space, not necessarily regular. Let V be a regular subspace of U. Then U is the direct sum*† *of V and $V^\perp$.*

Proof. Let $\mathbf{u} \in U$. Then $\mathbf{u}$ determines an element of $\mathrm{Hom}(V, k)$ by

$$\mathbf{v} \to \phi(\mathbf{u}, \mathbf{v}).$$

By the definition of regularity there is a uniquely defined $\mathbf{w} = \mathbf{w}(\mathbf{u}) \in V$ such that

$$\phi(\mathbf{w}, \mathbf{v}) = \phi(\mathbf{u}, \mathbf{v}) \qquad (\text{all } \mathbf{v} \in V).$$

Hence

$$\mathbf{u} = \mathbf{w} + (\mathbf{u} - \mathbf{w})$$

where

$$\mathbf{w} \in V, \qquad \mathbf{u} - \mathbf{w} \in V^\perp.$$

Conversely, if

$$\mathbf{u} = \mathbf{s} + \mathbf{t}, \qquad \mathbf{s} \in V, \quad \mathbf{t} \in V^\perp$$

we have

$$\phi(\mathbf{s}, \mathbf{v}) = \phi(\mathbf{u}, \mathbf{v}) = \phi(\mathbf{w}, \mathbf{v}) \qquad (\text{all } \mathbf{v} \in V)$$

and so (regularity again) $\mathbf{s} = \mathbf{w}$.

Definition. A basis $\mathbf{u}_1, \ldots, \mathbf{u}_n$ of a quadratic space is said to be *normal* if

$$\phi(\mathbf{u}_i, \mathbf{u}_j) = 0 \qquad (i \neq j). \tag{1.8}$$

LEMMA 1.4. *Every quadratic space has a normal basis.*

† We say that a vector space U is the *direct sum* of subspaces V, W if every $\mathbf{u} \in U$ is uniquely of the shape $\mathbf{u} = \mathbf{v} + \mathbf{w}$ with $\mathbf{v} \in V$, $\mathbf{w} \in W$. If U, ϕ is a quadratic space, the direct sum is *orthogonal* if $\phi(\mathbf{v}, \mathbf{w}) = 0$ for all $\mathbf{v} \in V$, $\mathbf{w} \in W$. As the direct sums which we have to consider will always be orthogonal, we shall often omit the adjective.

Proof. If ϕ is identically 0, every basis is normal. Otherwise, by Lemma 1.1 there is a $\mathbf{u}_1 \in U$ such that $\phi(\mathbf{u}_1) \neq 0$. The 1-dimensional space V spanned by $\mathbf{u}_1$ is thus regular and U is the direct sum of V and $V^\perp$. By induction on the dimension there is a normal basis $\mathbf{u}_2, \ldots, \mathbf{u}_n$ of $V^\perp$ and $\mathbf{u}_1, \ldots, \mathbf{u}_n$ is the required normal basis of U.

Note. In the usual geometry of euclidean spaces over the reals one usually works with *orthonormal* bases, i.e. bases for which $\phi(\mathbf{u}_j) = 1$ holds for all j in addition to (1.8). In the more general context discussed here orthonormal bases do not usually exist and we shall seldom, if ever, have occasion to refer to them.

LEMMA 1.5. *If* $\mathbf{u}_1, \ldots, \mathbf{u}_n$ *is a normal basis for* U, ϕ *then a necessary and sufficient condition that* U, ϕ *be regular is that*

$$\phi(\mathbf{u}_j) \neq 0 \qquad (1 \leqslant j \leqslant n)$$

Proof. See Lemma 1.2.

We shall mainly be concerned with the case when U, ϕ is regular.

LEMMA 1.6. *Suppose that* U, ϕ *is regular. Then*

$$(V^\perp)^\perp = V$$

for every linear subspace V *of* U.

Proof. Clearly

$$V \subset (V^\perp)^\perp.$$

By the definition of regularity we have

$$\dim V + \dim V^\perp = \dim U.$$

Hence

$$\dim (V^\perp)^\perp = \dim V,$$

and the result follows.

LEMMA 1.7. *Let* V *be a linear subspace of the regular quadratic space* U, ϕ. *Then the following statements are equivalent.*

(i) $V \cap V^\perp = \{\mathbf{0}\}$.

(ii) V *is regular*

(iii) $V^\perp$ *is regular*.

Proof. After Lemma 1.6 we need only prove the equivalence of (i) and (ii). This follows directly from the definition of regularity.

2. ISOTROPIC SPACES

We say that a quadratic space U, ϕ represents $b \in k$ if there is a $\mathbf{b} \in U$ such that $\phi(\mathbf{b}) = b$. We say that 0 is represented *non-trivially* if there is a $\mathbf{b} \neq \mathbf{0}$ with $\phi(\mathbf{b}) = 0$. A regular space is said to be *isotropic* if it represents 0 nontrivially, otherwise it is *anisotropic*.

An example of an isotropic space is a *hyperbolic plane*. This is a 2-dimensional space U, ϕ with a basis $\mathbf{u}_1, \mathbf{u}_2$ such that

$$\phi(\mathbf{u}_1) = \phi(\mathbf{u}_2) = 0 \qquad \phi(\mathbf{u}_1, \mathbf{u}_2) = 1. \tag{2.1}$$

LEMMA 2.1. *Every isotropic quadratic space U, ϕ contains a hyperbolic plane.*

Proof. By the definition of isotropy there is a $\mathbf{u}_1 \neq \mathbf{0}$ with $\phi(\mathbf{u}_1) = 0$. Since isotropy implies regularity there is a $\mathbf{w}$ such that $\phi(\mathbf{u}_1, \mathbf{w}) \neq 0$ and without loss of generality

$$\phi(\mathbf{u}_1, \mathbf{w}) = 1. \tag{2.2}$$

Put $\mathbf{u}_2 = \mathbf{w} + \lambda \mathbf{u}_1$ where $\lambda \in k$ is defined by

$$\begin{aligned} \phi(\mathbf{u}_2) &= \phi(\mathbf{w}) + 2\lambda\phi(\mathbf{w}, \mathbf{u}_1) + \phi(\mathbf{u}_1) \\ &= \phi(\mathbf{w}) + 2\lambda \\ &= 0. \end{aligned}$$

Then $\mathbf{u}_1, \mathbf{u}_2$ span a hyperbolic plane $H \subset U$.

Remark. Since $\mathbf{w}$ is any point of U satisfying (2.2), we shall later find the algorithm in the above proof very handy for constructing points $\mathbf{u}_2$ with $\phi(\mathbf{u}_2) = 0$ and with further desirable properties.

COROLLARY 1. *U is the direct sum of H and $H^{\perp}$.*

Proof. Lemma 1.3.

Definition. We say that the regular space U, ϕ is *universal* if it represents every non-zero element of k.

COROLLARY 2. *An isotropic space is universal.*

Proof. For a hyperbolic plane is clearly universal.

LEMMA 2.2. *Let $k = \mathbf{F}_p$ be the finite field of p elements ($p \neq 2$, prime). Then every 2-dimensional regular space U, ϕ is universal.*

Proof. Let $\mathbf{u}_1, \mathbf{u}_2$ be a normal basis, say

$$\phi(\mathbf{u}_1) = a_1, \qquad \phi(\mathbf{u}_2) = a_2 \qquad \phi(\mathbf{u}_1, \mathbf{u}_2) = 0$$

where $a_1 \neq 0, a_2 \neq 0$ by regularity. Then for $b \neq 0$ we have to show that there exists $c_1, c_2 \in \mathbf{F}_p$ such that

$$a_1c_1^2 + a_2c_2^2 = b.$$

We write this in the shape

$$a_1c_1^2 = b - a_2c_2^2. \tag{2.3}$$

Let S, T be the set of values taken respectively by $a_1c_1^2$ and $b - a_2c_2^2$ as c_1, c_2 run through $\mathbf{F}_p$. Then S, T each have $\frac{1}{2}(p+1)$ elements and so $S \cap T$ is not empty, i.e. (2.3) has a solution as required.

COROLLARY. *U, ϕ can be universal without being isotropic.*

Proof. For we can take $a_1 = 1$ and choose a_2 so that $-a_2$ is not a square in $\mathbf{F}_p$.

Finally, we prove a technical lemma which we shall use repeatedly. We use the language of forms rather than that of quadratic spaces since it will be quoted in that context.

LEMMA 2.3. *Let k be any field of characteristic $\neq 2$ and let $f(\mathbf{x}), g(\mathbf{y})$ be regular forms in the variables $\mathbf{x} = (x_1, \ldots, x_n)$ and $\mathbf{y} = (y_1, \ldots, y_m)$ respectively. Suppose that $f(\mathbf{x}) - g(\mathbf{y})$ is isotropic as a form in the $m + n$ variables $(x_1, \ldots, x_n, y_1, \ldots, y_m)$. Then there is a $b \neq 0$ which is represented both by f and g.*

Proof. By hypothesis there are $\mathbf{a} \in k^n, \mathbf{b} \in k^m$ not both zero such that

$$f(\mathbf{a}) = g(\mathbf{b}).$$

If $f(\mathbf{a}) \neq 0$ we are done. If $f(\mathbf{a}) = 0$ but $\mathbf{a} \neq \mathbf{0}$ then the form $f(\mathbf{x})$ is isotropic and so universal by Lemma 2.1, Corollary 2. Let b be any non-zero element of k represented by $g(\mathbf{y})$. Then b is represented by $f(\mathbf{x})$ and we are done. If $\mathbf{a} = \mathbf{0}$, then $\mathbf{b} \neq \mathbf{0}$ and we argue similarly.

3. NORMAL BASES

We have already seen (Lemma 1.4) that every quadratic space has a normal basis. It will often be convenient to define properties of the quadratic space in terms of a normal basis and then we shall have to show that the definition is independent of the particular choice of basis. In this context the following result will be vital.

LEMMA 3.1. *Let U, ϕ be a regular quadratic space and let A: $\mathbf{a}_1, \ldots, \mathbf{a}_n$ and B: $\mathbf{b}_1, \ldots, \mathbf{b}_n$ be any pair of normal bases. Then there is a sequence of* *normal bases*

$$C_t: \mathbf{c}_{t1}, \mathbf{c}_{t2}, \ldots, \mathbf{c}_{tn} \qquad (1 \leqslant t \leqslant T)$$

such that

(i) $C_1 = A, C_T = B$; *and*

(ii) *For each* t *in* $1 \leqslant t < T$ *the bases* C_t, C_{t+1} *have at least* $n-2$ *elements in common in the sense that*

$$\mathbf{c}_{tj} = \mathbf{c}_{(t+1)j}$$

for at least $n-2$ *values of* j.

Proof. We note first that the assertion is true if $\mathbf{b}_1, \ldots, \mathbf{b}_n$ is a permutation of $\mathbf{a}_1, \ldots, \mathbf{a}_n$, since every permutation is a product of transpositions. Hence it will be sufficient to disregard the order in which the elements in the various bases occur.

Secondly it will be enough to find a sequence of bases C_t leading from A to some normal basis B^*: $\mathbf{b}_1, \mathbf{b}_2^*, \ldots, \mathbf{b}_n^*$ with some elements $\mathbf{b}_2^*, \ldots, \mathbf{b}_n^*$ of U. For then the 1-dimensional space V spanned by $\mathbf{b}_1$ is regular (e.g. Lemma 1.5) and hence so is $V^\perp$ (Lemma 1.7). By induction on the dimension n of U we may therefore suppose that there is a sequence of bases of the required sort of $V^\perp$ going from $\mathbf{b}_2^*, \ldots, \mathbf{b}_n^*$ to $\mathbf{b}_2, \ldots, \mathbf{b}_n$. On adjoining $\mathbf{b}_1$ to these we get a sequence of bases of U going from B^* to B.

Put

$$\mathbf{b}_1 = \sum_{1 \leqslant j \leqslant n} s_j \mathbf{a}_j.$$

We shall now show how to construct a sequence of bases from A to some B^* using induction on the number J of non-zero s_j. After a permutation of the base A we may suppose that

$$s_j \neq 0 \qquad (1 \leqslant j \leqslant J)$$
$$s_j = 0 \qquad (j > J).$$

and will use induction on J.

First case $J = 1$. We can take

$$\mathbf{b}_j^* = \mathbf{a}_j \qquad (2 \leqslant j \leqslant n)$$

and the sequence from A to B^* consists only of A and B^*.

Second case. Suppose that $J \geqslant 2$ and that

$$\phi(s_1\mathbf{a}_1 + s_2\mathbf{a}_2) \neq 0.$$

We can choose non-zero t_1, t_2 such that

$$0 = \phi(s_1\mathbf{a}_1 + s_2\mathbf{a}_2, t_1\mathbf{a}_1 + t_2\mathbf{a}_2)$$
$$= s_1 t_1 \phi(\mathbf{a}_1) + s_2 t_2 \phi(\mathbf{a}_2);$$

for example

$$t_1 = s_2\phi(\mathbf{a}_2), \qquad t_2 = -s_1\phi(\mathbf{a}_1).$$

Put

$$\mathbf{a}_1^* = s_1\mathbf{a}_1 + s_2\mathbf{a}_2$$
$$\mathbf{a}_2^* = t_1\mathbf{a}_1 + t_2\mathbf{a}_2$$
$$\mathbf{a}_j^* = \mathbf{a}_j \qquad (2 < j \leqslant n).$$

Then the bases A and A^*: $\mathbf{a}_1^*, \ldots, \mathbf{a}_n^*$ have $n - 2$ elements in common. Since

$$\mathbf{b}_1 = \mathbf{a}_1^* + \sum_{2<j\leqslant J} s_j \mathbf{a}_j^*$$

we are reduced to the case $J - 1$ and use induction on J.

We note that the second case certainly applies when $J = 2$ since then

$$\phi(s_1\mathbf{a}_1 + s_2\mathbf{a}_2) = \phi(\mathbf{b}_1) \neq 0.$$

Third case $J > 2$. If there are indices i,j with $1 \leqslant i < j \leqslant J$ and

$$\phi(s_i\mathbf{a}_i + s_j\mathbf{a}_j) \neq 0$$

we can permute the a's and reduce ourselves to the second case. Hence we have to deal only with the possibility that

$$0 = \phi(s_1\mathbf{a}_1 + s_2\mathbf{a}_2) = s_1^2\phi(\mathbf{a}_1) + s_2^2\phi(\mathbf{a}_2)$$
$$0 = \phi(s_2\mathbf{a}_2 + s_3\mathbf{a}_3) = s_2^2\phi(\mathbf{a}_2) + s_3^2\phi(\mathbf{a}_3)$$
$$0 = \phi(s_3\mathbf{a}_3 + s_1\mathbf{a}_1) = s_3^2\phi(\mathbf{a}_3) + s_1^2\phi(\mathbf{a}_1).$$

But then

$$s_1^2\phi(\mathbf{a}_1) = s_2^2\phi(\mathbf{a}_2) = s_3^2\phi(\mathbf{a}_3) = 0.$$

Contradiction.

4. ISOMETRIES AND AUTOMETRIES

Let U_1, ϕ_1 and U_2, ϕ_2 be quadratic spaces over the same field k. An isomorphism

$$\sigma: \ U_1 \to U_2$$

of k-linear spaces is called an *isometry* if it also preserves the quadratic space structure in the sense that

$$\phi_2(\sigma\mathbf{u}) = \phi_1(\mathbf{u}) \qquad (\text{all } \mathbf{u} \in U_1).$$

Two quadratic spaces are said to be *isometric* (or "in the same *equivalence class*") if there is an isometry from one to the other. Clearly two quadratic spaces are isometric if and only if they correspond to the same equivalence class of quadratic forms in the sense of Section 1.

Although the term is not (yet) standard we shall call an isometry of U, ϕ with itself an *autometry*. The autometries of U, ϕ form a group under composition, the *orthogonal group* $O(U) = O$.

LEMMA 4.1. *Let σ be an autometry of a regular space U, ϕ. Then* $\det(\sigma) = \pm 1$.

Proof. Let $\mathbf{u}_1, \ldots, \mathbf{u}_n$ be any basis of U. Then

$$\phi(\sigma\mathbf{u}_i, \sigma\mathbf{u}_j) = \phi(\mathbf{u}_i, \mathbf{u}_j) \qquad (\text{all } i, j).$$

On expressing the $\sigma\mathbf{u}_i, \sigma\mathbf{u}_j$ in terms of $\mathbf{u}_i, \mathbf{u}_j$ and using the bilinearity of ϕ we deduce that

$$(\det \sigma)^2 \det_{1 \leqslant i \leqslant n,\, 1 \leqslant j \leqslant n}(\phi(\mathbf{u}_i, \mathbf{u}_j)) = \det(\phi(\mathbf{u}_i, \mathbf{u}_j)).$$

But $\det(\phi(\mathbf{u}_i, \mathbf{u}_j)) \neq 0$ by Lemma 1.2 and the result follows.

If $\det \sigma = +1$ we say that the autometry σ is *proper*, otherwise *improper*. The proper autometries form a group, the *proper orthogonal* group $O^+(U)$. Since, as we shall see in a moment, improper autometries always exist, the index of $O^+(U)$ in $O(U)$ is precisely 2 and the set $O^-(U)$ of improper autometries is a coset.

Let V be any regular subspace of U, ϕ so that (Lemma 1.3) U is the direct sum of V and $V^\perp$. There is a uniquely defined linear map $\sigma \colon U \to U$ such that

$$\sigma\mathbf{u} = -\mathbf{u} \qquad (\mathbf{u} \in V)$$
$$\sigma\mathbf{u} = \mathbf{u} \qquad (\mathbf{u} \in V^\perp).$$

Clearly σ is an autometry. In particular, if $\mathbf{v} \in U$ has $\phi(\mathbf{v}) \neq 0$ we can take for V the 1-dimensional subspace spanned by $\mathbf{v}$. We denote by $\tau_\mathbf{v}$ the autometry such that

$$\left.\begin{aligned} \tau_\mathbf{v}\mathbf{v} &= -\mathbf{v} \\ \tau_\mathbf{v}\mathbf{u} &= \mathbf{u} \quad \text{if} \quad \phi(\mathbf{u}, \mathbf{v}) = 0. \end{aligned}\right\} \tag{4.1}$$

Clearly

$$\det \tau_\mathbf{v} = -1. \tag{4.2}$$

The $\tau_\mathbf{v}$ are called *symmetries*†. It is easy to see that (4.1) is equivalent to

$$\tau_\mathbf{v}\mathbf{u} = \mathbf{u} - \frac{2\phi(\mathbf{u}, \mathbf{v})}{\phi(\mathbf{v})}\mathbf{v}. \tag{4.3}$$

LEMMA 4.2. *Let $\mathbf{u}, \mathbf{v} \in U$ with $\phi(\mathbf{u}) = \phi(\mathbf{v})$ and $\phi(\mathbf{u} - \mathbf{v}) \neq 0$. Then*

$$\tau_{\mathbf{u}-\mathbf{v}}\mathbf{u} = \mathbf{v}.$$

† In ordinary n-dimensional space they correspond to reflexions in $(n - 1)$-dimensional hyperplanes.

Proof. We use (4.3) with $\mathbf{u} - \mathbf{v}$ instead of $\mathbf{v}$. Evaluation of the expressions occurring gives

$$\phi(\mathbf{u}, \mathbf{u} - \mathbf{v}) = \phi(\mathbf{u}) - \phi(\mathbf{u}, \mathbf{v})$$

and

$$\begin{aligned}\phi(\mathbf{u} - \mathbf{v}, \mathbf{u} - \mathbf{v}) &= \phi(\mathbf{u}) - 2\phi(\mathbf{u}, \mathbf{v}) + \phi(\mathbf{v}) \\ &= 2\{\phi(\mathbf{u}) - \phi(\mathbf{u}, \mathbf{v})\}\end{aligned}$$

and so

$$\begin{aligned}\tau_{\mathbf{u}-\mathbf{v}}\mathbf{u} &= \mathbf{u} - (\mathbf{u} - \mathbf{v}) \\ &= \mathbf{v},\end{aligned}$$

as required.

COROLLARY 1. *If* $\mathbf{u}, \mathbf{v} \in U$ *with* $\phi(\mathbf{u}) = \phi(\mathbf{v}) \neq 0$ *then there is an autometry* σ *such that* $\sigma\mathbf{u} = \mathbf{v}$, *and* σ *is either a symmetry or the product of two symmetries.*

Proof. If $\phi(\mathbf{u} - \mathbf{v}) \neq 0$ we can take $\sigma = \tau_{\mathbf{u}-\mathbf{v}}$. If $\phi(\mathbf{u} + \mathbf{v}) \neq 0$, then

$$\tau_{\mathbf{u}+\mathbf{v}}\mathbf{u} = -\mathbf{v},$$

and so we can take $\sigma = \tau_{\mathbf{v}}\tau_{\mathbf{u}+\mathbf{v}}$. At least one of these cases must be possible because

$$\begin{aligned}\phi(\mathbf{u} + \mathbf{v}) + \phi(\mathbf{u} - \mathbf{v}) &= 2\phi(\mathbf{u}) + 2\phi(\mathbf{v}) \\ &= 4\phi(\mathbf{u}) \\ &\neq 0.\end{aligned}$$

COROLLARY 2. *If* U, ϕ *is regular and* $n = \dim U > 1$ *we can suppose that the* σ *of Corollary* 1 *is the product of precisely two symmetries.*

Proof. For there is a $\mathbf{w}$ normal to $\mathbf{u}$ with $\phi(\mathbf{w}) \neq 0$. If $\phi(\mathbf{u} - \mathbf{v}) \neq 0$ we can take $\sigma = \tau_{\mathbf{u}-\mathbf{v}}\tau_{\mathbf{w}}$.

LEMMA 4.3. *Suppose that* U, ϕ *is regular. Then every autometry is a product of symmetries.*

Note. The proof given here shows that $2n$ symmetries suffice, where $n = \dim U$. A more elaborate argument [see Example 8] shows that n symmetries are enough.

Proof. Let ρ be any autometry of U and let $\mathbf{v}$ be any element of U with $\phi(\mathbf{v}) \neq 0$. Then

$$\phi(\rho\mathbf{v}) = \phi(\mathbf{v}) \neq 0$$

and so by Lemma 4.2, Corollary there is a product σ of symmetries such that

$$(\sigma\rho)\mathbf{v} = \mathbf{v}.$$

Since $\sigma\rho$ is an autometry of U which maps $\mathbf{v}$ into $\mathbf{v}$, it maps the space $V^{\perp}$

of vectors normal to $\mathbf{v}$ onto itself. By induction on the dimension of U we may assume that there is a sequence of vectors $\mathbf{w}(1), \ldots, \mathbf{w}(T)$ of $V^{\perp}$ such that

$$\sigma^* \text{ (say)} = \tau_{\mathbf{w}(1)} \tau_{\mathbf{w}(2)} \cdots \tau_{\mathbf{w}(T)}$$

and $\sigma\rho$ have the same action on $V^{\perp}$. But the $\sigma_{\mathbf{w}(t)}$ leave $\mathbf{v}$ fixed (since $\mathbf{w}(t) \in V^{\perp}$) and so both σ^* and $\sigma\rho$ leave $\mathbf{v}$ fixed. Hence $\sigma\rho = \sigma^*$ and so $\rho = \sigma^{-1}\sigma^*$ is a product of symmetries, as required.

The following theorem due to Witt will prove absolutely indispensible and shows the power of his quadratic space approach to the theory of quadratic forms.

THEOREM 4.1. *Let V_1, V_2 be subspaces of U, ϕ. Suppose that there is an isometry ρ: $V_1 \to V_2$ and that V_1 (and hence V_2) is regular. Then there is an autometry σ of U, ϕ which coincides with ρ on V_1.*

Note. It is not required that U be regular, though that is the case of interest.

Proof. By Lemma 1.1 there is a $\mathbf{v} \in V_1$ with $\phi(\mathbf{v}) \neq 0$. By Lemma 4.2, Corollary 1, there is an autometry λ of U such that $\lambda(\rho\mathbf{v}) = \mathbf{v}$. On taking $\lambda\rho, \lambda V_2$ instead of ρ, V_2 respectively, we may thus suppose without loss of generality that $\mathbf{v} \in V_1 \cap V_2$ and $\rho\mathbf{v} = \mathbf{v}$. If $\dim V_1 = 1$, then we are done. If not, we shall use induction on $\dim V_1$. Let U^* be the orthogonal complement of the 1-dimensional space spanned by $\mathbf{v}$ and put

$$V_j^* = U^* \cap V_j \qquad (j = 1, 2).$$

Then $\rho V_1^* = V_2^*$. By the induction hypothesis, there is an autometry σ^* of U^* which coincides with ρ on V_1^*. Clearly the autometry σ of U defined by $\sigma\mathbf{v} = \mathbf{v}$ and $\sigma\mathbf{u} = \sigma^*\mathbf{u}$ for $\mathbf{u} \in U^*$ has the required properties.

COROLLARY 1 ("Witt's Lemma"). *Let U_1, ϕ_1 and U_2, ϕ_2 be isometric quadratic spaces and let $V_j \subset U_j$ $(j = 1, 2)$ also be isometric. Suppose that V_1 (and so V_2) is regular. Denote by $V_j^{\perp}$ the orthogonal complement of V_j in U_j. Then $V_1^{\perp}$ and $V_2^{\perp}$ are isometric.*

Proof. By hypothesis there is an isometry μ: $U_2 \to U_1$ and by taking $\mu U_2, \mu V_2$ for U_2, V_2 respectively, we may suppose that $U_1 = U_2 = U$, $\phi_1 = \phi_2 = \phi$ (say). By the Theorem, there is an autometry σ of U such that $\sigma V_1 = V_2$. Then $\sigma V_1^{\perp} = V_2^{\perp}$ and we are done.

Because of the importance of the preceding Corollary, we enunciate explicitly the interpretation in terms of quadratic forms.

COROLLARY 2. *Let $\mathbf{x} = (x_1, \ldots, x_l)$ and $\mathbf{y} = (y_1, \ldots, y_m)$ be two independent sets of variables and let $f_j(\mathbf{x}), g_j(\mathbf{y})$ be quadratic forms $(j = 1, 2)$. Suppose* *that*

(i) $f_1(\mathbf{x})$ *and* $f_2(\mathbf{x})$ *are equivalent and regular.*

(ii) $f_1(\mathbf{x}) + g_1(\mathbf{y})$ *and* $f_2(\mathbf{x}) + g_2(\mathbf{y})$ *are equivalent as forms in the* $l + m$ *variables* $(x_1, \ldots, x_l, y_1, \ldots, y_m)$.

Then $g_1(\mathbf{y})$ *and* $g_2(\mathbf{y})$ *are equivalent.*

Proof. Let U_1 be a k-vector space with basis $\mathbf{u}_1, \ldots, \mathbf{u}_{l+m}$, define ϕ_1 by

$$\phi_1(x_1\mathbf{u}_1 + \ldots + x_l\mathbf{u}_l + y_1\mathbf{u}_{l+1} + \ldots + y_m\mathbf{u}_{l+m}) = f_1(\mathbf{x}) + g_1(\mathbf{y}),$$

and let $V_1 \subset U_1$ be spanned by $\mathbf{u}_1, \ldots, \mathbf{u}_l$. Define U_2, ϕ_2, V_2 similarly but with f_2, g_2 instead of f_1 and g_1. Then the hypotheses and conclusion of Corollary 2 are the translations of the hypotheses and conclusion of Corollary 1.

5. THE GROTHENDIECK AND WITT GROUPS

The topics discussed in this section will play only a peripheral role in this book but they are important in the theory of quadratic forms over general rings and fields.

From now on in this section all quadratic spaces are regular and k is fixed. Let U_1, ϕ_1 and U_2, ϕ_2 be two regular quadratic spaces over k. Let $U = U_1 \oplus U_2$ be the direct sum and define ϕ on U by

$$\phi(\mathbf{u}_1 \oplus \mathbf{u}_2) = \phi_1(\mathbf{u}_1) + \phi_2(\mathbf{u}_2) \qquad (\mathbf{u}_j \in U_j).$$

Then U, ϕ is a quadratic space and clearly the equivalence class (= isometry class) of U, ϕ depends only on the equivalence classes of U_1, ϕ_1 and U_2, ϕ_2. Let S be the set of all equivalence classes of regular quadratic spaces. If $U_1, \phi_1; U_2, \phi_2$ and U, ϕ above are respectively in the classes s_1, s_2 and s, then we write

$$s = s_1 + s_2.$$

Clearly

$$(s_1 + s_2) + s_3 = s_1 + (s_2 + s_3);$$

that is $+$ gives S a semigroup structure. The semigroup is abelian, that is

$$s_1 + s_2 = s_2 + s_1$$

since clearly $U_1 \oplus U_2$ is equivalent to $U_2 \oplus U_1$. Further, by "Witt's Lemma" (Theorem 4.1, Corollary),

$$s_1 + s_3 = s_2 + s_3$$

for any $s_1, s_2, s_3 \in S$ implies that

$$s_1 = s_2.$$

We have thus proved

LEMMA 5.1. *The set S of equivalence classes of quadratic spaces with the operation + defined above is an abelian semigroup with cancellation.*

It is convenient in this Section to adopt the definition of quadratic space which includes the 0-dimensional quadratic space (cf. the introductory remarks in Section 1). Its equivalence class, which we may denote by $0_S = 0$, provides a zero for the semigroup $S: 0 + s = s + 0 = s$ for all $s \in S$.

LEMMA 5.2. *S is generated (as abelian semigroup with 0) by the classes of the 1-dimensional spaces.*

Proof. For if U, ϕ has the normal basis $\mathbf{u}_1, \ldots, \mathbf{u}_n$ then its class s satisfies $s = s_1 + \ldots + s_n$, where s_j is the class of the 1-dimensional quadratic space spanned by $\mathbf{u}_j$.

By the dimension of an element s of S we shall mean the dimension of any quadratic space U, ϕ in the equivalence class s, and similarly for the determinant. Since U, ϕ is regular the determinant of an $s \in S$ is an element of $k^*/(k^*)^2$. [Here, as always k^* is the group of non-zero elements of k under multiplication.]

For a later purpose we shall need

LEMMA 5.3. *There is an automorphism $s \to s^*$ of the semigroup S defined as follows. If U, ϕ is in s then $U, -\phi$ is in s^*. Further*

$$s + s^* = nh, \tag{5.1}$$

where $h \in S$ is the class of the hyperbolic plane and n is the dimension of s.

Proof. The operation * is clearly well defined and an automorphism. To prove (5.1) it is enough, after Lemma 5.2, to suppose that s is 1-dimensional, say the class of the 1-dimensional space U, with basis $\mathbf{u}$ and $\phi(\mathbf{u}) = a \neq 0$. Then $s + s^*$ is the class of the 2-dimensional space V, ψ (say) with normal basis $\mathbf{u}, \mathbf{u}^*$ (say) and $\psi(\mathbf{u}) = a$, $\psi(\mathbf{u}^*) = -a$. This is isotropic and so is a hyperbolic plane.

We now apply the following piece of "abstract nonsense".

LEMMA 5.4. *Let S be an abelian semigroup with cancellation. Then there is an abelian group G and a semigroup homomorphism α: $S \to G$ with the following property:*

P. *Let H be any abelian group and β: $S \to H$ a semigroup homomorphism.*

Then there is a unique group homomorphism $\gamma\colon G \to H$ *such that*

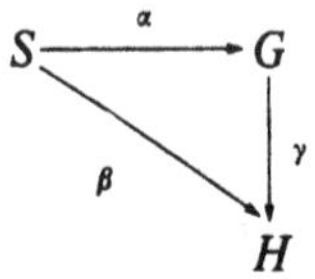

(5.2)

commutes.

The group G is uniquely defined by the above property up to isomorphism. Finally, $\alpha\colon S \to G$ *is an injection.*

Note. A mundane case of Lemma 5.4 is when S is the multiplicative semigroup of rational integers and G is the group of non-zero rationals.

Proof. (Sketch). This is a typical "universal mapping property" and as usual the uniqueness is routine granted the existence. Indeed if $\alpha_j\colon S \to G_j$ $(j = 1, 2)$ both have the property P then (taking $H = G_2$, and G_1 respectively) there are unique group homomorphisms γ_1, γ_2 such that

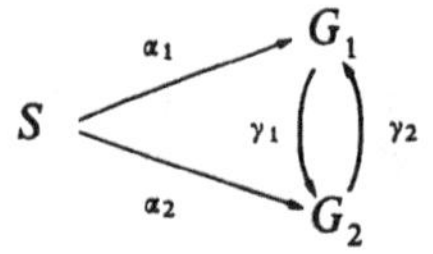

commutes. It follows that γ_1, γ_2 are isomorphisms.

To prove existence denote by $S \times S$ the set of pairs (s, s') with $s, s' \in S$. Then $S \times S$ has a semigroup structure defined by

$$(s_1, s_1') + (s_2, s_2') = (s_1 + s_2, s_1' + s_2').$$

It is easily verified that $S \times S$ is a semigroup with cancellation.

We write $(s_1, s_1') \equiv (s_2, s_2')$ if $s_1 + s_2' = s_1' + s_2$. It is readily verified that $\equiv$ is an equivalence relation in the technical sense (reflexive, symmetric, transitive) and that it respects addition in the semigroup $S \times S$. Hence the set G of classes for $\equiv$ has a semigroup structure induced by that of $S \times S$. We now show that this structure makes G a group. In fact the equivalence class $0 \in G$ consisting of the pairs (s, s) $(s \in S)$ satisfies $0 + g = g + 0 = g$ for all $g \in G$: and if $(s, s') \in g$ then the class $(-g)$ to which (s', s) belongs satisfies $(-g) + g = 0$. The group axioms are thus satisfied.

The map $\alpha\colon S \to G$ which takes s into the class of $(s, 0)$ is an injection.

Finally, any semigroup homomorphism $\beta\colon S \to H$ into any abelian group H extends to $S \times S$ by mapping (s, s') into $\beta s - \beta s'$: and this induces a group homomorphism $G \to H$, which clearly has all the required properties. This concludes the proof.

When Lemma 5.4 is applied to the semigroup S of equivalence classes of regular quadratic spaces we obtain the *Grothendieck group* $G = G(k)$ of the field k. By the lemma, the definitions of dimension and determinant on S extend to give group homomorphisms

$$\dim\colon G \to \mathbf{Z} \tag{5.3}$$

$$\det\colon G \to k^*/(k^*)^2 \tag{5.4}$$

which we shall continue to call the dimension and determinant respectively.

We now introduce the Witt group $W = W(k)$. This is both more concrete and historically older than the Grothendieck group but it is convenient now to introduce it via the latter. Let $W \subset S$ denote the set of equivalence classes of anisotropic forms. The zero-dimensional form is, by definition, anisotropic and so $0 \in W$. Let s be any element of S. Then either $s \in W$ or by Lemma 2.1, Corollary, we have $s = s' + h$ where h is the class of the hyperbolic plane and $s' \in S$. It follows by induction that

$$s = w + mh \tag{5.5}$$

where $w = w(s) \in W$ and $m = m(s)$ is a non-negative integer. In particular if $w_1, w_2 \in W$ we have $w_1 + w_2 = w_3 + m_{12}h$ for some w_3 and some non-negative integer m_{12}. Let us write $w_3 = w_1 \underset{W}{+} w_2$. Then $\underset{W}{+}$ gives the set W the structure of an abelian semigroup with zero 0.

Further, if w^* is defined for $w \in W$ as in Lemma 5.3, we have $w + w^* = nh$ for some $n \in \mathbf{Z}$, and so $w \underset{W}{+} w^* = 0$. Hence W with the operation $\underset{W}{+}$ is an abelian group, the *Witt group* $W(k)$.

The map $s \to w$ given by (5.5) gives a semigroup homomorphism $S \to W$. This extends to a group homomorphism $G \to W$ whose kernel is clearly generated by h. This gives

LEMMA 5.5. $W = G/K$ *where* K *is the subgroup of* G *generated by* h.

Since the hyperbolic plane h has dimension 2, we have a homomorphism

$$\dim_W\colon W \to \mathbf{Z}/2\mathbf{Z} \tag{5.6}$$

on W which we call dimension [this is not to be confused with the dimension $\dim(w)$ of $w \in W$ considered as an element of S]. Further, since $\det(h) = (-1)(k^*)^2$ we can define a determinant map only as†:

$$\det_W\colon W \to k^*/K_W \tag{5.7}$$

where K_W is the subgroup of k^* generated by -1 and $(k^*)^2$.

† Some authors introduce a modified determinant on S (and G): $\det^*(s) = (-1)^{n(n-1)/2}\det(s)$, where $n = \dim(s)$. This induces a well-defined $\det^*_W$ on W but neither $\det^*$ not $\det^*_W$ is a homomorphism.

We are now in a position to describe the Grothendieck group $G(k)$ and the Witt group $W(k)$ in terms of generators and relations. If $a \in k^*$ we write $\langle a \rangle$ for the equivalence class of the 1-dimensional regular quadratic space U, ϕ with a basis $\mathbf{u}_1$ such that $\phi(\mathbf{u}_1) = a$.

LEMMA 5.6. *$G(k)$ is the abelian group given by the generators $\langle a \rangle$, $a \in k^*$ with the relations*

(i) $\langle ab^2 \rangle = \langle a \rangle \qquad (a, b \in k^*)$
(ii) $\langle a \rangle + \langle b \rangle = \langle a + b \rangle + \langle ab(a+b) \rangle$ for $a, b \in k^*$ *with* $a + b \in k^*$.

Proof. By Lemma 5.2 the $\langle a \rangle$ generate G. We shall check first that the relations (i), (ii) hold and then show that all other relations follow from them.

The relation (i) expresses the fact that the 1-dimensional space U, ϕ with basis $\mathbf{u}_1, \phi(\mathbf{u}_1) = a$ also has the basis $b\mathbf{u}_1$ where $\phi(b\mathbf{u}_1) = b^2\phi(\mathbf{u}_1) = b^2 a$.

The left-hand side of (ii) is the equivalence class of the 2-dimensional space U, ϕ with normal basis $\mathbf{u}_1, \mathbf{u}_2$ with $\phi(\mathbf{u}_1) = a, \phi(\mathbf{u}_2) = b$. Provided that $a + b \neq 0$, another normal basis is $\mathbf{u}_1 + \mathbf{u}_2$ and $b\mathbf{u}_1 - a\mathbf{u}_2$ where $\phi(\mathbf{u}_1 + \mathbf{u}_2) = a + b$ and $\phi(b\mathbf{u}_1 - a\mathbf{u}_2) = ab(a + b)$. [Note that the determinants on both sides of (ii) are the same element of $k^*/(k^*)^2$ which gives a check.]

We now verify that there are no further relations. Any relation can be put in the shape

$$\sum_{j=1}^{n} \langle a_j \rangle = \sum_{j=1}^{n} \langle c_j \rangle \tag{5.8}$$

with $a_j, c_j \in k^*$. This expresses the fact that there is a regular quadratic space U, ϕ with two normal bases $\mathbf{u}_1, \ldots, \mathbf{u}_n$ and $\mathbf{v}_1, \ldots, \mathbf{v}_n$ (say) such that $\phi(\mathbf{u}_j) = a_j$ and $\phi(\mathbf{v}_j) = c_j$. Suppose first that $n = 1$. Then $\mathbf{v}_1 = b\mathbf{u}_1$ for some $b \in k^*$ and we have a case of relation (i). Now let $n = 2$. Then $\mathbf{v}_1 = b_1\mathbf{u}_1 + b_2\mathbf{u}_2$. If either $b_1 = 0$ or $b_2 = 0$ our relation (5.8) is a consequence of relation (i), and, possibly, of the fact that $G(k)$ is given as abelian. Otherwise, on using relation (i) we may suppose without loss of generality that $b_1 = b_2 = 1$. Then $0 \neq \phi(\mathbf{v}_1) = \phi(\mathbf{u}_1) + \phi(\mathbf{u}_2)$; and we have a case of (ii).

Finally, suppose that $n > 2$. By Lemma 3.1 we can get from the basis $\mathbf{u}_1, \ldots, \mathbf{u}_n$ to the basis $\mathbf{v}_1, \ldots, \mathbf{v}_n$ through a sequence of normal bases where each time we change at most two basis vectors. Hence the identity (5.8) must be a consequence of those with $n = 1$ or $n = 2$. This concludes the proof.

COROLLARY. *The Witt group $W(k)$ is given by the generators $\langle a \rangle$ with $a \in k^*$ and the relations*

$(\mathrm{i})_W \quad \langle ab^2 \rangle \underset{W}{=} \langle a \rangle \qquad (a, b \in k^*)$

$(\mathrm{ii})_W$ $\langle a\rangle \underset{W}{+} \langle b\rangle = \langle a \underset{W}{+} b\rangle + \langle ab(a+b)\rangle \qquad (a, b, a+b \in k^*)$

$(\mathrm{iii})_W$ $\langle 1\rangle \underset{W}{+} \langle -1\rangle = 0.$

Proof. $(\mathrm{i})_W$, $(\mathrm{ii})_W$ reproduce (i) and (ii) with $\underset{W}{+}$ instead of $+$. The Corollary now follows from Lemma 5.5 since (in G)

$$\langle 1\rangle + \langle -1\rangle = h.$$

We conclude this section by finding $G(k)$ and $W(k)$ in two special cases.

LEMMA 5.7. *Let* $\mathbf{R}$ *be the real field. Then* $G(\mathbf{R})$ *is the free abelian group on the two generators* $\langle 1\rangle$ *and* $\langle -1\rangle$.

Proof. Clear. This is effectively "Sylvester's Law of Inertia".

COROLLARY. $W(\mathbf{R})$ *is free on the single generator* $\langle 1\rangle$.

LEMMA 5.8. *Let* $\mathbf{F}_p$ *be the field of* p *elements* ($p \neq 2$, *prime*). *Let* $\alpha \in \mathbf{F}_p^*$, $\alpha \notin (\mathbf{F}_p^*)^2$. *Then* $G(\mathbf{F}_p)$ *is generated by* $\langle 1\rangle$ *and* $\langle \alpha\rangle$ *with the relation* $2\langle 1\rangle = 2\langle \alpha\rangle$.

Proof. $\mathbf{F}_p^*/(\mathbf{F}_p^*)^2$ is of order 2 and so there are precisely the two 1-dimensional regular quadratic spaces $\langle 1\rangle$ and $\langle \alpha\rangle$. By Lemma 2.2 the form $\langle 1\rangle + \langle 1\rangle$ is universal and so represents α. Hence $\langle 1\rangle + \langle 1\rangle = \langle \alpha\rangle + \langle \alpha\rangle$ and this can be the only relation since any other would have the false implication $\langle 1\rangle = \langle \alpha\rangle$.

COROLLARY. $W(\mathbf{F}_p)$ *is of order* 4. *It is non-cyclic if* $-1 \in (\mathbf{F}_p^*)^2$, *otherwise cyclic.*

6. SINGULAR FORMS

In this section we indicate briefly how the theory extends to singular quadratic spaces and to singular quadratic forms.

Let V, ϕ be a possibly singular quadratic space of dimension n over a field k (of characteristic $\neq 2$, as always). The *radical* V_0 of V, ϕ consists of the $\mathbf{a} \in V$ such that $\phi(\mathbf{a}, \mathbf{b}) = 0$ for all $\mathbf{b} \in V$. If $\mathbf{a}_1, \mathbf{a}_2 \in V_0$ then clearly $u_1\mathbf{a}_1 + u_2\mathbf{a}_2 \in V_0$ for all $u_1, u_2 \in V$. Hence V_0 is a linear subspace of V. The space V, ϕ is regular, by the definition given in Section 1, precisely when the radical V_0 consists only of the zero vector.

If $\mathbf{b}_1 - \mathbf{b}_2 \in V_0$, then clearly $\phi(\mathbf{b}_1) = \phi(\mathbf{b}_2)$. Hence ϕ gives a well-defined function $\bar{\phi}$ (say) on the quotient space $\bar{V} = V \bmod V_0$. It is readily verified that $\bar{V}, \bar{\phi}$ is a k-quadratic space. Further, $\bar{V}, \bar{\phi}$ is regular: for if $\bar{\mathbf{a}} \in \bar{V}$ is such that $\bar{\phi}(\bar{\mathbf{a}}, \bar{\mathbf{b}}) = 0$ for all $\bar{\mathbf{b}} \in \bar{V}$ then $\phi(\mathbf{a}, \mathbf{b}) = 0$ for any $\mathbf{a} \in \bar{\mathbf{a}}$ and all $\mathbf{b} \in V$, so $\mathbf{a} \in V_0$ and $\bar{\mathbf{a}} = 0$.

Let $\bar{\mathbf{e}}_1, \ldots, \bar{\mathbf{e}}_{n-r}$ be a basis of $\bar{V}$, let $\mathbf{e}_{n-r+1}, \ldots, \mathbf{e}_n$ be a basis of V_0, where $r = \dim V_0$, and let $\mathbf{e}_1, \ldots, \mathbf{e}_{n-r}$ be any representatives in V of $\bar{\mathbf{e}}_1, \ldots, \bar{\mathbf{e}}_{n-r}$. Then clearly $\mathbf{e}_1, \ldots, \mathbf{e}_n$ is a basis of V. Let V_1 be the subspace of V spanned by $\mathbf{e}_1, \ldots, \mathbf{e}_{n-r}$ so V is the direct sum of V_0 and V_1. Then ϕ restricted to V_1 gives it a quadratic space structure isomorphic to $\bar{V}, \bar{\phi}$ and so regular. We have thus proved

LEMMA 6.1. *Any quadratic space* V, ϕ *can be expressed as a direct sum* $V_0 \oplus V_1$, *where* V_0 *is the radical and* ϕ *restricted to* V_1 *gives a regular quadratic space.*

We can reformulate this in terms of quadratic forms as:

LEMMA 6.2. *Every quadratic form* $f(x_1, \ldots, x_n)$ *over* k *is equivalent to a form*

$$g(x_1, \ldots, x_n) = h(x_1, \ldots, x_{n-r})$$

for some r, *where* h *is a regular quadratic form in some number* $n - r \leqslant n$ *of variables.*

NOTES

The potentialities of the point of view adopted here were first recognized by Witt (1936). For generalizations see Artin (1957) and Dieudonné (1955).

Section 5. One may also define a tensor product U, ϕ of two quadratic spaces U_1, ϕ_1 and U_2, ϕ_2 by putting

$$U = U_1 \otimes U_2$$

and

$$\phi(\mathbf{u}_1 \otimes \mathbf{u}_2, \mathbf{v}_1 \otimes \mathbf{v}_2) = \phi_1(\mathbf{u}_1, \mathbf{v}_1)\,\phi_2(\mathbf{u}_2, \mathbf{v}_2)$$

for

$$\mathbf{u}_j, \mathbf{v}_j \in U_j \qquad (j = 1, 2).$$

This gives the Grothendieck and Witt groups a ring structure. See, for example, Lam (1973), Lorenz (1970), Milnor and Husemoller (1973), Scharlau (1969).

EXAMPLES

1. Find $a_1, a_2, a_3 \in \mathbf{Q}$ such that $f(a_1, a_2, a_3) = 5$, where $f(\mathbf{x}) = 3x_1^2 - 2x_2^2 - x_3^2$.
[*Hint*: $f(1, 1, 1) = 0$.]

2. Let $f(\mathbf{x})$ be a regular isotropic form in the variables $\mathbf{x} = (x_1, \ldots, x_n)$ over the field k. Show that there are $a_1, \ldots, a_n, b_1, \ldots, b_n \in k$ such that

$$f(a_1 + b_1 t, a_2 + b_2 t, \ldots, a_n + b_n t) = t$$

identically in the variable t.

3. Let $b \in k^*$ be fixed. Let A be the set of $a \in k^*$ which are represented over k by the binary form $x_1^2 + bx_2^2$. Show that A is a subgroup of k^*. [k^* is the multiplicative group of non-zero elements of k].

[*Hint*: cf. proof of Lemma 2.1 of Chapter 3.]

4. Let $a, b \in k^*$. Show that the form

$$x_1^2 + ax_2^2 + bx_3^2$$

is isotropic if and only if the form

$$x_1^2 + ax_1^2 + bx_3^2 + abx_4^2$$

is isotropic.

[*Hint*: see previous example.]

5. If σ is any autometry of U, ϕ and $\mathbf{a}, \mathbf{b} \in U$, then

$$\phi(\sigma\mathbf{a} - \mathbf{a}, \mathbf{b}) + \phi(\mathbf{a}, \sigma\mathbf{b} - \mathbf{b}) + \phi(\sigma\mathbf{a} - \mathbf{a}, \sigma\mathbf{b} - \mathbf{b}) = 0.$$

6. Let U, ϕ be a regular quadratic space of dimension n and suppose that ϕ vanishes identically on a subspace V of dimension r, Show that $2r \leqslant n$ and that U is the direct sum of r hyperbolic planes $H_1, \ldots, H_r$ and a regular subspace W of dimension $n - 2r$, all mutually orthogonal.

7. Let U, ϕ be a regular quadratic space of dimension $n = 2r$ and let V be a subspace of dimension r on which ϕ vanishes completely.

(i) Show that U is the direct sum of V and of a subspace W of U of dimension r on which ϕ also vanishes identically.

(ii) If $\mathbf{v}_1, \ldots, \mathbf{v}_r$ is a basis of V, show that W has a basis $\mathbf{w}_1, \ldots, \mathbf{w}_r$ such that

$$\phi(\mathbf{v}_i, \mathbf{w}_j) = \begin{cases} 1 & \text{if } i = j \\ 0 & \text{otherwise.} \end{cases}$$

(iii) Let W^* be any subspace of U having the properties prescribed for W in (i). Show that W^* has a basis $\mathbf{w}_1^*, \ldots, \mathbf{w}_r^*$ given by

$$\mathbf{w}_i^* = \mathbf{w}_i + \sum_j l_{ij}\mathbf{v}_j$$

where $l_{ij} + l_{ji} = 0$.

(iv) Let σ be an autometry of U such that $\sigma V = V$. Show that σ is proper (i.e. $\det \sigma = +1$).

[*Hint*. Reduce to the case when also $\sigma W = W$.]

8. Show that an autometry σ of a regular quadratic space U, ϕ of dimension n can be expressed as the product of at most n symmetries by filling in details of the following argument:

(i) if there is an anisotropic $\mathbf{a}$ with $\sigma\mathbf{a} = \mathbf{a}$ use induction on n.

(ii) if there is an anisotropic $\mathbf{a}$ such that $\mathbf{b} = \sigma\mathbf{a} - \mathbf{a}$ is also anisotropic, consider $\sigma\tau_{\mathbf{b}}$ and use induction.

(iii) Hence can suppose from now on (A) that $\sigma\mathbf{a} = \mathbf{a}$ implies $\mathbf{a}$ isotropic and (B) that $\sigma\mathbf{a} - \mathbf{a}$ is isotropic whenever $\mathbf{a}$ is anisotropic.

(iv) If $\mathbf{a}$ is isotropic, there is a $\mathbf{b}$ such that $\mathbf{a} \pm \mathbf{b}$ are both anisotropic. Hence $\sigma\mathbf{a} - \mathbf{a}$ is isotropic.

[*Hint*. (iiiB).]

(v) Hence ϕ vanishes identically on

$$V = \{\sigma\mathbf{a} - \mathbf{a} : \mathbf{a} \in U\}$$

and $\sigma V = V$.

(vi) If $\mathbf{b} \in V^{\perp}$ then $\phi(\mathbf{a}, \sigma\mathbf{b} - \mathbf{b}) = 0$ for all $\mathbf{a} \in U$ and so $\sigma\mathbf{b} = \mathbf{b}$.

[*Hint*. Example 5.]

(vii) Hence ϕ vanishes identically on $V^{\perp}$ by (iiiA).

(viii) Hence $\dim U = 2 \dim V = 2r$ (say).

(ix) Hence $\sigma \in O^{+}(V)$.

[*Hint*. Example 7.]

(x) Let τ be any symmetry. Then $\tau\sigma$ does not satisfy the conditions (iii) and so is the product of at most $n = 2r$ symmetries.

(xi) From determinantal considerations $\tau\sigma$ is the product of at most $n - 1$ symmetries and so σ requires at most n symmetries.

[For crib, see Artin (1957) pp 129–30, or O'Meara (1963) pp 102–103.]

9. Let $\sigma \in O^{+}(V)$ where $n = \dim V$ is odd. Show that there is an $\mathbf{a} \neq \mathbf{0}$ in V such that $\sigma\mathbf{a} = \mathbf{a}$.

[*Hint*. Show that σ is the product of $<n$ symmetries, cf. also example 11.]

10. If $\sigma \in O(V)$ can be expressed as the product of $n = \dim V$ symmetries show that it can be expressed as such a product with the first symmetry being fixed arbitrarily.

11. (i) Let $\phi(\lambda) = \det(\mathbf{T} - \lambda\mathbf{I})$ where $\mathbf{T}$ is a proper automorph of a regular quadratic form f in n variables. Show that $\phi(\lambda) = (-\lambda)^n \phi(\lambda^{-1})$.

(ii) If n is odd deduce that $\phi(1) = 0$ and so obtain another solution of example 9.

[*Hint.* $\mathbf{T}' = \mathbf{F}\mathbf{T}^{-1}\mathbf{F}^{-1}$, where $\mathbf{T}'$ is the transpose and $\mathbf{F}$ is the matrix of f.]

12. Show that the order of the multiplicative group of $n \times n$ matrices with elements in the field of p elements is given by

$$(p^n - 1)(p^n - p)\dots(p^n - p^{n-1}).$$

[*Hint.* A matrix $\mathbf{T}$ is determined by the $\mathbf{t}_j = \mathbf{T}\mathbf{e}_j$. Further, $\mathbf{T}$ is non-singular provided that $\mathbf{t}_1 \neq \mathbf{0}$ and $\mathbf{t}_j$ is linearly independent of $\mathbf{t}_1, \dots, \mathbf{t}_{j-1}$ for each $j > 1$.]

13. In this example k is the field of p elements, where p is an odd prime. Let f be a regular quadratic form over k and denote by N, R, A respectively (a) the number of representations of 0 by f (including the trivial representation $f(\mathbf{0}) = 0$), (b) the number of representations of 1 by f and (c) the order of the group of proper automorphs of f.

(i) Show that N, R, A depend only on the dimension n of f and the value of the determinant d as an element of $k^*/(k^*)^2$. Denote them respectively by $N_n(d), R_n(d), A_n(d)$.

(ii) Show that

$$N_2(d) = \begin{cases} 2p - 1 & \text{if } d \in (k^*)^2 \\ 1 & \text{if } d \notin (k^*)^2. \end{cases}$$

(iii) For $n \geqslant 3$ show that

$$N_n(d) = p^{n-2}(p - 1) + pN_{n-2}(-d).$$

[*Hint.* Consider forms $f = 2x_1x_2 + g(x_3, \dots, x_n)$.]

(iv) Deduce that

$$N_{2m+1}(d) = p^{2m} \qquad (m \geqslant 0)$$

and that

$$N_{2m}(d) = p^{2m-1} + (p - 1)p^{m-1}\chi((-)^m d) \qquad (m > 0)$$

where

$$\chi(a) = \begin{cases} 1 & \text{if } a \in (k^*)^2 \\ -1 & \text{otherwise.} \end{cases}$$

(v) Show that

$$R_{2m}(d) = p^{2m-1} - p^{m-1}\chi((-)^m d) \qquad (m > 0)$$

and

$$R_{2m+1}(d) = p^{2m} + p^m\chi((-)^m d) \qquad (m \geqslant 0).$$

[*Hint*. Consider the representations of 0 by $-x_0^2 + f(x_1, \ldots, x_n)$ distinguishing between $x_0 = 0$ and $x_0 \neq 0$.]

(vi) Show that

$$A_{2m+1}(d) = p^{m^2}(p^2 - 1)(p^4 - 1)\ldots(p^{2m} - 1)$$

and

$$A_{2m}(d) = p^{m(m-1)}(p^2 - 1)(p^4 - 1)\ldots(p^{2m-2} - 1)(p^m - \varepsilon)$$

where

$$\varepsilon = \chi((-)^m d).$$

[*Hint*. Show that $A_n(d) = R_n(d)\,A_{n-1}(d)$.]

14. Let V be a quadratic space of dimension n over the real field $\mathbf{R}$. Define a topology on $O(V)$ as follows: Let $\mathbf{b}_1, \ldots, \mathbf{b}_n$ be a basis of V. To each $\sigma \in O(V)$ there corresponds the point of $\mathbf{R}^{n^2}$ with coordinates s_{ij} where $\sigma\mathbf{b}_i = \Sigma_j s_{ij}\mathbf{b}_j$ $(s_{ij} \in \mathbf{R})$.

(i) Show that this topology on $O(V)$ is independent of the choice of basis $\mathbf{b}_1, \ldots, \mathbf{b}_n$.

(ii) Show that $O^+(V)$ is both open and closed in $O(V)$.

(iii) Show that $O^+(V)$ is connected.

[*Hint*. Each $\sigma \in O^+(V)$ is the product of expressions $\tau_{\mathbf{a}}\tau_{\mathbf{b}}$. Consider what happens as $\mathbf{b}$ tends to $\mathbf{a}$.]

15. Let $f(\mathbf{x})$ be a regular quadratic form over a field k of characteristic $\neq 2$. Suppose that f is isotropic over K, where K is an algebraic extension of k of odd degree. Show that f is isotropic over k.

[*Hint*. Suppose not and let the degree m of K/k be minimal for a counter-example, say $K = k(\theta)$, where θ has the irreducible equation $F(t) = 0$. Let $f(\boldsymbol{\alpha}) = 0$, where $\alpha_j = \sum_{0 \leqslant r < m} a_{jr}\theta^r$ $(a_{jr} \in k)$. Put $A_j(t) = \sum_r a_{jr}t^r$. Show that

$$f(A_1(t), \ldots, A_n(t))$$

is divisible by an irreducible polynomial $G(t)$ of odd degree $< m$. Deduce that f is isotropic over $L = k(\lambda)$, where $G(\lambda) = 0$.]

16. Let U, ϕ be an anisotropic quadratic space over the field k of characteristic $\neq 2$. Let $K = k(\delta)$ where $\delta^2 = D \in k^*$ but $\notin (k^*)^2$. Suppose that $\phi(\mathbf{u} + \mathbf{v}\delta) = 0$, where $\mathbf{u}, \mathbf{v} \in U$ are not both $\mathbf{0}$, and let $V \subset U$ be the k-subspace of U spanned by $\mathbf{u}$ and $\mathbf{v}$. Show that $\dim V = 2$ and that ϕ induces on V a regular form of determinant $-D(k^*)^2$.

17. Let U, ϕ be a regular quadratic space over a field k of characteristic $\neq 2$. Suppose that $\dim U = 4$, that $d(\phi) = d(k^*)^2$ and that U, ϕ is isotropic over $K = k(\delta)$, where $\delta^2 = d$. Show that U, ϕ is isotropic over k.

[*Hint*. We can suppose $K \neq k$. With the notation of the previous example show that $V^\perp$ is a hyperbolic plane.]

18. Deduce from the relations for the Witt group (Lemma 5.6, Corollary) that

$$\langle a \rangle \underset{W}{+} \langle -a \rangle = 0$$

for all $a \in k^*$.

19. Let t be transcendental over the field k and let $f(\mathbf{x}) = g(\mathbf{x}) + th(\mathbf{x})$, where $g(\mathbf{x})$ and $h(\mathbf{x})$ are quadratic forms over k. Show that the two following statements are equivalent:

(i) $f(\mathbf{x})$ is isotropic over $k(t)$.
(ii) there is an $\mathbf{a} \neq \mathbf{0}$ in k^n such that $g(\mathbf{a}) = h(\mathbf{a}) = 0$.

[*Hint*. Let $\mathbf{b} = \mathbf{c} + t^m\mathbf{d}$ for some $m > 0$, where $\mathbf{d} \in k^n$, the elements of $\mathbf{c}$ are polynomials in t of degree $< m$, and $f(\mathbf{b}) = 0, f(\mathbf{d}) \neq 0$. Show that $\mathbf{b}^* = \tau_{\mathbf{d}}\mathbf{b}$ satisfies $f(\mathbf{b}^*) = 0$, $\mathbf{b}^* \neq \mathbf{0}$, and that the elements of $\mathbf{b}^*$ are polynomials of degree $< m$.

Note. See Brumer (1978). This is essentially a special case of the method of Aubry (1912), cf. Chapter 6, Section 8 and Example 6.]

20. Let $g(\mathbf{x})$, $h(\mathbf{x})$ be quadratic forms with coefficients in the field k and let K be an algebraic extension of k of odd degree. Suppose that there is an $\mathbf{A} \neq \mathbf{0}$ in K^n such that $g(\mathbf{A}) = h(\mathbf{A}) = 0$. Show that there is an $\mathbf{a} \neq \mathbf{0}$ in k^n such that $g(\mathbf{a}) = h(\mathbf{a}) = 0$.

[*Hint*. Examples 15, 19. See Brumer (1978).]

CHAPTER 3

p-Adic Numbers

1. INTRODUCTION

It was the introduction of p-adic numbers by Hensel and their exploitation in the context of quadratic forms by his pupil Hasse which gave rise to the elegant formulation of the theory of quadratic forms over the rationals given here. The p-adic numbers play an increasingly central role in the theory of numbers and in mathematics generally. In this Section we introduce the p-adics from scratch for those who have the misfortune of not having met them before. In Section 2 we introduce Hilbert's Norm Residue Symbol and in Section 3 we discuss the link between behaviour over the rationals and over the various p-adic fields.

Throughout, our purpose is to prove the minimum that will be necessary for our purposes. Practically all the results of this chapter generalize to algebraic number fields and the corresponding local fields but the proofs are often substantially more sophisticated. [See, for example, O'Meara (1963), Serre (1962) or Cassels and Fröhlich (1967) (especially the exercises at the end).]

Let k be any field. A real-valued function $|a|$ $(a \in k)$ is called a *valuation*† if it has the following three properties:

(i) $|a| \geqslant 0$ and $|a| = 0$ only for $a = 0$.

(ii) $|ab| = |a|\,|b| \qquad (a, b \in k)$.

(iii) $|a + b| \leqslant |a| + |b| \qquad (a, b \in k)$.

For example if k is the rational field $\mathbf{Q}$, then the ordinary absolute value has these properties. To distinguish the absolute value from the other valuations we may encounter, we denote it by $|\ |_\infty$.

Let p be any prime number which we keep fixed. Then any $r \in \mathbf{Q}$ other

† Strictly, a rank 1 valuation: but we shall not encounter any others.

than 0 can be put in the shape

$$r = p^\rho u/v \tag{1.1}$$

where ρ, u, v are integers, $p \nmid u$, $p \nmid v$ and ρ may be positive, negative or zero. By the uniqueness of factorization ρ is uniquely determined by p and r. We write

$$|r|_p = p^{-\rho}$$

and complete the definition by putting

$$|0|_p = 0.$$

Then $|\ |_p$ is also a valuation. The properties (i), (ii) above are readily verified. We show now that not merely is (iii) true but that $|\ | = |\ |_p$ satisfies the stronger condition:

(iii)′ $|a + b| \leqslant \max\{|a|, |b|\}$.

For suppose that

$$a = p^\alpha u/v, \qquad b = p^\beta x/y$$

where u, v, x, y are integers prime to p. We may suppose without loss of generality that $|a|_p \leqslant |b|_p$; that is $\beta \leqslant \alpha$. Then

$$\begin{aligned} a + b &= p^\beta(p^{\alpha-\beta}uy + vx)/vy \\ &= p^{\beta+\delta}z/vy, \end{aligned}$$

where $p^{\alpha-\beta}uy + vx = p^\delta z$ and $p \nmid z$ and $\delta \geqslant 0$. Here $p \nmid vy$ and so

$$|a + b|_p = p^{-\beta-\delta} \leqslant p^{-\beta} = |b|_p = \max\{|a|_p, |b|_p\}.$$

We call (iii) the "*triangle inequality*" and (iii)′ is the "*ultrametric inequality*". A valuation which satisfies the ultrametric inequality is *non-archimedean*. The triangle inequality implies that a valuation $|\ |$ on a field k gives a metric, the distance between $a, b \in k$ being $|a - b|$. When the ultrametric inequality holds this metric has somewhat paradoxical properties: for example, with the obvious definitions every point of a circular disc can be taken as the centre.

Before going on we note that it follows immediately from (i) and (ii) that

$$|-1| = |1| = 1$$

and so

$$|-a| = |a| \qquad (\text{all } a \in k).$$

Further we have

LEMMA 1.1. *Suppose that* $|\ |$ *is non-archimedean. Then*

$$|a + b| = \max\{|a|, |b|\}$$

whenever $|a|$, $|b|$ *are unequal.*

Proof. For suppose that $|a| < |b|$ and that $|a+b| < |b|$. Then $b = a + b + (-a)$ and so

$$|b| \leqslant \max\{|a+b|, |a|\} < |b|.$$

Contradiction! We now recall the definitions leading up to the notation of completeness in the context of a field k with valuation $|\;|$. First, a sequence a_j $(j = 1, 2, \ldots)$ of elements of k has the *limit* b if for any $\varepsilon > 0$ there is an $n_0 = n_0(\varepsilon)$ such that $|a_j - b| < \varepsilon$ whenever $n \geqslant n_0$. Clearly a sequence can have at most one limit (with respect to a fixed valuation $|\;|$). A sequence which has a limit is *convergent*. A sequence a_j is a *fundamental sequence* (Cauchy sequence) if for every $\varepsilon > 0$ there is an $u_0(\varepsilon)$ such that $|a_i - a_j| < \varepsilon$ whenever $i, j > n_0(\varepsilon)$. Clearly every convergent sequence is a fundamental sequence. The field k is said to be *complete* (with respect to the valuation) if every fundamental sequence is convergent.

Every field with a valuation can be embedded in a complete field, its *completion* with respect to the valuation. For example the real field $\mathbf{R}$ is the completion of $\mathbf{Q}$ with respect to the absolute value $|\;|_\infty$. Similarly the *p-adic field* $\mathbf{Q}_p$ is the completion of $\mathbf{Q}$ with respect to $|\;|_p$. We shall sometimes write $\mathbf{R} = \mathbf{Q}_\infty$ for the completion with respect to "the infinite prime" ∞ which we have used as a label for the ordinary absolute value. The analogy is only partial, but we shall see that it is very useful.

[There is a parallel case in which the analogy is more complete. Let k_0 be any algebraically closed field and let $k = k_0(t)$, where t is transcendental over k_0. Let $c \in k_0$. Every $f(t) \in k_0(t)$ other than 0 can be written in the shape

$$f(t) = (x - c)^{\rho(c)} f_1(t)/f_2(t)$$

where $f_j(t) \in k_0[t]$, $f_j(c) \neq 0$ $(j = 1, 2)$ and $\rho(c)$ is positive negative or zero. We can write

$$|f(t)|_c = e^{-\rho(c)}$$

where $e > 1$ is some constant. Similarly if

$$f(t) = g_1(t)/g_2(t)$$

where $g_j(t) \in k_0[t]$ and $\rho(\infty) = \deg g_2 - \deg g_1$ we have the infinite valuation

$$|f(t)|_\infty = e^{-\rho(\infty)}$$

Then all the $|\;|_c$ and $|\;|_\infty$ are non-archimedean and it is not difficult to show that any valuation of $k_0(t)$ which is "trivial" on k_0 is "equivalent" to one of the $|\;|_c$ or $|\;|_\infty$ (with appropriate senses for "trivial" and "equivalent"). But now $|\;|_\infty$ is completely analogous to the $|\;|_c$. Indeed which valuation bears the label "∞" depends on the choice of the generator t of $k_0(t)/k_0$. If we take another generator t^* (say) $= 1/t$ (or, more generally, $t^* = (at + b)/(ct + d)$: $a, b, c, d \in k_0$) then another valuation becomes the valuation "at infinity".]

Lemma 1.2. *There is a field* $\mathbf{Q}_p$ *containing the rational field* $\mathbf{Q}$ *and a non-archimedean valuation* $| \, |_p$ *on* $\mathbf{Q}_p$ *with the following properties.*

(i) *On* $\mathbf{Q}$ *the valuation* $| \, |_p$ *is the same as that defined above.*

(ii) $\mathbf{Q}_p$ *is complete with respect to* $| \, |_p$.

(iii) $\mathbf{Q}_p$ *is the closure of* $\mathbf{Q}$ *with respect to (the topology defined by)* $| \, |_p$.

Further, $\mathbf{Q}_p$ *is unique up to isomorphism.*

Note. Uniqueness is trivial. Indeed let K be any field containing $\mathbf{Q}$ and let $| \, |$ be a valuation on K such that (i) K is complete and (ii) $| \, |$ coincides with $| \, |_p$ on $\mathbf{Q}$. Then the closure C (say) of $\mathbf{Q}$ with respect to K is clearly isomorphic to $\mathbf{Q}_p$. Below we shall give a formal proof of the existence of $\mathbf{Q}_p$. As we shall never refer to the details of the proof again the trusting reader may skip to the enunciation of Lemma 1.3.

Proof. After the Note we need prove only existence. This is a special case of the general theorem about the existence of the completion of a valued field but we take advantage of some simplifications permitted by our special case.

Let $\mathscr{F}$ denote the set of fundamental sequences $\mathbf{a} = \{a_1, a_2, \ldots\}$, as defined above, in the special case when the field k is $\mathbf{Q}$ and $| \, |$ is the p-adic valuation $| \, |_p$. Then $\mathscr{F}$ can be made into a ring by defining $+$, $-$, $\times$ element-wise:

$$\{a_1, a_2, \ldots, a_j, \ldots\} \overset{+}{\underset{\times}{-}} \{b_1, b_2, \ldots, b_j, \ldots\}$$
$$= \left\{a_1 \overset{+}{\underset{\times}{-}} b_1, \ldots, a_j \overset{+}{\underset{\times}{-}} b_j, \ldots\right\}.$$

A sequence $\mathbf{a} = \{a_1, a_2, \ldots\}$ is *null* if it has 0 as its limit. The set $\mathscr{N}$ of null sequences is clearly an ideal in $\mathscr{F}$.

Suppose that $\mathbf{a} \in \mathscr{F}$, $\mathbf{a} \notin \mathscr{N}$. Then there is some $\delta > 0$ (depending on $\mathbf{a}$) such that $|a_j|_p \geqslant \delta$ infinitely often. By the definition of $\mathscr{F}$ there is some $n_0 = n_0(\varepsilon)$ such that $|a_i - a_j|_p < \delta$ for all $i, j \geqslant n_0$. On picking j so that $|a_j|_p \geqslant \delta$, we see by Lemma 1.1 that $|a_i|_p = |a_j|_p$ for all $i \geqslant n_0$. We denote this common value by $\|\mathbf{a}\|$. If $\mathbf{a} \in \mathscr{N}$ we put $\|\mathbf{a}\| = 0$. It is then easily verified that

(i)″ $\|\mathbf{a}\| \geqslant 0$, with equality only when $\mathbf{a} \in \mathscr{N}$

(ii)″ $\|\mathbf{ab}\| = \|\mathbf{a}\| \, \|\mathbf{b}\|$

(iii)″ $\|\mathbf{a} + \mathbf{b}\| \leqslant \max \|\mathbf{a}\|, \|\mathbf{b}\|$

(iv)″ if $\mathbf{a} - \mathbf{b} \in \mathscr{N}$ then $\|\mathbf{a}\| = \|\mathbf{b}\|$.

The next step is to show that $\mathscr{N}$ is actually a maximal ideal in $\mathscr{F}$. For let $\mathscr{I}$ be an ideal strictly greater than $\mathscr{N}$, so $\mathscr{I}$ contains a $\mathbf{c} = \{c_1, c_2, \ldots\} \notin \mathscr{N}$.

We may suppose that all the c_j are non-zero since replacing the zeros in the sequence by (say) 1 adds a null-sequence to $\mathbf{c}$. Then the sequence $\mathbf{c}^{-1} = \{c_1^{-1}, c_2^{-1}, \ldots, c_j^{-1}, \ldots\}$ is well defined. It is a fundamental sequence since

$$|c_i^{-1} - c_j^{-1}| = \frac{|c_i - c_j|}{|c_i|\,|c_j|}$$
$$= \frac{|c_i - c_j|}{\|\mathbf{c}\|^2}$$

if i, j are large enough and since $\mathbf{c}$ is a fundamental sequence by definition. Since $\mathbf{c}^{-1} \in \mathscr{F}$ and $\mathbf{c} \in \mathscr{T}$, the ideal $\mathscr{T}$ contains $\mathbf{c}^{-1}\mathbf{c} = \{1, 1, \ldots\}$. Hence $\mathscr{T} = \mathscr{F}$; and so $\mathscr{N}$ is maximal.

Since $\mathscr{N}$ is a maximal ideal, the quotient ring $\mathscr{F}/\mathscr{N}$ is a field, the field $\mathbf{Q}_p$ of p-adic numbers which we are seeking. For any $\alpha \in \mathscr{F}/\mathscr{N}$ we write temporarily $|\alpha|' = \|\mathbf{a}\|$ for $\mathbf{a} \in \alpha$. By (iv)″ this is a valid definition and by (i)″, (ii)″, (iii)″ the function $|\ |'$ is a non-archimedean valuation on $\mathbf{Q}_p$.

For any $r \in \mathbf{Q}$ we denote by $\lambda(r) \in \mathbf{Q}_p$ the class modulo $\mathscr{N}$ of the fundamental sequence $\{r, r, \ldots\}$. Clearly

$$r \to \lambda(r)$$

is an embedding† of $\mathbf{Q}$ in $\mathbf{Q}_p$ and $|\lambda(r)|' = |r|_p$. We may therefore identify $\mathbf{Q}$ with $\lambda(\mathbf{Q}) \subset \mathbf{Q}_p$ and we shall write $|\alpha|_p$ for $|\alpha|'$ for all $\alpha \in \mathbf{Q}_p$. With this identification any fundamental sequence $\mathbf{a} = \{a_1, a_2, \ldots\}$ of elements of $\mathbf{Q}$ has the limit $\alpha \in \mathbf{Q}_p$ where $\mathbf{a} \in \alpha$: that is $\mathbf{Q}_p$ is the closure of $\mathbf{Q}$ with respect to $|\ |_p$.

It remains to show that $\mathbf{Q}_p$ is complete. Let $\boldsymbol{\alpha} = \{\alpha_1, \ldots, \alpha_j, \ldots\}$ be a fundamental sequence of elements of $\mathbf{Q}_p$. By the previous remark there are $a_j^* \in \mathbf{Q}$ such that‡ $|a_j^* - \alpha_j|_p < 2^{-j}$. Then the sequence $\mathbf{a}^* = \{a_1^*, \ldots, a_j^*, \ldots\}$ is fundamental, and so defines an $\alpha^* \in \mathbf{Q}_p$ which is clearly the limit of $\boldsymbol{\alpha}$. This completes the proof of Lemma 1.2.

LEMMA 1.3. *$|\alpha|_p$ is a power of p for§ $\alpha \in \mathbf{Q}_p^*$.*

Proof. Since $\mathbf{Q}_p$ is the closure of $\mathbf{Q}$ there is an $a \in \mathbf{Q}$ such that $|\alpha - a|_p < |a|_p$. Then $|\alpha|_p = |a|_p$.

Definition. The $\alpha \in \mathbf{Q}_p$ with $|\alpha|_p \leqslant 1$ are the *p-adic integers* $\mathbf{Z}_p$. Clearly $\mathbf{Z}_p$ is a ring.

† For it is certainly a ring homomorphism. A ring homomorphism of one field k into another field K is either an embedding or it maps the whole of k into $0 \in K$. And the second alternative certainly does not happen here.

‡ Here 2^{-j} may be replaced by any sequence of positive real numbers which tends to 0.

§ Recall that k^* is the group of non-zero elements of a field k.

Definition. The set U_p of $\alpha \in \mathbf{Q}_p$ with $|\alpha|_p = 1$ is a group under multiplication. They are the *p-adic units.*

It is also convenient to introduce the notion of convergence of infinite sums in $\mathbf{Q}_p$. We say that

$$\sum_{n=1}^{\infty} \beta_n \qquad (\beta_n \in \mathbf{Q}_p) \tag{1.3}$$

is convergent if the partial sums

$$\sigma_N = \sum_{n=1}^{N} \beta_n$$

tend to a limit. The limit is then said to be the value of the infinite sum.

LEMMA 1.4. *A necessary and sufficient condition that* (1.3) *converge is that* $\beta_n \to 0$ (*in the p-adic sense, of course*).

Proof. If the series is convergent then σ_n is a fundamental sequence and $\beta_n = \sigma_n - \sigma_{n-1}$ tends to 0. Conversely if $m > n$ then

$$\begin{aligned} |\sigma_m - \sigma_n|_p &= |\beta_{n+1} + \beta_{n+2} + \dots + \beta_m| \\ &\leqslant \max_{j>n} |\beta_j| \\ &\to 0 \qquad (n \to \infty). \end{aligned}$$

[Here, of course, $\to 0$ and $\to \infty$ are meant in the ordinary sense.]

LEMMA 1.5. *Every* $\alpha \in \mathbf{Z}_p$ *can be put uniquely in the shape*

$$\alpha = \sum_{j=0}^{\infty} a_j p^j \tag{1.4}$$

where each a_j *is in the set* $\{0, 1, 2, \dots, p-1\}$.

Proof. (1.4) converges by Lemma 1.4 and the sum is clearly in $\mathbf{Z}_p$. Conversely let $\alpha \in \mathbf{Z}_p$. Then there is a $c \in \mathbf{Q}$ such that $|\alpha - c|_p \leqslant 1/p$. Then c is of the shape $c = u/v$ where $u, v \in \mathbf{Z}$ and $p \nmid v$. We can therefore pick an $a_0 \in \{0, 1, \dots, p-1\}$ such that $u \equiv a_0 v \pmod{p}$: that is $|c - a_0|_p \leqslant 1/p$. Hence $\alpha = a_0 + p\alpha_1$ where

$$|\alpha_1|_p = \frac{|\alpha - a_0|_p}{|p|_p} \leqslant 1.$$

We now repeat the process with α_1 to get a_1, etc.

LEMMA 1.6. *Let* $\alpha \in \mathbf{Q}_p^*$, *and suppose that*

$$|\alpha - 1|_p \leqslant 1/p \qquad (p \neq 2)$$
$$\leqslant \tfrac{1}{8} \qquad (p = 2)$$

Then $\alpha \in (\mathbf{Q}_p^*)^2$.

Proof. We have to show that there is a $\xi \in \mathbf{Q}_p^*$ such that $\xi^2 = \alpha$.
Suppose we have a ξ_n such that

$$|\xi_n^2 - \alpha| \leqslant p^{-n}.$$

We attempt to find ξ_{n+1} in the shape $\xi_n + \delta_n$:

$$(\xi_n + \delta_n)^2 - \alpha = \xi_n^2 - \alpha + 2\xi_n\delta_n + \delta_n^2$$
$$= \delta_n^2$$

if

$$\delta_n = \frac{\alpha - \xi_n^2}{2\xi_n}.$$

Here

$$|\delta_n| = \frac{|\alpha - \xi_n^2|}{|2|\,|\xi_n|} = \begin{cases} |\alpha - \xi_n^2| & (p \neq 2) \\ 2|\alpha - \xi_n^2| & (p = 2) \end{cases}$$

and so

$$|\delta_n|^2 \leqslant \begin{cases} p^{-2n} \leqslant p^{-n-1} & (p \neq 2, n \geqslant 1) \\ 4p^{-2n} \leqslant p^{-n-1} & (p = 2, n \geqslant 3). \end{cases}$$

The hypotheses of the lemma state that we can take $\xi_1 = 1$ $(p \neq 2)$, $\xi_3 = 1$ $(p = 2)$. In either case we get a fundamental sequence ξ_n which tends to a limit ξ which must satisfy $\xi^2 = \alpha$.

COROLLARY. (i) $\mathbf{Q}_2^*/(\mathbf{Q}_2^*)^2$ *is of order* 8, *generators* $2, -1, 5$.

(ii) *for* $p \neq 2$ *let* r *be a quadratic nonresidue.*

Then $\mathbf{Q}_p^*/(\mathbf{Q}_p^*)^2$ *is of order* 4, *generators* p, r.

Proof. (i) Let $\alpha \in \mathbf{Q}_2^*$. Then $\alpha = 2^n\beta$ for some β with $|\beta| = 1$. Then $(-1)^u 5^v \beta = \gamma$ (say) $\equiv 1 \pmod{2^3}$ for some $u, v = 0$ or 1. Further γ with $|\gamma|_2 = 1$ is a square only if $\gamma \equiv 1 \pmod{2^3}$.

(ii) An $\alpha \in \mathbf{Q}_p^*$ is of the shape $\alpha = p^n\beta$ with $|\beta|_p = 1$. Hence $\beta \equiv a^2 \pmod{p}$ or $\beta \equiv ra^2 \pmod{p}$ for some $a \in \mathbf{Z}$. Then $a^{-2}\beta \equiv 1$ or $r^{-1}a^{-2}\beta \equiv 1 \pmod{p}$ and the rest is easy.

We conclude with an application to quadratic forms.

LEMMA 1.7. *Let $p \neq 2$ and let*

$$f(\mathbf{x}) = a_1x_1^2 + a_2x_2^2 + a_3x_3^2$$

with

$$|a_1|_p = |a_2|_p = |a_3|_p.$$

Then $f(\mathbf{x})$ is isotropic.

Proof. Without loss of generality we may suppose that

$$|a_1|_p = |a_2|_p = |a_3|_p = 1.$$

By Lemma 2.2 of Chapter 2 there exist $u_1, u_2 \in \mathbf{Z}_p$ such that

$$a_1u_1^2 + a_2u_2^2 \equiv -a_3 \pmod{p}.$$

Then

$$a_1u_1^2 + a_2u_2^2 + a_3u_3^2 = 0$$

for some $u_3 \in \mathbf{Z}_p, u_3 \equiv 1 \pmod{p}$.

COROLLARY. *A form in $n \geqslant 5$ variables over $\mathbf{Q}_p$ is isotropic.*

Proof. Without loss of generality

$$f(\mathbf{x}) = \sum_{1 \leqslant i \leqslant n} a_i x_i^2$$

is diagonal. By a substitution $x_i \to c_i x_i$ $(c_i \in \mathbf{Q}_p^*)$ we can ensure that three of the $|a_i|_p$ are equal.

Note. The Corollary (but not the Lemma) is true also for $p = 2$. [Lemma 2.7 of Chapter 4.]

2. NORM RESIDUE SYMBOL

In this section we allow the possibility $p = \infty$ (so $\mathbf{Q}_\infty = \mathbf{R}$). The *Hilbert Norm Residue Symbol*

$$(a, b) = \left(\frac{a, b}{p}\right)$$

for $a, b \in \mathbf{Q}_p^*$ is defined by

$$(a, b) = \begin{cases} 1 \text{ if } ax^2 + by^2 - z^2 \text{ is isotropic} \\ -1 \text{ otherwise.} \end{cases}$$

Clearly (a, b) depends only on a, b modulo squares.

There is an asymmetric form of this definition, namely that $(a, b) = 1$ if and only if

$$a = z^2 - by^2 \tag{2.1}$$

for some $y, z \in \mathbf{Q}_p$. For suppose $ax^2 + by^2 - z^2 = 0$ with $x, y, z \in \mathbf{Q}_p$ not all zero. If $x \neq 0$, then (2.1) has a solution by homogeneity. If, however, $x = 0$, then the form $z^2 - by^2$ is isotropic, so universal and again we are done. [cf. Chapter 2, Lemma 2.3].

LEMMA 2.1. *The Norm Residue Symbol has the following properties:*

(i) $(a, b) = (b, a)$.

(ii) $(a_1 a_2, b) = (a_1, b)(a_2, b)$.

(ii)′ $(a, b_1 b_2) = (a, b_1)(a, b_2)$.

(iii) *If* $(a, b) = 1$ *for all* b, *then* $a \in (\mathbf{Q}_p^*)^2$.

(iii)′ *If* $(a, b) = 1$ *for all* a, *then* $b \in (\mathbf{Q}_p^*)^2$.

(iv) $(a, -a) = 1$ *for all* a.

(v) *If* $p \neq 2, \infty$ *and* $|a|_p = |b|_p = 1$, *then* $(a, b) = 1$.

Proof. Here (i) and (iv) are immediate from the definition. It follows from (i) that (ii) and (ii)′ are equivalent and similarly for (iii) and (iii)′. Finally, (v) is an immediate consequence of Lemma 1.7.

We now show that for fixed b the a with $(a, b) = 1$ form a group under multiplication. Such a have the shape (2.1), so let

$$a_j = z_j^2 - by_j^2 \qquad (j = 1, 2).$$

Then

$$a_1 a_2 = (z_1 z_2 + b y_1 y_2)^2 - b(z_1 y_2 + z_2 y_1)^2.$$

To complete the proof of (ii) we would have to show that the group of $a \in \mathbf{Q}_p^*$ of the shape (2.1) is either the whole of $\mathbf{Q}_p^*$ or of index 2. From our present, aggressively lowbrow, point of view this has to be regarded as a "fact of life" which can be read off from the explicit tables given below for (a, b). Similarly (iii) and (iii)′ are immediate from the tables. *Vom höheren Standpunkt aus* the lemma is a very special case of a very general phenomenon. The equation (2.1) states that a is a norm for the quadratic extension $\mathbf{Q}_p(\sqrt{b})/\mathbf{Q}_p$, whence the name Norm Residue symbol. For all this see, for example, O'Meara (1963), Serre (1962) or, alternatively, Chapter IV of Cassels and Fröhlich (1967).

To complete the proof it remains to write down tables of (a, b). We distinguish three cases (α) p an odd prime (β) $p = 2$ and (γ) $p = \infty$.

(α) p an odd prime. Here the group $\mathbf{Q}_p^*/(\mathbf{Q}_p^*)^2$ is generated by p and r, where $r \in \mathbf{Z}$, and r is not congruent to a square mod p.

b \ a	1	r	p	pr
1	+1	+1	+1	+1
r	+1	+1	−1	−1
p	+1	−1	ε	$-\varepsilon$
pr	+1	−1	$-\varepsilon$	ε

Here $\varepsilon = 1$ if $-1 \in (\mathbf{Q}_p^*)^2$ [that is, if $p \equiv 1 \pmod 4$] and $\varepsilon = -1$ otherwise.

To check this table one notes first that $(a,b) = 1$ when $|a|_p = |b|_p = 1$ by Lemma 1.7. Again, if $ax^2 + by^2 - z^2$ is isotropic, then there is by homogeneity a solution $x_0, y_0, z_0 \in \mathbf{Q}_p$ of $ax_0^2 + by_0^2 - z_0^2 = 0$ with $\max\{|x_0|, |y_0|, |z_0|\} = 1$. For example, consider $a = p, b = r$. If $|y_0| < 1$, $|z_0| < 1$ we should have $|ry_0^2 - z_0^2|_p \leqslant p^{-2}$ and so $|px_0^2| \leqslant p^{-2}$ and $|x_0| < 1$ contrary to the normalization. Hence $\max(|y_0|, |z_0|) = 1$ and $|ry_0^2 - z_0^2| = |px_0^2| < 1$; a contradiction. Hence $(p, r) = -1$. All the other entries in the table may be checked similarly.

(β) $p = 2$. Here $\mathbf{Q}_2^*/(\mathbf{Q}_2^*)^2$ is generated by 2, 5, −1.

b \ a	1	5	−1	−5	2	10	−2	−10
1	+1	+1	+1	+1	+1	+1	+1	+1
5	+1	+1	+1	+1	−1	−1	−1	−1
−1	+1	+1	−1	−1	+1	+1	−1	−1
−5	+1	+1	−1	−1	−1	−1	+1	+1
2	+1	−1	+1	−1	+1	−1	+1	−1
10	+1	−1	+1	−1	−1	+1	−1	+1
−2	+1	−1	−1	+1	+1	−1	−1	+1
−10	+1	−1	−1	+1	−1	+1	+1	−1

This is also readily checked. Note that if x_0 is a 2-adic unit then $x_0^2 \equiv 1 \pmod{2^3}$.

(γ) $p = \infty$. Here $\mathbf{Q}_\infty^*/(\mathbf{Q}_\infty^*)^2$ is generated by -1.

$b \backslash a$	1	-1
1	$+1$	$+1$
-1	$+1$	-1

3. LOCAL AND GLOBAL

It is customary to call the $\mathbf{Q}_p$ the "local" fields while $\mathbf{Q}$ is the "global" field. The main theme in the first part of this book is the relationship between "local" and "global" behaviour of quadratic forms. In this section we first show that behaviour of elements of $\mathbf{Q}$ in a *finite* set of $\mathbf{Q}_p$ is quite independent. We then give two formulae which relate the behaviour of elements of $\mathbf{Q}$ in *all* $\mathbf{Q}_p$. These will be basic in later chapters.

LEMMA 3.1 ("Strong Approximation Theorem"). *Let P be a finite set of primes $p \neq \infty$. For each $p \in P$ let $z_p \in \mathbf{Z}_p$ be given arbitrarily and let $\varepsilon > 0$ be arbitrarily small. Then there is a $z \in \mathbf{Z}$ such that*

$$|z - z_p|_p < \varepsilon \qquad (\textit{all } p \in P).$$

Proof. This is essentially just a reformulation of the "Chinese Remainder Theorem". Since $\mathbf{Z}$ is dense in $\mathbf{Z}_p$ with respect to $|\ |_p$ (by Lemma 1.5) there are $z_p^* \in \mathbf{Z}$ such that

$$|z_p - z_p^*|_p < \varepsilon.$$

Now by the Chinese Remainder Theorem† there is a $z \in \mathbf{Z}$ such that

$$z \equiv z_p^* \pmod{p^{m(p)}} \qquad (\text{all } p \in P)$$

where the $m(p)$ are positive integers so large that

$$p^{m(p)}\varepsilon > 1.$$

Then

$$|z - z_p^*|_p \leqslant p^{-m(p)} < \varepsilon$$

and we are done.

† The Chinese Remainder Theorem states that if u_j $(1 \leqslant j \leqslant J)$ are non-zero integers coprime in pairs and if a_j are integers then there is an integer b such that

$$b \equiv a_j \pmod{u_j} \qquad (1 \leqslant j \leqslant J).$$

It is to be found in every elementary textbook.

LEMMA 3.2. ("Weak Approximation Theorem"). *Let P be a finite set of primes p, including possibly $p = \infty$. Let $a_p \in \mathbf{Q}_p$ be given for all $p \in P$ and let $\varepsilon > 0$ be given arbitrarily small. Then there is an $a \in \mathbf{Q}$ such that*

$$|a - a_p|_p < \varepsilon \qquad (\textit{all } p \in P). \tag{3.1}$$

Note. Lemma 3.2 is a particular case of a general theorem about the independence of valuations on fields [see, for example, Chapter II of Cassels and Fröhlich (1967)] while Lemma 3.1 is much more special. However, from our purely utilitarian viewpoint, it is convenient to deduce Lemma 3.2 from Lemma 3.1.

Proof. For $p \in P, p \neq \infty$ let $n(p) \geqslant 0$ be an integer such that $p^{n(p)}a_p \in \mathbf{Z}_p$. Let

$$h = \prod_{\substack{p \in P \\ \neq \infty}} p^{n(p)}.$$

Put

$$z_p = ha_p$$

for all p (including $p = \infty$ if $\infty \in P$). Then $z_p \in \mathbf{Z}_p$ $(p \neq \infty)$. Let $\eta > 0$ be given, arbitrarily small. By Lemma 3.1 there is $z \in \mathbf{Z}$ such that

$$|z - z_p|_p < \eta \qquad (\text{all } p \in P, p \neq \infty). \tag{3.2}$$

If $\infty \notin P$ we are done since we can take $a = z/h$ and can choose η so that $|h|_p\varepsilon > \eta$: so suppose that $\infty \in P$. For $p \in P, p \neq \infty$ choose $m(p) \geqslant 1$ so that $p^{m(p)}\eta > 1$. Put

$$A = \prod_{\substack{p \in P \\ \neq \infty}} p^{m(p)}. \tag{3.3}$$

Now choose a positive integer B prime to A [that is, prime to the $p \neq \infty$ in P] such that

$$A/B < \eta. \tag{3.4}$$

Put

$$b = z + rA/B$$

for an $r \in \mathbf{Z}$ to be chosen later. Then

$$|b - z_p|_p < \eta \qquad (p \in P, \neq \infty)$$

by (3.2) and (3.3). Further by (3.4) we can choose the integer r so that

$$|b - z_\infty|_\infty < \eta.$$

On putting $a = b/h$ and choosing η suitably, as before, we have the conclusion of the Lemma.

It is convenient to introduce the convention that a statement holds for *almost all* primes p if it fails for at most a finite set of primes.

LEMMA 3.3. *Let* $a \in \mathbf{Q}^*$. *Then* $|a|_p = 1$ *for almost all primes p and*

$$\prod_{\substack{\text{all } p \\ \text{inc } \infty}} |a|_p = 1.$$

Proof. We have

$$a = \pm \prod_{p \neq \infty} p^{\alpha(p)}$$

where $\alpha(p) \in \mathbf{Z}$ and $\alpha(p) = 0$ for almost all p (in the sense just defined). Then

$$|a|_p = p^{-\alpha(p)} \qquad (p \neq \infty)$$

and

$$|a|_\infty = \prod_{p \neq \infty} p^{\alpha(p)}.$$

LEMMA 3.4 ("Product Formula for the Norm Residue Symbol"). *Let* $a, b \in \mathbf{Q}^*$. *Then* $\left(\frac{a, b}{p}\right) = 1$ *for almost all p and*

$$\prod_{\substack{\text{all } p \\ \text{inc } \infty}} \left(\frac{a, b}{p}\right) = 1. \tag{3.5}$$

Proof. We shall deduce this from the Law of Quadratic Reciprocity and conversely it is easy to deduce Quadratic Reciprocity from Lemma 3.4.

By Lemma 3.3 we have $|a|_p = |b|_p = 1$ for almost all p and then $\left(\frac{a, b}{p}\right) = 1$ by Lemma 2.1(v), provided that $p \neq 2, \infty$.

Denote the left-hand side of (3.5) by $f(a, b)$. By Lemma 2.1 we have

$$f(a_1 a_2, b) = f(a_1, b) f(a_2, b)$$
$$f(a, b_1 b_2) = f(a, b_1) f(a, b_2).$$

Hence it is enough to show that $f(a, b) = 1$ when a and b run through a set of generators of the group $\mathbf{Q}^*$:

$$-1, 2, q = \text{odd prime}.$$

First case. $a = q_1, b = q_2$ where $q_1 \neq q_2$ odd primes. Then

$$\begin{aligned} \left(\frac{q_1, q_2}{p}\right) &= 1 && (p \text{ odd}, p \neq q_1, q_2) \\ &= 1 && (p = \infty) \\ &= (q_1/q_2) && (p = q_2) \\ &= (q_2/q_1) && (p = q_1) \\ &= (-1)^{(q_1 - 1)(q_2 - 1)/4} && (p = 2). \end{aligned}$$

Here (q_1/q_2) is the ordinary symbol of quadratic reciprocity. The statement

$$f(q_1, q_2) = \prod_{\substack{\text{all } p \\ \text{inc } \infty}} \left(\frac{q_1, q_2}{p}\right) = 1$$

is thus equivalent to the law of quadratic reciprocity.

Second case. $a = b = -1$. We have

$$\left(\frac{-1, -1}{p}\right) = -1 \qquad (p = 2, \infty)$$
$$= +1 \qquad \text{(otherwise)}$$

and again $f(-1, -1) = 1$.

Other cases. Are dealt with similarly.

4. HENSEL'S LEMMA

We conclude this chapter by returning to the purely local situation. The solution of equations by successive approximation, which in the real case bears the name of Newton, is also applicable in p-adic fields, where it is generically referred to as "Hensel's Lemma". We give below the simplest variant. It is given here primarily for background interest as we shall have little use for it. Lemma 1.6, which is a special case, will usually do all that we need.

LEMMA 4.1. *Let $f(x) \in \mathbf{Z}_p[x]$ be a polynomial in the single variable x and suppose that there exists an $a \in \mathbf{Z}_p$ such that*

$$|f(a)|_p < |f'(a)|_p^2, \tag{4.1}$$

where $f'(x)$ denotes the (formal) derivative with respect to x. Then there is a $b \in \mathbf{Z}_p$ such that $f(b) = 0$.

Note. Lemma 1.6 is the case $f(x) = x^2 - \alpha, a = 1$.

Proof. We have an identity

$$f(x + y) = f(x) + f_1(x)y + f_2(x)y^2 + \dots \tag{4.2}$$

in the pair of variables x, y, where

$$f_j(x) \in \mathbf{Z}_p[x] \tag{4.3}$$

and

$$f_1(x) = f'(x).$$

Put $x = a, y = d$ in (4.2), where d is chosen so that the linear terms in y vanish, that is

$$d = -f(a)/f'(a), \tag{4.4}$$

so

$$d \in \mathbf{Z}_p$$

by (4.1). Then $f_j(a) \in \mathbf{Z}_p$ by (4.3) and so (4.2) gives

$$\begin{aligned} |f(a+d)|_p &\leqslant \max_{j \geqslant 2} |f_j(a) d^j|_p \\ &\leqslant |d|_p^2 \\ &= |f(a)|_p^2 / |f'(a)|_p^2 \\ &< |f(a)|_p \end{aligned} \tag{4.5}$$

by (4.1). Further, by using the analogue of (4.2) for $f'(x)$ it is easy to see that

$$|f'(a+d) - f'(a)|_p < |f'(a)|_p,$$

and so

$$|f'(a+d)|_p = |f'(a)|_p. \tag{4.6}$$

We now construct an infinite sequence $a = a_0, a_1, \ldots$ of elements of $\mathbf{Z}_p$ by

$$\begin{aligned} a_{j+1} &= a_j + d_j, \\ d_j &= -f(a_j)/f'(a_j). \end{aligned} \tag{4.7}$$

Then

$$|f'(a_j)|_p = |f'(a)|_p \qquad \text{(all } j\text{)}$$

by (4.6), and so

$$|f(a_0)|_p > |f(a_1)|_p > |f(a_2)|_p > \ldots . \tag{4.8}$$

Hence

$$d_j \to 0$$

p-adically by (4.7) and a_j converges to a limit $b \in \mathbf{Z}_p$. Finally $f(a_j) \to 0$ by (4.8), so $f(b) = 0$ as required.

NOTES

Section 1. There are many applications of p-adic numbers to diophantine problems but no really satisfactory account. The best are Mordell (1969), Chapter 23 and (more sophisticated) in Borevič and Šafarevič (1966). See also Cassels (1976). There are introductions to p-adic numbers from widely differing points of view in Bachman (1964), Mahler (1973) and Koblitz (1977). For the general theory of local fields, see Serre (1962).

Sections 2 and 3. The Hilbert Norm Residue Symbol generalizes to any finite extension k of a p-adic field. This generalization is itself the special case $K = k(\sqrt{b})$ of a symbol (K, a) where $a \in k^*$ and K is an abelian extension of k [i.e. K is a normal extension whose galois group is abelian]. See Hasse (1926), Cassels and Fröhlich (1967), Exercise 2, or, for the local theory only, Serre (1962). All that is required for quadratic forms over an arbitrary number-field is developed from first principles in O'Meara (1963).

There are proofs of the Law of Quadratic Reciprocity in almost any elementary book on the theory of numbers, e.g. Hardy and Wright (1938), Chapter 6.

EXAMPLES

1. Elements a_1, a_2 of $\mathbf{Z}_3$ are defined by

$$5a_1 + 1 = 0; \qquad a_2^2 = 7, \qquad |a_2 - 1|_3 < 1.$$

Find $b_1, b_2 \in \mathbf{Z}$ such that

$$|b_j - a_j|_3 \leqslant 3^{-3} \qquad (j = 1, 2).$$

2. For what $a \in \mathbf{Z}$ is $5x^2 = a$ soluble in $\mathbf{Z}_7$?, in $\mathbf{Q}_7$?

3. Find $a, b \in \mathbf{Z}$ such that

$$|a^2 + 6b^2 - 5|_5 \leqslant 5^{-4}.$$

4. If $c \in \mathbf{Z}_p$ satisfies $|c|_p < 1$, show that

$$\frac{1}{1+c} = 1 - c + c^2 - c^3 + \dots .$$

Hence or otherwise find $a \in \mathbf{Z}$ such that

$$|\tfrac{1}{4} - a|_5 \leqslant 5^{-10}.$$

5. For positive integer n define

$$\binom{a}{n} = \frac{a(a-1)\dots(a-n+1)}{n!}.$$

Show that $\binom{a}{n} \in \mathbf{Z}_p$ whenever $a \in \mathbf{Z}_p$.

[*Hint.* Approximate p-adically to a by positive integers.]

6. Let $p \neq 2$ and let $c \in \mathbf{Q}_p$ satisfy $|c|_p < 1$.
Show that

$$1 + \binom{\frac{1}{2}}{1}c + \binom{\frac{1}{2}}{2}c^2 + \dots + \binom{\frac{1}{2}}{n}c^n + \dots$$

converges to a square root of $1 + c$.

Hence or otherwise find a $d \in \mathbf{Q}$ such that

$$|d^2 - 11|_5 \leqslant 5^{-10}.$$

7. By an appropriate modification of the method of the preceding question, find $u \in \mathbf{Z}$ such that

$$|u^2 + 7|_2 \leqslant 2^{-10}.$$

8. Let $p \neq 2$ and let $a \in \mathbf{Z}_p$ satisfy $| \quad - 1|_p < 1$. Show that $p|a^p - 1|_p = |a - 1|_p$. What is the corresponding result for $p = 2$?

9. For $a \in \mathbf{Q}_p^*$, $a \neq 0, 1$ show that

$$\left(\frac{a+1, \quad -a}{p}\right) = +1$$

and

$$\left(\frac{a, \quad a}{p}\right) = \left(\frac{a, \quad -1}{p}\right).$$

10. Let $a + b + c = 0$ where $a, b, c \in \mathbf{Q}_p^*$. Show that

$$\left(\frac{-1, \quad -abc}{p}\right) = \left(\frac{a, \quad b}{p}\right)\left(\frac{b, \quad c}{p}\right)\left(\frac{c, \quad a}{p}\right).$$

11. Let $p = 2m + 1$ be an odd prime, let a, b be p-adic units and let $u, v \in \mathbf{Z}$.

Show that

$$\left(\frac{ap^u, bp^v}{p}\right) \equiv a^{vm} b^{um} (-1)^{uvm} \pmod p.$$

12. Let $a \in \mathbf{Q}_p$ and suppose that

$$|a - 1|_p \leqslant \begin{cases} p^{-1} & \text{if } p \neq 3 \\ 3^{-2} & \text{if } p = 3. \end{cases}$$

Show that $a \in \mathbf{Q}_p^3$.

[*Hint*. Hensel's Lemma.]

13. Let $f(x) = x^2 + bx + c$ with $b, c \in \mathbf{Z}_p$ and suppose that $|b^2 - 4c|_p = 1$. Suppose that $f(v) \equiv 0 \pmod p$ for some $v \in \mathbf{Z}_p$. Show that there is a $u \in \mathbf{Z}_p$ such that $f(u) = 0$ and $u \equiv v \pmod p$.

[*Hint*. Hensel's Lemma.]

14. Let $p > 2$ be prime and

(i) Let $\mathbf{A}$ be an $n \times n$ matrix with elements in $\mathbf{Z}_p$ and suppose that $\mathbf{A} \equiv \mathbf{I} \pmod p$ but $\mathbf{A} \neq \mathbf{I}$. Show that $\mathbf{A}$ is of infinite order in the multiplicative group of $n \times n$ matrices.

[*Hint*. Show that $\mathbf{A}^q \neq \mathbf{I}$ for prime q and use induction. It is necessary to distinguish $q = p$ and $q \neq p$.]

(ii) Let G be a finite group of $n \times n$ matrices with elements in $\mathbf{Z}_p$. Show that the order of G divides

$$(p^n - 1)(p^n - p)(p^n - p^2)\dots(p^n - p^{n-1}).$$

[*Hint.* Consider the matrices in G modulo p: cf. Chapter 2, example 12.]

15. Let $\mathbf{A}$ be an $n \times n$ matrix with elements in $\mathbf{Z}_2$.

(i) If $\mathbf{A} \equiv \mathbf{I}$ (mod 4) but $\mathbf{A} \neq \mathbf{I}$ show that $\mathbf{A}$ has infinite order.

(ii) If $\mathbf{A} \equiv \mathbf{I}$ (mod 2) show that either $\mathbf{A}^2 = \mathbf{I}$ or $\mathbf{A}$ has infinite order.

(iii) Let H be a finite group of matrices $\mathbf{A} \equiv \mathbf{I}$ (mod 2). Show that H has order 2^m for some $m \leqslant n$.

[*Hint.* For H is a group of $n \times n$ matrices of exponent 2 and so abelian. Choose a basis for $\mathbf{Q}_2^n$ consisting of the common eigenvectors of the $\mathbf{A} \in H$.]

(iv) Let G be a finite group of matrices $\mathbf{A}$. Show that the order of $\mathbf{G}$ divides

$$2^n(2^n - 1)(2^n - 2)\dots(2^n - 2^{n-1}).$$

[*Hint*: previous example.]

16. (i) Let G be a finite group of $n \times n$ matrices with elements in $\mathbf{Q}$. Show that the order g of G divides

$$g^*(n) = \prod_{q \text{ prime}} q^{\beta(q)}$$

where

$$\beta(2) = n + 2[n/2] + [n/2^2] + [n/2^3] + \dots$$

and

$$\beta(q) = [n/(q-1)] + [n/q(q-1)] + [n/q^2(q-1)] + \dots.$$

(ii) If all the elements of G have determinant $+1$ show, further, that g divides $\frac{1}{2}g^*(n)$.

[*Hint.* For G has elements in $\mathbf{Z}_p$ for all p greater than some p_0. For q odd determine p, by Dirichlet's theorem about primes in arithmetic progressions so that p is a primitive root modulo q^2 and apply example 14(ii). For $q = 2$ let $p \equiv 3$ (mod 8).

Note. This result is best possible, see W. Burnside, *Theory of groups of finite order* (2nd ed., Cambridge 1911) pp. 479–484 or N. Bourbaki *Éléments de Mathematique* Fasc. **37** (Groupes et algébres de Lie, Chaps. II, III)pp. 272–274.]

The following examples are designed to illustrate the scope of p-adic methods and are unrelated to the main theme of the book.

17. Let $f(x) = ax^2 + bx + c$ where $a, b, c \in \mathbf{Z}$, $a \neq 0$. Show that there are infinitely many primes p for which there is a $u_p \in \mathbf{Z}_p$ with $f(u_p) = 0$.
[*Note*. The corresponding result is true for polynomials of any degree. For a further generalization see Cassels (1976).]

[*Hint*. On replacing x by x/a we may suppose that $a = 1$. If $c = 0$ result is trivial so suppose $c \neq 0$. If $b^2 = 4c$, then $f(x)$ is a perfect square, so we may suppose that $d = b^2 - 4c \neq 0$. Let e be any integer divisible by all the primes dividing $2cd$. Then $f(e^N) \to \infty$ as $N \to \infty$ but $|f(e^N)|_p$ is bounded below for $p|e$. Hence $f(e^N)$ is divisible by some prime $q \nmid e$ when N is large enough. Apply example 13 with $q = p$.]

18. For positive integer m show that

$$|m!|_p = p^{-M}$$

where

$$M = \sum_{j>0} [m/p^j].$$

Deduce that

$$M < m/(p-1).$$

19. Let $a_{i,j} \in \mathbf{Q}_p$ be given for $i, j = 0, 1, 2, \ldots$ and suppose that $a_{i,j} \to 0$ p-adically as $i, j \to \infty$ independently in the usual sense. Show that

$$\sum_i \left(\sum_j a_{i,j}\right) = \sum_j \left(\sum_i a_{i,j}\right).$$

20. Let $b \in \mathbf{Q}_p$ satisfy

$$|b| \leqslant \begin{cases} p^{-1} & \text{if } p \neq 2 \\ \frac{1}{4} & \text{if } p = 2. \end{cases}$$

Show that there is a sequence $c_1, c_2, \ldots, c_n, \ldots$ of elements of $\mathbf{Z}_p$ such that

$$(1+b)^m = 1 + c_1 m + c_2 m^2 + \ldots + c_n m^n + \ldots$$

for all positive integers m.

[*Hint*.

$$(1+b)^m = \sum_{s=0}^{\infty} \frac{m(m-1)\ldots(m-s+1)}{s!} b^s.$$

Use two previous examples to justify rearranging in powers of m.]

21. (Strassmann's Theorem). Let $c_0, c_1, c_2, \ldots, c_n, \ldots$ be elements of $\mathbf{Z}_p$, not all 0. Suppose that $c_n \to 0$ p-adically and put

$$f(x) = c_0 + c_1 x + \ldots + c_n x^n + \ldots$$

for $x \in \mathbf{Z}_p$. Show that there are only a finite number of $a \in \mathbf{Z}_p$ such that $f(a) = 0$.

More precisely, define $N \geqslant 0$ by

$$|c_N|_p = \max_n |c_n|_p$$

$$|c_n|_p < |c_N|_p \qquad (\text{all } n > N).$$

Show that there are at most N solutions of $f(a) = 0$.

[*Hint*. If $f(a) = 0$ show that

$$f(x) = \Sigma c_n(x^n - a^n) = (x - a)g(x),$$

where $g(x)$ has similar properties to $f(x)$ but with $N - 1$ instead of N.]

22. Let $b, c, v \in \mathbf{Z}$ and let the sequence $u_0, u_1, u_2, \ldots$ of elements of $\mathbf{Z}$ be defined by the recurrence relation

$$u_n = bu_{n-1} + cu_{n-2} \qquad (n \geqslant 2).$$

Show that either there are only finitely many n such that $u_n = v$ or that $u_n = v$ for all the n in an arithmetic progression.

[*Note*. This is a special case of a theorem of Mahler which was generalized by Lech. See discussion in Cassels (1976).]

[*Hint*. If $c = 0$ or if $x^2 - bx - c$ has equal roots this is trivial. Otherwise

$$u_n = \lambda\alpha^n + \mu\beta^n$$

where α, β are the roots of $x^2 - bx - c$ and λ, μ are determined by $\lambda + \mu = u_0, \lambda\alpha + \mu\beta = u_1$. By example 17 we can find a prime $p \neq 2$ such that α, β are p-adic units and $\lambda, \mu \in \mathbf{Z}_p$. Put $A = \alpha^{p-1}$, $B = \beta^{p-1}$ so $A \equiv B \equiv 1 \pmod p$. Hence for fixed $r, 0 \leqslant r < p - 1$ we have

$$u_{r+(p-1)s} = \sum_{j=0}^{\infty} e_j s^j$$

with $e_j \in \mathbf{Z}_p$ by example 20. Now apply example 21 to $f(s) = u_{r+(p-1)s} - v$.]

23. Define a sequence u_n by

$$u_0 = 0, \qquad u_1 = 1$$

and

$$u_n = u_{n-1} - 2u_{n-2} \qquad (n \geqslant 2).$$

Show that the only solutions of $u_n = \pm 1$ are $n = 1, 2, 3, 5, 13$.

[*Hint*. The polynomial $x^2 - x + 2$ has roots in $\mathbf{Q}_{11}$.]

[*Note*. The solutions of $u_n = \pm 1$ may be shown to correspond to the solutions of the equation $x^2 + 7 = 2^{n+2}$, which was first treated by Nagell using $\mathbf{Q}_7$: but his proof does not follow the paradigm of the previous example. See Mordell (1969) Chapter 23, Theorem 6.]

24. (von Staudt's Theorem, Witt's Proof).

(i) Define the Bernoulli numbers by

$$\frac{x}{e^x - 1} = B_0 + \frac{B_1}{1!}x + \ldots + \frac{B_k}{k!}x^k + \ldots.$$

Show that $B_0 = 1$, $B_1 = -\frac{1}{2}$ and that $B_k = 0$ for all odd $k > 1$.

(ii) (Euler–Maclaurin Sum Formula). Show that

$$S_k(n) = 1^k + 2^k + \ldots + (n-1)^k$$

is given by

$$S_k(n) = \sum_{r=0}^{k} \binom{k}{r} \frac{B_r}{k+1-r} n^{k+1-r}.$$

[*Hint*. Consider the identity

$$1 + e^x + \ldots + e^{(n-1)x} = \frac{e^{nx}-1}{x} \cdot \frac{x}{e^x - 1}.]$$

(iii) Let p be any prime. Deduce that

$$B_k = \lim_{n \to 0} n^{-1} S_k(n)$$

where the limit is in the p-adic sense.

(iv) If $p = 2$, suppose that k is even. Show that

$$|p^{-m-1} S_k(p^{m+1}) - p^{-m} S_k(p^m)|_p \leqslant 1.$$

[*Hint*. Take the range of summation in the definition of $S_k(p^{m+1})$ in the shape $up^m + v$ where $0 \leqslant u < p$ and $0 \leqslant v < p^m$.]

(v) Deduce from (iii) and (iv) that

$$|B_k - p^{-1} S_k(p)|_p \leqslant 1;$$

and hence that

$$B_k + p^{-1} \in \mathbf{Z}_p \qquad ((p-1)|k)$$

$$B_k \in \mathbf{Z}_p \qquad \text{(otherwise)}.$$

(vi) Hence

$$B_k + \sum_{\substack{p \text{ prime} \\ (p-1)|k}} p^{-1} \in \mathbf{Z}.$$

CHAPTER 4

Quadratic Forms Over Local Fields

1. INTRODUCTION

In this chapter we give the theory of quadratic forms over a local field $\mathbf{Q}_p$. Unless explicitly excluded we allow the possibility $p = \infty$, so $\mathbf{Q}_p = \mathbf{R}$.

Let f be a regular form in n variables over $\mathbf{Q}_p$. It is equivalent over $\mathbf{Q}_p$ to a diagonal form

$$a_1x_1^2 + \ldots + a_nx_n^2 \qquad (a_j \in \mathbf{Q}_p^*).$$

We shall see in Section 2 that

$$c_p(f) = c(f) = \prod_{i<j} (a_i, a_j)$$

depends only on the equivalence class of f and not the particular choice of the diagonal form in that class. Here, of course,

$$(a_i, a_j) = \left(\frac{a_i, a_j}{p}\right)$$

is the Hilbert Norm Residue Symbol. The invariant $c_p(f) = \pm 1$ just defined is the *Hasse–Minkowski invariant*.† We already know two invariants of an equivalence class of a regular form f, which are defined for any field k: the dimension $n(f)$ (= numbers of variables) and the determinant‡ $d(f) \in k^*/(k^*)^2$.

THEOREM 1.1. *Suppose $p \neq \infty$. Then $n(f)$, $d(f)$, $c(f)$ is a complete set of invariants of the equivalence class of f.*

In other words, if $n(f_1) = n(f_2)$, $d(f_1) = d(f_2)$ and $c(f_1) = c(f_2)$ then f_1 and f_2 are equivalent.

For completeness we recall the familiar situation when $p = \infty$.

† *Warning*. There are several definitions of the Hasse-Minkowski invariant in the literature. Each differs from the other by a factor depending on the number of variables and the determinant.

‡ Since we are interested in f only up to equivalence, we regard $d(f)$ as an element of $k^*/(k^*)^2$ rather than as an element of k^*.

THEOREM 1.2. *A complete set of invariants of the equivalence class of a regular form f over* $\mathbf{R} = \mathbf{Q}_\infty$ *is the pair* $n(f)$, $s(f)$ *where* $s(f)$ *is the number of negative coefficients in a diagonal form equivalent to f. Further*

$$d(f) = (-1)^s(\mathbf{Q}_\infty^*)^2$$

and

$$c_\infty(f) = (-1)^{s(s-1)/2}.$$

The values of $d(f)$, c_∞ follow at once from the definitions. The rest has already been proved (Lemma 5.7 of Chapter 2).

THEOREM 1.3. *If* $n = 1$, *then* $c(f) = 1$. *If* $n = 2$ *and* $d(f) = -(\mathbf{Q}_p^*)^2$ *then* $c(f) = 1$.

Suppose that $p \neq \infty$. *Then the triplet* $\{n(f), d(f), c(f)\}$ *runs through all values not excluded by the previous paragraph.*

In Section 2 we shall prove the assertions made above. We shall also characterize the isotropic forms in terms of their invariants.

2. THE PROOFS

LEMMA 2.1. *Let g be a regular binary quadratic form over* $\mathbf{Q}_p$. *Then the* $b \in \mathbf{Q}_p^*$ *which are represented by g are precisely those for which*†

$$(b, -d(g)) = \varepsilon, \tag{2.1}$$

where $\varepsilon = \pm 1$ *depends only on g.*

Proof. Without loss of generality

$$g(\mathbf{x}) = a_1 x_1^2 + a_2 x_2^2$$

is diagonal. It represents b if and only if

$$a_1 x_1^2 + a_2 x_2^2 - b x_3^2$$

is isotropic [Chapter 2, Lemma 2.3], that is if

$$(a_1/b)x_1^2 + (a_2/b)x_2^2 - x_3^2$$

is isotropic. By the definition of the Hilbert Norm Residue Symbol the condition for this is

$$(a_1/b, a_2/b) = +1.$$

† $d(g)$ is defined only up to squares. The Hilbert Norm Residue Symbol (a, b) depends only on a and b modulo squares and so "by abuse of notation" the symbol on the left-hand side of (2.1) has an obvious meaning.

By the multiplicative properties of the Norm Residue Symbol (Lemma 2.1 of Chapter 3) the left-hand side of this is

$$(b, a_2)(b, a_1)(b, b)(a_1, a_2).$$

By (iv) of the Lemma just cited we have $(b, b) = (b, -1)$ and so

$$\begin{aligned}(b, a_2)(b, a_1)(b, b) &= (b, -a_1a_2)\\ &= (b, -d(g)).\end{aligned}$$

The condition for representability is thus

$$(b, -d(g)) = (a_1, a_2). \tag{2.2}$$

COROLLARY. *Suppose that the binary form g is equivalent to $a_1x_1^2 + a_2x_2^2$. Then (a_1, a_2) depends only on g and not on the particular diagonalization.*

LEMMA 2.2. *Suppose that the regular diagonal forms*

$$\sum_{j=1}^{n} a_jx_j^2$$

and

$$\sum_{j=1}^{n} b_jx_j^2$$

are equivalent. Then

$$\prod_{i<j}(a_i, a_j) = \prod_{i<j}(b_i, b_j).$$

Proof. This is true for $n = 1$ since an empty product is 1 by definition and true for $n = 2$ by the preceding corollary. We therefore suppose $n > 2$. By Lemma 3.1 of Chapter 2 we can get from one diagonal form to an equivalent diagonal form through a chain of diagonal forms, changing at most two basis vectors at a time. Hence we may suppose that $a_i \neq b_i$ for at most two values of i. Further, $\prod_{i<j}(a_i, a_j)$ is independent of the order of $a_1, \ldots, a_n$ and so we may suppose that

$$a_i = b_i \qquad (i > 2)$$

and that $a_1x_1^2 + a_2x_2^2$ is equivalent to $b_1x_1^2 + b_2x_2^2$. In particular

$$a_1a_2 \in b_1b_2(\mathbf{Q}_p^*)^2$$

and

$$(a_1, a_2) = (b_1, b_2)$$

by the previous Corollary. But then

$$\begin{aligned}\prod_{i<j}(a_i, a_j) &= (a_1, a_2)\prod_{j>2}(a_1a_2, a_j)\prod_{2<i<j}(a_i, a_j)\\ &= (b_1, b_2)\prod_{j>2}(b_1b_2, b_j)\prod_{2<i<j}(b_i, b_j)\\ &= \prod_{i<j}(b_i, b_j);\end{aligned}$$

as asserted.

COROLLARY 1. *The value of the Hasse–Minkowski invariant $c(f)$, as defined in Section 1 does indeed depend only on f.*

COROLLARY 2. *Suppose that $n(f) = 2$. Then a necessary and sufficient condition that $b \in \mathbf{Q}_p^*$ be represented by f is that*

$$(b, -d(f)) = c(f).$$

Proof. Follows from (2.1) (with g for f).

Let f, g be two regular quadratic forms. By $f + g$ we shall mean the form $f(\mathbf{x}) + g(\mathbf{y})$ where $\mathbf{x}$, $\mathbf{y}$ are independent sets of variables. Then $f + g$ is clearly regular. [In quadratic space terms this corresponds to taking the direct sum of two quadratic spaces. Compare Theorem 4.1, Corollary 2, of Chapter 2.]

LEMMA 2.3. *Let f, g be regular forms and let $f + g$ be defined as above. Then*

$$\text{(i)}\ n(f + g) = n(f) + n(g)$$
$$\text{(ii)}\ d(f + g) = d(f)\, d(g)$$
$$\text{(iii)}\ c(f + g) = (d(f), d(g))\, c(f)\, c(g)$$

where (a, b) is the Norm Residue Symbol and $n(\ ), d(\ ), c(\ )$ denote the dimension, determinant and Hasse–Minkowski invariant respectively.

Proof. (i) and (ii) have already been proved in Chapter 2 for forms over a general field k, so only (iii) remains. Without loss of generality f and g are diagonal:

$$f = \sum_{i=1}^{n} a_i x_i^2, \qquad g = \sum_{j=1}^{m} b_j y_j^2$$

with $n = n(f)$, $m = n(g)$. Then

$$c(f + g) = \prod_{i<j} (a_i, a_j) \prod_{i<j} (b_i, b_j) \prod_{i,j} (a_i, b_j)$$
$$= c(f)\, c(g) \left(\prod_i a_i, \prod_j b_j\right)$$
$$= c(f)\, c(g)\, (d(f), d(g));$$

as required.

We now consider when a form f is isotropic. The next lemma is quite general and quoted only for completeness.

LEMMA 2.4. *A regular binary form f over a field k is isotropic if and only if*

$$-d(f) \in (k^*)^2$$

Proof. For the corresponding quadratic space must be a hyperbolic plane and f is equivalent to $x_1^2 - x_2^2$ (or to $x_1 x_2$).

LEMMA 2.5. *A necessary and sufficient condition that the regular ternary form f be isotropic is that*

$$c(f) = (-1, -d(f)). \tag{2.3}$$

Proof. For without loss of generality f is diagonal:

$$f = a_1x_1^2 + a_2x_2^2 + a_3x_3^2.$$

By the definition of the Norm Residue Symbol, f is isotropic if and only if

$$(-a_1/a_3, -a_2/a_3) = 1.$$

The rest of the proof follows by manipulation of the norm residue symbol as in Lemma 2.1. [If f is isotropic it contains a hyperbolic plane and so is equivalent to $x_1^2 - x_2^2 - d_0x_3^2$ for some d_0. This shows that an isotropic form satisfies (2.3) but the converse requires a further argument like that given above.]

LEMMA 2.6. *A necessary and sufficient condition that the quaternary form f be anisotropic is that it satisfies both the following conditions:*

(i) $d(f) \in (\mathbf{Q}_p^*)^2$

(ii) $c(f) = -(-1, -1)$.

Proof. Without loss of generality f is diagonal. We write it in the shape

$$f(\mathbf{x}, \mathbf{y}) = g(\mathbf{x}) - h(\mathbf{y})$$

where

$$g(\mathbf{x}) = a_1x_1^2 + a_2x_2^2$$
$$h(\mathbf{y}) = b_1y_1^2 + b_2y_2^2.$$

Then by Lemma 2.3 of Chapter 2 f is isotropic if and only if there is an $e \in \mathbf{Q}_p^*$ which is represented by both g and h. By Lemma 2.1 a necessary and sufficient condition for e to be represented by both g and h is

$$(e, -a_1a_2) = (a_1, a_2)$$
$$(e, -b_1b_2) = (b_1, b_2).$$

The set of e satisfying each of these conditions lies in precisely half of the classes of $\mathbf{Q}_p^*$ mod $(\mathbf{Q}_p^*)^2$. Hence f is anisotropic if and only if these two sets of e are complementary. By the properties of the Norm Residue Symbol (Lemma 2.1 of Chapter 3) this is possible if and only if

$$a_1a_2(\mathbf{Q}_p^*)^2 = b_1b_2(\mathbf{Q}_p^*)^2 \tag{2.4}$$

and

$$(a_1, a_2) = -(b_1, b_2). \tag{2.5}$$

Here (2.4) is equivalent to

$$d(f) = a_1 a_2 b_1 b_2 (\mathbf{Q}_p^*)^2 = (\mathbf{Q}_p^*)^2.$$

It is now readily verified using the multiplicative properties of the Norm Residue Symbol that (2.5) is equivalent to $c(f) = -(-1, -1)$; as required.

Note. Once Theorem 1.1 has been proved, the above lemma will show that there is precisely one anisotropic class of quaternary quadratic forms over $\mathbf{Q}_p$. A form in this class is easily seen to be

$$x_1^2 - ax_2^2 - bx_3^2 + abx_4^2$$

where

$$(a, b) = -1.$$

In any case we have the

COROLLARY. *A regular ternary form f represents all except possibly one of the cosets of* $\mathbf{Q}_p^*$ mod $(\mathbf{Q}_p^*)^2$.

Proof. For if $f(x_1, x_2, x_3)$ does not represent e, then the form $g(x_1, \ldots, x_4) = f(x_1, x_2, x_3) - ex_4^2$ must be anisotropic [Lemma 2.3 of Chapter 2]. In particular

$$d(g) = -ed(f) = (\mathbf{Q}_p^*)^2$$

by the Lemma.

LEMMA 2.7. *Suppose that* $p \neq \infty$. *Then every regular form f over* $\mathbf{Q}_p$ *of dimension* $\geqslant 5$ *is isotropic.*

Proof. Without loss of generality f is diagonal and the dimension is exactly 5. If $f = \sum_{1 \leqslant i \leqslant 5} a_i x_i^2$ it is enough to show that there is an $e \in \mathbf{Q}_p^*$ which is represented both by

$$a_1 x_1^2 + a_2 x_2^2 + a_3 x_3^3$$

and by

$$-a_4 x_4^2 - a_5 x_5^2.$$

By the last Corollary, the ternary form represents all except possibly one of the classes of $\mathbf{Q}_p^*$ mod $(\mathbf{Q}_p^*)^2$. By Lemma 2.1 the binary form represents at least half of the classes of $\mathbf{Q}_p^*$ mod $(\mathbf{Q}_p^*)^2$. Since $p \neq \infty$ the order of $\mathbf{Q}_p^*/(\mathbf{Q}_p^*)^2$ is greater than 2. Hence the result.

Note. We have already proved this lemma more simply in the case $p \neq 2$ (Lemma 1.7, Corollary, of Chapter 3). If we had merely aimed at Lemma 2.7 it might have been more in keeping with the philistine approach adopted in this book to give an *ad hoc* proof for $p = 2$. But the above proof (which follows Fröhlich (1967) at a respectful distance) has given additional information on the way. It has the immense virtue that it works in any finite algebraic extension of $\mathbf{Q}_p$.

Proof of Theorem 1.1. This asserts that if $p \neq \infty$, then $n(f)$, $d(f)$, $c(f)$ is a complete set of invariants for the equivalence class of f.

Let then f_1, f_2 be regular forms with

$$n(f_1) = n(f_2)$$
$$d(f_1) = d(f_2)$$
$$c(f_1) = c(f_2).$$

We want to show that f_1 is equivalent to f_2. This is trivial for $n = 1$ so suppose that $n > 1$. The first step is to show that there is a $b \in \mathbf{Q}_p^*$ which is represented both by f_1 and by f_2. For $n = 2$ this follows from Lemma 2.2, Corollary 2 (which asserts that, for $n = 2$, the set of elements of $\mathbf{Q}_p$ represented by f is determined by $d(f)$ and $c(f)$). For $n > 2$ it follows from Lemma 2.7, which asserts that the form $f_1(\mathbf{x}) - f_2(\mathbf{y})$ in the $2n \geqslant 5$ variables $(\mathbf{x}, \mathbf{y})$ is isotropic, and from Lemma 2.3 of Chapter 2.

In both cases f_j $(j = 1, 2)$ is equivalent to a form

$$bx_1^2 + g_j(x_2, \ldots, x_n) \qquad (j = 1, 2).$$

It is readily verified from Lemma 2.3 that $n(g_j)$, $d(g_j)$ and $c(g_j)$ are uniquely determined by $n(f_j)$, $d(f_j)$, $c(f_j)$ and b. In particular

$$n(g_1) = n(g_2) = n - 1$$
$$d(g_1) = d(g_2)$$
$$c(g_1) = c(g_2).$$

By the induction hypothesis g_1, g_2 are equivalent and hence so are f_1 and f_2.

Proof of Theorem 1.3. This is now trivial. Suppose that $n = 2$ and that $d \in \mathbf{Q}_p^*$ is given. Then the form

$$f = ax_1^2 + (d/a)x_2^2$$

has

$$d(f) = d(\mathbf{Q}_p^*)^2$$

and

$$c(f) = (a, -d).$$

This proves the statements about $n = 2$.

Now let $n = 3$ and let $d \in \mathbf{Q}_p^*$ be given as before. Pick $a \in \mathbf{Q}_p^*$ so that $-ad \notin (\mathbf{Q}_p^*)^2$. Consider the form

$$f(x_1, x_2, x_3) = ax_1^2 + g(x_2, x_3),$$

where $d(g) \in ad(\mathbf{Q}_p^*)^2$. Then $d(f) = d(\mathbf{Q}_p^*)^2$ and

$$c(f) = (a, ad)\, c(g).$$

By the case $n = 2$ we can choose g so that $c(g)$ is either $+1$ or -1 and so also $c(f)$ can have both signs.

The induction to $n > 3$ is obvious.

We conclude this section by a technical lemma which will be needed later. It shows that if a form $f(\mathbf{x})$ is isotropic over $\mathbf{Q}_p$ then there are solutions $\mathbf{c}$ of $f(\mathbf{c}) = 0$ with further desirable properties (cf. the Remark after the proof of Lemma 2.1 of Chapter 2). It is clear that the argument can be further extended.

LEMMA 2.8. *Let $f(x_1, \ldots, x_n)$ be a regular form over $\mathbf{Q}_p$ in $n \geqslant 3$ variables and let*

$$h(\mathbf{x}) = h_1 x_1 + \ldots + h_n x_n$$

be a linear form. Suppose that not all h_j are 0. Let $\mathbf{b} \in \mathbf{Q}_p^n$ be a solution of $f(\mathbf{b}) = 0$. Then there is a solution $\mathbf{c} \in \mathbf{Q}_p^n$ of $f(\mathbf{c}) = 0$, $h(\mathbf{c}) \neq 0$ in every neighbourhood of $\mathbf{b}$.

Proof. We may suppose without loss of generality that $\mathbf{b} = (1, 0, \ldots, 0)$ and, after a linear transformation on $x_2, \ldots, x_n$, that

$$f(\mathbf{x}) = 2f_{12}x_1x_2 + f(0, x_2, \ldots, x_n)$$

with

$$f_{12} \neq 0.$$

After a further substitution

$$x_1 \to x_1 + \text{linear form in } x_2, \ldots, x_n$$

we have

$$f(\mathbf{x}) = 2f_{12}x_1x_2 + g(x_3, \ldots, x_n)$$

where g is a regular form in $n - 2$ variables.

If $h_1 \neq 0$ there is nothing to prove, so we may suppose that

$$h_1 = 0.$$

Suppose, now, that at least one of $h_3, \ldots, h_n$ is non-zero. Choose $d_3, \ldots, d_n \in \mathbf{Q}_p$ such that

$$h_3 d_3 + \ldots + h_n d_n \neq 0.$$

Then for any $\lambda \in \mathbf{Q}_p$ the point $\mathbf{b}_\lambda$ with coordinates

$$(1, -\lambda^2 g(d_3, \ldots, d_n)/2f_{12}, \lambda d_3, \ldots, \lambda d_n)$$

satisfies $f(\mathbf{b}_\lambda) = 0$. It also satisfies $h(\mathbf{b}_\lambda) \neq 0$ if λ is small enough and so does what is required.

Finally, suppose that $h_3 = \ldots = h_n = 0$, so $h_2 \neq 0$. Pick $d_3, \ldots, d_n \in \mathbf{Q}_p$ such that $g(d_3, \ldots, d_n) \neq 0$. Then $\mathbf{b}_\lambda$ satisfies $f(\mathbf{b}_\lambda) = 0$ and $h(\mathbf{b}_\lambda) \neq 0$ if $\lambda \neq 0$.

3. THE WITT GROUP

It is an easy exercise to compute the Witt group $W(\mathbf{Q}_p)$ from first principles. The following result, however, introduces ideas which we shall require later in discussing $W(\mathbf{Q})$ (Section 11 of Chapter 6).

LEMMA 3.1. *Suppose that p is odd. Then $W(\mathbf{Q}_p)$ is isomorphic to the direct sum of two copies of $W(\mathbf{F}_p)$, where $\mathbf{F}_p$ is the field of p elements.*

Proof. We shall in this proof use small Latin letters to denote elements of $\mathbf{Q}_p$ and small Greek letters near the beginning of the alphabet for elements of $\mathbf{F}_p$. The group of units of $\mathbf{Q}_p$ is U. The residue class map $\mathbf{Z}_p \to \mathbf{F}_p$ will be written $a \to \bar{a}$. We know that it sets up an isomorphism between U/U^2 and $\mathbf{F}^*/(\mathbf{F}^*)^2$. In Chapter 2 we used $\underset{w}{+}$ for addition in the Witt group but here we shall use $+$ as there can be no confusion with other groups. By Lemma 5.6, Corollary, of Chapter 2 $W(\mathbf{Q}_p)$ is generated by symbols $\langle a \rangle$ $(a \in \mathbf{Q}_p^*)$ subject to the relations

(i) $\langle ab^2 \rangle = \langle a \rangle \qquad (a, b \in \mathbf{Q}_p^*)$

(ii) $\langle a \rangle + \langle b \rangle = \langle a + b \rangle + \langle ab(a + b) \rangle \qquad (a, b, a + b \in \mathbf{Q}_p^*)$.

(iii) $\langle 1 \rangle + \langle -1 \rangle = 0$.

It is a consequence of these axioms that

(iv) $\langle a \rangle + \langle -a \rangle = 0 \qquad (a \in \mathbf{Q}_p^*)$.

We first define a map

$$\omega_0 \colon W(\mathbf{F}_p) \to W(\mathbf{Q}_p) \tag{3.1}$$

by

$$\omega_0 \langle \alpha \rangle = \langle u \rangle \qquad (\alpha \in \mathbf{F}_p^*) \tag{3.2}$$

where u is any element of U such that $\bar{u} = \alpha$. If u_1 is any other element of U such that $\bar{u}_1 = \alpha$ then $u_1 = uv^2$ for some $v \in U$, so $\langle u_1 \rangle = \langle u \rangle$. Hence the map is uniquely defined on our set of generators for $W(\mathbf{F}_p)$. To show, however, that the map ω_0 is well-defined, we must show that it respects the relations in $W(\mathbf{F}_p)$: but this is easy. For let $\alpha, \beta \in \mathbf{F}_p$ and let $u, v \in U$ satisfy $\bar{u} = \alpha, \bar{v} = \beta$. Then the analogues of the relations (i), (ii), (iii) for $\mathbf{F}_p$ (with α, β for a, b respectively) are mapped by ω_0 into the relations (i), (ii), (iii) with u, v for a, b.

It is rather more remarkable that there is a map

$$\psi_0 \colon W(\mathbf{Q}_p) \to W(\mathbf{F}_p) \tag{3.3}$$

in the opposite direction. This is defined as follows. Every $a \in \mathbf{Q}_p^*$ is uniquely of the shape

$$a = p^l u \qquad (l \in \mathbf{Z}, u \in U). \tag{3.4}$$

Then we put

$$\psi_0\langle a\rangle = \begin{cases} 0 & \text{if } l \text{ is odd} \\ \langle \bar{u}\rangle & \text{if } l \text{ is even.} \end{cases} \tag{3.5}$$

Again we have to check that the relations (i), (ii), (iii) are preserved and the only one which might cause trouble is (ii). We distinguish three cases.

(I) $|a|_p$ and $|b|_p$ are distinct, say $|b|_p < |a_p|$. Then $a + b = ac^2$ for some $c \in \mathbf{Q}_p^*$. Then both sides of (ii) are mapped by ψ_0 into visibly the same element of $W(\mathbf{F}_p)$.

(II) $|a|_p = |b|_p > |a + b|_p$. Then $b = -ac^2$ for some $c \in \mathbf{Q}_p$ and both sides of (ii) are mapped into 0 or into an expression of the type $\langle\alpha\rangle + \langle-\alpha\rangle$, which is 0 by the analogue of (iv) for $W(\mathbf{F}_p)$.

(III) $|a|_p = |b|_p = |a + b|_p = p^{-l}$ (say). If l is odd, every term in (ii) is mapped by ψ_0 into 0. If, however, l is even then (ii) is mapped into a relation (ii) for $W(\mathbf{F}_p)$.

Hence ψ_0 is well-defined. Clearly

$$\psi_0\omega_0 = \text{identity on } W(\mathbf{F}_p). \tag{3.6}$$

We can define further maps. First

$$\omega_1 \colon W(\mathbf{F}_p) \to W(\mathbf{Q}_p) \tag{3.7}$$

is given by

$$\omega_1\langle\alpha\rangle = \langle pu\rangle \qquad (u \in U, \bar{u} = \alpha). \tag{3.8}$$

Secondly

$$\psi_1 \colon W(\mathbf{Q}_p) \to W(\mathbf{F}_p) \tag{3.9}$$

is given by

$$\psi_1\langle a\rangle = \begin{cases} 0 & \text{if } l \text{ is even} \\ \langle \bar{u}\rangle & \text{if } l \text{ is odd,} \end{cases} \tag{3.10}$$

where l, u are given by (3.4). It is readily verified that ω_1, ψ_1 are well-defined and

$$\psi_1\omega_1 = \text{identity on } W(\mathbf{F}_p). \tag{3.11}$$

Finally, it is readily checked using (3.6) and (3.11) that the map Ψ which takes $\langle a \rangle \in W(\mathbf{Q}_p)$ into the pair $(\psi_0\langle a \rangle, \psi_1\langle a \rangle)$ is an isomorphism between $W(\mathbf{Q}_p)$ and the direct sum of two copies of $W(\mathbf{F}_p)$. This completes the proof.

For completeness we prove

LEMMA 3.2. ($p = 2$). *The group* $W(\mathbf{Q}_2)$ *is the direct sum of a cyclic group of order* 8 *and two cyclic groups of order* 2.

Proof. Put

$$\langle 1 \rangle = \varepsilon$$
$$\langle 2 \rangle = \varepsilon + \lambda$$
$$\langle 5 \rangle = \varepsilon + \mu$$
$$\langle 10 \rangle = \varepsilon + \nu$$

so

$$\lambda \neq 0, \qquad \mu \neq 0, \qquad \nu \neq 0.$$

Then

$$\langle 1 \rangle + \langle 1 \rangle = \langle 2 \rangle + \langle 2 \rangle \qquad ,$$
$$\langle 1 \rangle + \langle 1 \rangle = \langle 1 \rangle + \langle 4 \rangle = \langle 5 \rangle + \langle 5 \rangle$$
$$\langle 1 \rangle + \langle 1 \rangle = \langle 1 \rangle + \langle 9 \rangle = \langle 10 \rangle + \langle 10 \rangle,$$

so

$$2\lambda = 2\mu = 2\nu = 0.$$

Further

$$\langle 1 \rangle + \langle 2 \rangle = \langle 3 \rangle + \langle 6 \rangle$$
$$= -\langle 5 \rangle - \langle 10 \rangle$$

so

$$4\varepsilon = \lambda + \mu + \nu,$$

and hence

$$8\varepsilon = 0.$$

We have

$$4\varepsilon \neq 0,$$

since $x_1^2 + x_2^2 + x_3^2 + x_4^2$ is not isotropic in $\mathbf{Q}_2$. Further,

$$\langle 2 \rangle + \langle 5 \rangle \neq \langle 1 \rangle + \langle 1 \rangle$$

(by consideration of determinants on both sides), so

$$\lambda + \mu \neq 0.$$

Similarly

$$\mu + \nu \neq 0, \qquad \nu + \lambda \neq 0.$$

Hence $W(\mathbf{Q}_2)$ is generated by $\varepsilon, \lambda, \mu$ and the only relations between them are

$$8\varepsilon = 2\lambda = 2\mu = 0.$$

NOTES

Sections 1, 2. All this generalizes without trouble to finite extensions of p-adic fields, granted the relevant properties of the Hilbert Norm Residue Symbol. See O'Meara (1963) or Fröhlich (1967):

Section 3. Lemma 3.1 was proved by Springer (1955) in the general context of local fields for which the residue class field is of characteristic $\neq 2$: his results for characteristic 2 are less complete.

In general the properties of quadratic forms over finite extensions of $\mathbf{Q}_2$ are complicated and difficult to manage. See O'Meara (1963).

EXAMPLES

1. For what primes p are the following forms isotropic over $\mathbf{Q}_p$?

(i) $5x_1^2 - x_2^2 - 3x_3^2$

(ii) $x_1^2 + x_2^2 + 7x_3^2 + 5x_4^2$.

2. For each of the following pairs of forms find the primes p for which they are equivalent over $\mathbf{Q}_p$:

(i) $3x_1^2 + 7x_2^2$ and $x_1^2 + 84x_2^2$

(ii) $x_1^2 - 3x_2^2 + 15x_3^2$ and $3x_1^2 - 5x_2^2 + 3x_3^2$.

(iii) $x_1^2 - 5x_2^2 + 3x_3^2 - 7x_4^2$ and $x_1^2 - x_2^2 + x_3^2 - x_4^2$.

3. Let $b \in \mathbf{Q}_p^*$, where $p \neq \infty$. Show that the forms $x_1^2 + x_2^2 + x_3^2 + x_4^2$ and $b(x_1^2 + x_2^2 + x_3^2 + x_4^2)$ are equivalent over $\mathbf{Q}_p$. What happens for $p = \infty$?

4. Let $m \in \mathbf{Q}_p^*, m \neq -1$. Show that $c_p(f) = 1$, where

$$f(\mathbf{x}) = (m + 1)(x_1^2 + x_2^2 + x_3^2) + mx_4^2.$$

For $p \neq \infty$ deduce that f is $\mathbf{Q}_p$—equivalent to $x_1^2 + x_2^2 + x_3^2 + m(m + 1)x_4^2$. What happens when $p = \infty$?

CHAPTER 5

Tools from the Geometry of Numbers

1. INTRODUCTION

A great advantage in confining attention to quadratic forms over the rational field, rather than working over an algebraic number field, is that one of the key theorems (the "Strong Hasse Principle") can be very simply proved using ideas from the Geometry of Numbers. The appropriate generalization is true for algebraic number-fields but there is (as yet) no correspondingly easy proof: indeed a very substantial portion of O'Meara's book (1963) is taken up with the requisite preliminary material.

It would, in fact, be possible to modify the argument used in the proof of the Strong Hasse Principle so as to avoid the use of the Geometry of Numbers and, instead, use an argument of Hermite which we give later in the book. We have preferred to use the Geometry of Numbers partly because it makes the proof more transparent and partly because the Geometry of Numbers will also be useful elsewhere.

In Section 2 the necessary results from the Geometry of Numbers will be developed in the shape in which we shall require them. This shape is not quite the usual one and in Section 3, which will not be required later, we put Section 2 into the setting of the standard approach and also discuss the relationship to the argument of Hermite mentioned above.

2. THE TOOLS

In this section we shall be concerned with sets $\mathscr{S}$ in n-dimensional space $\mathbf{R}^n$. In the applications these will be extremely simple sets: ellipsoids or parallelepipeds: in the formal proofs, however, we shall require them only to be Lebesgue-measurable. We shall speak of the measure of $\mathscr{S}$ as its volume and denote it by $v(\mathscr{S})$.

$\mathbf{R}^n$ is given with a fixed co-ordinate system, so that its elements are sets of n real numbers, say $\mathbf{r} = (r_1, \ldots, r_n)$. We use the habitual notation for vector addition and for multiplication by elements of $\mathbf{R}$: in particular the origin is $\mathbf{0} = (0, \ldots, 0)$. By $\mathbf{Z}^n$ we mean the set of $\mathbf{u} = (u_1, \ldots, u_n) \in \mathbf{R}^n$ for which $u_j \in \mathbf{Z}\,(1 \leqslant j \leqslant n)$.

THEOREM 2.1 (Blichfeld). *Let k be a positive integer and let $\mathscr{S}$ be a set in $\mathbf{R}^n$ with*

$$v(\mathscr{S}) > k.$$

Then there are $k + 1$ distinct points $\mathbf{s}_0, \mathbf{s}_1, \ldots, \mathbf{s}_k \in \mathscr{S}$ such that

$$\mathbf{s}_i - \mathbf{s}_j \in \mathbf{Z}^n \qquad (0 \leqslant i \leqslant j \leqslant k).$$

Note. This is the continuous analogue of the following generalization of Dirichlet's Schubfachprinzip: if m papers are filed in n pigeon-holes and $m > kn$ then at least one pigeon-hole must contain at least $k + 1$ papers.

Blichfeld's original theorem is the case $k = 1$.

Proof. For $\mathbf{u} \in \mathbf{Z}^n$ denote by $\mathscr{S}(\mathbf{u})$ the set of points $\mathbf{s}$ of $\mathscr{S}$ which lie in the cube

$$u_j \leqslant s_j < u_j + 1 \qquad (1 \leqslant j \leqslant n).$$

The $\mathscr{S}(\mathbf{u})$ are disjoint and their union is $\mathscr{S}$, so

$$v(\mathscr{S}) = \sum_{\mathbf{u}} v(\mathscr{S}(\mathbf{u})).$$

For each $\mathbf{u}$ let $\mathscr{S}^*(\mathbf{u})$ be the set of

$$\mathbf{s} - \mathbf{u}, \qquad \mathbf{s} \in \mathscr{S}(\mathbf{u}).$$

Then

$$\mathscr{S}^*(\mathbf{u}) \subset \mathscr{W}, \tag{2.1}$$

where $\mathscr{W}$ is the unit cube:

$$\mathscr{W} : 0 \leqslant x_j < 1 \qquad (1 \leqslant j \leqslant n).$$

Further,

$$v(\mathscr{S}^*(\mathbf{u})) = v(\mathscr{S}(\mathbf{u}))$$

and so

$$\begin{aligned} \sum_{\mathbf{u}} v(\mathscr{S}^*(\mathbf{u})) &= \Sigma v(\mathscr{S}(\mathbf{u})) \\ &= v(\mathscr{S}) \\ &> k. \end{aligned} \tag{2.2}$$

It follows from (2.1) and (2.2) that at least one point $\mathbf{w}$ (say) of $\mathscr{W}$ must be contained in at least $k + 1$ of the $\mathscr{S}^*(\mathbf{u})$, say for $\mathbf{u} = \mathbf{u}_0, \ldots, \mathbf{u}_k$. Then

$$\mathbf{s}_j = \mathbf{w} + \mathbf{u}_j \qquad (0 \leqslant j \leqslant k) \tag{2.3}$$

clearly satisfy the conclusions of the Theorem.

There is an alternative formulation of the first part of the above proof which may be found more perspicuous. Let $\sigma(\mathbf{x})$ be the characteristic function of $\mathscr{S}$, that is

$$\sigma(\mathbf{x}) = \begin{cases} 1 & \text{if } \mathbf{x} \in \mathscr{S} \\ 0 & \text{otherwise.} \end{cases}$$

Then

$$\int \cdots \int_{\mathbf{R}^n} \sigma(\mathbf{x}) \, dx_1 \ldots dx_n = v(\mathscr{S}).$$

Hence

$$\begin{aligned} \int \cdots \int_{\mathscr{W}} \left(\sum_{\mathbf{u}} \sigma(\mathbf{x} + \mathbf{u}) \right) dx_1 \ldots dx_n &= \int \cdots \int_{\mathbf{R}^n} \sigma(\mathbf{x}) \, dx_1 \ldots dx_n \\ &= v(\mathscr{S}) \\ &> k. \end{aligned}$$

There is thus a $\mathbf{w} \in \mathscr{W}$ such that

$$\sum_{\mathbf{u}} \sigma(\mathbf{w} + \mathbf{u}) > k$$

and so

$$\geqslant k + 1$$

since both sides are integers. We can now put (2.3) as before where $\sigma(\mathbf{w} + \mathbf{u}_j) = 1$.

We owe to Minkowski the recognition of the importance of convexity in this context. We recall that a set $\mathscr{C}$ is *convex* if

$$\lambda_1 \mathbf{c}_1 + \lambda_2 \mathbf{c}_2 \in \mathscr{C} \tag{2.4}$$

whenever

$$\lambda_1 \geqslant 0, \qquad \lambda_2 \geqslant 0, \qquad \lambda_1 + \lambda_2 = 1$$

and

$$\mathbf{c}_1, \mathbf{c}_2 \in \mathscr{C}.$$

We say that a set $\mathscr{S}$ is *symmetric* (about $\mathbf{0}$) if $-\mathbf{s} \in \mathscr{S}$ whenever $\mathbf{s} \in \mathscr{S}$. Minkowski's celebrated "convex body Theorem" in its simplest formulation is

THEOREM 2.2 (Minkowski). *Let $\mathscr{C}$ be a convex symmetric set with*

$$v(\mathscr{C}) > 2^n.$$

Then $\mathscr{C}$ contains a $\mathbf{u} \in \mathbf{Z}^n$ other than $\mathbf{0}$.

Proof. Let $\mathscr{S} = \frac{1}{2}\mathscr{C}$, that is the set of $\frac{1}{2}\mathbf{c}, \mathbf{c} \in \mathscr{C}$. Then

$$v(\mathscr{S}) > 1.$$

Hence by the case $k = 1$ of Theorem 2.1 there are two distinct points $\frac{1}{2}\mathbf{c}_0$ and $\frac{1}{2}\mathbf{c}_1$ of $\frac{1}{2}\mathscr{C}$ such that

$$\mathbf{u}\ (\text{say}) = \tfrac{1}{2}\mathbf{c}_0 - \tfrac{1}{2}\mathbf{c}_1 \in \mathbf{Z}^n.$$

But then $-\mathbf{c}_1 \in \mathscr{C}$ by the hypothesis that $\mathscr{C}$ is symmetric, and so

$$\mathbf{u} = \tfrac{1}{2}\mathbf{c}_0 + \tfrac{1}{2}(-\mathbf{c}_1) \in \mathscr{C}$$

by the convexity ($\lambda_1 = \lambda_2 = \frac{1}{2}$ in (2.4)). This concludes the proof.
We shall want the following generalization.

THEOREM 2.3. *Let Λ be the set of $\mathbf{u} \in \mathbf{Z}^n$ satisfying a finite set of congruences*

$$h_{i1}u_1 + \dots + h_{in}u_n \equiv 0 \pmod{m_i} \qquad (1 \leqslant i \leqslant I),$$

where

$$h_{ij} \in \mathbf{Z}, \qquad m_i \in \mathbf{Z}, \qquad m_i \geqslant 1.$$

Put

$$m = m_1 \dots m_I.$$

Let $\mathscr{C}$ be a convex symmetric set with

$$v(\mathscr{C}) > 2^n m.$$

Then $\mathscr{C}$ contains a $\mathbf{u} \in \Lambda$ other than $\mathbf{0}$.

Proof. We have

$$v(\tfrac{1}{2}\mathscr{C}) > m$$

and so by Theorem 2.1 there are $m + 1$ distinct points $\frac{1}{2}\mathbf{c}_0, \dots, \frac{1}{2}\mathbf{c}_m$ such that

$$\tfrac{1}{2}\mathbf{c}_i - \tfrac{1}{2}\mathbf{c}_j \in \mathbf{Z}^n \qquad (0 \leqslant i \leqslant j \leqslant m).$$

The set $\mathbf{Z}^n$ is an abelian group under addition and clearly Λ is a subgroup of index $\leqslant m$. Hence two of the

$$\mathbf{v}_i\ (\text{say}) = \tfrac{1}{2}\mathbf{c}_i - \tfrac{1}{2}\mathbf{c}_0 \in \mathbf{Z}^n \qquad (0 \leqslant i \leqslant m)$$

must be in the same coset of $\mathbf{Z}^n \bmod \Lambda$ (Schubfachprinzip). Suppose then that $\mathbf{v}_k$ and $\mathbf{v}_l (l \neq k)$ are in the same coset. Then

$$\mathbf{u}\ (\text{say}) = \mathbf{v}_k - \mathbf{v}_l \in \Lambda.$$

On the other hand,

$$\mathbf{u} = \tfrac{1}{2}\mathbf{c}_k - \tfrac{1}{2}\mathbf{c}_l \in \mathscr{C}$$

on using the same argument as in the proof of Theorem 2.2.

Let us now revert to Theorem 2.2. When the set $\mathscr{C}$ is a parallelepiped the condition $v(\mathscr{C}) > 2^n$ can be relaxed.

THEOREM 2.4 (Minkowski's Linear Forms Theorem). *Let* $L_j(\mathbf{x})$ $(1 \leqslant j \leqslant n)$ *be real linear forms in the n variables* $\mathbf{x} = (x_1, \ldots, x_n)$ *of determinant* $D \neq 0$. *Let* $t_j > 0$ $(1 \leqslant j \leqslant n)$ *satisfy*

$$t_1 \ldots t_n \geqslant |D|. \tag{2.5}$$

Then there is an $\mathbf{a} \in \mathbf{Z}^n$ *other than* $\mathbf{0}$ *such that*

$$|L_1(\mathbf{a})| \leqslant t_1 \tag{2.6}$$

and

$$|L_j(\mathbf{a})| < t_j \qquad (2 \leqslant j \leqslant n). \tag{2.7}$$

Proof. Here $|\,|$ denotes the ordinary absolute value. Let $0 < \varepsilon < 1$. The set $\mathscr{C}_\varepsilon$ given by

$$\begin{aligned} |L_1(\mathbf{x})| &< t_1(1 + \varepsilon) \\ |L_j(\mathbf{x})| &< t_j \qquad (2 \leqslant j \leqslant n) \end{aligned}$$

has volume

$$\begin{aligned} v(\mathscr{C}_\varepsilon) &= 2^n |D|^{-1} t_1 \ldots t_n (1 + \varepsilon) \\ &\geqslant 2^n(1 + \varepsilon) \\ &> 2^n \end{aligned}$$

as is readily verified by introducing new variables $y_j = L_j(\mathbf{x})(1 \leqslant j \leqslant n)$. Further $\mathscr{C}_\varepsilon$, being an n-dimensional parallelepiped, is convex and symmetric. Hence $\mathscr{C}_\varepsilon$ contains an $\mathbf{a}_\varepsilon \neq \mathbf{0}$, $\mathbf{a}_\varepsilon \in \mathbf{Z}^n$. But $\mathscr{C}_\varepsilon \subset \mathscr{C}_1$ and $\mathscr{C}_1$, being bounded, contains only a finite number of elements $\mathbf{a} \neq \mathbf{0}$ of $\mathbf{Z}^n$. We have just shown that there is one of these in every $\mathscr{C}_\varepsilon$ and so there must be one in $\bigcap_\varepsilon \mathscr{C}_\varepsilon$: and this is the set defined by (2.6) and (2.7).

Note. A similar argument shows that the conditions $v(\mathscr{C}) > 2^n$ in Theorem 2.2 can be relaxed to $v(\mathscr{C}) \geqslant 2^n$ whenever $\mathscr{C}$ is bounded and closed (i.e. compact): one considers the sets $(1 + \varepsilon)\mathscr{C}$ with $\varepsilon > 0$.

We shall need a result on simultaneous approximation which is effectively a special case of Theorem 2.4.

COROLLARY. *Let* $n \geqslant 1, M > 1$ *be integers and let* $\theta_1, \ldots, \theta_n$ *be real numbers. Then there is an integer m in*

$$0 < m < M \tag{2.8}$$

and integers $l_1, \ldots, l_n$ *such that*

$$|m\theta_j - l_j| \leqslant M^{-1/n} \qquad (1 \leqslant j \leqslant n). \tag{2.9}$$

Proof. We apply the Theorem to the $n + 1$ linear forms

$$\begin{aligned} L_j(\mathbf{x}) &= \theta_j x_{n+1} - x_j \qquad (1 \leqslant j \leqslant n) \\ L_{n+1}(\mathbf{x}) &= x_{n+1}. \end{aligned}$$

Then $D = \pm 1$. Take $t_j = M^{-1/n}$ $(1 \leqslant j \leqslant n)$; $t_{n+1} = M$. By the Theorem (with $n+1$ for n and relaxing (2.7)) there is an $\mathbf{a} = (l_1, \ldots, l_n, m) \neq \mathbf{0}$ in $\mathbf{Z}^{n+1}$ such that

$$|\theta_j m - l_j| \leqslant M^{-1/n} \qquad (1 \leqslant j \leqslant n) \tag{2.10}$$

$$|m| < M. \tag{2.11}$$

If $m = 0$, the inequalities (2.10) imply $l_j = 0$ since $M > 1$, contradicting $\mathbf{a} \neq \mathbf{0}$. Hence on taking $-l_j$, $-m$ for l_j, m if need be we may suppose that $m > 0$, Then (2.10), (2.11) give the conclusions of the Corollary.

3. BACKGROUND

The results of this section will not be required later. In it we set the results of Section 2 in a wider framework.

The usual treatment of the geometry of numbers (e.g. Cassels (1959)) depends on the notion of a lattice Γ in $\mathbf{R}^n$. This is the set of points

$$u_1 \mathbf{b}_1 + \ldots + u_n \mathbf{b}_n$$

where $\mathbf{b}_1, \ldots, \mathbf{b}_n$ is a set of n linearly independent elements of $\mathbf{R}^n$ (a basis of Γ) and $u_1, \ldots, u_n$ run through $\mathbf{Z}$. It is easy to see that Γ is a subgroup of $\mathbf{R}^n$ considered as a group under addition and that Γ is discrete in the ordinary topology on $\mathbf{R}^n$. Conversely it can be shown that every discrete subgroup of $\mathbf{R}^n$ which contains n linearly independent points is a lattice. The number

$$d(\Gamma) = |\det(\mathbf{b}_1, \ldots, \mathbf{b}_n)| \tag{3.1}$$

is independent of the choice of basis $\mathbf{b}_1, \ldots, \mathbf{b}_n$ and is called the *determinant* of the lattice Γ.

In this terminology Theorem 2.2 at once implies

THEOREM 3.1. *Let Γ be a lattice and $\mathscr{C}$ a convex symmetric set in $\mathbf{R}^n$ with*

$$v(\mathscr{C}) > 2^n d(\Gamma). \tag{3.2}$$

Then $\mathscr{C}$ contains a point of Γ other than $\mathbf{0}$.

For one has only to introduce new co-ordinates $\mathbf{y} = (y_1, \ldots, y_n)$ by

$$\mathbf{x} = y_1 \mathbf{b}_1 + \ldots + y_n \mathbf{b}_n$$

where $\mathbf{b}_1, \ldots, \mathbf{b}_n$ is a basis and to note that the Jacobian of the transformation from $\mathbf{x}$ to $\mathbf{y}$ is $\pm d(\Gamma)$.

In this setting Theorem 2.3 is no more general than Theorem 2.2. Indeed every subgroup Λ of finite index in a lattice Γ is itself a lattice. This follows at once from the characterization of a lattice given above: alternatively it is an immediate consequence of Lemma 3.3 of Chapter 7.

Now let

$$\mathbf{x} = u_1\mathbf{b}_1 + \ldots + u_n\mathbf{b}_n \tag{3.3}$$

be a point of a lattice Γ with basis $\mathbf{b}_1, \ldots, \mathbf{b}_n$. Then

$$x_1^2 + \ldots + x_n^2 = g(u_1, \ldots, u_n) \tag{3.4}$$

is a positive definite form in $u_1, \ldots, u_n$ with real coefficients and

$$d(g) = \{d(\Gamma)\}^2. \tag{3.5}$$

Conversely, every positive definite form g can be obtained in this way, e.g. by "completing the square" successively. Let m be the minimum taken by g for integral $\mathbf{u} \neq \mathbf{0}$. Then there are no points of Γ other than $\mathbf{0}$ in the convex symmetric set

$$\mathscr{C}: x_1^2 + \ldots + x_n^2 < m. \tag{3.6}$$

By Theorem 3.1 we must have

$$v(\mathscr{C}) \leqslant 2^n d(\Gamma).$$

But

$$v(\mathscr{C}) = m^{n/2} V_n,$$

where V_n is the volume of the unit sphere. It follows that

$$m \leqslant C_n(d(g))^{1/n} \tag{3.7}$$

where

$$C_n = 4V_n^{-2/n}. \tag{3.8}$$

In fact, before Minkowski had invented the Geometry of Numbers, Hermite gave a simple proof of (3.7) with

$$C_n = (4/3)^{(n-1)/2} \tag{3.9}$$

instead of (3.8). We shall reproduce his argument in Section 3 of Chapter 9 as part of a proof of a more general theorem. Minkowski's estimate is much better (i.e. smaller) than Hermite's for large values of n but Hermite's is better for $n \leqslant 8$. Indeed for $n = 2$ the value (3.9) is best-possible, as is shown by the form $u_1^2 + u_1u_2 + u_2^2$. A great deal of work has been done on possible values of C_n and the best are now known for $n \leqslant 8$. For all this see Cassels (1959) or, better, Milnor and Husemoller (1973).

It follows, then that whenever we apply Theorem 2.2 or 2.3 to ellipsoids, we could obtain stronger numerical bounds by using information about possible values of C_n. This remark applies, for example, to Theorem 4.1, Corollary, of Chapter 6.

NOTES

The Geometry of Numbers is the creation of Hermann Minkowski, and he exploited it to great effect in the theory of numbers. There are recent accounts in Cassels (1959) and Lekkerkerker (1969).

EXAMPLES

1. (i) Let m be a positive integer and suppose that there is an integer r such that $r^2 \equiv -1 \pmod{m}$. Let Λ be the set of $(x, y) \in \mathbf{Z}^2$ such that $x \equiv ry \pmod{m}$. Show that $x^2 + y^2 \equiv 0 \pmod{m}$ for all $x, y \in \Lambda$. Show, further that there is an $\mathbf{a} = (a, b) \neq \mathbf{0}$ of Λ with $a^2 + b^2 < 2m$.. Deduce that $a^2 + b^2 = m$.

(ii) Show that the conditions of (i) are satisfied when m is an odd prime $\equiv 1 \pmod{4}$.

(iii) More generally, show that the conditions of (i) are satisfied whenever $m = m_0$ or $m = 2m_0$ and all the prime factors of m_0 are $\equiv 1 \pmod{4}$.

2. (i) Let m be a positive integer and suppose that there are integers r, s such that $r^2 + s^2 + 1 \equiv 0 \pmod{m}$. Let Λ be the set of $(x, y, z, w) \in \mathbf{Z}^4$ such that

$$x \equiv rz + sw, \qquad y \equiv sz - rw \pmod{m}.$$

Show that there is a point $(a, b, c, d) \neq \mathbf{0}$ of Λ for which $a^2 + b^2 + c^2 + d^2 < 2m$. Deduce that $m = a^2 + b^2 + c^2 + d^2$.

(ii) Show that the conditions of (i) are satisfied whenever m is odd or twice an odd integer. Deduce that every positive integer is the sum of four squares.

CHAPTER 6

Quadratic Forms over the Rationals

1. INTRODUCTION

The theme of this chapter is the relation between the behaviour of quadratic forms over the rational field $\mathbf{Q}$ and over the "local" fields $\mathbf{Q}_p$ (where $\mathbf{Q}_\infty = \mathbf{R}$). General theorems enable us to reduce questions over $\mathbf{Q}$ to questions over the $\mathbf{Q}_p$.

THEOREM 1.1 (Strong Hasse Principle). *Let f be a regular quadratic form over $\mathbf{Q}$. A necessary and sufficient condition for f to be isotropic over $\mathbf{Q}$ is that it is isotropic over all $\mathbf{Q}_p$ (inc. $p = \infty$).*

The obvious analogue of this theorem is true for any algebraic number field instead of $\mathbf{Q}$: the formulation and proof are due to Hasse (1923, 1923a; 1924, 1924a, b). We shall, however, as usual confine attention to $\mathbf{Q}$: in particular when the number n of variables is 3 we shall give a proof using the ideas of the geometry of numbers which does not, apparently, generalize. For $n \geqslant 4$ we shall use Dirichlet's theorem about the existence of primes in arithmetic progressions. This gives very perspicuous proofs but may be felt to introduce an alien element. In Chapter 14 we shall show, following the ideas of Gauss in his *Disquisitiones*, that it is possible to develop the theory without invoking Dirichlet's theorem.

The proof of Theorem 1.1 will occupy Sections 3, 4 and 5. We prove at once here two important corollaries.

COROLLARY 1 (Meyer's Theorem). *An indefinite regular form f in $n \geqslant 5$ variables is isotropic.*

Proof. For f is isotropic over $\mathbf{Q}_\infty = \mathbf{R}$ by hypothesis, and, by Lemma 2.7 of Chapter 4, every regular form over $\mathbf{Q}_p$ in $\geqslant 5$ variables is isotropic for $p \neq \infty$.

COROLLARY 2. *Let f be a regular form over* $\mathbf{Q}$ *and let* $e \in \mathbf{Q}^*$ *be given. Suppose that f represents e in every* $\mathbf{Q}_p$ *(including* $p = \infty$*). Then f represents e in* $\mathbf{Q}$.

Proof. For the representability of e by $f(x_1, \ldots, x_n)$ is equivalent to the isotropy of $f(x_1, \ldots, x_n) - ex_{n+1}^2$ in $n + 1$ variables. [Chapter 2, Lemma 2.3]

Another major theorem is

THEOREM 1.2 (Weak Hasse Principle). *Two regular forms over* $\mathbf{Q}$ *which are equivalent over every* $\mathbf{Q}_p$ *(including* $p = \infty$*) are equivalent over* $\mathbf{Q}$.

We shall give the deduction of Theorem 1.2 from Theorem 1.1 in Section 2. It is perhaps worth noting that there are situations generally analogous to that considered in this chapter where the Weak Hasse Principle holds but the Strong Hasse Principle fails [e.g. when the ground field is $k = \mathbf{R}(t)$, t being transcendental, cf. the last section of Cassels, Ellison and Pfister (1971) and Hsia (1973)].

In Sections 3, 4, 5 and 6 we shall prove the Strong Hasse Principle for $n \leqslant 2$, $n = 3$, $n = 4$ and $n \geqslant 5$ respectively.

Another general problem is this: given a collection of forms f_p defined over $\mathbf{Q}_p$ for all primes p, does there exist a form f defined over $\mathbf{Q}$ which is equivalent to each f_p over $\mathbf{Q}_p$? There is a very satisfactory solution to this problem but we require some preliminary discussion.

Since $\mathbf{Q} \subset \mathbf{Q}_p$ we can define the Hasse–Minkowski invariant $c_p(f)$ of a regular form defined over $\mathbf{Q}$.

LEMMA 1.1. $c_p(f) = 1$ *for almost all p and* $\prod_{\substack{\text{all } p \\ \text{inc } \infty}} c_p(f) = 1$.

Note. We recall that "almost all" p means all except, possibly, finitely many.

Proof. We may suppose without loss of generality that f is diagonal

$$f(\mathbf{x}) = a_1 x_1^2 + \ldots + a_n x_n^2.$$

Then

$$c_p(f) = \prod_{i<j} \left(\frac{a_i, a_j}{p} \right).$$

Lemma 1.1 now follows from the corresponding properties of the Hilbert Norm Residue Symbol (Lemma 3.4 of Chapter 3).

The next theorem shows that Lemma 1.1 is the only relation between the $c_p(f)$.

THEOREM 1.3. *Let $n \geqslant 2$ and $d_0 \in \mathbf{Q}^*$ be given. For all p (including $p = \infty$) let $f_p \in \mathbf{Q}_p[x_1, \ldots, x_n]$ be a regular form and suppose that* †

(i) $d(f_p) = d_0(\mathbf{Q}_p^*)^2$

(ii) $c(f_p) = 1$ *for almost all p and* $\prod_{\substack{\text{all } p \\ \text{inc } \infty}} c_p(f_p) = 1$.

Then there is an $f \in \mathbf{Q}[x_1, \ldots, x_n]$ with $d(f) = d_0(\mathbf{Q}^*)^2$ which is equivalent over $\mathbf{Q}_p$ to f_p for every p.

In Section 7 we shall give a brief proof of Theorem 1.3 using Dirichlet's theorem about primes in arithmetic progression. We shall later (Chapter 14) give an elementary but more complicated proof. In Sections 8, 9 and 10, we give two amplifications of the Strong Hasse Principle of a rather minor nature and an application to the theory of projective planes. Finally, in Section 11, we determine the Witt group $W(\mathbf{Q})$. The proof simultaneously gives a proof of the Weak Hasse Principle which is both simple and elementary (in the sense that it does not use Dirichlet's theorem about primes in arithmetic progressions).

2. THE WEAK HASSE PRINCIPLE

In this section we show that the Strong Hasse Principle (Theorem 1.1) implies the Weak Hasse Principle (Theorem 1.2).

Let $f(\mathbf{x}), g(\mathbf{x})$ be two regular forms in n variables with rational coefficients and suppose that they are equivalent over all $\mathbf{Q}_p$ (including $p = \infty$). We have to show that f, g are equivalent over $\mathbf{Q}$.

For each p the $\mathbf{Q}_p$-equivalent forms f and g represent over $\mathbf{Q}_p$, the same elements of $\mathbf{Q}_p$. In particular the form $f(\mathbf{x}) - g(\mathbf{y})$ in the $2n$ variables $(\mathbf{x}, \mathbf{y})$ is isotropic over $\mathbf{Q}_p$. By Theorem 1.1, which we are supposing to be true, the form $f(\mathbf{x}) - g(\mathbf{y})$ must be isotropic over $\mathbf{Q}$. By Lemma 2.3 of Chapter 2 there is thus an $e \neq 0$ which is represented both by f and g over $\mathbf{Q}$.

It follows that the forms f, g are equivalent over $\mathbf{Q}$ to forms

$$\langle e \rangle + f_1, \qquad \langle e \rangle + g_1 \tag{2.1}$$

where $\langle e \rangle$ is the 1-dimensional form ex^2 and where f_1, g_1 are regular forms in $(n-1)$ variables. By hypothesis the forms f, g are equivalent over every $\mathbf{Q}_p$ and so, by Witt's Lemma (Theorem 4.1, Corollary 1, of Chapter 2) the forms f_1, g_1 are equivalent over every $\mathbf{Q}_p$. We now use induction on the dimension n of the forms f, g. Since f_1, g_1 are in $(n-1)$ variables and

† In this Chapter we shall often be concerned with a form f only up to k-equivalence, where k is $\mathbf{Q}$ or $\mathbf{Q}_p$. In these circumstances we shall regard $d(f)$ as an element of $k^*/(k^*)^2$ rather than as an element of k.

satisfy the hypotheses of the weak Hasse principle we may suppose that they satisfy the conclusions of the principle: that is, that f_1, g_1 are equivalent over $\mathbf{Q}$. But then the forms (2.1) are equivalent over $\mathbf{Q}$, that is f and g are equivalent. This concludes the proof.

3. THE STRONG HASSE PRINCIPLE, $n \leq 2$

Let $f(\mathbf{x})$ be a regular quadratic form with rational coefficients which is isotropic over all $\mathbf{Q}_p$ (including $p = \infty$). The Strong Hasse Principle asserts that f is isotropic over $\mathbf{Q}$. The truth of this statement depends only on the equivalence class of f so we may suppose, for example, that f is diagonal. The truth of the statement is also unaffected by replacing f with af where $a \in \mathbf{Q}^*$.

The principle is vacuous when the dimension n of f is 1. When $n = 2$ by the remarks which have just been made we need consider only f of the shape

$$x_1^2 - bx_2^2 \qquad b \in \mathbf{Q}^*.$$

This form is isotropic over $\mathbf{Q}_p$ if and only if $b \in (\mathbf{Q}_p^*)^2$. To verify the Strong Hasse Principle for $n = 2$ it is thus enough to verify the following.

LEMMA 3.1. *Let $b \in \mathbf{Q}^*$ and suppose that $b \in (\mathbf{Q}_p^*)^2$ for all p (including $p = \infty$). Then $b \in (\mathbf{Q}^*)^2$.*

Proof. We have

$$b = \pm \prod_p p^{\alpha(p)}$$

where $\alpha(p) \in \mathbf{Z}$ and almost all $\alpha(p)$ are 0. Since $b \in (\mathbf{Q}_p^*)^2$ we see that $\alpha(p)$ is even. Finally the fact that $b \in (\mathbf{Q}_\infty^*)^2$ implies that the $\pm$ sign is, in fact, $+$. [The lemma remains correct if $p = \infty$ is excluded: it is enough to note that $-1 \notin (\mathbf{Q}_2^*)^2$.]

4. THE STRONG HASSE PRINCIPLE, $n = 3$

LEMMA 4.1. *Let f be any regular ternary quadratic form over $\mathbf{Q}$. Then there is an $a \in \mathbf{Q}^*$ such that af is rationally equivalent to a form*

$$g(\mathbf{x}) = a_1x_1^2 + a_2x_2^2 + a_3x_3^2 \qquad (4.1)$$

where $a_1, a_2, a_3 \in \mathbf{Z}$ and $a_1a_2a_3$ is squarefree.

Proof. We may suppose without loss of generality that f is diagonal and that the coefficients are integers, say

$$f(\mathbf{x}) = b_1x_1^2 + b_2x_2^2 + b_3x_3^2.$$

If b_1 is not squarefree, say $b_1 = b_1'c^2$, then on replacing x_1 by $c^{-1}x_1$ we may replace b_1 by b_1'. If there is a prime p such that $p|b_1, p|b_2$ say $b_1 = pb_1', b_2 = pb_2'$ then we have

$$p^{-1}f(x_1, x_2, px_3) = b_1'x_1^2 + b_2'x_2^2 + b_3'x_3^2,$$

where $b_3' = pb_3$. Both the operations just described decrease the absolute value of the non-zero integer $b_1b_2b_3$. Hence after a finite number of steps we reach a form (4.1) of the type described.

LEMMA 4.2. *Let $g(\mathbf{x})$ be as in Lemma 4.1. Let p be an odd prime $p|a_3$ and suppose that $g(\mathbf{x})$ is isotropic in $\mathbf{Q}_p$. Then there is an $r \in \mathbf{Z}$ such that $a_1r^2 + a_2 \equiv 0 \pmod p$.*

Proof. By hypothesis there are $t_1, t_2, t_3 \in \mathbf{Q}_p$ not all 0 such that $g(t_1, t_2, t_3) = 0$. By homogeneity we may suppose that

$$\max\{|t_1|_p, |t_2|_p, |t_3|_p\} = 1.$$

Since $\Sigma a_jt_j^2 = 0$, at least two of the $|a_jt_j^2|_p$ must be equal. If $|a_jt_j^2|_p = p^{-\mu(j)}$, clearly $\mu(1)$, $\mu(2)$ are even but $\mu(3)$ is odd. Hence the only possibility is

$$1 = |a_1t_1^2|_p = |a_2t_2^2|_p > |a_3t_3^2|_p.$$

The required result now follows on taking $\Sigma a_jt_j^2$ modulo p.

LEMMA 4.3. *Let $g(\mathbf{x})$ be as in Lemma 4.1 and suppose that $g(\mathbf{x})$ is isotropic over $\mathbf{Q}_2$.*

(i) *Suppose that $2 \nmid a_1a_2a_3$. Then after permuting the suffices* 1, 2, 3 *if need be we may suppose that*

$$a_1 + a_2 \equiv 0 \pmod 4.$$

(ii) *Suppose that $2|a_3$. Then there is an $s = 0$ or 1 such that*

$$a_1 + a_2 + a_3s^2 \equiv 0 \pmod 8.$$

Proof. As in the proof of Lemma 4.2 we may suppose without loss of generality that there are $t_j \in \mathbf{Q}_2$ such that $\sum a_jt_j^2 = 0$ and

$$1 = |a_1t_1^2|_2 = |a_2t_2^2|_2 > |a_3t_3^2|_2.$$

The result now follows since $t^2 \equiv 1 \pmod 8$ whenever $|t|_2 = 1$.

THEOREM 4.1 (Legendre). *Let*

$$g(\mathbf{x}) = a_1x_1^2 + a_2x_2^2 + a_3x_3^2 \tag{4.2}$$

where $a_1, a_2, a_3 \in \mathbf{Z}$ *aod* $a_1a_2a_3$ *is squarefree. Suppose that the following conditions are satisfied*

(i) *if* p *is an odd prime dividing* $a_1a_2a_3$, *say* $p|a_3$ *then there is an integer* r_p such *that*

$$a_1r_p^2 + a_2 \equiv 0 \pmod{p}. \tag{4.3}$$

(ii) *if* $2|a_1a_2a_3$, *say* $2|a_3$ *then*

$$a_1 + a_2 + a_3s^2 \equiv 0 \pmod{8} \tag{4.4}$$

where $s = 0$ or 1.

(iii) *if* $2 \nmid a_1a_2a_3$ *then, on permuting* a_1, a_2, a_3 *if need be,*

$$a_1 + a_2 \equiv 0 \pmod{4}. \tag{4.5}$$

Then there are $b_1, b_2, b_3 \in \mathbf{Z}$, *not all* 0, *such that*

$$a_1b_1^2 + a_2b_2^2 + a_3b_3^2 = 0. \tag{4.6}$$

Proof. We shall apply Theorem 2.3 of Chapter 5 and will denote by Λ the set of $\mathbf{z} = (z_1, z_3, z_3) \in \mathbf{Z}^3$ satisfying a number of congruences which we now describe.

(i) Let p be an odd prime, $p|a_3$. We impose the condition

$$z_1 \equiv r_pz_2 \pmod{p} \tag{4.7}$$

where r_p is given by (i) of the enunciation. Then

$$\begin{aligned} g(\mathbf{z}) &= a_1z_1^2 + a_2z_2^2 + a_3z_3^2 \\ &\equiv a_1z_1^2 + a_2z_2^2 \pmod{p} \\ &\equiv (a_1r_p^2 + a_2)z_2^2 \pmod{p} \\ &\equiv 0 \pmod{p}. \end{aligned}$$

(ii) Suppose that $2|a_3$. We impose the conditions

$$\left.\begin{aligned} z_1 &\equiv z_2 \pmod{4} \\ z_3 &\equiv sz_2 \pmod{2} \end{aligned}\right\} \tag{4.8$_1$}$$

where $s = 0$ or 1 is given by (ii) of the enunciation. It is readily verified (considering separately the cases z_2 odd, $z_2 \equiv 2 \pmod{4}$ and $z_2 \equiv 0 \pmod{4}$) that these congruences imply

$$f(\mathbf{z}) \equiv 0 \pmod{8}.$$

(iii) Suppose that $2 \nmid a_1a_2a_3$. By (iii) of the enunciation we may suppose that $a_1 + a_2 \equiv 0 \pmod 4$. We impose the conditions

$$\left.\begin{aligned} z_1 &\equiv z_2 \pmod 2 \\ z_3 &\equiv 0 \pmod 2. \end{aligned}\right\} \tag{4.8$_2$}$$

Taking all these conditions together we have a collection of congruences

$$L_j(\mathbf{z}) \equiv 0 \pmod{m_j}$$

where L_j is a linear form with integral coefficients and where

$$m\,(\text{say}) = \prod_j m_j = \pm 4a_1a_2a_3.$$

Together they imply that

$$g(\mathbf{z}) \equiv 0 \pmod{4a_1a_2a_3}.$$

We now apply Theorem 2.3 of Chapter 5 to the set Λ of (z_1, z_2, z_3) satisfying the congruences, so

$$m = |4a_1a_2a_3|,$$

where in the rest of this proof we shall write $|\;|$ for the absolute value $|\;|_\infty$. For $\mathscr{C}$ we take the ellipsoid

$$|a_1|x_1^2 + |a_2|x_2^2 + |a_3|x_3^2 < 4|a_1a_2a_3|.$$

This has volume

$$\begin{aligned} v(\mathscr{C}) &= \frac{\pi}{3} \cdot 2^3 \cdot |4a_1a_2a_3| \\ &> 2^3 m. \end{aligned}$$

There is thus a $\mathbf{b} \in \Lambda$, $\mathbf{b} \neq \mathbf{0}$ in $\mathscr{C}$.

Since $\mathbf{b} \in \Lambda$ we have

$$a_1b_2^2 + a_2b_2^2 + a_3b_3^2 \equiv 0 \pmod{4|a_1a_2a_3|}.$$

Since $\mathbf{b} \in \mathscr{C}$ we have

$$\begin{aligned} |a_1b_1^2 + a_2b_2^2 + a_3b_3^2| &\leqslant |a_1|b_1^2 + |a_2|b_2^2 + |a_3|b_3^2 \\ &< 4|a_1a_2a_3|. \end{aligned}$$

It follows that

$$a_1b_1^2 + a_2b_2^2 + a_3b_3^2 = 0,$$

as required

In the proof of the Theorem we have also proved

COROLLARY 1. *There is a solution of* $g(\mathbf{b}) = 0$ *with*

$$|a_1|b_1^2 + |a_2|b^2 + |a_3|b_3^2 < 4|a_1 a_2 a_3|.$$

It is implicit in the proof that the constant 4 on the right-hand side can be improved somewhat and there is a considerable literature on the "smallest" solutions of $f(\mathbf{b}) = 0$ with various measures of the "size" of $\mathbf{b}$. Forms of the type $x_1^2 + x_2^2 - px_3^2$ with prime $p \equiv 1 \pmod 4$ show that the 4 in the corollary cannot be replaced by 2.

We note that in the proof of Theorem 4.1 we did not need to suppose that $g(\mathbf{x})$ is indefinite (i.e. isotropic at $\mathbf{R} = \mathbf{Q}_\infty$), so Theorem 4.1 is stronger than the strong Hasse theorem in this case. In fact we have

COROLLARY 2. *Let* $f(\mathbf{x})$ *be a regular ternary form over* $\mathbf{Q}$. *Suppose that* f *is isotropic over* $\mathbf{Q}_p$ *for all* p *(including* ∞*) with one possible exception* p_0 *(which may be either* ∞ *or a finite prime). Then* f *is isotropic over* $\mathbf{Q}$.

Note. In the theorem actually proved by Legendre $p_0 = 2$. Thus in the enunciation of Theorem 4.1 one may omit conditions (ii) and (iii) provided that one adds the condition that $g(\mathbf{x})$ be indefinite.

Proof. It is enough to consider the $g(\mathbf{x})$ given by (4.1). By the definition of the Hilbert Norm Residue Symbol $g(\mathbf{x})$ is isotropic over $\mathbf{Q}_p$ precisely when

$$\left(\frac{-a_1 a_3, -a_2 a_3}{p}\right) = 1. \tag{4.9}$$

The product formula for the Norm Residue Symbol (Lemma 3.4 of Chapter 3) states that the number of p (inc ∞) for which (4.9) fails is even. In particular if (4.2) holds for all $p \neq p_0$ then it holds for $p = p_0$ as well. The Corollary now follows from Theorem 4.1 on recalling Lemmas 4.2 and 4.3.

We conclude this section with a historical note. Theorem 4.1 implies some of the consequences of the law of quadratic reciprocity. Thus suppose, if possible, that odd (positive) primes q_1, q_2 exist such that

(α) $q_1 \equiv 3 \pmod 4$.

(β) $-q_1$ is a quadratic residue of q_2.

(γ) $-q_2$ is a quadratic residue of q_1.

Then the hypotheses of Theorem 4.1 would be satisfied with $a_1 = q_1$, $a_2 = q_2, a_3 = 1$. The conclusion of the Theorem cannot, however, hold since $q_1 x_1^2 + q_2 x_2^2 + x_3^2$ is definite. It follows that (α), (β), (γ) cannot all hold simultaneously. Special cases of the quadratic reciprocity law had been proved in this sort of way (as well as in other ways) before Gauss discovered the first complete proof.

5. THE STRONG HASSE PRINCIPLE, $n = 4$

Here we shall make use of Dirichlet's theorem on primes in arithmetic progressions:

THEOREM 5.1 (Dirichlet). *Let $l > 0$ and m be a coprime pair of integers. There are infinitely many primes $p \equiv m \pmod{l}$.*

For a proof see almost any text on analytic number theory. We shall need only the

COROLLARY. *Let P be a finite set of primes p (possibly $\infty \in P$). For $p \in P$ let $t_p \in \mathbf{Q}_p^*$ be given. Then there is a $t \in \mathbf{Q}^*$ such that*

(i) $t \in t_p(\mathbf{Q}_p^*)^2$ *(all $p \in P$).*

(ii) $|t|_p = 1$ *for all $p \notin P, p \neq \infty$*

except possibly for one $p = p_0$ (say).

Proof. For $p \in P, p \neq \infty$ let $t_p = p^{\alpha(p)} s_p$ where $|s_p|_p = 1$. We shall take t in the shape

$$t = \pm p_0 \prod_{\substack{p \in P \\ \neq \infty}} p^{\alpha(p)}$$

where $p_0 \notin P$ is a prime. If $\infty \in P$ we choose the sign $\pm$ to be that of t_∞; otherwise we choose it arbitrarily. We now select the prime $p_0 > 0$ to satisfy the congruences

$$p^{-\alpha(p)} t \equiv s_p \begin{cases} \bmod p & \text{if } p \neq 2 \\ \bmod 8 & \text{if } p = 2. \end{cases}$$

Then t clearly has all the required properties.

We now proceed to the proof of the Strong Hasse Principle for $n = 4$. It is enough to consider forms f of the type

$$f(\mathbf{x}) = a_1 x_1^2 + a_2 x_2^2 + a_3 x_3^2 + a_4 x_4^2$$

with

$$a_j \in \mathbf{Z} \qquad (1 \leqslant j \leqslant 4).$$

Let P consist of $p = \infty$ together with all the finite primes p dividing $2a_1a_2a_3a_4$. We are supposing that $f(\mathbf{x})$ is isotropic for all p (including ∞). Hence there are b_{jp} $(1 \leqslant j \leqslant 4)$ not all 0 and t_p, all in $\mathbf{Q}_p$ such that

$$a_1 b_{1p}^2 + a_2 b_{2p}^2 = t_p$$
$$a_3 b_{3p}^2 + a_4 b_{4p}^2 = -t_p.$$

By Chapter 2, Lemma 2.3 we may suppose without loss of generality that $t_p \neq 0$. With these t_p let t and p_0 be as given by Theorem 5.1, Corollary, and consider the two ternary forms

$$\left.\begin{aligned} a_1x_1^2 + a_2x_2^2 - tu^2 \\ a_3x_3^2 + a_4x_4^2 + tv^2 \end{aligned}\right\} \tag{5.1}$$

in the variables (x_1, x_2, u) and (x_3, x_4, v). They are isotropic for $p \in P$ by the construction of t. For $p \notin P$, $p \neq p_0$ the forms (5.1) are isotropic by Lemma 1.7 of Chapter 3 since $|a_j|_p = 1\ (1 \leqslant j \leqslant 4)$ and $|t|_p = 1$. By Corollary 2 to Theorem 4.1 it follows that the forms (5.1) are isotropic over $\mathbf{Q}$. There are thus $b_1, b_2, b_3, b_4 \in \mathbf{Q}$ such that

$$\left.\begin{aligned} a_1b_1^2 + a_2b_2^2 - t = 0 \\ a_3b_3^2 + a_4b_4^2 + t = 0, \end{aligned}\right\}$$

and so

$$a_1b_1^2 + a_2b_2^2 + a_3b_3^2 + a_4b_4^2 = 0,$$

as required.

6. THE STRONG HASSE PRINCIPLE, $n \geq 5$

It is enough to consider forms

$$a_1x_1^2 + \ldots + a_nx_n^2$$

with

$$a_j \in \mathbf{Z} \qquad (1 \leqslant j \leqslant n).$$

Let P consist of $p = \infty$ together with the finite primes dividing $2a_1 \ldots a_n$. As in Section 5 there are $t_p \in \mathbf{Q}_p^*$ for $p \in P$ such that

$$a_1x_1^2 + a_2x_2^2 - t_pu^2$$

and

$$a_3x_3^2 + \ldots + a_nx_n^2 + t_pv^2$$

are both isotropic over $\mathbf{Q}_p$. Let

$$t_p = a_1b_{1p}^2 + a_2b_{2p}^2 \qquad (b_{1p}, b_{2p} \in \mathbf{Q}_p).$$

By the Weak Approximation Theorem (Lemma 3.2 of Chapter 3) we can find $b_1, b_2 \in \mathbf{Q}$ so close to b_{1p}, b_{2p} in $\mathbf{Q}_p$ for $p \in P$ that

$$t = a_1b_1^2 + a_2b_2^2 \tag{6.1}$$

satisfies

$$t \in t_p(\mathbf{Q}_p^*)^2 \qquad (p \in P) \tag{6.2}$$

By (6.2) the form

$$a_3x_3^2 + \ldots + a_nx_n^2 + tv^2 \tag{6.3}$$

is isotropic for all $p \in P$. It is isotropic also for $p \notin P$: indeed for $p \notin P$ we have $|a_3|_p = \ldots = |a_n|_p = 1$ and since $n \geqslant 5$ the form $a_3x_3^2 + \ldots + a_nx_n^2$ is already isotropic over $\mathbf{Q}_p$ by Lemma 1.7 of Chapter 3. We now use induction and suppose that the Strong Hasse Principle has already been proved for forms of dimension less than n. Then (6.3) is isotropic over $\mathbf{Q}$ and on making the substitution (6.1) it follows that $a_1x_1^2 + \ldots + a_nx_n^2$ is isotropic over $\mathbf{Q}$, as required.

7. AN EXISTENCE THEOREM

In this section we prove Theorem 1.3. This asserts that if we are given forms f_p in $n \geqslant 2$ variables over $\mathbf{Q}_p$ with the "same" determinant $d_0 \in \mathbf{Q}^*/(\mathbf{Q}^*)^2$ [in the sense that $d(f_p) = d_0(\mathbf{Q}_p^*)^2$] and such that $\prod_{p \text{ inc } \infty} c_p(f_p) = 1$, then there is a rational form f which is equivalent to each f_p over $\mathbf{Q}_p$.

We first prove this result for $n = 2$. Let P be a finite set of primes p such that

(i) $\infty \in P$; $2 \in P$

(ii) if $|d_0|_p \neq 1$ then $p \in P$

(iii) if $c_p(f_p) = -1$ then $p \in P$.

For each $p \in P$ let $t_p \in \mathbf{Q}_p^*$ be represented by f_p. By the Corollary to Dirichlet's Theorem 5.1 there is a $p_0 \notin P$ and a $t \in \mathbf{Q}^*$ such that $t \in t_p(\mathbf{Q}_p^*)^2$ $(p \in P)$ and $|t|_p = 1$ for $p \notin P, p \neq p_0$.

Put

$$f(x_1, x_2) = tx_1^2 + (d_0/t)x_2^2.$$

Then for $p \in P$ we have

$$\begin{aligned} c_p(f) &= \left(\frac{t, -d_0}{p}\right) \\ &= \left(\frac{t_p, -d_0}{p}\right) \\ &= c_p(f_p) \qquad (p \in P). \end{aligned}$$

For $p \notin P, p \neq p_0$ we have

$$c_p(f) = \left(\frac{t, -d_0}{p}\right) = +1 = c_p(f_p).$$

Finally for $p = p_0$ the product formula $\prod c_p(f) = 1$ together with the hypothesis $\prod c_p(f_p) = 1$ give

$$c_{p_0}(f) = 1 = c_{p_0}(f_{p_0}).$$

This concludes the proof when $n = 2$.

Now suppose that $n \geqslant 3$. Define the set P by (i), (ii), (iii) above. For $p \in P$ let $t_p \in \mathbf{Q}_p^*$ be represented by f_p over $\mathbf{Q}_p$. Let t be any element of $\mathbf{Q}^*$ such that $t/t_p \in (\mathbf{Q}_p^*)^2$ for all $p \in P$. For $p \notin P$ we have $|d_0|_p = 1, c_p(f_p) = 1$ and so f_p is isotropic: in particular f_p represents t (Lemma 2.1, Corollary 2, of Chapter 2). Hence for *all* p the form f_p is equivalent to

$$\langle t \rangle + g_p$$

for some form g_p over $\mathbf{Q}_p$ in $n - 1$ variables. We now show that the g_p satisfy the hypotheses of Theorem 1.3. In the first place

$$d(g_p) = (d_0/t)(\mathbf{Q}_p^*)^2.$$

Secondly

$$\begin{aligned} c_p(f_p) &= c_p(\langle t \rangle + g_p) \\ &= \left(\frac{t, d(g_p)}{p}\right) c_p(g_p) \\ &= \left(\frac{t, d_0/t}{p}\right) c_p(g_p). \end{aligned}$$

Hence $\prod c_p(f_p) = 1$ together with the product formula for the Norm Residue Symbol implies that

$$\prod c_p(g_p) = 1.$$

By induction on the dimension, there is a rational form g which is equivalent to each g_p over $\mathbf{Q}_p$. Then

$$f = \langle t \rangle + g$$

does what is required.

8. SIZE OF SOLUTIONS

In Section 4 we not only showed that a ternary form f which is everywhere locally isotropic is isotropic over $\mathbf{Q}$ but we showed that there is in a sense a solution of $f(\mathbf{a}) = 0$ which is "not too big". The proofs in Section 5 and 6 for $n > 3$ did not give a such a bound and could not be reframed so as to give a reasonable one. Here we show by a quite different argument that if a rational form f is isotropic then there is always a solution of $f(\mathbf{a}) = 0$ which is not too big.

LEMMA 8.1. *Let*

$$f(\mathbf{x}) = \Sigma f_{ij} x_i x_j \in \mathbf{Z}[x_1, \ldots, x_n] \tag{8.1}$$

be an isotropic form in n variables. Then there is an

$$\mathbf{a} \in \mathbf{Z}^n, \mathbf{a} \neq \mathbf{0} \tag{8.2}$$

with

$$f(\mathbf{a}) = 0 \tag{8.3}$$

such that

$$\max_{1 \leqslant j \leqslant n} |a_j| \leqslant (3F)^{(n-1)/2}, \tag{8.4}$$

where

$$F = \sum_{i,j} |f_{ij}|. \tag{8.5}$$

Here $|\ |$ denotes the absolute value.

Note. Although the constant 3 in (8.4) could doubtless be improved, the exponent $(n - 1)/2$ cannot. For let b be a large positive integer and consider

$$f(\mathbf{x}) = x_1^2 - \sum_{j=2}^{n} (x_j - bx_{j-1})^2. \tag{8.6}$$

Here

$$F = n + 2(n - 1)b + (n - 1)b^2. \tag{8.7}$$

Suppose that $f(\mathbf{a}) = 0$. Then clearly $a_1 \neq 0$ and

$$a_n = \lambda_n + \lambda_{n-1} b + \ldots + \lambda_2 b^{n-2} + a_1 b^{n-1}, \tag{8.8}$$

where

$$\lambda_n = a_n - ba_{n-1},$$

so

$$\lambda_n^2 + \lambda_{n-1}^2 + \ldots + \lambda_2^2 = a_1^2. \tag{8.9}$$

It is easy to check that these equations imply that

$$\begin{aligned} |a_n| &\geqslant -|a_1|b^{n-2} + |a_1|b^{n-1} \\ &\geqslant b^{n-1} - b^{n-2}. \end{aligned} \tag{8.10}$$

Proof of Lemma. We take for $\mathbf{a} \in \mathbf{Z}^n, \mathbf{a} \neq \mathbf{0}$ a solution of $f(\mathbf{a}) = 0$ for which

$$\|\mathbf{a}\| \text{ (say)} = \max_{1 \leqslant j \leqslant n} |a_j| \tag{8.11}$$

is minimal. If (8.4) is false we shall find an $\mathbf{a}^* \in \mathbf{Z}^n, \neq \mathbf{0}$ for which $f(\mathbf{a}^*) = 0$, $\|\mathbf{a}^*\| < \|\mathbf{a}\|$: which would contradict the minimality of $\|\mathbf{a}\|$.

We may suppose by permuting the indices and taking $-\mathbf{a}$ for $\mathbf{a}$ if need be that

$$a_1 = \max_j |a_j|. \tag{8.12}$$

If $a_1 = 1$ there is nothing to prove, so we suppose that

$$a_1 \geqslant 2. \tag{8.13}$$

Let $\theta_2, \ldots, \theta_n$ be any real numbers. By a theorem on diophantine approximation (Theorem 2.4, Corollary, of Chapter 5) there are integers $b_1, \ldots, b_n$ such that

$$0 < b_1 < a_1 \tag{8.14}$$

and

$$|b_1\theta_j - b_j| \leqslant a_1^{-1/(n-1)} \qquad (2 \leqslant j \leqslant n). \tag{8.15}$$

We apply this with

$$\theta_j = a_j/a_1. \tag{8.16}$$

Then

$$\begin{aligned} |b_j| &\leqslant |b_1\theta_j| + a_1^{-1/(n-1)} \\ &\leqslant b_1 + a_1^{-1/(n-1)} \\ &< b_1 + 1; \end{aligned}$$

and so

$$\|\mathbf{b}\| = \max_j |b_j| = b_1 < \|\mathbf{a}\|. \tag{8.17}$$

The minimality of $\|\mathbf{a}\|$ now implies that

$$f(\mathbf{b}) \neq 0. \tag{8.18}$$

We now choose $\lambda, \mu \in \mathbf{Z}$ so that

$$\mathbf{a}^* = \lambda\mathbf{a} + \mu\mathbf{b} \tag{8.19}$$

satisfies

$$f(\mathbf{a}^*) = 0. \tag{8.20}$$

We have

$$\begin{aligned} f(\mathbf{a}^*) &= \lambda^2 f(\mathbf{a}) + 2\lambda\mu f(\mathbf{a}, \mathbf{b}) + \mu^2 f(\mathbf{b}) \\ &= 2\lambda\mu f(\mathbf{a}, \mathbf{b}) + \mu^2 f(\mathbf{b}); \end{aligned}$$

so it is enough to choose

$$\lambda = f(\mathbf{b}) \in \mathbf{Z}\,(\neq 0) \tag{8.21}$$

and

$$\mu = -2f(\mathbf{a}, \mathbf{b}) \in \mathbf{Z}. \tag{8.22}$$

We note that

$$\mathbf{a}^* \neq \mathbf{0} \tag{8.23}$$

since otherwise (8.19) would imply $f(\mathbf{b}) = f(\mathbf{a}) = 0$ contrary to (8.18).

By (8.16) we can write (8.15) in the shape

$$b_j = \phi a_j + \delta_j$$

where

$$\phi = b_1/a_1$$

and

$$\delta_1 = 0, \qquad |\delta_j| \leqslant a_1^{-1/(n-1)} \qquad (2 \leqslant j \leqslant n). \tag{8.24}$$

We shall express $\mathbf{a}^*$ in terms of $\mathbf{a}$ and $\boldsymbol{\delta}$ by eliminating $\mathbf{b}$.

We have

$$f(\mathbf{a}, \mathbf{b}) = f(\mathbf{a}, \phi\mathbf{a} + \boldsymbol{\delta}) = \phi f(\mathbf{a}) + f(\mathbf{a}, \boldsymbol{\delta})$$
$$= f(\mathbf{a}, \boldsymbol{\delta})$$

and

$$f(\mathbf{b}) = f(\phi\mathbf{a} + \boldsymbol{\delta})$$
$$= 2\phi f(\mathbf{a}, \boldsymbol{\delta}) + f(\boldsymbol{\delta}).$$

Hence by (8.19), (8.21), (8.22) we have

$$\mathbf{a}^* = f(\mathbf{b})\mathbf{a} - 2f(\mathbf{a}, \mathbf{b})\mathbf{b}$$
$$= \{2\phi f(\mathbf{a}, \boldsymbol{\delta}) + f(\boldsymbol{\delta})\}\, \mathbf{a} - 2f(\mathbf{a}, \boldsymbol{\delta})\{\phi\mathbf{a} + \boldsymbol{\delta}\}$$
$$= f(\boldsymbol{\delta})\,\mathbf{a} - 2f(\mathbf{a}, \boldsymbol{\delta})\,\boldsymbol{\delta}.$$

Estimating crudely and recalling the definition (8.5) of F it follows that

$$\|\mathbf{a}^*\| \leqslant 3F\|\mathbf{a}\| \;\|\boldsymbol{\delta}\|^2.$$

By the minimality of $\|\mathbf{a}\|$ we have, however,

$$\|\mathbf{a}^*\| \geqslant \|\mathbf{a}\|$$

and so

$$3F\|\boldsymbol{\delta}\|^2 \geqslant 1. \tag{8.25}$$

But by (8.24) we have

$$\|\boldsymbol{\delta}\| \leqslant a_1^{-1/(n-1)} = \|\mathbf{a}\|^{-1/(n-1)}.$$

Hence (8.25) implies

$$\|\mathbf{a}\| \leqslant (3F)^{(n-1)/2},$$

as asserted.

9. AN APPROXIMATION THEOREM

The object of this section is to prove

LEMMA 9.1. *Let $f(\mathbf{x})$ be an isotropic form over $\mathbf{Q}$ in $n \geqslant 3$ variables. Let $\varepsilon > 0$ be arbitrarily small and P be a finite set of primes p (possibly $\infty \in P$) and for $p \in P$ let $\mathbf{b}_p \in \mathbf{Q}_p^n$ be given with $f(\mathbf{b}_p) = 0$. Then there is a $\mathbf{b} \in \mathbf{Q}^n$*

with

$$f(\mathbf{b}) = 0$$

and

$$\|\mathbf{b} - \mathbf{b}_p\|_p < \varepsilon \qquad (\textit{all } p \in P).$$

Here for a vector $\mathbf{a} \in \mathbf{Q}_p^n$ *we have written*

$$\|\mathbf{a}\|_p = \max_j |a_j|_p.$$

We note at once the

COROLLARY. *Let* $g(\mathbf{x})$ *be a regular rational form in* $n \geqslant 2$ *variables and let* $a \in \mathbf{Q}^*$ *be represented rationally by* g. *For* $p \in P$ *let* $\mathbf{a}_p \in \mathbf{Q}_p^n$ *with* $g(\mathbf{a}_p) = a$. *Then there is an* $\mathbf{a} \in \mathbf{Q}^n$ *such that*

$$g(\mathbf{a}) = a$$

and

$$\|\mathbf{a} - \mathbf{a}_p\|_p < \varepsilon \qquad (\textit{all } p \in P).$$

Proof of Corollary. For we may consider

$$f(x_1, \ldots, x_{n+1}) = g(x_1, \ldots, x_n) - ax_{n+1}^2.$$

Proof of Lemma 9.1. By hypothesis there is a $\mathbf{c} \in \mathbf{Q}^n$, $\mathbf{c} \neq \mathbf{0}$ such that

$$f(\mathbf{c}) = 0.$$

Suppose, first, that

$$f(\mathbf{c}, \mathbf{b}_p) \neq 0 \tag{9.1}$$

for all $p \in P$ where, as usual, $f(\mathbf{x}, \mathbf{y})$ is the bilinear form belonging to the quadratic form $f(\mathbf{x})$. By the weak approximation Theorem (Lemma 3.2 of Chapter 3) we can choose a $\mathbf{d} \in \mathbf{Q}^n$ which is arbitrarily close p-adically to $\mathbf{b}_p$ for each $p \in P$. We now choose $\lambda, \mu \in \mathbf{Q}$ such that

$$f(\lambda\mathbf{c} + \mu\mathbf{d}) = 0.$$

Since $f(\mathbf{c}) = 0$ the condition for this is

$$2\lambda\mu f(\mathbf{c}, \mathbf{d}) + \mu^2 f(\mathbf{d}) = 0.$$

We may therefore take

$$\mu = 1: \ \lambda = -f(\mathbf{d})/2f(\mathbf{c}, \mathbf{d}).$$

As $\mathbf{d}$ approaches $\mathbf{b}_p$ in the p-adic sense we have

$$\begin{aligned}\lambda &\to -f(\mathbf{b}_p)/2f(\mathbf{c}, \mathbf{b}_p) \\ &= 0\end{aligned}$$

by (9.1). Hence

$$\lambda\mathbf{c} + \mathbf{d} \to \mathbf{b}_p.$$

Since the limits can be achieved simultaneously for all $p \in P$ this completes the proof of the Lemma subject to the restriction (9.1).

Finally, if (9.1) fails for some p then by Lemma 2.8 of Chapter 4 we can find $\mathbf{b}'_p$ arbitrarily close to $\mathbf{b}_p$ and with $f(\mathbf{c}, \mathbf{b}'_p) \neq 0$. We can then argue with $\mathbf{b}'_p$ instead of $\mathbf{b}_p$.

10. AN APPLICATION. FINITE PROJECTIVE PLANES

In this section we give a rather striking application of the preceding theory due to Bruck and Ryser (1949). A finite projective plane consists of two finite sets, one Π of "points" and one Λ of "lines". There is one relation of "a point lying on a line" and two axioms

(i) any two distinct points lie on precisely one line

(ii) any two lines meet in precisely one point.

If one excludes certain "degenerate" cases (e.g. only one line) it is easy to see that there is an integer n such that every line contains $n + 1$ points and every point lies on $n + 1$ lines. This integer n is called the *order* of the plane. There are then precisely

$$N = n^2 + n + 1$$

points and N lines in all.

One example of a finite projective plane is that constructed in the usual way over a finite field $\mathbf{F}$: in this case the order n is equal to the number of elements of $\mathbf{F}$ and so is a prime power. Other examples of finite projective planes have been found but in every known case the order is a prime power. It is not known, for example, whether finite projective planes of order 10 can exist, though thousands of dollars worth of computer time has been devoted to the investigation.

Perhaps the only general "non-existence proof" for projective planes is the following which shows, for example, that there are no projective planes of order 6. [There is a generalization to the non-existence of certain "block designs", see for example Marshall Hall (1954), Lecture 5.]

LEMMA 10.1 (Bruck and Ryser). *Let n be the order of a finite projective plane and suppose that*

$$n \equiv 1 \text{ or } 2 \pmod{4}. \tag{10.1}$$

Then every odd prime p which divides n to an odd power satisfies

$$p \equiv 1 \pmod{4}. \tag{10.2}$$

Proof. We take N variables x_j, one for each point π_j. With each line λ_k we associate the linear form

$$l_k = \Sigma x_j,$$

the sum being over the j for which π_j lies on λ_k. The axioms of a projective plane ensure that

$$\begin{aligned}\sum_{k=1}^{N} l_k^2 &= (n+1)\sum_j x_j^2 + \sum_{i \neq j} x_i x_j \\ &= g(\mathbf{x}) \text{ (say).}\end{aligned} \tag{10.3}$$

We can diagonalize $g(\mathbf{x})$, for example, by introducing new variables

$$\begin{aligned} y_1 &= x_2 + \ldots + x_N \\ y_j &= x_j + n^{-1} x_1 \qquad (j > 1) \end{aligned}$$

and find that it is rationally equivalent to

$$y_1^2 + n \sum_{j>1} y_j^2 = h(\mathbf{y}) \text{ (say).} \tag{10.4}$$

It follows from (10.3) that $g(\mathbf{x})$, and so $h(\mathbf{y})$, is rationally equivalent to

$$f(\mathbf{z}) = \sum_{n=1}^{N} z_j^2.$$

In particular f and h must have the same Hasse–Minkowski invariants:

$$\begin{aligned} c_p(h) &= c_p(f) \\ &= 1. \end{aligned} \tag{10.5}$$

On the other hand by (10.4) we have

$$\begin{aligned} c_p(h) &= \left(\frac{n, n}{p}\right)^{(N-1)(N-2)/2} \\ &= \left(\frac{n, n}{p}\right) \end{aligned} \tag{10.6}$$

since $N = n^2 + n + 1 \equiv 3 \pmod 4$ by (10.1).

From the general properties of the Hilbert Norm Residue Symbol we have

$$\left(\frac{n, n}{p}\right) = \left(\frac{n, -1}{p}\right). \tag{10.7}$$

Putting (10.5), (10.6) and (10.7) together we get

$$\left(\frac{n, -1}{p}\right) = 1. \tag{10.8}$$

Now let p be an odd prime dividing n to an odd power. Then (10.8) asserts that -1 is a quadratic residue of p, that is $p \equiv 1 \pmod 4$, as asserted.

11. THE WITT GROUP

In this section we determine the Witt group $W(\mathbf{Q})$. The argument will also give a simple elementary proof of the Weak Hasse Principle.

We recall (Lemma 5.6 of Chapter 2) that $W(\mathbf{Q})$ is generated by symbols $\langle a \rangle$ ($a \in \mathbf{Q}^*$) subject to the relations

(i) $\langle ab^2 \rangle = \langle a \rangle$

(ii) $\langle a \rangle + \langle b \rangle = \langle a + b \rangle + \langle ab(a + b) \rangle$

(iii) $\langle 1 \rangle + \langle -1 \rangle = 0$.

For each prime p (including ∞) there is the localization map

$$\lambda_p\colon\ W(\mathbf{Q}) \to W(\mathbf{Q}_p)$$

which maps $\langle a \rangle$ considered as an element of $W(\mathbf{Q})$ into $\langle a \rangle$ considered as an element of $W(\mathbf{Q}_p)$.

For odd primes p we write

$$W(p) = W(\mathbf{F}_p), \tag{11.1}$$

where $\mathbf{F}_p$ is the field of p elements. In Section 3 of Chapter 3 we introduced two maps ψ_0, ψ_1 from $W(\mathbf{Q}_p)$ to $W(p)$. It is ψ_1 which plays an important rôle here, and we denote its composition with the localization map by $\psi(p)$. The map

$$\psi(p)\colon\ W(\mathbf{Q}) \to W(p) \tag{11.2}$$

is thus defined as follows. Every $a \in \mathbf{Q}^*$ is uniquely of the shape

$$a = p^l u, \qquad |u|_p = 1 \tag{11.3}$$

and then

$$\psi(p)\langle a \rangle = \begin{cases} \langle \bar{u} \rangle & \text{if } l \text{ is odd} \\ 0 & \text{if } l \text{ is even.} \end{cases} \tag{11.4}$$

Here $\bar{u}$ is the image of u in the residue class map $\mathbf{Z}_p \to \mathbf{F}_p$.

We now define

$$W(2) = \mathbf{Z}/2\mathbf{Z} \tag{11.5}$$

(the cyclic group of order 2). The map $\psi(2)$ is defined to be

$$\psi(2)\langle a \rangle = l \pmod 2 \tag{11.6}$$

where

$$|a|_2 = 2^{-l}. \tag{11.7}$$

This map is well-defined since it clearly respects the relations (i), (ii), (iii).

Finally we put

$$W(\infty) = \mathbf{Z} \qquad (= W(\mathbf{R})) \tag{11.8}$$

and define the map

$$\psi(\infty)\langle a\rangle = \begin{cases} +1 & \text{if } a > 0 \\ -1 & \text{if } a < 0. \end{cases} \tag{11.9}$$

This is again well-defined.

We can now enunciate

THEOREM 11.1. *The maps $\psi(p)$ (including $p = \infty$) establish an isomorphism between $W(\mathbf{Q})$ and the direct sum of the $W(p)$.*

Before going on to the proof we note

COROLLARY 1. *Suppose that an element of $W(\mathbf{Q})$ is in the kernel of all the localization maps λ_p (including ∞). Then it is 0.*

For then it is certainly in the kernel of all the maps $\psi(p)$.

COROLLARY 2. *The Weak Hasse Principle holds.*

For let $f, g \in \mathbf{Q}[\mathbf{x}]$ be quadratic forms which are equivalent over all $\mathbf{Q}_p$ (including $p = \infty$). Then $f - g$, considered as an element of $W(\mathbf{Q})$, is in the kernel of all the localization maps. Hence it is the zero element of $W(\mathbf{Q})$.

Proof of Theorem 11.1. Let $p > 0$ be a prime. Let $P = P(p)$ (resp. P') denote the set of non-zero rational integers all of whose prime factors q satisfy $q \leqslant p$ (resp. $q < p$). Let $L(p)$ (resp. $L'(p)$) be the subgroup of $W(\mathbf{Q})$ generated by the $\langle a\rangle$ with $a \in P(p)$ (resp. $a \in P'(p)$). It is obvious that $W(\mathbf{Q})$ is the union of the $L(p)$.

If $p > 2$ let p' be the largest prime with $p' < p$. Then clearly $P'(p) = P(p')$ and $L'(p) = L(p')$. Further, $P'(2)$ consists only of ± 1, so $L'(2)$ is the subgroup of $W(\mathbf{Q})$ generated by $\langle 1\rangle$ and $\langle -1\rangle = -\langle 1\rangle$.

Clearly $\psi(\infty)$ gives an isomorphism of $L'(2)$ with $W(\infty) = \mathbf{Z}$.

For $p \neq \infty$ the map $\psi(p)$ is clearly trivial on $L'(p)$. Hence we have reduced the theorem to showing that $\psi(p)$ induces an isomorphism between $L(p)/L'(p)$ and $W(p)$ or, equivalently, that the kernel of $\psi(p)$ in its action on $L(p)$ is precisely $L'(p)$ (not larger).

First let $p = 2$. Then

$$\langle 2\rangle + \langle 2\rangle = \langle 1\rangle + \langle 1\rangle \in L'(2)$$

and

$$\langle -2\rangle = -\langle 2\rangle.$$

Hence the kernel of $\psi(2)$ is precisely $L'(2)$.

The case $p > 2$ is more complicated. Our first objective is to show that

$$\langle pu \rangle \equiv \langle pv \rangle \pmod{L'(p)} \tag{11.10}$$

whenever

$$u, v \in P', \qquad u \equiv v \pmod{p}. \tag{11.11}$$

Suppose, first, that u, v, w are integers in

$$-p < u, v, w < +p$$

and

$$v \equiv uw \pmod{p}.$$

Then

$$uw - v = pt$$

where

$$-p < t < +p.$$

Hence

$$\begin{aligned}\langle vp \rangle + \langle t \rangle &= \langle vp \rangle + \langle tp^2 \rangle \\ &= \langle (v + tp)\,p \rangle + \langle vt(v + tp) \rangle \\ &= \langle uwp \rangle + \langle vtuw \rangle\end{aligned}$$

and so

$$\langle vp \rangle \equiv \langle uwp \rangle \pmod{L'(p)}. \tag{11.12}$$

Now let u be any element of P'. Then either $-p < u < +p$ or there is a prime $q < p$ such that $u = qu_1$ for some $u_1 \in P'$ (or, of course, both). An easy induction on the absolute value of u using (11.12) shows then that (11.10) is true whenever (11.11) holds and, in addition $-p < v < p$. Finally, if v does not satisfy this additional restriction, let w be either of the two integers satisfying

$$-p < w < +p \qquad u \equiv v \equiv w \pmod{p}.$$

Then $\langle pu \rangle \equiv \langle pw \rangle \equiv \langle pv \rangle \pmod{L(p)}$. This completes the proof that (11.11) implies (11.10).

We have now to define a map

$$\nu\colon\ W(p) \to L(p)/L'(p).$$

We recall that $W(p) = W(\mathbf{F}_p)$ and will use $\alpha, \beta, \ldots$ to denote elements of $\mathbf{F}_p$. Then we put

$$\nu\langle \alpha \rangle = \langle ap \rangle \pmod{L'(p)} \qquad (\alpha \in \mathbf{F}_p^*)$$

where $a \in \mathbf{Z}$ belongs to $\alpha \in \mathbf{Z}/p\mathbf{Z} = \mathbf{F}_p$ and

$$-\tfrac{1}{2}p < a < \tfrac{1}{2}p.$$

To show that ν is well-defined we have to show that it preserves the relations corresponding to (i), (ii), (iii) for $W(\mathbf{F}_p)$ and the only one which causes trouble is (ii). Suppose that $\beta \in \mathbf{F}_p^*$ and b are related as are α, a. Then

$$-p < a + b < +p$$

so $a, b, a + b$ and $ab(a + b)$ are all in P'. We have

$$\begin{aligned}\langle ap\rangle + \langle bp\rangle &= \langle (a + b)\,p\rangle + \langle ab(a + b)\,p\rangle \\ &\equiv \langle cp\rangle + \langle dp\rangle \pmod{L'(p)}\end{aligned}$$

by (11.10), where

$$c \equiv a + b \pmod p \qquad -\tfrac{1}{2}p < c < \tfrac{1}{2}p$$

and

$$d \equiv ab(a + b) \pmod p \qquad -\tfrac{1}{2}p < d < \tfrac{1}{2}p.$$

Hence v respects the analogue of (ii) for $\mathbf{F}_p$ (written with α, β for a, b).

This shows that v is well defined. Since $\psi(p)\,v$ is the identity on $W(p) = W(\mathbf{F}_p)$ and $v\psi(p)$ is the identity on $L(p)/L'(p)$ it follows that $\psi(p)$ is an isomorphism of $L(p)/L'(p)$ with $W(p)$, as required.

NOTES

As was mentioned in Section 1, there are fields over which the analogue of the Weak Hasse Principle holds but that of the Strong Hasse principle does not. Over an algebraic number field k, however, the strong principle holds and hence so also does the weak. The use of the geometry of numbers does not, apparently, generalize and it is necessary to invoke quite deep results from the theory of algebraic numbers. We sketch the argument. [Hasse (1924a), O'Meara (1963).]

For $n = 3$ the Strong Hasse Principle is easily seen to be equivalent to the following. Let K/k be a quadratic extension. Then $b \in k^*$ is the norm of an element β of K provided that it is a norm "everywhere locally". The corresponding result is in fact true whenever K/k is an abelian extension [i.e. a normal extension whose galois group is abelian], though not for a general extension K/k. [Hasse (1931).] There is a proof of what is needed for quadratic forms in O'Meara (1963). An alternative but related approach to the Strong Hasse Principle for $n = 3$ is by the theory of algebras. The even Clifford algebra $C_0(f)$ of a ternary form f is a "generalized quaternion algebra". It is a full matrix algebra over the ground-field precisely when f is isotropic [cf. Notes to Chapter 10]. The Strong Hasse Principle for ternary quadratics over an algebraic number field is thus a special case of the Hasse principle for algebras [Deuring (1935), Chapter 7].

For $n = 4$ Hasse's original argument [Hasse 1924a)] used the argument of our Section 5 and the appropriate generalization of Dirichlet's theorem about primes in arithmetic progressions. However, granted the case $n = 3$ for general algebraic number-fields, there is the following elementary argument [Springer (1957), O'Meara (1963) p. 187, or Cassels and Fröhlich (1967), Exercise 4]. For quaternary forms f for which $d(f) \in k^{*2}$ one is reduced to the case $n = 3$ by Chapter 2, example 4. The general case reduces to this one over a quadratic extension K of k by the use of Chapter 2, example 17.

Finally, for $n \geqslant 5$ the argument of Section 6 generalizes readily to algebraic number fields.

The actual form of Theorem 4.1 proved by Legendre supposes that $a_1 a_2 a_3$ is square-free, and replaces the conditions (i), (ii), (iii) by the condition that there exist integers e_1, e_2, e_3 such that

$$\left.\begin{aligned} a_1 e_1^2 + a_2 &\equiv 0 \pmod{a_3} \\ a_2 e_2^2 + a_3 &\equiv 0 \pmod{a_1} \\ a_3 e_3^2 + a_1 &\equiv 0 \pmod{a_2} \end{aligned}\right\} \qquad (£)$$

together with the condition that $g(\mathbf{x})$ be indefinite. His proof [Legendre (1798) Part I, Section 4] uses induction. We sketch a variant which uses the integer

$$I = (\min_j |a_j|)(\max_j |a_j|).$$

If $I = 1$ then $a_j = \pm 1$ for all j and there is nothing to prove. Otherwise, without loss of generality,

$$|a_1| \leqslant |a_2| < |a_3|.$$

By (£) there is an $r \in \mathbf{Z}$ such that

$$a_1 r^2 + a_2 = a_3 A_3 \qquad |r| \leqslant |\tfrac{1}{2} a_3|$$

for some $A_3 \in \mathbf{Z}$. Put

$$A_2 = a_1 a_2,$$

and

$$X_1 = a_1 r x_1 - a_2 x_2, \qquad X_2 = x_1 + r x_2, \qquad X_3 = a_3 x_3.$$

Then identically

$$X_1^2 + A_2 X_2^2 + A_3 X_3^2 = a_3 A_3 (a_1 x_1^2 + a_2 x_2^2 + a_3 x_3^2).$$

There are integers a'_1, a'_2, a'_3, b such that

$$A_2 = a'_1 a'_2, \qquad A_3 = a'_1 a'_3 b^2$$

and $a'_1a'_2a'_3$ is square-free. The above identity shows that $g(\mathbf{x})$ is isotropic precisely when

$$g'(\mathbf{x}) = a'_1x_1^2 + a'_2x_2^2 + a'_3x_3^2$$

is isotropic. It may be shown that g' satisfies the hypotheses of Legendre's theorem whenever g does. Further it is easy to see that

$$I' \text{ (say)} = (\min_j |a'_j|)(\max_j |a'_j|) < I.$$

We thus have the required induction. Variants of this argument are given in several books.

Gauss (1801) Section 294 proves Legendre's theorem quite differently. He shows that (£) implies that $g(\mathbf{x})$ is integrally equivalent to a form $h(\mathbf{x}) = \Sigma h_{ij}x_ix_j$ with $d|h_{22}$, $d|h_{33}$ and $d|_{23}$, where $d = a_1a_2a_3$. Then $h(dx_1, x_2, x_3) = dk(\mathbf{x})$, where $k(\mathbf{x})$ is classically integral and of determinant 1. If $g(\mathbf{x})$ is indefinite then so is $k(\mathbf{x})$ and Gauss had already shown that an indefinite classically integral form of determinant $+1$ is integrally equivalent to $-x_1^2 + 2x_1x_2$. In particular, $k(\mathbf{x})$, and so $g(\mathbf{x})$, is isotropic. Gauss's proof of the existence of $h(\mathbf{x})$ is explicit but not very transparent, which perhaps explains why it is not often reproduced in the literature. However, the existence of $h(\mathbf{x})$ follows readily from Lemma 4.2 of Chapter 9. Alternatively, one can consider the lattice Γ consisting of the points $\mathbf{z} = (z_1, z_2, z_3)$ for which

$$\left.\begin{aligned} z_1 &\equiv e_1z_2 \pmod{a_3} \\ z_2 &\equiv e_2z_3 \pmod{a_1} \\ z_3 &\equiv e_3z_1 \pmod{a_2} \end{aligned}\right\} \qquad (\$)$$

where e_1, e_2, e_3 are the integers occurring in (£). Let $\mathbf{b}_1, \mathbf{b}_2, \mathbf{b}_3$ be a basis of Γ. The conditions (£) imply that

$$g(x_1\mathbf{b}_1 + x_2\mathbf{b}_2 + x_3\mathbf{b}_3) = a_1a_2a_3K(\mathbf{x}),$$

where $K(\mathbf{x})$ is classically integral and of determinant ± 1. We may now apply Gauss' argument to $K(\mathbf{x})$ instead of $k(\mathbf{x})$.

The proof of Theorem 4.1 given in the text has obvious similarities with that of Gauss. Let $\mathbf{c}_1, \mathbf{c}_2, \mathbf{c}_3$ be a basis of the lattice Λ of the text. Then

$$g(x_1\mathbf{c}_1 + x_2\mathbf{c}_2 + x_3\mathbf{c}_3) = 4a_1a_2a_3l(\mathbf{x}),$$

where $l(\mathbf{x})$ is an integer-valued form of determinant $\pm\frac{1}{4}$. One can then use the results of Chaper 9, Section 3, to show that $l(\mathbf{x})$ is integrally equivalent to $\pm(x_1^2 + x_2x_3)$. One can use the methods of Chapter 9, Section 3, to get the conclusion of Theorem 4.1, Corollary, and indeed with a better constant. Put

$$m(\mathbf{x}) = |a_1|x_1^2 + |a_2|x_2^2 + |a_3|x_3^2$$

and

$$m(x_1\mathbf{c}_1 + x_2\mathbf{c}_2 + x_3\mathbf{c}_3) = 4a_1a_2a_3\, M(x_1, x_2, x_3).$$

Then M is a definite quadratic form of determinant $\frac{1}{4}$ with rational coefficients. By Lemma 3.2 of Chapter 9, there is a $\mathbf{d} \in \mathbf{Z}^3$ other than $\mathbf{0}$ such that

$$M(\mathbf{d}) \leqslant (4/3)(1/4)^{1/3} < 1.$$

But $|l(\mathbf{x})| \leqslant M(\mathbf{x})$ for all $\mathbf{x} \in \mathbf{R}^3$, so $l(\mathbf{d}) = 0$: that is, we have a solution of $g(\mathbf{x}) = 0$, $\mathbf{x} \neq 0$, $\mathbf{x} \in \mathbf{Z}^3$ with

$$|a_1|x_1^2 + |a_2|x_2^2 + |a_3|x_3^2 \leqslant (4/3)(1/4)^{1/3}|4a_1a_2a_3|.$$

We note finally that the argument in the text (or the variant just discussed) does not appear to have an analogue for the lattice Γ defined by ($), at least not without introducing an extra twist. This explains why we have chosen to prove Theorem 4.1 rather than Legendre's original formulation.

Section 4. For estimates of the smallest solution of $a_1x_1^2 + a_2x_2^2 + a_3x_3^2 = 0$ see Kneser (1959) and, for the algebraic number case, Siegel (1973). [cf. also notes on Section 8.]

For a discussion of the extent to which the law of quadratic reciprocity can be deduced from Theorem 4.1, see Gauss (1801), Section 296.

Section 8. The first occurrence of the argument used in the proof of Lemma 8.1 appears to be in Aubry (1912) in the context of example 6. The counter-example (8.6) is due to Kneser [see Cassels (1955)]. For a generalization see Birch and Davenport (1958): and for a generalization to algebraic number-fields see Raghavan (1973). See also Davenport (1971).

For another application of Aubry's method, see Cassels (1964) and Pfister (1965), Section 3.

Section 9. Compare Milnor and Husemoller (1973) Chapter 4 and Conway (1973). See also Hsia (1973) and Waterhouse (1976, 1977).

EXAMPLES

1. Determine which of the following forms are isotropic over $\mathbf{Q}$.

(i) $x_1^2 + x_2^2 - 15(x_3^2 + x_4^2)$

(ii) $3x_1^2 + 2x_2^2 - 7x_3^2$

(iii) $3x_1^2 + 2x_2^2 - 11x_3^2$

(iv) $3x_1^2 + 2x_2^2 - 7x_3^2 + 2x_2x_3 + 2x_3x_1 + 2x_1x_2$.

2. Determine which of the following pairs of forms are equivalent over $\mathbf{Q}$ and give an explicit equivalence when one exists.

(i) $x_1^2 - 15x_2^2$ and $3x_1^2 - 5x_2^2$
(ii) $x_1^2 - 82x_2^2$ and $2x_1^2 - 41x_2^2$
(iii) $x_1^2 + x_2^2 + 16x_3^2$ and $2x_1^2 + 2x_2^2 + 5x_3^2 - 2x_2x_3 - 2x_1x_3$.

3. Let C denote the set of rational numbers c such that

$$f(\mathbf{x}) = 42x_1^2 - 20x_2^2 + 15x_3^2 - 18x_4^2$$

is rationally equivalent to a form of the type

$$c(x_1^2 + 7x_2^2) + g(x_3, x_4)$$

with a binary form g (which may, of course, depend on c). Show that there is a $b \in \mathbf{Q}^*$ such that the set bC of $bc(c \in C)$ is a subgroup of $\mathbf{Q}^*$ and so obtain a complete description of C.

4. Show that the forms $x_1^2 + x_2^2 + x_3^2 + x_4^2$ and $b(x_1^2 + x_2^2 + x_3^2 + x_4^2)$ are rationally equivalent for every rational $b > 0$.
[*Hint*. cf. example of Chapter 4.]

5. For what rational numbers m are the two following forms rationally equivalent?

$$(m + 1)(x_1^2 + x_2^2 + x_3^2) + mx_4^2$$
$$x_1^2 + x_2^2 + x_3^2 + m(m + 1)x_4^2.$$

[*Hint*. cf. example 4 of Chapter 4.]

6. Let c be a positive integer and let $f(\mathbf{x})$ be one of the following forms:

$$cx_1^2 - x_2^2 - x_3^2$$
$$cx_1^2 - x_2^2 - x_3^2 - x_4^2$$
$$cx_1^2 - x_2^2 - x_3^2 - x_4^2 - x_5^2.$$

If $f(\mathbf{x})$ is isotropic use the methods of Section 8 to prove the existence of an integral $\mathbf{a}$ with $f(\mathbf{a}) = 0$ and $a_1 = 1$. Deduce that c is the sum of n integral squares precisely when c is the sum of n rational squares ($n = 2, 3, 4$).

Deduce that every positive integer c is the sum of 4 integral squares and give necessary and sufficient conditions for c to be the sum of 2 or 3 integral squares.

[*Note*. This argument seems to have been first found by Aubry (1912) but was rediscovered on at least one later occasion. See also Serre (1970).]

7. Let G be a finite group of proper rational automorphs of a regular quadratic form $f \in \mathbf{Q}[x_1, \ldots, x_n]$. Show that the order g of G divides

$$g(n) = \prod_{q \text{ prime}} q^{\gamma(q)}$$

where

$$\gamma(2) = n - 1 + [n/2] + [n/2^2] + \ldots$$

and

$$\gamma(q) = [n/(q-1)] + [n/q(q-1)] + [n/q^2(q-1)] + \ldots \qquad (q > 2).$$

[*Hint.* Apply the hint of Chapter 3, example 16, to the result of Chapter 2, example 13.

Note. There is an improvement over the direct use of Chapter 3, example 16, only for $q = 2$ where we have gained $2^{[n/2]}$. For a connection between $g(n)$ and the Bernoulli numbers see Minkowski (1887).]

8. Let $f(\mathbf{x}), g(\mathbf{y})$ be forms with rational coefficients in n, m variables respectively. Suppose that

(i) $f(\mathbf{x})$ is regular.
(ii) $n \geqslant m + 3$.
(iii) f represents g over $\mathbf{R}$.

Show that f represents g over $\mathbf{Q}$.

Show, further, that there is no m for which (ii) may be replaced by $n \geqslant m + 2$.

[*Hint.* Without loss of generality g is diagonal. Use induction on m.]

9. Let $f(\mathbf{x})$, $g(\mathbf{x})$ be regular rational forms in n variables. Suppose that they are equivalent over $\mathbf{R}$ and over $\mathbf{Q}_p$ for all finite primes p, except possibly for a single prime p_0. Show that they are equivalent over $\mathbf{Q}$.

CHAPTER 7

Quadratic Forms over Integral Domains

1. INTRODUCTION

Let k be any field of characteristic $\neq 2$ and let I be any ring in k containing 1. The most common case will be when k is the quotient ring of I but it may be larger. We denote by U the *units* of I, that is the set of $u \in I$ such that $u^{-1} \in I$. Clearly U is a group under multiplication.

In Chapter 2 we saw how the k-equivalence class of a quadratic form

$$f(\mathbf{x}) \in k[\mathbf{x}]$$

could be studied in terms of a quadratic space V, ϕ. In Section 2 we shall introduce the notion of a lattice and show how the I-equivalence class of f corresponds to a lattice in V, ϕ. In Section 3 we deduce some general properties of lattices which will be needed later.

There is little, if anything, that is at all deep in this Chapter: its purpose is to set up the machinery for subsequent Chapters.

2. QUADRATIC FORMS AND LATTICES

Let V be an n-dimensional vector space over the field k. If $\mathbf{b}_1, \ldots, \mathbf{b}_n$ is any basis for V as a k-vector space we call the set of

$$r_1\mathbf{b}_1 + \ldots + r_n\mathbf{b}_n, \tag{2.1}$$

where $r_1, \ldots, r_n$ run through I, the *lattice* with *basis* $\mathbf{b}_1, \ldots, \mathbf{b}_n$. If we wish to emphasize the role of the ring I we speak of an I-lattice.

A lattice has many bases. Clearly a necessary and sufficient condition that $\mathbf{b}_1, \ldots, \mathbf{b}_n$ and $\mathbf{c}_1, \ldots, \mathbf{c}_n$ be bases of the same lattice is that there exist $r_{ij}, s_{ij} \in I$ such that

$$\mathbf{c}_i = \sum_j r_{ij}\mathbf{b}_j \tag{2.2}$$

and

$$\mathbf{b}_i = \sum_j s_{ij}\mathbf{c}_j. \tag{2.3}$$

On eliminating the $\mathbf{c}_i$ between (2.2) and (2.3) and using the fact that $\mathbf{b}_1, \ldots, \mathbf{b}_n$ are a vector-space basis of V we see that

$$\sum_j s_{ij}r_{jl} = \begin{cases} 1 & \text{if } i = l \\ 0 & \text{otherwise.} \end{cases}$$

Hence

$$\det(s_{ij})\det(r_{ij}) = 1.$$

In particular $\det(r_{ij})$ lies in the group U of units of I. Conversely, if $\det(r_{ij}) \in U$, then we can solve (2.2) for the $\mathbf{b}_j$ in terms of the $\mathbf{c}_j$ and obtain expressions of the type (2.3) with $s_{ij} \in I$. We have thus proved

LEMMA 2.1. *Let Λ be the lattice with basis $\mathbf{b}_1, \ldots, \mathbf{b}_n$ and let $\mathbf{c}_1, \ldots, \mathbf{c}_n$ be elements of Λ. If the $\mathbf{c}_i$ are given by (2.2) then a necessary and sufficient condition that $\mathbf{c}_1, \ldots, \mathbf{c}_n$ be a basis for Λ is that*

$$\det(r_{ij}) \in U. \tag{2.4}$$

Now let

$$f(x_1, \ldots, x_n) \in k[x_1, \ldots, x_n]$$

be a quadratic form. In Chapter 2 we introduced the quadratic space V, ϕ with basis $\mathbf{b}_1, \ldots, \mathbf{b}_n$, where

$$f(x_1, \ldots, x_n) = \phi(x_1\mathbf{b}_1 + \ldots + x_n\mathbf{b}_n).$$

Let Λ be the lattice with basis $\mathbf{b}_1, \ldots, \mathbf{b}_n$. If $\mathbf{c}_1, \ldots, \mathbf{c}_n$ is another basis of Λ, put

$$g(x_1, \ldots, x_n) = \phi(x_1\mathbf{c}_1 + \ldots + x_n\mathbf{c}_n).$$

Then

$$g(x_1, \ldots, x_n) = f(\sum_i x_i r_{i1}, \sum_i x_i r_{i2}, \ldots, \sum_i x_i r_{in}).$$

By (2.4) the forms g and f are I-equivalent. Conversely if g is I-equivalent to f we can reverse the process. We have thus set up the required correspondence between I-equivalence classes of k-valued quadratic forms and I-lattices in quadratic spaces. Of course two distinct lattices can correspond to the same class of forms (cf. Chapter 11).

3. LATTICES

In this section we prove some general results about lattices. Except at the beginning we shall suppose that I is a *principal ideal domain* (i.e. that every ideal in I is general by a single element). In the applications I will be either the ring $\mathbf{Z}$ of rational integers or the ring $\mathbf{Z}_p$ of p-adic integers.

Let Λ, Γ be lattices in the same n-dimensional space V with respective bases $\mathbf{b}_1, \ldots, \mathbf{b}_n$ and $\mathbf{e}_1, \ldots, \mathbf{e}_n$. Since both these lattice-bases are bases of V as a k-vector space, we have

$$\mathbf{e}_i = \sum_j t_{ij}\mathbf{b}_j \qquad t_{ij} \in k \tag{3.1}$$

with

$$d \text{ (say)} = \det(t_{ij}) \neq 0. \tag{3.2}$$

The value of d depends not merely on Λ and Γ but on the choice of bases. By Lemma 2.1 another choice of basis for either Λ or Γ multiplies d by an element of U. We denote the image of d in k^*/U by $d(\Gamma/\Lambda)$ and call it the *relative determinant*.

If Λ is a lattice and l is a non-zero element of k we denote by $l\Lambda$ the set of $l\mathbf{a}$, $\mathbf{a} \in \Lambda$. Clearly $l\Lambda$ is a lattice. It depends only on the class of l in k^*/U.

LEMMA 3.1. *Suppose that Λ, Γ are lattices and that*† $\Gamma \subset \Lambda$. *Then* $d(\Gamma/\Lambda) \in I/U$ *and*

$$d(\Gamma/\Lambda)\,\Lambda \subset \Gamma. \tag{3.3}$$

Proof. We retain the notation used in (3.1). Since $\mathbf{e}_i \in \Gamma \subset \Lambda$ we have $t_{ij} \in I$ and so $d \in I$ by (3.2). Hence $d(\Gamma/\Lambda) \in I/U$.

On solving (3.1) for the $\mathbf{b}_j$ we have

$$d\mathbf{b}_j = \sum_i s_{ji}\mathbf{e}_i$$

for some $s_{ji} \in I$ and so $d\Lambda \subset \Gamma$.

LEMMA 3.2. *Suppose that I is a principal ideal domain and let Λ, Γ be lattices with $\Gamma \subset \Lambda$. Let $\mathbf{b}_1, \ldots, \mathbf{b}_n$ be any basis of Λ. Then there is a basis $\mathbf{c}_1, \ldots, \mathbf{c}_n$ of Γ of the shape*

$$\left.\begin{aligned} \mathbf{c}_1 &= t_{11}\mathbf{b}_1 \\ \mathbf{c}_2 &= t_{21}\mathbf{b}_1 + t_{22}\mathbf{b}_2 \\ &\vdots \\ \mathbf{c}_n &= t_{n1}\mathbf{b}_1 + \ldots + t_{nn}\mathbf{b}_n \end{aligned}\right\} \tag{3.4}$$

with

$$t_{ij} \in I, \qquad t_{ii} \neq 0. \tag{3.5}$$

Note. Here the basis $\mathbf{b}_j$ is given in advance. For the case when it is allowed to vary, see Theorem 5.1 of Chapter 11.

† We say that Γ is a *sublattice* of Λ. Note that this term implies that Λ and Γ span the same vector space.

Proof. By Lemma 3.1 there is a

$$d \in I, \qquad d \neq 0 \tag{3.6}$$

such that

$$d\Lambda \subset \Gamma. \tag{3.7}$$

For fixed j let $\Gamma^{(j)}$ be the set of $\mathbf{a} \in \Gamma$ of the shape

$$\mathbf{a} = s_1\mathbf{b}_1 + \ldots + s_j\mathbf{b}_j. \tag{3.8}$$

If $\mathbf{a}^{(1)}, \mathbf{a}^{(2)} \in \Gamma^{(j)}$ then $r_1\mathbf{a}^{(1)} + r_2\mathbf{a}^{(2)} \in \Gamma^{(j)}$ for any $r_1, r_2 \in I$. Hence the set S_j of s_j occurring in (3.8) is an I-ideal. Since $d\mathbf{b}_j \in d\Lambda \subset \Gamma$, we have $d \in S_j$ and so S_j is not the zero ideal. By hypothesis S_j is principal. Hence we can find a

$$\mathbf{c}_j = t_{j1}\mathbf{b}_1 + \ldots + t_{jj}\mathbf{b}_j \in \Gamma$$

such that t_{jj} is a generator of S_j. We shall show that $\mathbf{c}_1, \ldots, \mathbf{c}_n$ is a basis for Γ.

Let

$$\mathbf{a} = s_1\mathbf{b}_1 + \ldots + s_n\mathbf{b}_n$$

be any element of Γ. Then $s_n \in S_n$ so $s_n = v_n t_{nn}$ for some $v_n \in I$. Hence

$$\mathbf{a} - v_n\mathbf{c}_n = s'_1\mathbf{b}_1 + \ldots + s'_{n-1}\mathbf{b}_{n-1}$$

for some $s'_1, \ldots, s'_{n-1} \in I$. We can repeat the argument:

$$s'_{n-1} \in S_{n-1}, \qquad \text{say} \quad s'_{n-1} = v_{n-1}t_{n-1,n-1}.$$

Then

$$\mathbf{a} - v_n\mathbf{c}_n - v_{n-1}\mathbf{c}_{n-1} = s''_1\mathbf{b}_1 + \ldots + s''_{n-2}\mathbf{b}_{n-2},$$

and so on. Ultimately we obtain

$$\mathbf{a} = v_n\mathbf{c}_n + v_{n-1}\mathbf{c}_{n-1} + \ldots + v_1\mathbf{c}_1.$$

for some $v_1, \ldots, v_n \in I$, as required.

A similar argument proves

LEMMA 3.3. *Suppose that* $I = \mathbf{Z}$. *Let* Λ *be a lattice and let* Γ *be a subgroup of finite index of* Λ, *considered as a group under addition. Then* Γ *is a lattice.*

Proof. For $d\Lambda \in \Gamma$ where d is now the group index. We can argue now precisely as in the proof of Lemma 3.2 to find $\mathbf{c}_1, \ldots, \mathbf{c}_n$ which are a basis for Γ considered either as a group or as a $\mathbf{Z}$-lattice.

LEMMA 3.4. *Suppose that* I *is a principal ideal domain. Let* $\mathbf{c}_1, \ldots, \mathbf{c}_J$ *be linearly independent elements of* Λ. *Then there is a basis* $\mathbf{b}_1, \ldots, \mathbf{b}_n$ *of* Λ *such that*

$$\left.\begin{aligned} \mathbf{c}_1 &= s_{11}\mathbf{b}_1 \\ \mathbf{c}_2 &= s_{21}\mathbf{b}_1 + s_{22}\mathbf{b}_2 \\ &\vdots \\ \mathbf{c}_J &= s_{J1}\mathbf{b}_1 + \ldots + s_{JJ}\mathbf{b}_J \end{aligned}\right\} \tag{3.9}$$

for some

$$s_{ij} \in I, \qquad s_{ii} \neq 0. \tag{3.10}$$

Note. We need not specify whether $\mathbf{c}_1, \ldots, \mathbf{c}_J$ are linearly independent over I or over k. Any elements of a lattice Λ which are linearly independent over I are necessary linearly independent over k.

Proof. We can pick $\mathbf{c}_{J+1}, \ldots, \mathbf{c}_n \in \Lambda$ so that $\mathbf{c}_1, \ldots, \mathbf{c}_n$ are linearly independent. Let Γ be the lattice with basis $\mathbf{c}_1, \ldots, \mathbf{c}_n$. By Lemma 3.1 we have $d\Lambda \subset \Gamma$ for some $d \in I, d \neq 0$ and so by Lemma 3.2 (with $d\Lambda, \Gamma$ for Γ, Λ respectively) there is a basis $\mathbf{b}_1, \ldots, \mathbf{b}_n$ of Λ such that

$$\begin{aligned} d\mathbf{b}_1 &= t_{11}\mathbf{c}_1 \\ d\mathbf{b}_2 &= t_{21}\mathbf{c}_1 + t_{22}\mathbf{c}_2 \\ &\vdots \\ d\mathbf{b}_n &= t_{n1}\mathbf{c}_1 + \ldots + t_{nn}\mathbf{c}_n \end{aligned}$$

with $t_{ij} \in I, t_{jj} \neq 0$.

On solving for the $\mathbf{c}_j$ in terms of the $\mathbf{b}_j$ we obtain expressions of the type (3.9) where in the first place we know only that $s_{ij} \in k$. But $\mathbf{b}_1, \ldots, \mathbf{b}_n$ is a basis for Λ and $\mathbf{c}_j \in \Lambda$, so the s_{ij} are actually in I. Further

$$s_{jj} = d/t_{jj} \neq 0.$$

This concludes the proof.

THEOREM 3.1. *Let I be a principal ideal domain contained in the field k. Exclude the trivial case when $k = I$. Let $\mathbf{e}_1, \ldots, \mathbf{e}_n$ be any I-basis of a lattice Λ and let $\mathbf{c}_1, \ldots, \mathbf{c}_J$ be elements of Λ. Then the following three statements are equivalent*:

(i) *there exist $\mathbf{c}_{J+1}, \ldots, \mathbf{c}_n$ such that $\mathbf{c}_1, \ldots, \mathbf{c}_n$ is a basis for Λ.*

(ii) *let*

$$\mathbf{c}_j = \sum_{i=1}^{n} r_{ji}\mathbf{e}_i \qquad (1 \leqslant j \leqslant J) \tag{3.11}$$

be the expression of the $\mathbf{c}_j$ in terms of the given basis $\mathbf{e}_i$. Then the set of determinants of the $J \times J$ submatrices of the $J \times n$ matrix r_{ji} is coprime.

(iii) *if*

$$\mathbf{a} = v_1\mathbf{c}_1 + \ldots + v_J\mathbf{c}_J \in \Lambda \tag{3.12}$$

with $v_1, \ldots, v_J \in k$ *then necessarily* $v_j \in I$ $(1 \leqslant j \leqslant J)$.

Note (to ii). If I is any ring containing 1 we say that a set f_l $(1 \leqslant l \leqslant L)$ of elements of I is *coprime*† if the ideal generated by them is the whole of I: that is if there exist g_l $(1 \leqslant l \leqslant L)$ such that

$$\sum_l f_l g_l = 1.$$

When I is a principal ideal ring (which we are supposing) then this is equivalent to the statement that if h divides f_l in I for $1 \leqslant l \leqslant L$ then h is a unit of I.

Proof. We first note that we may suppose that $\mathbf{c}_1, \ldots, \mathbf{c}_J$ are linearly independent. Indeed (i), (ii) both clearly imply linear independence and so does (iii) since we have excluded the trivial case $k = I$. [The only place where we have used this exclusion.]

We now introduce a fourth condition

(iv) let $\mathbf{b}_1, \ldots, \mathbf{b}_n$ be the basis given by Lemma 3.4 and let $s_{ij} \in I$ be given by (3.9). Then s_{jj} is a unit of I for $1 \leqslant j \leqslant J$.

For linearly independent $\mathbf{c}_1, \ldots, \mathbf{c}_J$ we shall show that (i), (ii), (iii), (iv) are all equivalent by proving the following implications

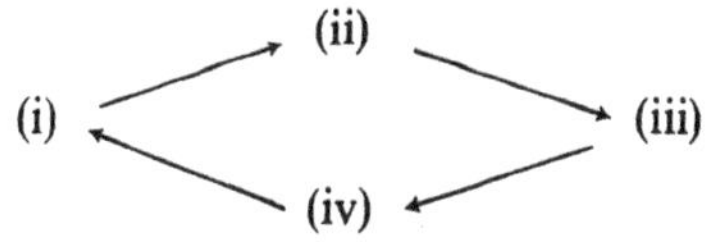

There are, of course, other logical ways of proving the equivalence: in particular (i) → (iii) is trivial.

(iii) → (iv). For if s_{JJ} is not a unit the choice $\mathbf{a} = \mathbf{b}_J$ contradicts (iii).

(iv) → (i). For if $s_{11}, \ldots, s_{JJ}$ are all units, then $\mathbf{c}_1, \ldots, \mathbf{c}_J, \mathbf{b}_{J+1}, \ldots, \mathbf{b}_n$ is clearly a basis for Λ.

(i) → (ii). For let $\mathbf{c}_{J+1}, \ldots, \mathbf{c}_n$ exist and extend the definition of r_{ji} so that

$$\mathbf{c}_j = \sum_{i=1}^{n} r_{ji}\mathbf{e}_i \qquad (1 \leqslant j \leqslant n).$$

By Lemma 2.1

$$\det_{\substack{1 \leqslant j \leqslant n \\ 1 \leqslant i \leqslant n}} (r_{ji}) = u \qquad \text{(say)}$$

is a unit of I. On expanding the determinant by the first J and the last

† This is often called comaximal.

$n - J$ rows (Laplace expansion) we have

$$u = \sum R_M R'_{M'}$$

where R_M runs through the determinants of the $J \times J$ matrices M formed from the first J rows and $R'_{M'}$ is the determinant of the "complementary" $(n - J) \times (n - J)$ matrix. Since $R'_{M'} \in I$ this implies that the R_M are coprime; that is, that (ii) holds.

(ii) → (iii). On substituting (3.11) in (3.12) we have

$$\mathbf{a} = w_1\mathbf{e}_1 + \ldots + w_n\mathbf{e}_n$$

where

$$w_i = \sum_{1 \leqslant j \leqslant J} v_j r_{ji}. \tag{3.13}$$

Here $r_{ji} \in I$ and (3.12) implies that $w_i \in I$ $(1 \leqslant i \leqslant n)$. We can pick any J out of the n equations (3.13) and on solving for the v_j obtain

$$R_M v_j \in I \qquad (1 \leqslant j \leqslant J)$$

where R_M is the determinant of the appropriate $J \times J$ submatrix M of r_{ji}. But condition (ii) asserts that there are $\lambda_M \in I$ such that

$$\sum_M \lambda_M R_M = 1.$$

Hence

$$v_j = \sum_M \lambda_M (R_M v_j) \in I \qquad (1 \leqslant j \leqslant J).$$

This is what is asserted by (iii).

This concludes the proof of the Theorem.

The case $J = 1$ is especially important. We shall say that a vector $\mathbf{c} \in \Lambda$ is *primitive* if there is a basis $\mathbf{c} = \mathbf{c}_1, \mathbf{c}_2, \ldots, \mathbf{c}_n$ of Λ.

COROLLARY. *Let*

$$\mathbf{c} = c_1\mathbf{e}_1 + \ldots + c_n\mathbf{e}_n \in \Lambda,$$

where $\mathbf{e}_1, \ldots, \mathbf{e}_n$ *is any basis. A necessary and sufficient condition for* $\mathbf{c}$ *to be primitive is that* $c_1, \ldots, c_n$ *should be coprime.*

4. SINGULAR FORMS

In this section we indicate briefly how the theory of lattices in a singular quadratic space and so of singular quadratic forms can be reduced to the regular case. We consider only the situation when the ring I is a principal ideal ring and then the position is very similar to that over fields, see Section 6 of Chapter 2.

Let then Λ be an I-lattice in a k-vector space V, ϕ, where I is a principal ideal ring. The *radical* Λ_0 of Λ consists of the $\mathbf{a} \in \Lambda$ such that $\phi(\mathbf{a}, \mathbf{b}) = 0$ for all $\mathbf{b} \in \Lambda$. Let r be the maximum number of elements of Λ_0 which can be linearly independent. If $\mathbf{c}_1, \ldots, \mathbf{c}_r$ is a linearly independent set of elements of Λ_0 then by Lemma 3.4 there is a basis $\mathbf{b}_1, \ldots, \mathbf{b}_n$ of Λ such that

$$\begin{aligned} \mathbf{c}_1 &= s_{11}\mathbf{b}_1 \\ \mathbf{c}_2 &= s_{21}\mathbf{b}_1 + s_{22}\mathbf{b}_2 \\ &\vdots \\ \mathbf{c}_r &= s_{r1}\mathbf{b}_1 + \ldots + s_{rr}\mathbf{b}_r \end{aligned}$$

where $s_{ij} \in I$ and $s_{ii} \neq 0$ $(1 \leqslant i \leqslant r)$. It is easy to see that $\mathbf{b}_i \in \Lambda_0$ $(1 \leqslant i \leqslant r)$ and so, by the maximality of r, the radical Λ_0 is precisely the sub-I-module of Λ spanned by $\mathbf{b}_1, \ldots, \mathbf{b}_r$; that is, the set of

$$u_1\mathbf{b}_1 + \ldots + u_r\mathbf{b}_r \qquad (u_i \in I, 1 \leqslant i \leqslant r).$$

Let Λ_1 be the submodule of Λ spanned by $\mathbf{b}_{r+1}, \ldots, \mathbf{b}_n$. Clearly the quadratic form induced by ϕ on Λ_1 is regular. Interpreting this in terms of forms we have

Lemma 4.1. *Suppose that I is a principal ideal domain. Then any singular quadratic form is I-equivalent to a regular form in a smaller number of variables.*

In conclusion, we note that the conclusion of Lemma 4.1 does not necessarily hold if I is not a principal ideal domain, e.g. $I = \mathbf{Z}[\sqrt{-5}]$. For example the form

$$\{(1 + 2\sqrt{-5})x_1 + 3x_2\}^2$$

is not $\mathbf{Z}[\sqrt{-5}]$—equivalent to a form in 1 variable.

NOTES

When the integral domain I is not a principal ideal domain we must distinguish between two generalizations of the notion of a lattice Λ.

(i) the definition (2.1) adopted in the text: that is, Λ is a free I-module which spans V as a k-vector space.

(ii) any finitely-generated I-module which spans V as a k-vector space.

When I is a PID the notions (i), (ii) coincide but otherwise they need not even for $n = 1$. Let V be the 1-dimensional vector space with basis $\mathbf{b}_1$ and let J be a non-principal ideal of I. Then the set of $s\mathbf{b}_1, s \in J$ is not a lattice in sense (i) but it is in sense (ii) if J is finitely generated.

The definition (i) is natural in connection with quadratic forms but the definition (ii) can be more natural in working with lattices. The counter-example at the end of Section 4 evaporates in the context of definition (ii).

When I is a Dedekind domain (e.g. the ring of integers of an algebraic number field) there is a generalization of the notion of basis to lattices Λ in the sense (ii): there are linearly independent elements $\mathbf{b}_1, \ldots, \mathbf{b}_n$ of V and an ideal J of I such that Λ is precisely the set of

$$s_1\mathbf{b}_1 + s_2\mathbf{b}_2 + \ldots + s_n\mathbf{b}_n,$$

where

$$s_j \in I \qquad (1 \leqslant j < n)$$
$$s_n \in J.$$

EXAMPLES

Examples 1, 2 and 3 introduce concepts which are important in the classical theory but which are not needed in this book.

1. Let I be a principal ideal domain. For a classically integral form $f(x_1, \ldots, x_n)$ and for integer r $(1 < r < n)$, define $t_r(f) \in k^*/U$ to be the greatest common divisor of the $r \times r$ minors of the corresponding matrix $\mathbf{F}$. Show that $t_r(f)$ depends only on r and the equivalence class of f.

Show also that $t'_r(f)$ has the same property, where the definition is the same as that of $t_r(f)$ except that the non-principal minors are multiplied by 2 before the g.c.d. is taken.

2. Let I, f be as in example 1. Suppose, further, that f is primitive and regular. Let F_{ij} $(1 \leqslant i, j \leqslant n)$ be the elements of the adjoint of $\mathbf{F}$ and let $f^*_{ij} \in I, s \in I$ be such that

$$F_{ij} = sf^*_{ij},$$
$$\text{g.c.d.}\,(f^*_{ij}) = 1.$$

Put $f^*(\mathbf{x}) = \Sigma f^*_{ij}x_i x_j$, so f^* is determined by f up to multiplication by a unit.

(i) If f, g are equivalent, show that g^* is equivalent to uf^* for some unit u of I.

(ii) Show that $(f^*)^* = vf$ for some unit v.

3. Notation as in example 2. Suppose that $I = \mathbf{Z}$ and that f is positive definite. Show that f^* can be chosen so that it is positive definite and that then it is uniquely determined by f.

CHAPTER 8

Integral p-Adic Forms

1. INTRODUCTION

In this chapter p is always an ordinary prime and not the symbol ∞. The object is to discuss the behaviour of $\mathbf{Q}_p$-valued forms under $\mathbf{Z}_p$-equivalence ("p-adic integral equivalence"). In particular the results apply to integral p-adic forms, where we use the classical definition of integrality:

$$f(\mathbf{x}) = \sum_{i,j=1}^{n} f_{ij}x_i x_j$$

with

$$f_{ij} = f_{ji} \in \mathbf{Z}_p$$

(cf. Chapter 1). The two definitions of integral form make a difference, of course, only when $p = 2$.

We say that f is *primitive* if

$$\max_{i,j} |f_{ij}| = 1;$$

that is when f is integral but $p^{-1}f$ is not. In the case $p = 2$ we say that f is *properly primitive* if it is primitive and

$$\max_{i} |f_{ii}| = 1.$$

If f is primitive but not properly so, then we say that it is *improperly primitive*. Thus $x_1^2 + 2x_1x_2$ is properly primitive but $2x_1^2 + 2x_1x_2$ is improperly primitive.

The next section, Section 2, is preparatory and discusses the special properties of lattices over $\mathbf{Z}_p$.

In Section 3 we first prove that every $\mathbf{Q}_p$-valued form has an integral automorph of determinant -1: this will play an important role later in the discussion of genera. We then restrict attention to $p \neq 2$ and obtain a very pleasant set $\mathscr{C}$ of *canonical forms* for $\mathbf{Z}_p$-equivalence: that is every $\mathbf{Q}_p$-valued form is $\mathbf{Z}_p$- equivalent to precisely one form of $\mathscr{C}$. An important tool is a partial generalization of "Witt's Lemma" (Theorem 4.1, Corollary, of Chapter 2), which of course, as it stands, applies only to fields.

In Section 4 we attempt to extend the results of Section 3 to $p = 2$. This is much less satisfactory both because the facts are less elegant and because the arguments are more complicated. Fortunately the results of this section are not needed later. When we come to apply the results of this chapter to "global" ($\mathbf{Z}$-) equivalence, there will always be a finite set of "bad" primes which are treated specially; and 2 will always be "bad".

An important property of integral p-adic forms is that two forms whose coefficients are sufficiently close p-adically are actually $\mathbf{Z}_p$-equivalent. We give a simple proof of this by a "Hensel's Lemma" argument and then give a slightly stronger result whose proof depends on Sections 3 and 4.

2. BASES OF $\mathbf{Z}_p^n$

By $\mathbf{Z}_p^n$ we mean of course the set of $\mathbf{a} = (a_1, \ldots, a_n)$ with $a_j \in \mathbf{Z}_p$. We consider $\mathbf{Z}_p^n$ as a lattice in $\mathbf{Q}_p^n$.

LEMMA 2.1. *Let* $\mathbf{c}_1, \ldots, \mathbf{c}_J$ *be linearly independent elements of* $\mathbf{Z}_p^n$. *Then the following three statements are equivalent:*

(i) *there exist* $\mathbf{c}_{J+1}, \ldots, \mathbf{c}_n \in \mathbf{Z}_p^n$ *so that* $\mathbf{c}_1, \ldots, \mathbf{c}_n$ *is a basis.*

(ii) *The* $n \times J$ *matrix* $\mathbf{c}_1\mathbf{c}_2 \ldots \mathbf{c}_J$ *contains a* $J \times J$ *minor whose determinant is a p-adic unit.*

(iii) *If,*

$$\mathbf{a} = v_1\mathbf{c}_1 + \ldots + v_J\mathbf{c}_J \in \mathbf{Z}_p^n \tag{2.1}$$

for some $v_1, \ldots, v_J \in \mathbf{Q}_p$, *then necessarily* $v_1, \ldots, v_J \in \mathbf{Z}_p$.

Proof. The ring $\mathbf{Z}_p$ is a principal ideal domain, the only ideals being those generated by p^m $(m = 0, 1, 2, \ldots)$. This Lemma is thus a special case of Theorem 3.1 of Chapter 7.

In accordance with the terminology of Chapter 7 we say that $\mathbf{a} \in \mathbf{Z}_p^n$ is *primitive* if $\max_j |a_j|_p = 1$. The Corollary to Theorem 3.1 of Chapter 7 gives:

COROLLARY. *A necessary and sufficient condition that a vector* $\mathbf{a} \in \mathbf{Z}_p^n$ *form a part of some basis is that* $\mathbf{a}$ *be primitive.*

3. CANONICAL FORMS

In this section we consider the properties of $\mathbf{Q}_p$-valued forms under $\mathbf{Z}_p$-equivalence. At first we permit all values of p but as soon as $p = 2$ would cause difficulties we make the restriction $p \neq 2$ and leave $p = 2$ for Section 4.

Let

$$f(\mathbf{x}) \in \mathbf{Q}_p[\mathbf{x}] \tag{3.1}$$

be a regular (= non-singular) quadratic form in n variables. Clearly $|f(\mathbf{a})|$ for $\mathbf{a} \in \mathbf{Z}_p^n$ is bounded above. Further

$$2f(\mathbf{a}, \mathbf{b}) = f(\mathbf{a} + \mathbf{b}) - f(\mathbf{a}) - f(\mathbf{b}) \tag{3.2}$$

and so

$$\sup_{\mathbf{a}, \mathbf{b}} |2f(\mathbf{a}, \mathbf{b})| \leqslant \sup_{\mathbf{a}} |f(\mathbf{a})| \tag{3.3}$$

where the suprema are taken over

$$\mathbf{a} \in \mathbf{Z}_p^n, \qquad \mathbf{b} \in \mathbf{Z}_p^n. \tag{3.4}$$

Since the p-adic valuation takes only a discrete set of values, it is clear that the suprema on both sides of (3.3) are attained. In particular there is some $\mathbf{c} \in \mathbf{Z}_p^n$ such that

$$|f(\mathbf{c})| = \sup_{\mathbf{a}} |f(\mathbf{a})|. \tag{3.5}$$

Then $\mathbf{c}$ must be primitive since if $\mathbf{c} = p\mathbf{b}$ with $\mathbf{b} \in \mathbf{Z}_p^n$ then we should have $|f(\mathbf{b})| = p^2 |f(\mathbf{c})|$. This gives

LEMMA 3.1. *$f(\mathbf{x})$ is $\mathbf{Z}_p$-equivalent to a form $g(\mathbf{x})$ with*

$$|g(1, 0, \ldots, 0)| = \sup_{\mathbf{a} \in \mathbf{Z}_p^n} |g(\mathbf{a})|. \tag{3.6}$$

Proof. By Lemma 2.1 there is a basis

$$\mathbf{c} = \mathbf{c}_1, \mathbf{c}_2, \ldots, \mathbf{c}_n$$

of $\mathbf{Z}_p^n$ where $\mathbf{c}$ is given by (3.5).

We recall (Chapter 2, Section 4) that if $\mathbf{b} \in \mathbf{Q}_p^n$, $f(\mathbf{b}) \neq 0$ then the symmetry

$$\tau_{\mathbf{b}}: \mathbf{x} \to \mathbf{x} - \frac{2f(\mathbf{b}, \mathbf{x})}{f(\mathbf{b})} \mathbf{b} \tag{3.7}$$

is a $\mathbf{Q}_p$-automorphism of f.

LEMMA 3.2. *Suppose that*

$$|f(\mathbf{b})| = \sup_{\mathbf{a} \in \mathbf{Z}_p^n} |f(\mathbf{a})|. \tag{3.8}$$

Then $\tau_{\mathbf{b}}$ is a $\mathbf{Z}_p$-automorphism of f.

Proof. Since $(\tau_{\mathbf{b}})^2$ is the identity, it will be enough to show that $\tau_{\mathbf{b}}\mathbf{a} \in \mathbf{Z}_p^n$ for all $\mathbf{a} \in \mathbf{Z}_p^n$. By (3.3) and (3.8) we have

$$|2f(\mathbf{a}, \mathbf{b})| \leq |f(\mathbf{b})|,$$

that is

$$\frac{2f(\mathbf{a}, \mathbf{b})}{f(\mathbf{b})} \in \mathbf{Z}_p.$$

Then (3.7) gives $\tau_{\mathbf{b}}\mathbf{a} \in \mathbf{Z}_p^n$ at once.

The following Corollary will be important later.

COROLLARY. *$f(\mathbf{x})$ has a $\mathbf{Z}_p$-automorph σ with* $\det(\sigma) = -1$.

Proof. For we can take $\sigma = \tau_{\mathbf{b}}$.

We now confine attention to

$$p \neq 2.$$

LEMMA 3.3 ($p \neq 2$). *Suppose that* $\mathbf{a}_1, \mathbf{a}_2 \in \mathbf{Z}_p^n$ *satisfy*

$$f(\mathbf{a}_1) = f(\mathbf{a}_2)$$

and that

$$|f(\mathbf{a}_1)| = |f(\mathbf{a}_2)| = \sup_{\mathbf{a}} |f(\mathbf{a})|.$$

Then there is an integral automorph σ of f such that

$$\sigma\mathbf{a}_1 = \mathbf{a}_2.$$

Proof. (cf. "Witt's Lemma"). We have

$$\begin{aligned} |f(\mathbf{a}_1 + \mathbf{a}_2) + f(\mathbf{a}_1 - \mathbf{a}_2)| &= |2f(\mathbf{a}_1) + 2f(\mathbf{a}_2)| \\ &= |4f(\mathbf{a}_1)| \\ &= \sup_{\mathbf{a}} |f(\mathbf{a})| \end{aligned}$$

since $p \neq 2$. Hence we have at least one of the two cases

$$\text{(i)}\ |f(\mathbf{a}_1 - \mathbf{a}_2)| = \sup_{\mathbf{a}} |f(\mathbf{a})|$$

$$\text{(ii)}\ |f(\mathbf{a}_1 + \mathbf{a}_2)| = \sup_{\mathbf{a}} |f(\mathbf{a})|.$$

In case (i) we can apply Lemma 3.2 with $\mathbf{b} = \mathbf{a}_1 - \mathbf{a}_2$. The automorph $\sigma = \tau_{\mathbf{b}}$ is integral and $\tau_{\mathbf{b}}\mathbf{a}_1 = \mathbf{a}_2$. Similarly in case (ii) we can put $\mathbf{b} = \mathbf{a}_1 + \mathbf{a}_2$ and $\sigma = \tau_{\mathbf{b}}\tau_{\mathbf{a}_1}$.

COROLLARY 1 ($p \neq 2$). *Let σ be any integral automorph of $f(\mathbf{x})$. Then σ is the product of integral automorphs $\tau_{\mathbf{b}}$ with integral $\mathbf{b}$.*

Proof. For we can take $\mathbf{a}_2 = \sigma\mathbf{a}_1$, where $\mathbf{a}_1$ is any element of $\mathbf{Z}_p^n$ with

$$|f(\mathbf{a}_1)| = \sup_{\mathbf{a}}|f(\mathbf{a})|.$$

The Lemma gives a σ_1 which is a product of $\tau_{\mathbf{b}}$'s of the specified shape such that $\sigma_1\mathbf{a}_1 = \mathbf{a}_2$. Hence on taking $\sigma_1^{-1}\sigma$ for σ we may suppose that $\sigma\mathbf{a}_1 = \mathbf{a}_1$. Then σ is an automorph of the orthogonal complement of $\mathbf{a}_1$. Every $\mathbf{c} \in \mathbf{Z}_p^n$ is of the shape $u\mathbf{a}_1 + \mathbf{d}$, where $u \in \mathbf{Z}_p$ and $\mathbf{d} \in \mathbf{Z}_p^n$ is orthogonal to $\mathbf{a}_1$. Hence we can use induction on the dimension (cf. the proof of Lemma 4.3 of Chapter 2).

COROLLARY 2 ($p \neq 2$). *Suppose, further, that $f(\mathbf{x})$ is integral and that its determinant is a p-adic unit. Then any integral automorph σ of $f(\mathbf{x})$ is a product of $\tau_{\mathbf{b}}$'s where $\mathbf{b}$ is integral and $f(\mathbf{b})$ is a p-adic unit (so $\tau_{\mathbf{b}}$ is integral).*

Proof. As for Corollary 1.

The following Lemma will be superseded by Theorem 3.1.

LEMMA 3.4 ($p \neq 2$). *Let $u_1, \ldots, u_n$ be p-adic units. Then*

$$u_1x_1^2 + \ldots + u_nx_n^2$$

is integrally equivalent to

$$x_1^2 + \ldots + x_{n-1}^2 + ux_n^2$$

where $u = u_1 \ldots u_n$.

Proof ($n = 2$). By Lemma 2.2 of Chapter 2 there are p-adic integers a_1, a_2 such that

$$u_1a_1^2 + u_2a_2^2 \equiv 1 \pmod p$$

and (on interchanging the indices 1, 2 if need be) we may suppose that a_1 is a unit. Then by Lemma 1.6 of Chapter 3 and since $p \neq 2$, there is an $a_1^* \in \mathbf{Z}_p$ such that

$$u_1a_1^{*2} + u_2a_2^2 = 1.$$

We thus have a primitive representation of 1 by $u_1x_1^2 + u_2x_2^2$ and this extends to give an integral equivalence with $x_1^2 + u_1u_2x_2^2$.

($n > 2$). There is an obvious induction argument.

THEOREM 3.1. *Let $p \neq 2$ and let r be some fixed quadratic non-residue of p, that is*

$$|r| = 1 \qquad r \notin (\mathbf{Q}_p^*)^2. \tag{3.9}$$

For $\varepsilon = 0$ *or* 1 *and for* $m = 1, 2, \ldots$ *let* $h(\mathbf{y}) = h(\varepsilon, m, \mathbf{y})$ *be the form*

$$\left.\begin{aligned} h(\varepsilon, m, \mathbf{y}) &= y_1^2 + \ldots + y_m^2 \quad \text{if } \varepsilon = 0 \\ &= y_1^2 + \ldots + y_{m-1}^2 + r y_m^2 \quad \text{if } \varepsilon = 1 \end{aligned}\right\}. \tag{3.10}$$

Then every non-singular $f(\mathbf{x}) \in \mathbf{Q}_p(\mathbf{x})$ *is* $\mathbf{Z}_p$*-equivalent to a form*

$$g(\mathbf{x}) = \sum_{j=1}^{J} p^{e(j)} h(\varepsilon_j, m_j, \mathbf{y}^{(j)}) \tag{3.11}$$

for some J, some $e(j)$ *with*

$$e(1) < e(2) < \ldots < e(J) \tag{3.12}$$

and some ε_j, m_j *with*

$$m_1 + \ldots + m_J = n. \tag{3.13}$$

Here

$$\mathbf{x} = (\mathbf{y}^{(1)}, \mathbf{y}^{(2)}, \ldots, \mathbf{y}^{(J)}). \tag{3.14}$$

Further, no two distinct forms (3.11) *are* $\mathbf{Z}_p$*-equivalent.*

We note at once the useful

COROLLARY ($p \neq 2$). *Suppose that* f_1, f_2 *are integral forms and that* $d(f_1) = d(f_2)$ *is a unit. Then* f_1, f_2 *are equivalent.*

Proof of Theorem. We shall first show that f is $\mathbf{Z}_p$-equivalent to a form (3.11) and then show that it is unique.

By Lemma 3.1 the form f is $\mathbf{Z}_p$-equivalent to a form

$$f(\mathbf{x}) = \sum_{i,j} f_{ij} x_i x_j$$

where

$$|f_{ii}| \leqslant |f_{11}|$$

and by (3.3) and since $p \neq 2$

$$|f_{ij}| \leqslant |f_{11}|.$$

Hence

$$f(\mathbf{x}) = f_{11}\left(x_1 + \frac{f_{12}}{f_{11}} x_2 + \ldots + \frac{f_{1n}}{f_{11}} x_n\right)^2 + h(x_2, \ldots, x_n)$$

for some quadratic form h, where $f_{ij}/f_{11} \in \mathbf{Z}_p$. Hence $f(\mathbf{x})$ is $\mathbf{Z}_p$-equivalent to

$$f_{11} x_1^2 + h(x_2, \ldots, x_n).$$

By induction it follows that f is $\mathbf{Z}_p$-equivalent to a diagonal form

$$f_{11} x_1^2 + f_{22} x_2^2 + \ldots + f_{nn} x_n^2$$

with

$$|f_{11}| \geqslant |f_{22}| \geqslant \dots \geqslant |f_{nn}|.$$

On lumping together the terms with the same $|f_{ii}|$ we see that f is equivalent to a sum of the type

$$\sum_{j=1}^{J} p^{e(j)} g_j(\mathbf{y}^{(j)})$$

where $g_j(\mathbf{y}^{(j)})$ is of the type

$$u_1 y_1^2 + \dots + u_m y_m^2$$

for units $u_1, \dots, u_m$. By Lemma 3.4 this is integrally equivalent to

$$y_1^2 + \dots + y_{n-1}^2 + u y_m^2$$

for some unit u. But either $u = v^2$ or $u = rv^2$ for some unit v, where r is given by 3.9). This shows that f is equivalent to a form of the shape (3.11).

It remains to show that f determines (3.11) uniquely and this follows easily from the generalization of Witt's Lemma, Lemma 3.3. On taking $p^e f$ instead of f for suitable integer e we may suppose that f is a primitive integral form. Then the rule for obtaining (3.11) can be put:

"If f represents integrally the square of a unit then split off the form $\langle 1 \rangle$. Otherwise it will certainly represent r, so split off $\langle r \rangle$ (we recall that $\langle a \rangle$ is the 1-dimensional form ax^2)." By Lemma 3.3 any two representations of $\langle 1 \rangle$ or any two representations of $\langle r \rangle$ are equivalent under the group of integral automorphs. Uniqueness of (3.11) now follows by induction on the dimension n.

4. CANONICAL FORMS, $p = 2$.

The properties of $\mathbf{Q}_2$-valued forms under $\mathbf{Z}_2$-equivalence are really rather tiresome. Fortunately, as we explained in the introductory Section 1, a detailed knowledge is not needed for the discussion of "global" forms. Only the masochist is invited to read the rest of this section.

LEMMA 4.1 ($p = 2$). *Every regular quadratic form taking values in $\mathbf{Q}_2$ is $\mathbf{Z}_2$-equivalent to a sum of forms of the following types in distinct variables:*

$$2^e x^2,\ 2^e(3x^2),\quad 2^e(5x^2),\quad 2^e(7x^2) \qquad (e \in \mathbf{Z}) \tag{4.1}$$

$$2^e(2x_1x_2) \tag{4.2}$$

$$2^e(2x_1^2 + 2x_1x_2 + 2x_2^2). \tag{4.3}$$

Note. It is not asserted that the representation is unique. Indeed there are equivalences between sums of forms (4.1), (4.2), (4.3) exemplified by the following ($\sim$ means $\mathbf{Z}_2$-equivalence):

$$x_1^2 + x_2^2 \sim 5x_1^2 + 5x_2^2 \tag{4.4}$$

$$x_1^2 + 2x_2^2 \sim 3x_1^2 + 6x_2^2 \tag{4.5}$$

$$x_1^2 + 4x_2^2 \sim 5x_1^2 + 20x_2^2 \tag{4.6}$$

$$ux_1^2 + 2x_2x_3 \sim ux_1^2 + x_2^2 - x_3^2 \qquad (|u| = 1) \tag{4.7}$$

$$ux_1^2 + (2x_2^2 + 2x_2x_3 + 2x_3^2) \sim u(3x_1^2 - x_2^2 - x_3^2) \qquad (|u| = 1) \tag{4.8}$$

$$(2x_1^2 + 2x_1x_2 + 2x_2^2) + (2x_3^2 + 2x_3x_4 + 2x_4^2) \sim 2x_1x_2 + 2x_3x_4. \tag{4.9}$$

Proof. Let

$$\sup|f(\mathbf{a}, \mathbf{b})| = 2^{-e} \tag{4.10}$$

when $\mathbf{a}, \mathbf{b}$ runs through $\mathbf{Z}_2^n$. As in the case $p \neq 2$, this supremum is attained and then $\mathbf{a}, \mathbf{b}$ are primitive. We now distinguish two cases.

(i) There is a $\mathbf{b} \in \mathbf{Z}_2^n$ such that $|f(\mathbf{b})| = 2^{-e}$. This is similar to the case $p \neq 2$. There is a basis $\mathbf{b} = \mathbf{b}_1, \mathbf{b}_2, \ldots, \mathbf{b}_n$ of $\mathbf{Z}_2^n$ and

$$f(\mathbf{b}_1, \mathbf{b}_j) = 0 \qquad (j > 1).$$

Hence

$$f(y_1\mathbf{b}_1 + y_2\mathbf{b}_2 + \ldots + y_n\mathbf{b}_n) = 2^e uy_1^2 + g(y_2, \ldots, y_n)$$

where u is a unit. Further, u is of one of the shapes v^2, $3v^2$, $5v^2$, $7v^2$ where v is a unit.

(ii)
$$|f(\mathbf{a})| < 2^{-e} \tag{4.11}$$

for all $\mathbf{a} \in \mathbf{Z}_2^n$. By hypothesis there are primitive $\mathbf{b}_1, \mathbf{b}_2 \in \mathbf{Z}_2^n$ such that

$$|f(\mathbf{b}_1, \mathbf{b}_2)| = 2^{-e}. \tag{4.12}$$

By (4.11) we have, in particular

$$|f(\mathbf{b}_1)| < 2^{-e} \qquad |f(\mathbf{b}_2)| < 2^{-e}. \tag{4.13}$$

Suppose first that $\mathbf{c}$ (say) $= \frac{1}{2}(\mathbf{b}_1 + \mathbf{b}_2) \in \mathbf{Z}_2^n$. Then by (4.12), (4.13) we have:

$$\begin{aligned} |f(\mathbf{b}_1, \mathbf{c})| &= |\tfrac{1}{2}f(\mathbf{b}_1) + \tfrac{1}{2}f(\mathbf{b}_1, \mathbf{b}_2)| \\ &= |\tfrac{1}{2}f(\mathbf{b}_1, \mathbf{b}_2)| \\ &> 2^{-e} \end{aligned}$$

in contradiction to the definition of e. Hence $\frac{1}{2}(\mathbf{b}_1 + \mathbf{b}_2) \notin \mathbf{Z}_2^n$. It follows from Lemma 2.1 that $\mathbf{b}_1, \mathbf{b}_2$ can be extended to a basis $\mathbf{b}_1, \ldots, \mathbf{b}_n$ of $\mathbf{Z}_2^n$.

We now show that if $\mathbf{d} \in \mathbf{Z}_2^n$ then there are $\lambda_1, \lambda_2 \in \mathbf{Z}_2$ such that

$$\mathbf{d}^* = \mathbf{d} - \lambda_1 \mathbf{b}_1 - \lambda_2 \mathbf{b}_2$$

satisfies

$$f(\mathbf{b}_1, \mathbf{d}^*) = f(\mathbf{b}_2, \mathbf{d}^*) = 0.$$

Indeed λ_1, λ_2 are determined by

$$\lambda_1 f(\mathbf{b}_1) + \lambda_2 f(\mathbf{b}_1, \mathbf{b}_2) = f(\mathbf{b}_1, \mathbf{d})$$
$$\lambda_1 f(\mathbf{b}_1, \mathbf{b}_2) + \lambda_2 f(\mathbf{b}) = f(\mathbf{b}_2, \mathbf{d}).$$

It follows from (4.10), (4.12), (4.13) that the solution λ_1, λ_2 is in $\mathbf{Z}_2$.

On applying this last remark with $\mathbf{d} = \mathbf{b}_j$ $(j > 2)$, where $\mathbf{b}_1, \ldots, \mathbf{b}_n$ is the basis constructed earlier, we obtain a new basis with

$$f(\mathbf{b}_1, \mathbf{b}_j) = f(\mathbf{b}_2, \mathbf{b}_j) = 0 \qquad (j > 2).$$

Then

$$f(y_1 \mathbf{b}_1 + \ldots + y_n \mathbf{b}_n) = 2^e h(y_1, y_2) + g(y_3, \ldots, y_n)$$

for some forms h and g and, further

$$h(y_1, y_2) = h_{11} y_1^2 + 2h_{12} y_1 y_2 + h_{22} y_2^2$$

with

$$|h_{12}| = 1, \qquad |h_{11}| < 1, \qquad |h_{22}| < 1.$$

But now the reader can readily verify that $h(y_1, y_2)$ is $\mathbf{Z}_2$-equivalent to

$$2y_1 y_2$$

or

$$2y_1^2 + 2y_1 y_2 + 2y_2^2$$

according as

$$h_{12}^2 - h_{11} h_{22} \equiv 1 \pmod 8$$
$$\text{or} \equiv 5 \pmod 8.$$

This conclues the proof.

We note at once the

COROLLARY. *Let f, g be two improperly primitive integral forms and suppose that $d(f) = d(g)$ is a unit. Then f, g are $\mathbf{Z}_2$-equivalent.*

Proof. For f is equivalent to a sum of forms of the type $2x_1 x_2$ and $2x_1^2 + 2x_1 x_2 + 2x_2^2$. By (4.9) we may suppose that all but one of these forms is of the shape $2x_1 x_2$. The last is of the shape $2x_1 x_2$ or $2x_1^2 + 2x_1 x_2 + 2x_2^2$ according as

$$d(f) \equiv \pm 1 \pmod 8$$
$$\text{or } \pm 5 \pmod 8.$$

We do not attempt to specify a unique canonical form [see Jones (1944), Pall (1945) or Watson (1976a)]: that is more a job for a parliamentary draftsman than for a mathematician. We shall however prove an appropriate generalization of "Witt's Lemma". Here again there are complications: the equivalence (4.7) shows that the obvious generalization fails.

LEMMA 4.2 ($p = 2$). *Let u be a 2-adic unit and let $g_1(\mathbf{x})$ and $g_2(\mathbf{x})$ be classically integral 2-adic forms. Suppose that*

either (i) *both g_1 and g_2 are properly primitive*

or (ii) *neither g_1 nor g_2 is properly primitive.*

If,

$$\langle u \rangle + g_1 \sim \langle u \rangle + g_2 \tag{4.14}$$

then

$$g_1 \sim g_2. \tag{4.15}$$

Note. We recall that $\langle u \rangle$ is the 1-dimensional form ux_1^2. The notation of (4.14) implies that $\langle u \rangle$ and g_1 have different variables: cf. Chapter 2, Section 5.

Proof. (cf. Lemma 3.3). The equivalence (4.14) means that the left-hand side and the right-hand side correspond to different choices of basis for a lattice Λ in a quadratic space V, ϕ. There are two vectors $\mathbf{b}, \mathbf{c} \in \Lambda$ with

$$\phi(\mathbf{b}) = \phi(\mathbf{c}) = u.$$

In order to prove (4.15) it will be enough to find an autometry σ of V, ϕ with $\sigma\Lambda = \Lambda$ and such that $\sigma\mathbf{b} = \mathbf{c}$. We distinguish some cases:

(i) Suppose that $\phi(\mathbf{b}, \mathbf{c}) \equiv 0 \pmod 2$. Then

$$\begin{aligned}\phi(\mathbf{b} - \mathbf{c}) &= 2u - 2\phi(\mathbf{b}, \mathbf{c}) \\ &\equiv 2 \pmod 4.\end{aligned}$$

Then $\sigma = \tau_{\mathbf{b}-\mathbf{c}}$ does what is required since for $\mathbf{a} \in \Lambda$ we have

$$\tau_{\mathbf{b}-\mathbf{c}}\mathbf{a} = \mathbf{a} - 2\frac{\phi(\mathbf{a}, \mathbf{b} - \mathbf{c})}{\phi(\mathbf{b} - \mathbf{c})}(\mathbf{b} - \mathbf{c}) \tag{4.16}$$

where

$$\frac{2\phi(\mathbf{a}, \mathbf{b} - \mathbf{c})}{\phi(\mathbf{b} - \mathbf{c})} \in \mathbf{Z}_2,$$

and so

$$\tau_{\mathbf{b}-\mathbf{c}}\mathbf{a} \in \Lambda.$$

(ii) Otherwise

$$\phi(\mathbf{b}, \mathbf{c}) \equiv 1 \pmod 2$$

and

$$\begin{aligned}\phi(\mathbf{b} \pm \mathbf{c}) &= 2u \pm 2\phi(\mathbf{b}, \mathbf{c}) \\ &\equiv 4 \pmod 8\end{aligned}$$

for one or other choice of sign $\pm$. On taking $-\mathbf{c}$ for $\mathbf{c}$ if necessary we may suppose that

$$\phi(\mathbf{b} - \mathbf{c}) \equiv 4 \pmod 8. \tag{4.17}$$

Now suppose that

$$\phi(\mathbf{b} - \mathbf{c}, \mathbf{a}) \equiv 0 \pmod 2 \qquad (\text{all } \mathbf{a} \in \Lambda).$$

By (4.16) and (4.17) we have

$$\tau_{\mathbf{b}-\mathbf{c}}\mathbf{a} \in \Lambda \qquad (\text{all } \mathbf{a} \in \Lambda),$$

as required.

There remains the case when

$$\phi(\mathbf{b} - \mathbf{c}, \mathbf{d}) \equiv 1 \pmod 2$$

for some $\mathbf{d} \in \Lambda$. Put

$$\begin{aligned}\mathbf{b}_1 &= \mathbf{b} \\ \mathbf{b}_2 &= \mathbf{b} - \mathbf{c} \\ \mathbf{b}_3 &= \mathbf{d}.\end{aligned}$$

Then

$$[\phi(\mathbf{b}_i, \mathbf{b}_j)] \equiv \begin{bmatrix} 1 & 0 & ? \\ 0 & 0 & 1 \\ ? & 1 & ? \end{bmatrix} \pmod 2 \tag{4.18}$$

where ? denotes a value about which we have no information. Hence in any case

$$\det(\phi(\mathbf{b}_i, \mathbf{b}_j))_{i, j = 1, 2, 3}$$

is a 2-adic unit. Since ϕ is integral on Λ, this implies that $\mathbf{b}_1, \mathbf{b}_2, \mathbf{b}_3, \ldots$ can be extended to a basis of Λ. It also implies that for any $\mathbf{a} \in \Lambda$ there are $\lambda_1, \lambda_2, \lambda_3, \in \mathbf{Z}_2$ such that

$$\phi(\mathbf{b}_j, \mathbf{a} - \lambda_1\mathbf{b}_1 - \lambda_2\mathbf{b}_2 - \lambda_3\mathbf{b}_3) = 0 \qquad (j = 1, 2, 3).$$

On putting these two remarks together, we see that there is a basis

$$\mathbf{b}_1, \mathbf{b}_2, \mathbf{b}_3, \ldots, \mathbf{b}_n$$

such that

$$\phi(\mathbf{b}_i, \mathbf{b}_j) = 0 \qquad (i = 1, 2, 3, j > 3).$$

In the language of forms this means that there are binary forms h_1, h_2 and an $(n-3)$-ary form k such that g_1, g_2 of (4.14) satisfy

$$g_1 \sim h_1 + k$$
$$g_2 \sim h_2 + k.$$

Further

$$\langle u\rangle + h_1 \sim \langle u\rangle + h_2 \sim l \text{ (say)},$$

where the form l is given by

$$l(y_1, y_2, y_3) = \phi(y_1\mathbf{b}_1 + y_2\mathbf{b}_2 + y_3\mathbf{b}_3).$$

A detailed consideration of the orthogonal complements of $\mathbf{b} = \mathbf{b}_1$ and $\mathbf{c} = \mathbf{b}_1 - \mathbf{b}_2$ in the space spanned by $\mathbf{b}_1$, $\mathbf{b}_2$ and $\mathbf{b}_3$ using (4.18) shows that

$$h_1 \sim 2x_1x_2$$
$$h_2 \sim x_1^2 - x_2^2$$

(or vice versa). If k is not properly primitive the form $g_1 \sim h_1 + k$ is not properly primitive but $g_2 \sim h_2 + k$ is: and then g_1 and g_2 are not equivalent. If however k is properly primitive we have $k \sim \langle v\rangle + k_1$ for some form k_1 and for some unit v. Then

$$h_1 + k \sim h_2 + k$$

by (4.7). This concludes the proof.

COROLLARY. *Suppose that g_1 and g_2 are integral forms but not properly primitive. If*

$$h + g_1 \sim h + g_2 \tag{4.19}$$

where

$$h(x_1, x_2) = 2x_1x_2 \tag{4.20}$$

or

$$h(x_1, x_2) = 2x_1^2 + 2x_1x_2 + 2x_2^2 \tag{4.21}$$

then

$$g_1 \sim g_2.$$

Proof. For (4.19) implies that

$$\langle 1\rangle + h + g_1 \sim \langle 1\rangle + h + g_2.$$

But

$$\langle 1\rangle + h \sim \langle 1\rangle + \langle 1\rangle + \langle -1\rangle \quad \text{or} \quad \langle 3\rangle + \langle -1\rangle + \langle -1\rangle$$

in case (4.20) or (4.21) respectively. The Corollary now follows by a triple application of the Lemma.

5. APPROXIMATION THEOREMS

Let $d \in \mathbf{Z}_p$, $d \neq 0$ and define δ by $|d| = p^{-\delta}$. Let

$$\left.\begin{aligned}\lambda &= 1 \quad \text{if } p = 2\\ &= 0 \text{ otherwise}\end{aligned}\right\}. \tag{5.1}$$

Let

$$f(\mathbf{x}) = \Sigma f_{ij} x_i x_j \tag{5.2}$$

be a classically integral form, i.e.

$$f_{ij} \in \mathbf{Z}_p \tag{5.3}$$

and suppose that

$$d(f) = d. \tag{5.4}$$

Let $g(\mathbf{x})$ be another integral form.

LEMMA 5.1. *Suppose that*

$$g_{ij} \equiv f_{ij} \pmod{p^{\delta+2\lambda+1}}. \tag{5.5}$$

Then the forms f and g are $\mathbf{Z}_p$-equivalent.

LEMMA 5.2. *Suppose that*

$$d(g) = du^2 \tag{5.6}$$

for some unit u and that

$$g_{ij} \equiv f_{ij} \pmod{p^{\delta+2\lambda}}. \tag{5.7}$$

Then f and g are $\mathbf{Z}_p$-equivalent.

Note. Lemma 5.1 follows from Lemma 5.2 since (5.5) implies that

$$d(g) \equiv d(f) \pmod{p^{\delta+2\lambda+1}}$$

and hence (5.6). However, we give a simple proof of Lemma 5.1 independent of Sections 3 and 4.

Proof of Lemma 5.1. Let $\mathbf{F}, \mathbf{G}$ be the symmetric matrices corresponding to f and g. We have to show the existence of a matrix $\mathbf{T}$ of unit determinant such that $\mathbf{T}'\mathbf{F}\mathbf{T} = \mathbf{G}$. Here the prime (′) denotes the transpose. We construct $\mathbf{T}$ by successive approximation as in Hensel's Lemma. If $\mathbf{I}$ denotes the unit matrix we have

$$(\mathbf{I}+\mathbf{S})'\mathbf{F}(\mathbf{I}+\mathbf{S}) = \mathbf{F} + \mathbf{S}'\mathbf{F} + \mathbf{F}\mathbf{S} + \mathbf{S}'\mathbf{F}\mathbf{S}.$$

Put

$$\mathbf{S} = \tfrac{1}{2}\mathbf{F}^{-1}(\mathbf{G}-\mathbf{F}) \tag{5.8}$$

and suppose that

$$\mathbf{G} \equiv \mathbf{F} \pmod{p^{\mu}} \tag{5.9}$$

where

$$\mu \geqslant \delta + 2\lambda + 1.$$

Then

$$\begin{aligned} 2\{\det(\mathbf{F})\}\mathbf{S} &= (\operatorname{adj}(\mathbf{F}))(\mathbf{G} - \mathbf{F}) \\ &= \mathbf{0} \pmod{p^{\mu}} \end{aligned}$$

since the adjoint adj(**F**) of the integral matrix **F** is integral. Hence

$$\mathbf{S} \equiv \mathbf{0} \pmod{p^{\mu-\delta-\lambda}}. \tag{5.10}$$

Put

$$\mathbf{F}_1 = (\mathbf{I} + \mathbf{S})'\mathbf{F}(\mathbf{I} + \mathbf{S}).$$

Then

$$\begin{aligned} \mathbf{G} - \mathbf{F}_1 &= -\mathbf{S}'\mathbf{F}\mathbf{S} \\ &= -\tfrac{1}{2}\mathbf{S}'(\mathbf{G} - \mathbf{F}) \end{aligned}$$

by (8). Hence

$$\mathbf{G} - \mathbf{F}_1 \equiv \mathbf{0} \pmod{p^{2\mu-\delta-2\lambda}}$$

by (5.9) and (5.10). Since $2\mu - \delta - 2\lambda > \mu$, we may replace $\mathbf{F}$ by $\mathbf{F}_1$ and continue the induction.

Proof of Lemma 5.2. Since we are concerned with $\mathbf{Z}_p$-equivalence we may suppose without loss of generality that f is in the canonical form given by Theorem 3.1 (for $p \neq 2$) or Lemma 4.1 (for $p = 2$). The proof is now almost immediate for $p \neq 2$ but somewhat more difficult for $p = 2$.

NOTES

The material of this chapter generalizes without trouble to finite extensions of $\mathbf{Q}_p$ with $p \neq 2$. The finite extensions of $\mathbf{Q}_2$ are much more troublesome. See O'Meara (1963).

EXAMPLES

1. Let $f(\mathbf{x}) = 5x^2 + 2xy + 5y^2$. Find diagonal forms which are $\mathbf{Z}_p$-equivalent to f for $p = 2, 3, 5$.

2. Over which rings $\mathbf{Z}_p$ are the following pairs of forms equivalent?
   (i) $2x_1x_2$ and $x_1^2 - x_2^2$
   (ii) $2x_1x_2$ and $x_1^2 + x_2^2$

(iii) $5x_1^2 + 13x_2^2 + 11x_3^2 + 2x_2x_3 + 2x_3x_1 + 16x_1x_2$
and $x_1^2 + 16x_2^2 + 10x_3^2 + 14x_2x_3 - 4x_3x_1 + 2x_1x_2$

(iv) $x_1^2 + 2x_2^2 + 6x_3^2 + 6x_2x_3$
and $2x_1^2 + 3x_2^2 + 5x_3^2$.

3. Let f be an improperly primitive $\mathbf{Z}_2$-form in n variables. Prove the following statements:

(i) if $d(f)$ is a unit then $2|n$ and f is $\mathbf{Z}_2$-equivalent to a sum of forms $2xy$ or $2x^2 + 2xy + 2y^2$.

(ii) if $d(f) = 1$, then $4|n$ and f is $\mathbf{Z}_2$-equivalent to a sum of forms $2xy$.

(iii) if $d(f) = 1$ and $c_2(f) = 1$, then $8|n$.

4. Let $f(\mathbf{x})$ be a classically integral $\mathbf{Z}_2$-quadratic form in n variables and suppose that $d(f)$ is a unit.

(i) Show that there is a $\mathbf{b} \in \mathbf{Z}_2^n$ such that

$$f(\mathbf{a}) \equiv f(\mathbf{a}, \mathbf{b}) \pmod 2$$

for every $\mathbf{a} \in \mathbf{Z}_2^n$ and that $\mathbf{b}$ is unique mod 2.

(ii) Show that $f(\mathbf{b})$ is uniquely determined mod 8.

(iii) Show that $f(\mathbf{b}) \equiv n \pmod 2$ and that

$$f(\mathbf{b}) - n \equiv d(f) - 1 \pmod 4$$

(iv) Show that

$$c_2(f) = (-1)^s$$

where s is any element of $\mathbf{Z}$ such that

$$4s \equiv f(\mathbf{b}) - d(f) - n + 1 \pmod 8$$

[*Hint*. Lemma 4.1.]

5. Let f_1, f_2 be two forms as specified in the previous example in the same number n of variables. Let $\mathbf{b}_1, \mathbf{b}_2$ be the corresponding elements of $\mathbf{Z}_2^n$. Show that f_1, f_2 are $\mathbf{Q}_2$-equivalent if and only if the two following conditions are satisfied:

$$d(f_1) \equiv d(f_2) \pmod 8 \qquad (\alpha)$$

$$f_1(\mathbf{b}_1) \equiv f_2(\mathbf{b}_2) \pmod 8. \qquad (\beta)$$

[*Note*. For background see Cassels (1962).]

6. Let V, ϕ be a regular quadratic space over $\mathbf{Q}_p$. A $\mathbf{Z}_p$-lattice $\Lambda \subset V$ is (classically) maximal when (i) ϕ induces on Λ a (classically) integral quadratic form and (ii) if Γ is a $\mathbf{Z}_p$-lattice in V properly containing Λ then ϕ does not induce an integral form on Γ.

(i) Show that every lattice Λ is contained in at least one maximal lattice.

(ii) Let $p \neq 2$ and suppose that Λ is maximal. Show that Λ has a basis $\mathbf{b}_1, \ldots, \mathbf{b}_n$ such that

$$\phi(\Sigma x_j \mathbf{b}_j) = \Sigma c_j x_j^2$$

where $|c_j| = 1$ or p^{-1} for all j and $|c_j| = p^{-1}$ for at most two values of j.

(iii) State and prove the corresponding statement for $p = 2$.

7. As the previous question except with the non-classical definition of integrality (Λ is integral if $\phi(\mathbf{c}) \in \mathbf{Z}_p$ for all $\mathbf{c} \in \Lambda$).

8. (i) Let Λ be a $\mathbf{Z}_p$-lattice in the regular $\mathbf{Q}_p$-quadratic space V, ϕ and suppose that $\phi(\mathbf{c}) \in \mathbf{Z}_p$ (all $\mathbf{c} \in \Lambda$). Let Γ be the set of $\mathbf{d} \in \Lambda$ such that

$$\phi(\mathbf{c} + \mathbf{d}) \equiv \phi(\mathbf{c}) \pmod{p} \qquad (\text{all } \mathbf{c} \in \Lambda).$$

Show that (a) Γ is a lattice

(b) $\phi(\mathbf{d}) \in p\mathbf{Z}_p$ (all $\mathbf{d} \in \Gamma$)

and (c) $\Lambda \supset \Gamma \supset p\Lambda$.

(ii) Let Δ be obtained from Γ as Γ was from Λ but with $p^{-1}\phi(\mathbf{x})$ instead of $\phi(\mathbf{x})$: that is Δ consists of the $\mathbf{e} \in \Gamma$ such that

$$\phi(\mathbf{d} + \mathbf{e}) \equiv \phi(\mathbf{d}) \pmod{p^2} \qquad (\text{all } \mathbf{d} \in \Gamma).$$

Show that

$$\Delta \supset p\Lambda.$$

(iii) Put $\Lambda_0 = \Lambda$ and $\Lambda_1 = p^{-1}\Delta$.
Show that one obtains by recursion a sequence $\Lambda_0, \Lambda_1, \Lambda_2, \ldots$ of lattices such that

$$\Lambda_{m+1} \supset \Lambda_m \qquad (\text{all } m)$$

and ϕ is integer-valued on each Λ_m. Deduce that there is some m such that $\Lambda_{m+1} = \Lambda_m$.

(iv) Suppose that $p \neq 2$ and that $\Lambda = \Lambda_0 = \Lambda_1$. Show that Λ has a basis $\mathbf{b}_1, \ldots, \mathbf{b}_n$ such that

$$\phi(\Sigma x_j \mathbf{b}_j) = \Sigma c_j x_j^2$$

where $|c_j| = 1$ or p^{-1} for each j.

(v) Consider the case when $\Lambda = \Lambda_0 = \Lambda_1$ and $p = 2$.

CHAPTER 9

Integral Forms over the Rational Integers

1. INTRODUCTION

In this chapter we discuss the behaviour forms under **Z**-equivalence. We shall be primarily concerned with integral forms but our initial remarks deal with **Q**-valued forms in general. We recall from the general theory (Chapter 1) that two such forms f, g are **Z**-equivalent if there is a square integral matrix $\mathbf{T}$ such that

$$g(\mathbf{x}) = f(\mathbf{T}\mathbf{x}) \tag{1.1}$$

and $\det \mathbf{T}$ is an invertible element of **Z**. The only invertible elements of **Z** are ± 1. We shall say that the equivalence between f and g given by (1.1) is *proper* if $\det \mathbf{T} = +1$ and *improper* if $\det \mathbf{T} = -1$. Clearly proper equivalence is an equivalence relation and we can speak of the *proper equivalence class* consisting of all forms which are properly equivalent to a given form. Of course improper equivalence is not an equivalence relation.

We recall that the determinant $d(f)$ of the form $\Sigma f_{ij} x_i x_j$ is given by

$$d(f) = \det(f_{ij}) \qquad (1 \leqslant i \leqslant n,\ 1 \leqslant j \leqslant n).$$

From the above we have

LEMMA 1.1. *Two* **Z**-*equivalent forms have the same determinant.*

An equivalence of the form f with itself is an integral *automorph* (or automorphism) of f; so again we can distinguish between *proper automorphs* (with $\det +1$) and *improper automorphs* (of $\det -1$). Both the set $O(f)$ of automorphs and the set $O^+(f)$ of proper automorphs of f form a group under composition. Either no improper automorphs exist [and then $O^+(f) = O(f)$] or $O^+(f)$ is of index 2 in $O(f)$. Both cases can occur: for example

$$f_1(x, y) = x^2 + 5y^2$$

has the improper automorph

$$x \to x \qquad y \to -y$$

but

$$f_2(x, y) = 3x^2 + 2xy + 101y^2$$

has no improper automorphs as the reader can easily verify. [For the only integral representations of 3 are given by $f_2(\mathbf{a}) = 3$ with $\mathbf{a} = \pm(1, 0)$. If $\mathbf{T}$ is an automorph then $f_2(\mathbf{T}(1,0)) = \pm(1,0)$ and then $\mathbf{T} = \pm\mathbf{I}$].

If f has an improper automorph then clearly the equivalence class and the proper equivalence class to which f belongs will coincide. If, however, f has no improper automorphs then the equivalence class of f splits into two proper equivalence classes.

We shall be considering integral automorphs for their own sake in Chapter 13.

We shall be primarily concerned with integral forms and use the classical definition: that is

$$f(\mathbf{x}) = \Sigma f_{ij} x_i x_j \qquad (f_{ij} = f_{ji})$$

is integral when

$$f_{ij} \in \mathbf{Z} \qquad (\text{all } i, j).$$

The form f is *primitive* when

$$\text{g.c.d.}(f_{ij}) = 1.$$

It is *properly primitive* if at least one of the f_{ii} is odd, otherwise *improperly primitive*. If f is improperly primitive, then $\frac{1}{2}f(\mathbf{x}) \in \mathbf{Z}[\mathbf{x}]$ but $\frac{1}{2}f(\mathbf{x})$ is not an integral form.

With these definitions we now review the contents of this Chapter. In Section 2 there are preliminaries about bases of $\mathbf{Z}^n$ and the way they can be made to approximate bases of the $\mathbf{Z}_p^n$. In Section 3 we prove the following important finiteness theorem:

THEOREM 1.1. *Let the integers $n \geqslant 1$ and $d \neq 0$ be given. Then there are only finitely many equivalence classes of integral quadratic forms f in n variables $\mathbf{x} = (x_1, \ldots, x_n)$ and with*

$$d(f) = d.$$

In accordance with our general philosophy we shall consider the "global" properties of rational integral forms (i.e. their properties over $\mathbf{Z}$) in relation to the corresponding "local" properties (i.e. their properties over the $\mathbf{Z}_p$). We shall say that two non-singular rational integral forms are in the same *genus* (plural: genera) if they are $\mathbf{Z}_p$-equivalent for every p including $p = \infty$ (with the usual convention that $\mathbf{Z}_\infty = \mathbf{R}$). Two forms which are globally (i.e.

$\mathbf{Z}-$) equivalent are clearly in the same genus. However the converse statement is in general false. For example the forms

$$f_1(x, y) = x^2 + 82y^2$$
$$f_2(x, y) = 2x^2 + 41y^2,$$

which are clearly not globally equivalent, are in the same genus. Indeed they are $\mathbf{Z}_2$-equivalent because $41 \in (\mathbf{Q}_2^*)^2$ and $\mathbf{Z}_{41}$-equivalent because $2 \in (\mathbf{Q}_{41}^*)^2$. They are clearly $\mathbf{Z}_\infty$-equivalent and they are $\mathbf{Z}_p$-equivalent for all $p \neq 2, 41, \infty$ by Lemma 3.4 (or Theorem 3.1) of Chapter 8.

Similarly there exist indefinite forms which are in the same genus but are not equivalent, but the proof of inequivalence requires the theory of automorphs. Thus in Section 3 of Chapter 13 we shall show that $x^2 - 82y^2$ and $2x^2 - 41y^2$ are inequivalent, though they are in the same genus for similar reasons to those given above. Indefinite genera in $n \geqslant 3$ variables are, however, usually (but not always) of one class: criteria for this will be given in Chapter 11.

In Section 4 we discuss the elementary consequences of the notion of a genus. In the introduction we mention only two of these. Two forms in the same genus have the same determinant: and so by Theorem 1.1 a genus contains only a finite number of equivalence classes. Secondly if f, g are in the same genus and M is any positive integer, then there is a form f^* properly equivalent to f such that, in an obvious notation

$$f_{ij}^* \equiv g_{ij} \pmod{M} \qquad (\text{all } i, j).$$

This second property emphasizes that it is impossible to distinguish two equivalence classes in the same genus by congruence properties alone.

In Section 5 we discuss the existence of genera with prescribed local properties. It turns out that no further conditions are required than those we considered in discussing the corresponding problem for rational forms and rational equivalence. More explicitly we have

THEOREM 1.2. *Let rational integers* $n \geqslant 1$ *and* $d \neq 0$ *be given. For all* p *(including* $p = \infty$*) let* $f_p(\mathbf{x})$ *be a* $\mathbf{Z}_p$*-integral form of determinant* d *in the variables* $\mathbf{x} = (x_1, \ldots, x_n)$. *Suppose that there exists a rational form* $g(\mathbf{x})$ *which is* $\mathbf{Q}_p$*-equivalent to* $f_p(\mathbf{x})$ *for each* p. *Then there is a* $\mathbf{Z}$*-integral form* $f(\mathbf{x})$ *which is* $\mathbf{Z}_p$*-equivalent to* $f_p(\mathbf{x})$ *for each* p.

The proof of the above theorem gives as a by-product the following equally important theorem about integral representations.

THEOREM 1.3. *Let* $f(\mathbf{x})$ *be a regular integral form and let* $a \neq 0$ *be an integer which is represented by* f *over each* $\mathbf{Z}_p$ *(including* $p = \infty$*). Then* a *is represented over* $\mathbf{Z}$ *by some form* f^* *in the same genus as* f.

There is a corresponding result in which "represented" is replaced by "primitively represented" both times†. Note that the example $f(\mathbf{x}) = 2x_1^2 + 41x_2^2$ and $a = 1$ shows that we cannot necessarily have $f^* = f$. Theorem 1.3 is particularly useful when f is in a genus which contains only one class: for then it gives necessary and sufficient conditions for the representation of a by f. For example, Theorem 1.3 implies that every positive integer is representable by $x_1^2 + x_2^2 + x_3^2 + x_4^2$.

In Section 6 we discuss a quantitative form of Theorem 1.3. If $\mathbf{b}$ is an integral representation of a by the integral form f then

$$f(\mathbf{Tb}) = a$$

for any integral automorph $\mathbf{T}$ of f. We shall say that the representations by $\mathbf{b}$ and $\mathbf{Tb}$ are in the same *orbit*. In general there are infinitely many integral representations of a by f but it is not difficult to show that they fall into finitely many orbits. The number of such orbits is given by a formula involving the number of equivalence classes in some genera of forms in $n - 1$ variables (where n is the number of variables in f). We illustrate the theory by discussing the number of representations of a positive integer as the sum of 2, 3 or 4 squares. From the point of view of this book the results of Section 6 are rather a digression but they are fundamental in the analytic theory (see Appendix B).

In Section 7 we turn to another aspect of the theory of genera. We have seen that two forms in the same genus need not be integrally equivalent. They are, however, "very nearly" integrally equivalent. Let P be any nonempty set of primes $p \neq \infty$. Denote by $\mathbf{Z}^{(P)}$ the set of rational numbers which do not have any of the $p \in P$ in their denominators: that is

$$\mathbf{Z}^{(P)} = \{c \in \mathbf{Q}: \quad |c|_p \leqslant 1 \quad \text{for all} \quad p \in P\}.$$

Then $\mathbf{Z}^{(P)}$ is clearly a ring. We shall say that two regular integral forms f and g are *semi-equivalent* if they are equivalent over $\mathbf{Z}^{(P)}$ for every finite set P of primes and have

THEOREM 1.4. *Two regular integral forms are in the same genus if and only if they are semi-equivalent.*

The proof of the "if" is trivial and may be given here. For $p \neq \infty$ we have $\mathbf{Z}^{(\{p\})} \subset \mathbf{Z}_p$ where $\{p\}$ is the set consisting only of p. If two forms are semi-equivalent then they are $\mathbf{Z}^{(\{p\})}$-equivalent by definition and so are certainly $\mathbf{Z}_p$-equivalent. Further $\mathbf{Z}^{(P)} \subset \mathbf{R} = \mathbf{Z}_\infty$ for any P, so semi-equivalence implies $\mathbf{Z}_\infty$-equivalence. Hence semi-equivalent forms are $\mathbf{Z}_p$-

† A representation $\mathbf{b}$ of b by f is *primitive* if the vector $\mathbf{b}$ is primitive

equivalent for all p (including ∞): that is they are in the same genus. The proof of the "only if" must be left for Section 7: it is related to theorems about approximations on the orthogonal group which will be discussed more fully in Chapter 11.

The theorem we have about representations of integers (Theorem 1.3) tells us about representations by the forms of a genus but tells us nothing about representation by the individual equivalence classes of the genus, and we saw that this is necessarily the case. For indefinite forms in $n \geqslant 4$ variables, however, there is a surprisingly strong result which we shall prove in the final section of this chapter (Section 8):

THEOREM 1.5. *Let f be a regular indefinite integral form in $n \geqslant 4$ variables and let $a \neq 0$ be an integer. Suppose that a is represented by f over all $\mathbf{Z}_p$. Then a is represented by f over $\mathbf{Z}$.*

Further, let P be a finite set of primes $p \neq \infty$ and for $p \in P$ let $\mathbf{b}_p \in \mathbf{Z}_p^n$ be any representation of a. Then there is a representation $\mathbf{b} \in \mathbf{Z}^n$ of a such that $\mathbf{b}$ is arbitrarily close to $\mathbf{b}_p$ for every $p \in P$.

The proof of Theorem 1.5 uses Theorem 1.4 and is the most elaborate argument in this chapter. The result itself will play a key role when we come to study approximation theorems on the orthogonal and spin groups and the theory of spinor-genera in Chapters 10 and 11.

We conclude this introductory section by bringing into the open a logical point which has been glossed over in the previous discussion. Let f be a regular integral form and a an integer. When f is positive definite we can always with a finite amount of calculation decide whether f represents a or not: for the equation

$$f(x_1, x_2, \ldots, x_n) = a \tag{1.2}$$

represents an ellipsoid in n-dimensional real space. In particular (1.2) for any real numbers $x_1, \ldots, x_n$ implies that

$$|x_j| \leqslant A \qquad (1 \leqslant j \leqslant n) \tag{1.3}$$

where A can be given explicitly in terms of a and the coefficients of f. To decide whether f represents a we have thus only to look at the integral vectors satisfying (1.3). If, however, f is indefinite then nothing which has been discussed so far gives us any way of deciding with certainty whether or not f represents a.

There is a similar situation with regard to the equivalence of two forms f and g. Of course f and g are equivalent if we can find an integral transformation $\mathbf{T}$ such that $g(\mathbf{x}) = f(\mathbf{Tx})$: but how can we show that no such $\mathbf{T}$ exists? In logical parlance our definition of equivalence is not "*effective*".

The reader can readily construct for himself a procedure for deciding whether two definite forms are equivalent but until fairly recently no such procedure was known for indefinite forms. A striking illustration of this is given by a footnote on page 147 of Dickson (1930): neither he nor A. E. Ross (both enthusiastic calculators) could decide whether the ternary forms

$$x^2 - 3y^2 - 2yz - 23z^2$$
$$x^2 - 7y^2 - 6yz - 11z^2$$

are equivalent or not.

In fact an effective decision procedure for equivalence of indefinite forms in $\geqslant 3$ variables is given by the theory of spinor genera in Chapter 11. For indefinite forms in 2 variables an effective decision procedure is given by the theory of integral automorphs (Chapter 13, Section 3). Another decision procedure which works for all indefinite forms is given in Chapter 13, Section 12, where there is a further discussion of the topic.

2. BASES OF $\mathbf{Z}^n$

The main object of this section is to show that bases of $\mathbf{Z}_p^n$ for several p can be simultaneously approximated by a base of $\mathbf{Z}^n$ subject to the obviously necessary condition that the bases all have the same determinant $+1$ or -1. There will be some more results in the same vein in Chapter 11, Section 2.

First however, because of its importance we reformulate Theorem 3.1 of Chapter 7 for the ring $I = \mathbf{Z}$.

LEMMA 2.1. *Let* $\mathbf{c}_1, \ldots, \mathbf{c}_J$ *be elements of* $\mathbf{Z}^n$. *Then the three following statements are equivalent*

(i) *There exist* $\mathbf{c}_{J+1}, \ldots, \mathbf{c}_n$ *such that* $\mathbf{c}_1, \ldots, \mathbf{c}_n$ *is a basis for* $\mathbf{Z}^n$.

(ii) *The determinants of the* $J \times J$ *submatrices of the* $n \times J$ *matrix* $\mathbf{c}_1\mathbf{c}_2 \ldots \mathbf{c}_J$ *have no common divisor* >1.

(iii) *If*

$$\mathbf{a} = v_1\mathbf{c}_1 + \ldots + v_J\mathbf{c}_J \in \mathbf{Z}^n$$

with $v_1, \ldots, v_J \in \mathbf{Q}$ *then necessarily* $v_1, \ldots, v_J \in \mathbf{Z}$.

Following our reformulation of Chapter 7 we say that an element $\mathbf{c}$ of $\mathbf{Z}^n$ is *primitive* if there is a basis $\mathbf{c} = \mathbf{c}_1, \mathbf{c}_2, \ldots, \mathbf{c}_n$ and have the

COROLLARY. *A necessary and sufficient condition that* $\mathbf{c} = (c_1, \ldots, c_n) \in \mathbf{Z}^n$ *be primitive is that*

$$\text{g.c.d.}\,(c_1, \ldots, c_n) = 1$$

All the above is a special case of what was proved in Chapter 7 since $\mathbf{Z}$ is a principal ideal domain. We now proceed to the main result of this section.

THEOREM 2.1. *Let P be a finite set of primes $p \neq \infty$ and for each $p \in P$ let*

$$\mathbf{c}_1^{(p)}, \ldots, \mathbf{c}_n^{(p)} \tag{2.1}$$

be a basis of $\mathbf{Z}_p^n$ with

$$\det(\mathbf{c}_1^{(p)}, \ldots, \mathbf{c}_n^{(p)}) = 1. \tag{2.2}$$

Then for any $\varepsilon > 0$ there is a basis $\mathbf{c}_1, \ldots, \mathbf{c}_n$ of $\mathbf{Z}^n$ with

$$\det(\mathbf{c}_1, \ldots, \mathbf{c}_n) = 1 \tag{2.3}$$

such that

$$\|\mathbf{c}_j - \mathbf{c}_j^{(p)}\|_p < \varepsilon \qquad (1 \leqslant j \leqslant n, \quad \text{all} \quad p \in P). \tag{2.4}$$

Here we have used the notation

$$\|\mathbf{b}\|_p = \max|b_j|_p$$

for $\mathbf{b} = (b_1, \ldots, b_n) \in \mathbf{Q}_p^n$.

The proof depends on the following simple lemma.

LEMMA 2.2. *Let $K > 1$ be an integer and let $m_k^{(p)} \in \mathbf{Z}_p$ be given for $p \in P$ and $1 \leqslant k \leqslant K$ such that*

$$\max_k |m_k^{(p)}|_p = 1. \tag{2.5}$$

Then there are $m_k \in \mathbf{Z}$ such that

$$|m_k - m_k^{(p)}|_p < \varepsilon \qquad (1 \leqslant k \leqslant K, \quad \text{all} \quad p \in P) \tag{2.6}$$

and

$$\text{g.c.d.}(m_1, \ldots, m_K) = 1. \tag{2.7}$$

Proof of Lemma. By the Chinese Remainder Theorem we can pick $m_1 \neq 0$ such that (2.6) holds for $k = 1$. Let P^* be the set of primes p^* which divide m_1 but are not in P. Applying the Chinese Remainder Theorem again we can find an $m_2 \in \mathbf{Z}$ such that (2.6) holds for $k = 2$ and, in addition

$$p^* \nmid m_2 \qquad (\text{all } p^* \in P^*). \tag{2.8}$$

Now for $k > 2$ choose $m_k \in \mathbf{Z}$ to satisfy (2.6). The condition (2.5) ensures that $\text{g.c.d.}(m_1, \ldots, m_K)$ is not divisible by any $p \in P$: and (2.8) ensures that it is not divisible by any other prime.

Proof of Theorem. We shall construct $\mathbf{c}_1, \ldots, \mathbf{c}_n$ in order using the following induction hypothesis:

H_J: there is a basis

$$\mathbf{c}_1, \ldots, \mathbf{c}_{J-1}, \mathbf{b}_J, \ldots, \mathbf{b}_n \tag{2.9}$$

of $\mathbf{Z}^n$ such that (2.4) is true for all $j < J$.

The hypothesis H_0 is vacuous. Suppose H_J has been proved. Then we can express the $\mathbf{c}_J^{(p)}$ $(p \in P)$ in terms of the basis (2.9):

$$\mathbf{c}_J^{(p)} = l_1^{(p)}\mathbf{c}_1 + \ldots + l_{J-1}^{(p)}\mathbf{c}_{J-1} + m_J^{(p)}\mathbf{b}_J + \ldots + m_n^{(p)}\mathbf{b}_n \tag{2.10}$$

where

$$l_1^{(p)}, \ldots, l_{J-1}^{(p)}, m_J^{(p)}, \ldots, m_n^{(p)} \in \mathbf{Z}_p. \tag{2.11}$$

Further, by the p-adic analogue of Lemma 2.1 (= Lemma 2.1 of Chapter 8) we have

$$\max_{J \leqslant j \leqslant n} |m_j^{(p)}|_p = 1. \tag{2.12}$$

By the Chinese Remainder Theorem we can find $l_j \in \mathbf{Z}$ such that

$$|l_j - l_j^{(p)}|_p < \varepsilon \qquad (j < J, \quad \text{all} \quad p \in P). \tag{2.13}$$

Suppose now first that

$$J \neq n. \tag{2.14}$$

Then by (2.12) and by Lemma 2.2 with $K = n - J + 1$ we can find $m_j \in \mathbf{Z}$ $(J \leqslant j \leqslant n)$ such that

$$|m_j - m_j^{(p)}|_p < \varepsilon \qquad (J \leqslant j \leqslant n, \quad \text{all} \quad p \in P) \tag{2.15}$$

and

$$\text{g.c.d.}(m_J, \ldots, m_n) = 1. \tag{2.16}$$

Put

$$\mathbf{c}_J = l_1\mathbf{c}_1 + \ldots + l_{J-1}\mathbf{c}_{J-1} + m_J\mathbf{b}_J + \ldots + m_n\mathbf{b}_n. \tag{2.17}$$

Then (2.4) holds for $j = J$ by (2.10), (2.13) and (2.15). Further, by (2.16) and Lemma† 2.1 the vectors $\mathbf{c}_1, \ldots, \mathbf{c}_J$ can be extended to a basis $\mathbf{c}_1, \ldots, \mathbf{c}_J$, $\mathbf{b}'_{J+1}, \ldots, \mathbf{b}'_n$ (say) of $\mathbf{Z}^n$. We thus have statement H_{J+1}.

Now suppose that

$$J = n.$$

Here we have to use the hypothesis (2.2) of the enunciation. It implies that

$$m_n^{(p)} = m_n \text{ (say)} = \pm 1,$$

where m_n is independent of p. Now choose l_j $(j < n)$ to satisfy (2.13) and the proof runs as before to give the required basis $\mathbf{c}_1, \ldots, \mathbf{c}_n$.

3. THE FINITENESS THEOREM

In this section we prove Theorem 1.1 which asserts the finiteness of the number of equivalence classes of integral quadratic forms in n variables with given determinant $d \neq 0$. This follows almost immediately from

† The equivalence of (i) and (ii) when (2.9) is taken as basis for $\mathbf{Z}^n$. Alternatively one can go back to Theorem 3.5 of Chapter 7 and take

$$\mathbf{e}_j = \mathbf{c}_j \quad (j < J), \qquad \mathbf{e}_j = \mathbf{b}_j \quad (j \geqslant J).$$

LEMMA 3.1. *For each $n \geqslant 1$ there is a constant C_n with the following property:*

Let $f(\mathbf{x})$ be a regular integral quadratic form in n variables. Then there is an integral vector $\mathbf{a}$ *such that*

$$f(\mathbf{a}) \neq 0 \tag{3.1}$$

and

$$|f(\mathbf{a})| \leqslant C_n |d|^{1/n} \tag{3.2}$$

where

$$d = d(f) \tag{3.3}$$

is the determinant of f.

Here and throughout this section we write $|\ |$ for the absolute value $|\ |_\infty$. The condition (3.1) is vital: it would be easy to prove a result in which (3.1) was replaced by $\mathbf{a} \neq \mathbf{0}$ but that would not suffice for our purposes.

We shall actually prove something more precise than Lemma 3.1 (Lemma 3.2, below) and apply it to the study of some special cases. First, however, we shall show that Lemma 3.1 implies Theorem 1.1.

Proof of Theorem 1.1. We can suppose without loss of generality that the vector $\mathbf{a}$ given by Lemma 3.1 is primitive: for if $\mathbf{a} = c\mathbf{a}'$ for some integer $c > 1$ and integral vector $\mathbf{a}'$, then $|f(\mathbf{a}')| < |f(\mathbf{a})|$. Put

$$h = f(\mathbf{a}) \tag{3.4}$$

so h belongs to a finite set of non-zero integers depending only on m and d by (3.1) and (3.2). Since $\mathbf{a}$ is primitive, by Lemma 2.1, Corollary, there is a basis $\mathbf{a} = \mathbf{a}_1, \mathbf{a}_2, \ldots, \mathbf{a}_n$ of $\mathbf{Z}^n$ and so f is equivalent to a form

$$f^*(\mathbf{x}) = \Sigma f^*_{ij} x_i x_j$$

with

$$f^*_{11} = h.$$

On "completing the square" it follows that

$$hf^*(\mathbf{x}) = (hx_1 + f^*_{12}x_2 + \ldots + f^*_{1n}x_n)^2 + g(x_2, \ldots, x_n) \tag{3.5}$$

for some integral form g in $n - 1$ variables of determinant

$$d(g) = h^{n-2} d. \tag{3.6}$$

Thus we can suppose† that $g(x_2, \ldots, x_n)$ is equivalent to one of a finite set of forms. On applying an integral unimodular transformation to $x_2, \ldots, x_n$ we can suppose, indeed, that $g(x_2, \ldots, x_n)$ itself is one of a finite set. Now, keeping $x_2, \ldots, x_n$ fixed, we can make a substitution

$$x_1 \to x_1 + u_2 x_2 + \ldots + u_n x_n$$

with integral $u_2, \ldots, u_n$ to ensure that

$$|f^*_{ij}| \leqslant |h| \qquad (2 \leqslant j \leqslant n).$$

† Here one uses induction on n.

Hence the right-hand side of (3.5) is one of a finite set of forms depending only on n, d and h. Since the set of possible h is also bounded, it follows that $f^*(\mathbf{x})$ is one of a finite set, as required. This completes the proof of Theorem 1.1.

It remains to prove Lemma 3.1. This is an immediate consequence of

LEMMA 3.2. *For all pairs of integers* $r \geqslant 1$ *and* $s \geqslant 0$ *there is a constant* $\theta = \theta(r, s)$ *with the following property:*

Let $f(\mathbf{x})$ *be a regular integral quadratic form which is real-equivalent to*

$$\xi_1^2 + \dots + \xi_r^2 - \xi_{r+1}^2 - \dots - \xi_{r+s}^2 \qquad (r + s = n). \tag{3.7}$$

Then there is an integral vector $\mathbf{a}$ *such that*

$$0 < f(\mathbf{a}) \leqslant \theta |d|^{1/n}, \tag{3.8}$$

where

$$d = d(f). \tag{3.9}$$

Further we can take θ *defined by*

$$\theta^n = 3^s (4/3)^{n(n-1)/2}. \tag{3.10}$$

Note 1. The Lemma remains true if $f(\mathbf{x})$ is supposed only to have real coefficients. We use integrality only to ensure that M (defined below) is attained. For the general case see Blaney (1948).

Note 2. The examples $2xy$ and $2x^2 + 2xy + 2y^2$ show that the constant (3.10) cannot be improved for $n = 2$. The best values of the constants for $n = 3$ are also known, see Davenport (1949).

Proof. The result is certainly true with $\theta = 1$ when $n = 1$, so we may suppose that $n > 1$ and use induction.

Let M be the least strictly positive value taken by f for integral values of the variables. Then $M = f(\mathbf{a})$ where the integral vector $\mathbf{a}$ is clearly primitive. On taking an equivalent form for f we may thus suppose that

$$M = f(1, 0, \dots, 0).$$

Hence, as before

$$M f(\mathbf{x}) = (M x_1 + f_{12} x_2 + \dots + f_{1n} x_n)^2 + g(x_2, \dots, x_n) \tag{3.11}$$

where

$$d(g) = M^{n-2} d. \tag{3.12}$$

We now have two cases

(i) $r > 1$. Then using induction we can apply the Lemma to g with $(r - 1, s)$ instead of (r, s). There are thus integers $b_2, \dots, b_n$ such that

$$0 < g(b_2, \dots, b_n) \leqslant \theta_1 |M^{n-2} d|^{1/(n-1)} \tag{3.13}$$

where

$$\theta_1 = \theta(r-1, s). \tag{3.14}$$

We can choose an integer b_1 such that

$$|Mb_1 + f_{12}b_2 + \ldots + f_{1n}b_n| \leqslant \tfrac{1}{2}M.$$

Then $f(\mathbf{b}) > 0$, where $\mathbf{b} = (b_1, \ldots, b_n)$ and so by the minimality of M we have

$$f(\mathbf{b}) \geqslant M. \tag{3.15}$$

On the other hand, by (3.11), (3.13) and (3.14) we have

$$Mf(\mathbf{b}) \leqslant \tfrac{1}{4}M^2 + \theta_1|M^{n-2}d|^{1/n-1}. \tag{3.16}$$

On eliminating $f(\mathbf{b})$ between (3.15) and (3.16) one readily obtains

$$M^n \leqslant (4\theta_1/3)^{n-1}|d|$$

and so one can suppose that

$$\theta^n = (4\theta_1/3)^{n-1}. \tag{3.17}$$

(ii) $r = 1$. Then using induction we can apply the Lemma to $-g$ with $(s, 0)$ instead of (r, s). There are thus integers $b_2, \ldots, b_n$ such that

$$0 > g(b_2, \ldots, b_n) \geqslant -\theta_2|M^{n-2}d|^{1/(n-1)} \tag{3.18}$$

where

$$\theta_2 = \theta(s, 0). \tag{3.19}$$

We can now choose an integer b_1 so that

$$\tfrac{1}{2}M \leqslant |Mb_1 + f_{12}b_2 + \ldots + f_{1n}b_n| \leqslant M. \tag{3.20}$$

Then by (3.11), (3.18), (3.20) we have

$$f(\mathbf{b}) < M$$

and so

$$f(\mathbf{b}) \leqslant 0.$$

Arguing as in the previous case we obtain

$$M^n \leqslant (4\theta_2)^{n-1}|d|$$

and so we can suppose that

$$\theta^n = (4\theta_2)^{n-1}. \tag{3.21}$$

Finally (3.10) follows by induction from (3.17) and (3.21). This completes the proof of the Lemma.

COROLLARY 1. *Let $f(\mathbf{x})$ be a real-valued regular form in n variables. Then there is an integral vector* $\mathbf{a} \neq \mathbf{0}$ *such that*

$$|f(\mathbf{a})| \leqslant (4/3)^{(n-1)/2}|d|^{1/n}, \qquad d = d(f).$$

Note. It is not excluded that $f(\mathbf{a}) = 0$.

Proof. Suppose first that f is positive definite. Then the proof of the lemma applies since M is clearly attained and (as remarked already) this is the only place where the integrality hypothesis was used. Since (3.10) with $s = 0$ gives $\theta = (4/3)^{(n-1)/2}$, we have what we required.

In the general case we have

$$f(\mathbf{x}) = \xi_1^2 + \ldots + \xi_r^2 - \xi_{r+1}^2 - \ldots - \xi_{r+s}^2 \qquad (n = r + s)$$

for some real linear forms $\xi_1, \ldots, \xi_n$ in the variables $\mathbf{x}$. Put

$$g(\mathbf{x}) = \xi_1^2 + \ldots + \xi_r^2 + \xi_{r+1}^2 + \ldots + \xi_{r+s}^2.$$

Then g is positive definite,

$$d(g) = |d(f)|$$

and

$$|f(\mathbf{a})| \leqslant g(\mathbf{a})$$

for all real-valued vectors $\mathbf{a}$. The Corollary for f now follows from the corresponding result for the positive definite form g.

COROLLARY 2. *Let f be a positive definite integral form with $d(f) = 1$ and $n \leqslant 5$. Then f is equivalent to*

$$x_1^2 + \ldots + x_n^2.$$

Proof. For by the Lemma there is an integral vector $\mathbf{a}$ such that

$$0 < f(\mathbf{a}) \leqslant (4/3)^{(n-1)/2} \leqslant (4/3)^2 < 2,$$

and so

$$f(\mathbf{a}) = 1.$$

After an integral unimodular transformation we may suppose that $f(1, 0, \ldots, 0) = 1$ and after a further transformation

$$x_1 \to x_1 + u_2 x_2 + \ldots + u_n x_n, \quad x_j \to x_j \qquad (j \neq 1)$$

with integral $u_2, \ldots, u_n$ we have

$$f(\mathbf{x}) = x_1^2 + g(x_2, \ldots, x_n)$$

where g satisfies the hypotheses of the Corollary with $n - 1$ for n. Induction.

We conclude with a methodological remark. In general we are interested in the number of classes in a genus. As we shall prove in the next section forms in the same genus have the same determinant. They are also rationally equivalent and so either all isotropic or all anisotropic. If they are anisotropic then it may be better to use Corollary 1 to Lemma 3.2 rather than the Lemma itself. For example consider forms f in the genus of

$$x^2 + y^2 - 3z^2, \qquad (*)$$

which is anisotropic. By the Corollary there is an integral vector $\mathbf{a}$ with

$$|f(\mathbf{a})| \leqslant (4/3)3^{1/3} < 2$$

and so

$$f(\mathbf{a}) = \pm 1$$

as $f(\mathbf{a}) = 0$ is excluded. Hence f is equivalent to

$$\pm x_1^2 + g(x_2, x_3)$$

for some binary form g: and it is now easy to establish equivalence with (*).

4. GENERA: ELEMENTARY PROPERTIES

We recall that two regular integral quadratic forms f, g are said to be in the same genus if they are equivalent in every $\mathbf{Z}_p$ (including $p = \infty$), say

$$f(\mathbf{x}) = g(\mathbf{T}_p\mathbf{x}) \qquad (4.1)$$

where the matrix $\mathbf{T}_p$ has elements in $\mathbf{Z}_p$ and $\det \mathbf{T}_p$ is a unit of $\mathbf{Z}_p$ for all p. In this section we make some easy deductions from this definition.

LEMMA 4.1. *Two forms in the same genus have the same determinant.*

Note. This will justify us in speaking of the determinant of a genus, meaning the determinant of any form in it.

Proof. With the notation (4.1) we have

$$d(f) = (\det \mathbf{T}_p)^2\, d(g) \qquad (4.2)$$

In particular the rational number $d(f)/d(g)$ is a p-adic unit for all p, so must be ± 1. Finally (4.2) with $p = \infty$ shows that $d(f)$ and $d(g)$ have the same sign.

COROLLARY 1. *The number of equivalence classes in a genus is finite.*

Proof. Follows from the above Lemma and Theorem 1.1.

LEMMA 4.2. *Let P be a finite set of primes $p \neq \infty$ and let f be an integral quadratic form of determinant $d \neq 0$. For $p \in P$ let f_p be a $\mathbf{Z}_p$-integral form of determinant d which is $\mathbf{Z}_p$-equivalent to f. Then there is a form f^* which is properly equivalent to f and whose coefficients are arbitrarily close to those of each of the f_p in the p-adic topology.*

Proof. By hypothesis for each $p \in P$ there is a p-adic integral matrix $\mathbf{T}_p$ such that

$$f_p(\mathbf{x}) = f(\mathbf{T}_p \mathbf{x}). \tag{4.3}$$

Since

$$d(f_p) = d(f) = d \neq 0,$$

we have

$$(\det \mathbf{T}_p)^2 = 1$$

so

$$\det \mathbf{T}_p = \pm 1.$$

By Lemma 3.2, Corollary, of Chapter 8 there is an integral p-adic automorph $\mathbf{S}_p$ of f_p with $\det \mathbf{S}_p = -1$. On replacing $\mathbf{T}_p$ by $\mathbf{S}_p \mathbf{T}_p$ if necessary, we may thus suppose that (4.3) holds with

$$\det \mathbf{T}_p = +1.$$

Then by Theorem 2.1 there is a global integral matrix $\mathbf{T}$ with

$$\det \mathbf{T} = +1$$

which is arbitrarily close p-adically to each of the $\mathbf{T}_p$. Clearly

$$f^*(\mathbf{x}) = f(\mathbf{T}\mathbf{x})$$

will do what is required.

COROLLARY 1. *Let P be as above and let f, g be forms in the same genus. Then there is an f^* properly equivalent to f which is arbitrarily close to g in the p-adic sense for each $p \in P$.*

Note. We can express the conclusion differently: let $M > 1$ be an integer. Then there is an f^* such that the coefficients of f^* and g satisfy

$$f^*_{ij} \equiv g_{ij} \pmod{M}.$$

Proof. By Lemma 4.1 we may take $f_p = g$ for each $p \in P$ in the enunciation of the Lemma.

Finally, as a sort of converse to Lemma 4.2 we have

LEMMA 4.3. *Let* $f(\mathbf{x}), g(\mathbf{x})$ *be integral forms of determinant* $d \neq 0$. *Suppose that they are real-equivalent and that*

$$f_{ij} \equiv g_{ij} \pmod{4d} \qquad (\text{all } i, j).$$

Then f, g *are in the same genus.*

Proof. For then they are $\mathbf{Z}_p$-equivalent for every $p \neq \infty$ by Lemma 5.2 of Chapter 8.

5. EXISTENCE OF GENERA: REPRESENTATIONS

In this section we prove Theorem 1.2 about the existence of genera with prescribed local properties and Theorem 1.3 about representations by the forms of a genus. We then apply Theorem 1.3, as an example, to study the representation of integers as the sum of 2, 3 and 4 squares.

To prove Theorem 1.2 we must prove the following statement for every $n \geqslant 1$:

> θ_n: Let $d \neq 0$ be a given integer and for each $p \neq \infty$ let $f_p(\mathbf{x})$ be a $\mathbf{Z}_p$-integral form in $\mathbf{x} = (x_1, \ldots, x_n)$ with determinant d. Suppose that there exists a rational form $g(\mathbf{x})$ which is $\mathbf{Q}_p$-equivalent to $f_p(\mathbf{x})$ for each p. Then there is an integral form $f(\mathbf{x})$ which is $\mathbf{Z}_p$-equivalent to $f_p(\mathbf{x})$ for each p and $\mathbf{Q}$-equivalent to g.

Note. It is convenient to speak of f_p as given for all p, but by Theorem 3.1, Corollary, of Chapter 8 there is only one $\mathbf{Z}_p$-equivalence class of forms of determinant d if $p \nmid 2d$. Hence the conditions that $f(\mathbf{x})$ be $\mathbf{Z}_p$-equivalent to $f_p(\mathbf{x})$ imposes a restriction on f only for finitely many p.

The statement θ_1 is trivial, since then $f_p(\mathbf{x}) = dx_1^2$ for all p. We may therefore suppose that $n \geqslant 2$ and use induction on n. The key step is the following Lemma, which is rather more general than is strictly required for the proof of Theorem 1.2 since it will also furnish Theorem 1.3.

LEMMA 5.1. *Let* $n \geqslant 2$ *and suppose that* θ_{n-1} *holds. Let* $d, f_p(\mathbf{x}), g(\mathbf{x})$ *be as in the hypotheses of* θ_n *and let* $a \neq 0$ *be a rational integer which is represented primitively by each* f_p *over* $\mathbf{Z}_p$ *and by* $g(\mathbf{x})$ *over* $\mathbf{R} = \mathbf{Z}_\infty$. *Then there exists an* $f(\mathbf{x})$ *which has the properties required by* θ_n *and which represents* a *primitively over* $\mathbf{Z}$.

Note. Once we have completed the proof of Theorem 1.2, the condition that θ_{n-1} holds will become unnecessary.

Proof. On replacing $f_p(\mathbf{x})$ for each p by a $\mathbf{Z}_p$-equivalent form we may suppose without loss of generality that

$$f_p(1, 0, \ldots, 0) = a \qquad (\text{all } p). \tag{5.1}$$

Further, by the Strong Hasse Principle (cf. Theorem 1.1, Corollary 2, of Chapter 6), the form $g(\mathbf{x})$ represents a over $\mathbf{Q}$ and so, on replacing $g(\mathbf{x})$ by a $\mathbf{Q}$-equivalent form, we may suppose that

$$g(1, 0, \ldots, 0) = a. \tag{5.2}$$

On "completing the square" we have

$$af_p(\mathbf{x}) = (ax_1 + b_{2p}x_2 + \ldots + b_{np}x_n)^2 + f_p^*(x_2, \ldots, x_n) \tag{5.3}$$

for some

$$b_{2p}, \ldots, b_{np} \in \mathbf{Z}_p, \tag{5.4}$$

where f_p^* is a $\mathbf{Z}_p$-integral form in $n-1$ variables with determinant

$$d^* = a^{n-2}d. \tag{5.5}$$

Similarly

$$ag(\mathbf{x}) = (ax_1 + c_2x_2 + \ldots + c_nx_n)^2 + g^*(x_2, \ldots, x_n) \tag{5.6}$$

for some

$$c_2, \ldots, c_n \in \mathbf{Q} \tag{5.7}$$

and some rational form g^* in $n-1$ variables.

By (5.3) and (5.7) and since $f_p(\mathbf{x}), g(\mathbf{x})$ are $\mathbf{Q}_p$-equivalent by hypothesis, Witt's Lemma (Theorem 4.1, Corollary 2, of Chapter 2) implies that $f_p^*(x_2, \ldots, x_n)$ and $g^*(x_2, \ldots, x_n)$ are $\mathbf{Q}_p$-equivalent. Hence the f_p^* together with d^* and g^* satisfy the hypotheses of θ_{n-1}, which we are assuming to have been established. It follows that there is a global integral form $f^*(x_2, \ldots, x_n)$ of determinant d^* which is $\mathbf{Z}_p$-equivalent to $f_p^*(x_2, \ldots, x_n)$ for all p and $\mathbf{Q}$-equivalent to g^*.

On replacing f^* by an equivalent form we may suppose by Lemma 4.2 that f^* is arbitrarily close p-adically to f_p^* for all p dividing $2ad$. Further, by the Chinese Remainder Theorem we can find

$$b_2, \ldots, b_n \in \mathbf{Z}$$

which are arbitrarily close p-adically to

$$b_{2p}, \ldots, b_{np}$$

for $p \mid 2ad$. Define a quadratic form $f(\mathbf{x})$ by

$$af(\mathbf{x}) = (ax_1 + b_2x_2 + \ldots + b_nx_n)^2 + f^*(x_2, \ldots, x_n) \tag{5.8}$$

so $f(\mathbf{x})$ has determinant d and is arbitrarily close to $f_p(\mathbf{x})$ for all $p \mid 2ad$. The form $f(\mathbf{x})$ has *a priori* only rational coefficients since it is obtained on dividing an integral form by a. However, the coefficients of $f(\mathbf{x})$ will be integers, if f is made close enough to f_p for the $p|a$; which we shall suppose.

Since f is arbitrarily close to f_p in the p-adic sense for $p|2d$ we can certainly ensure that f and f_p are $\mathbf{Z}_p$-equivalent for those p by Lemma 5.1 (or Lemma 5.2) of Chapter 8. By the *Note* after the enunciation of θ_n, the forms f and f_p are automatically $\mathbf{Z}_p$-equivalent for all other p.

Finally, $f(\mathbf{x})$ is $\mathbf{Q}$-equivalent to $g(\mathbf{x})$ by (5.6), (5.8) and since $f^*(x_2, \ldots, x_n)$ and $g^*(x_2, \ldots, x_n)$ are $\mathbf{Q}$-equivalent.

Proof of Theorem 1.2. We have to prove θ_n for all n. After Lemma 5.1 it will be enough to show that an integer a exists satisfying the conditions of that Lemma.

Let b be any non-zero integer which is represented by g over $\mathbf{Q}$ and let P be the set of primes dividing $2db$. For $p \notin P$ it is trivial that b is primitively represented by f_p over $\mathbf{Z}_p$ [cf. *Note* after enunciation of θ_n]. For $p \in P$ the form f_p certainly represents b over $\mathbf{Q}_p$ because f_p and g are $\mathbf{Q}_p$-equivalent by hypothesis. If

$$b = f(\mathbf{b}_p), \qquad \mathbf{b}_p \in \mathbf{Q}_p^n$$

we may choose $\beta(p) \in \mathbf{Z}$ so that

$$p^{\beta(p)}\mathbf{b}_p$$

is a primitive element of $\mathbf{Z}_p^n$. Then

$$a = b \prod_{p \in P} p^{2\beta(p)}$$

does what is required.

We have now done everything to establish

THEOREM 5.1. *Let $f(\mathbf{x})$ be an integral form in n variables of determinant $d \neq 0$. Let $a \neq 0$ be an integer which is represented by $f(\mathbf{x})$ over $\mathbf{R}$ and primitively represented by $f(\mathbf{x})$ over $\mathbf{Z}_p$ for all $p|2d$ (if $n \geq 3$), all $p|2ad$ (if $n = 2$). Then a is primitively represented over $\mathbf{Z}$ by some form f^* in the same genus as f.*

Note. This clearly implies Theorem 1.3.

Proof. The proof just completed above shows that a is primitively represented by f over $\mathbf{Z}_p$ for all p. We can therefore apply Lemma 5.1 putting $g(\mathbf{x})$ and $f_p(\mathbf{x})$ for all p equal to the $f(\mathbf{x})$ of the present Theorem. Then $f^*(\mathbf{x})$ is the $f(\mathbf{x})$ whose existence is given by Lemma 5.1. This completes the proof.

Because of its usefulness we enunciate the trivial

COROLLARY. *Suppose, further, that f is in a genus of one class. Then we may suppose that $f^* = f$.*

By Lemma 3.2 Corollary 2 the forms $x_1^2 + \ldots + x_n^2$ are in genera of one class for $n \leqslant 5$. The above Corollary together with a simple p-adic study (which we omit) implies the following results:

(i) a positive integer a is represented primitively by $x_1^2 + x_2^2$ if and only if it is not divisible by 4 or by any prime $p \equiv 3 \pmod 4$.

(ii) a positive integer a is represented primitively by $x_1^2 + x_2^2 + x_3^2$ if and only if

$$a \equiv 1, 2, 3, 5, 6 \pmod 8.$$

(iii) a positive integer a is represented primitively by $x_1^2 + x_2^2 + x_3^2 + x_4^2$ if and only if $a \not\equiv 0 \pmod 8$.

Note that for (ii) and (iii) we have to consider only representation over $\mathbf{Z}_2$ but for the binary form $x_1^2 + x_2^2$ we have to consider $\mathbf{Z}_p$ for $p|a$ as well.

6. QUANTITATIVE STUDY OF REPRESENTATIONS

In this section we show how the methods of the preceding sections can be made to yield quantitative results about the numbers of representations of integers by integral forms. We shall not have occasion to use them later, except that we shall refer to them in our survey of analytic methods as applied to quadratic forms [Appendix B].

If $f(\mathbf{x})$ is a regular integral quadratic form we denote by $O^+(f) = O_Z^+(f)$ the group of proper integral automorphs of f: that is the set of integral matrices $\mathbf{T}$ with

$$\det(\mathbf{T}) = +1 \tag{6.1}$$

such that

$$f(\mathbf{Tx}) = f(\mathbf{x}). \tag{6.2}$$

We shall say that two integral vectors $\mathbf{b}, \mathbf{b}^*$ are in the same *orbit* (for $O^+(f)$) if there is a $\mathbf{T} \in O^+(f)$ such that

$$\mathbf{b}^* = \mathbf{Tb}. \tag{6.3}$$

This is clearly an equivalence relation. If $\mathbf{b}, \mathbf{b}^*$ are in the same orbit then clearly

$$f(\mathbf{b}^*) = f(\mathbf{b}): \tag{6.4}$$

but not, of course, *vice-versa*.

LEMMA 6.1. *Let $a \neq 0$ be any integer. Then the set of representations of a fall into finitely many orbits.*

Proof. It is clearly enough to consider primitive representations. Suppose that $\mathbf{b}$ is a primitive integral vector with $f(\mathbf{b}) = a$. The argument used in the proof of Theorem 1.1 near the beginning of Section 3 shows that there exists a basis

$$\mathbf{b} = \mathbf{b}_1, \mathbf{b}_2, \ldots, \mathbf{b}_n \tag{6.5}$$

of $\mathbf{Z}^n$ such that

$$h(\mathbf{x}) \text{ (say)} = f(x_1\mathbf{b}_1 + \ldots + x_n\mathbf{b}_n) \tag{6.6}$$

is one of a finite set of forms: and (on taking $-\mathbf{b}_n$ for $\mathbf{b}_n$ if need be) we may suppose that

$$\det(\mathbf{b}_1, \ldots, \mathbf{b}_n) = +1. \tag{6.7}$$

We shall show that two vectors $\mathbf{b}$ which give rise to the same form $h(\mathbf{x})$ in this way must belong to the same orbit. For suppose that

$$\mathbf{b}^* = \mathbf{b}_1^*, \ldots, \mathbf{b}_n^*$$

is another basis of $\mathbf{Z}^n$ such that

$$f(x_1\mathbf{b}_1^* + \ldots + x_n\mathbf{b}_n^*) = h(\mathbf{x})$$

and

$$\det(\mathbf{b}_1^*, \ldots, \mathbf{b}_n^*) = +1.$$

Then we can define the transformation $\mathbf{T}$ by

$$\mathbf{b}_j^* = \mathbf{T}\mathbf{b}_j \qquad (1 \leqslant j \leqslant n).$$

Clearly

$$f(\mathbf{T}\mathbf{x}) = f(\mathbf{x})$$

and $\mathbf{T}$ is integral with determinant $+1$, so

$$\mathbf{T} \in O^+(f).$$

Further, $\mathbf{b}^* = \mathbf{T}\mathbf{b}$, so $\mathbf{b}$ and $\mathbf{b}^*$ are in the same orbit. This completes the proof.

We now examine (6.6) more closely. If $\mathbf{b} = \mathbf{b}_1$ gives a primitive representation of $a \neq 0$ then, on "completing the square" in familiar fashion we obtain

$$ah(\mathbf{x}) = (ax_1 + c_2x_2 + \ldots + c_nx_n)^2 + g(x_2, \ldots, x_n) \tag{6.8}$$

where $c_2, \ldots, c_n$ are integers and g is a form in $n-1$ variables of determinant $a^{n-2}d$. It is clear that the class of g for proper equivalence is independent of the choice of $\mathbf{b}_2, \ldots, \mathbf{b}_n$ and depends only on $\mathbf{b} = \mathbf{b}_1$. We shall say that the representation of a by $\mathbf{b}$ is *associated with* the class of g. We intend to use this notion to classify and enumerate the representations of a by a form f or, more precisely, by the forms of a genus and will confine attention to the simplest case, namely when the forms are positive definite. The number of representations of a given integer a will then be finite. Before enunciating the main theorem we must introduce some new concepts and notation.

For positive definite forms f we shall denote the order of the finite group $O^+(f)$ by $o^+(f)$: it clearly depends only on the equivalence class of f. In general two forms in the same genus do not have automorphism groups of the same order. An important rôle is played by the *weight*† (German: Mass, sometimes anglicized as *mass* although the German word means "measure") $W(\mathscr{F})$ of the genus $\mathscr{F}$ defined by

$$W(\mathscr{F}) = \sum \frac{1}{o^+(f)} \tag{6.9}$$

where the sum is over representatives f of the proper equivalence classes in the genus.

Now let $\mathscr{F}$ be a given n-dimensional genus of determinant d, let $a \neq 0$ and let $g(\mathbf{x})$ be a given integral $(n-1)$-dimensional form of determinant $a^{n-2}d$. Consider the sets of integers $(c_2, \ldots, c_n) = \mathbf{c}$ such that $h(\mathbf{x})$ given by (6.8) is in $\mathscr{F}$. If we replace $\mathbf{c}$ by any set of integers congruent to them modulo a, then clearly the new h is equivalent to the old h and *a fortiori* is also in $\mathscr{F}$. Let ρ be the number of congruence classes of $\mathbf{c}$ modulo a for which $h \in \mathscr{F}$. Clearly ρ depends only on the equivalence class of g. Indeed ρ depends only on the genus of g since every class in the genus contains a form g^* which is arbitrarily close p-adically to g for all $p|2ad$ (by Lemma 4.2, Corollary 1): then the corresponding form h^* is arbitrarily close to h for these p and so in the same genus (by Lemma 4.3). We can therefore write

$$\rho = \rho(\mathscr{F}, \mathscr{G}), \tag{6.10}$$

where $\mathscr{G}$ is the genus of g. Clearly ρ can be determined by purely local ($=p$-adic) considerations.

Finally, we shall call $1/o^+(f)$ the *weight of a representation* by f. Denote by

$$\sigma(a, \mathscr{F}, \mathscr{G}) \tag{6.11}$$

the sum of the weights of all the representations of a by (representatives of) all classes of the genus $\mathscr{F}$ which are associated with forms in $\mathscr{G}$. We have now:

Lemma 6.2. *With the notation introduced above, we have*

$$\sigma(a, \mathscr{F}, \mathscr{G}) = \rho(\mathscr{F}, \mathscr{G}) W(\mathscr{G}), \tag{6.12}$$

where $W(\mathscr{G})$ is the weight of the genus $\mathscr{G}$.

Proof. Let $\mathbf{b}$ be a representation of a by some $f \in \mathscr{F}$. The *orbit* of $\mathbf{b}$ is the set of distinct vectors of the shape $\mathbf{Tb}$ where $\mathbf{T} \in O^+(f)$. The total number

† Some authors use $\Sigma 1/o(f)$, where the sum is over the wide equivalence classes and $o(f)$ is the total number of automorphs (proper and improper). This is half of (6.9). [*Proof*: exercise for the reader.]

$N(\mathbf{b})$ of elements of the orbit is thus $o^+(f)/o^+(\mathbf{b}, f)$ where $o^+(\mathbf{b}, f)$ is the order of the subgroup $O^+(\mathbf{b}, f)$ of $O^+(f)$ for which $\mathbf{Tb} = \mathbf{b}$ (the *stabilizer* of $\mathbf{b}$). Hence the number $N(\mathbf{b})$ of elements in the orbit is given by

$$N(\mathbf{b}) = \frac{o^+(f)}{o^+(\mathbf{b}, f)}. \tag{6.13}$$

It is convenient to introduce a change of notation. Put

$$n = m + 1 \tag{6.14}$$

and write

$$\mathbf{e}_j = \mathbf{b}_{j+1} \qquad (1 \leqslant j \leqslant m) \tag{6.15}$$

so that $\mathbf{b}, \mathbf{e}_1, \ldots, \mathbf{e}_m$ is a basis of $\mathbf{Z}^{m+1}$ with

$$\det(\mathbf{b}, \mathbf{e}_1, \ldots, \mathbf{e}_m) = +1. \tag{6.16}$$

Write

$$\mathbf{x} = (y, \mathbf{z}) \tag{6.17}$$

where

$$\mathbf{z} = (z_1, \ldots, z_m). \tag{6.18}$$

Then (6.8) becomes

$$ah(y, \mathbf{z}) = (ay + v_1 z_1 + \ldots + v_m z_m)^2 + g(\mathbf{z}) \tag{6.19}$$

where

$$v_j = c_{j+1} \qquad (1 \leqslant j \leqslant m). \tag{6.20}$$

Here $\mathbf{e}_1, \ldots, \mathbf{e}_m$ are to some extent arbitrary and we must investigate how far the shape of (6.19) depends on this choice. Another choice $\mathbf{e}_j^*$ (say) satisfying (6.16) must be of the shape

$$\mathbf{e}_j^* = u_j \mathbf{b} + \sum_k s_{kj} \mathbf{e}_k \qquad (1 \leqslant j \leqslant m) \tag{6.21}$$

where u_j, s_{kj} are rational integers satisfying

$$\det(s_{kj}) = 1. \tag{6.22}$$

If $\mathbf{x}^* = (y^*, \mathbf{z}^*)$ are co-ordinates in $\mathbf{Z}^{m+1}$ to the basis $\mathbf{b}, \mathbf{e}_1^*, \ldots, \mathbf{e}_m^*$ then we have identically

$$y^* \mathbf{b} + \sum_j z_j^* \mathbf{e}_j^* = y\mathbf{b} + \sum_j z_j \mathbf{e}_j. \tag{6.23}$$

Hence

$$y = y^* + \sum_k u_k z_k \tag{6.24}$$

and

$$z_j = \sum_k s_{jk} z_k^* \qquad (1 \leqslant j \leqslant m); \tag{6.25}$$

which we shall write as

$$\mathbf{z} = \mathbf{S}\mathbf{z}^*. \tag{6.26}$$

On substituting these values in (6.19) and distinguishing corresponding entities with respect to the new basis by an asterisk, we obtain

$$g^*(\mathbf{z}^*) = g(\mathbf{S}\mathbf{z}^*) \tag{6.27}$$

and

$$\mathbf{v}^* = a\mathbf{u} + \mathbf{S}'\mathbf{v}, \tag{6.28}$$

where $\mathbf{S}'$ is the transpose of $\mathbf{S}$.

The equation (6.27) confirms, what we knew already, that g^* is properly equivalent to g. In our algorithm we have selected one representative from each equivalence class and so we must have

$$g^* = g, \tag{6.29}$$

and, accordingly,

$$\mathbf{S} \in O^+(g). \tag{6.30}$$

Further, the transformation given by (6.24) and (6.26) is an automorphism of f if and only if $\mathbf{v}^* = \mathbf{v}$. Hence $\mathbf{S} \in O^+(g)$ can be extended by suitable choice of $\mathbf{u}$ to give an element of $O^+(\mathbf{b}, f)$ if and only if

$$\mathbf{S}'\mathbf{v} \equiv \mathbf{v} \pmod{a}. \tag{6.31}$$

Conversely, every element of $O^+(\mathbf{b}, f)$ arises in this way. It follows that the total number of distinct vectors $\mathbf{v}$ (mod a) which can correspond to a given $\mathbf{b}$ is

$$M(\mathbf{b}) = \frac{o^+(g)}{o^+(\mathbf{b}, f)}. \tag{6.32}$$

On eliminating $o^+(\mathbf{b}, f)$ between (6.13) and (6.32) we obtain

$$\begin{aligned}\frac{M(\mathbf{b})}{o^+(g)} &= \frac{1}{o^+(\mathbf{b}, f)} \\ &= \frac{N(\mathbf{b})}{o^+(f)}.\end{aligned} \tag{6.33}$$

We note that the right-hand side is the sum of the weights of all the representations in the orbit of $\mathbf{b}$.

Next we keep g fixed and let f, $\mathbf{b}$ run respectively through representatives of all the classes in the genus $\mathscr{F}$ and through representatives $\mathbf{b}$ of all the orbits of representations of a asssociated with g. By the definition of $\rho(\mathscr{F}, \mathscr{G})$ we have

$$\sum M(\mathbf{b}) = \rho(\mathscr{F}, \mathscr{G})$$

and so

$$\frac{\rho(\mathscr{F}, \mathscr{G})}{o^+(g)} = \sum_{\mathbf{b}} \frac{N(\mathbf{b})}{o^+(f)}$$

is the sum of the weights of the representations in all the orbits associated with g.

Finally, letting g run through representatives of all the classes in $\mathscr{G}$ and summing again, we obtain (6.12) on recalling that

$$W(\mathscr{G}) = \sum \frac{1}{o^+(g)}.$$

This concludes the proof of the Lemma.

One often wishes to use Lemma 6.2 when the genus $\mathscr{F}$ is fixed and the integer a varies. Then the factor $\rho(\mathscr{F}, \mathscr{G})$ in (6.12) is easy to manage but one requires a formula for the weight $W(\mathscr{G})$ of the genus $\mathscr{G}$ which, of course, varies when a varies. There are useful formulae for $W(\mathscr{G})$ but their proof, in general, requires the analytic theory of quadratic forms (see Appendix B).

In the rest of this section we illustrate the use of Lemma 6.2 to discuss the representation of integers as the sum of 2, 3 and 4 squares. The answer in the case of 3 squares must remain imperfect because the formula for the weight of the corresponding binary genus can be proved only by analytic means: but fortunately, and exceptionally, the weights of the ternary genera required in the treatment of sums of four squares can be obtained in a simple elementary manner. For simplicity we confine ourselves in the case of 3 and 4 squares to the representation of squarefree and odd squarefree numbers respectively.

LEMMA 6.3. *A necessary and sufficient condition for an integer $a > 0$ to be representable primitively as the sum of two squares is that it be divisible neither by 4 nor by a prime $p \equiv -1 \pmod 4$. If a satisfies this condition then the number of primitive representations is $2^{\alpha+2}$ where α is the number of distinct odd prime divisors of a.*

Note. The method of proof extends at once to primitive representations by any genus of definite binary forms. The reader should have no difficulty in making the generalization for himself or he can consult any sufficiently elementary textbook. [cf. also Appendix B, Section 2].

Proof. By Lemma 3.2, Corollary 2, the form $f(x_1, x_2) = x_1^2 + x_2^2$ is in a genus $\mathscr{F}$ of one class. Clearly $o^+(f) = 4$. The equation (6.8) takes the shape

$$ah(\mathbf{x}) = (ax_1 + cx_2)^2 + g(x_2)$$

where $h(\mathbf{x})$ is equivalent to $f(\mathbf{x})$ and so $g(x_2) = x_2^2$ on comparing determinants. Hence $o^+(g) = 1$. We have to find the number of solutions modulo a of the congruence

$$c^2 + 1 \equiv 0 \pmod a.$$

Clearly there are no solutions if $4|a$ or if a is divisible by a prime $p \equiv -1 \pmod 4$. If neither of these things happen, the number of solutions of the congruence is 2^α, where α is defined in the enunciation. The truth of the Lemma now follows at once from Lemma 6.2.

LEMMA 6.4. *Let a be a positive square-free integer and suppose that*

$$a \neq 3, \quad a \not\equiv 7 \pmod 8.$$

Then the number $v(a)$ *of representations of a as the sum of three squares satisfies*

$$v(a) = 12 \,.\, 2^m \,.\, N,$$

where m is the number of odd prime divisors of a and N is the number of proper equivalence classes in a certain genus of positive definite binary forms of determinant a.

Note 1. We shall see in Chapter 14 that the number of classes in any genus of primitive binary forms depends only on the determinant, so it is not necessary to specify the genus more precisely.

Note 2. There are corresponding results when a is not square-free but then we have to distinguish between primitive and imprimitive representations and must consider the possibility that the form g is not primitive.

Proof. The genus $\mathscr{F}$ of

$$f(x_1, x_2, x_3) = x_1^2 + x_2^2 + x_3^2$$

contains only one class by Lemma 3.2, Corollary 2.

The only representations of $+1$ by f are $(\pm 1, 0, 0)$, $(0, \pm 1, 0)$, $(0, 0, \pm 1)$. Any element of $O^+(f)$ must permute these and so it is readily verified that

$$o^+(f) = 24. \tag{6.34}$$

Let $g(\mathbf{x})$ be the binary form occurring in (6.19). Since every positive definite integral form of determinant 1 is equivalent to f, the conditions on g are

(i) g has determinant a.

(ii) g is positive definite.

(iii) there exist integers v_1, v_2 such that

$$(v_1 z_1 + v_2 z_2)^2 + g(z_1, z_2) \tag{6.35}$$

is divisible by a, in the sense that the quotient must be an integral form.

Since a is squarefree, condition (iii) is equivalent to the following for each $p|a$:

$(\text{iii})'_p$ there exist integers v_1, v_2 such that (6.35) is divisible by p.

We now look at the information these requirements give us on the Hasse–Minkowski symbols $c_p(g)$. The definiteness implies that

$$c_\infty(g) = +1. \tag{6.36}$$

If p is an odd prime and $p|a$ then the condition (iii)$_p$ is equivalent to the condition that g represent -1, and so

$$c_p(g) = \left(\frac{-1, -a}{p}\right) \qquad p|a, \quad p \neq 2. \tag{6.37}$$

For $p \nmid 2a$ we have automatically for any form of determinant a that

$$c_p(g) = +1 \qquad p \nmid 2a. \tag{6.38}$$

Finally, whether or not a is divisible by 2, the condition (iii)$'_2$ does not give any 2-adic condition on g. On using the product formula for the Hilbert Norm Residue Symbol and for the Hasse–Minkowski symbols we deduce however from (6.36), (6.37), (6.38) that

$$c_2(g) = \prod_{\substack{p \neq 2 \\ \text{inc.}\infty}} c_p(g) = \prod_{p \neq 2, \infty} \left(\frac{-1, -a}{p}\right) = \left(\frac{-1, -a}{2}\right)\left(\frac{-1, -a}{\infty}\right). \tag{6.39}$$

When $a \equiv 7 \pmod 8$, so $-a \in (\mathbf{Q}_2^*)^2$, then $c_2(g) = +1$ for any form of determinant a, but the right-hand side of (6.39) is $(+1)(-1) = -1$: a contradiction. This is fortunate, since it is trivial that no $a \equiv 7 \pmod 8$ can be the sum of three squares. Otherwise, by Theorem 1.2, there exist forms g satisfying (6.36) and (6.37). They belong to precisely one genus $\mathscr{G}$ since the fact that a is squarefree together with (6.36)–(6.39) determines g up to $\mathbf{Z}_p$-equivalence for every p.

Since $a \neq 3$, the only elements of $O^+(g)$ are $\pm \mathbf{I}$, so $o^+(g) = 2$. Hence

$$W(\mathscr{G}) = N/2 \tag{6.40}$$

where $W(\mathscr{G})$ is the weight of the genus and N is the number of proper classes in it.

If p is odd and $p|a$ the condition that (6.34) be divisible by p gives precisely two possibilities for (v_1, v_2) modulo p. On the other hand, if $2|a$ then (v_1, v_2) is uniquely determined modulo 2. Hence in the notation of Lemma 5.2 we have

$$\rho(\mathscr{F}, \mathscr{G}) = 2^m. \tag{6.41}$$

Finally, by (6.34) every representation has weight 1/24 and so

$$\sigma(a, \mathscr{F}, \mathscr{G}) = \nu(a)/24. \tag{6.42}$$

The Lemma now follows from Lemma 6.2 and the values (6.40), (6.41) and (6.42).

LEMMA 6.5. *Let $a > 0$ be odd and squarefree. Then the number of primitive representations of a by*

$$f(x_1, x_2, x_3, x_4) = x_1^2 + x_2^2 + x_3^2 + x_4^2 \tag{6.43}$$

is

$$8 \prod_{p|a} (p + 1), \tag{6.44}$$

where p, as always, denotes a prime.

Proof. As before by Lemma 3.2, Corollary 2, the form (6.3) is in a genus of 1 class. It is easily verified that

$$o^+(f) = 2^6 . 3. \tag{6.45}$$

In the notation of (6.12) it is readily verified that only one genus $\mathscr{G}$ can occur with representations of a by f, namely that of forms g which are (i) positive definite, (ii) classically integral of determinant $d(g) = a^2$ and (iii) $\mathbf{Z}_p$-equivalent to

$$-y_1^2 + 2ay_2y_3 \tag{6.46}$$

for every odd prime p. In the notation of Lemma 6.2, clearly $\rho(\mathscr{F}, \mathscr{G}) = 2^\alpha$, where α is the number of prime factors of a. Hence by Lemma 6.2 the proof of Lemma 6.5 will be complete once we have proved

LEMMA 6.6. *Let $a > 0$ be odd and squarefree and let $\mathscr{G}$ be the genus of positive definite forms g with $d(g) = a^2$ which are $\mathbf{Z}_p$-equivalent to* (6.46) *for all odd primes p. Then*

$$W(\mathscr{G}) = \frac{1}{24} \prod_{p|a} \left(\frac{p+1}{2} \right). \tag{6.47}$$

Proof.† It is convenient to work in terms of automorphs of forms rather than proper automorphs. This causes no trouble because $o(h) = 2o^+(h)$ for every ternary form h since it possesses the improper automorph $-\mathbf{I}$.

We represent the integral equivalence class (in the wide sense) of a form g in the genus $\mathscr{G}$ by a lattice Γ in a quadratic space V, ϕ, say

$$\phi(y_1\mathbf{e}_1 + y_2\mathbf{e}_2 + y_3\mathbf{e}_3) = g(y_1, y_2, y_3) \tag{6.48}$$

for a basis $\mathbf{e}_1, \mathbf{e}_2, \mathbf{e}_3$ of Γ. By hypothesis we may choose the basis $\mathbf{e}_1, \mathbf{e}_2, \mathbf{e}_3$ so that

$$g(y_1, y_2, y_3) \equiv -y_1^2 + 2ay_1y_2 \pmod{a^2}. \tag{6.49}$$

† Although the proof is intended to be self-contained, ideas underlying it may be clearer after Sections 1 and 2 of Chapter 11.

There are precisely 2^α sublattices Δ of Γ on which the value of ϕ is divisible by a^2 namely those consisting of the $y_1\mathbf{e}_1 + y_2\mathbf{e}_2 + y_3\mathbf{e}_3$ such that

$$\left.\begin{aligned} y_1 &\equiv 0 \pmod{a} \\ y_2 &\equiv 0 \pmod{a_2} \\ y_3 &= 0 \pmod{a_3} \end{aligned}\right\} \tag{6.50}$$

where $a = a_2 a_3$ is one of the 2^α factorizations of a into positive integers a_2, a_3.

Let Δ be one of the lattices just described. Then $a^{-1}\phi$ induces on Δ a positive definite classically integral form of determinant 1, and so equivalent to $z_1^2 + z_2^2 + z_3^2$. We therefore introduce a new quadratic space W, ψ and a lattice Λ in W with basis $\mathbf{b}_1, \mathbf{b}_2, \mathbf{b}_3$ such that

$$\psi(z_1\mathbf{b}_1 + z_2\mathbf{b}_2 + z_3\mathbf{b}_3) = z_1^2 + z_2^2 + z_3^2. \tag{6.51}$$

We have, then, just shown that there is an isometry

$$\varepsilon\colon\ V, \phi \to W, \psi \tag{6.52}$$

such that

$$\varepsilon\Delta = a\Lambda. \tag{6.53}$$

The isometry ε is determined by Δ only up to an autometry of Λ and so there are precisely $o(\Lambda)$ distinct ε for which (6.56) holds.

There are, however, 2^α possible choices of Δ. It follows that there are precisely‡ $2^\alpha o(\Lambda)$ autometries (6.52) such that

$$a\Lambda \subset \varepsilon\Gamma \subset \Lambda. \tag{6.54}$$

Although the ε are distinct, the images $\varepsilon\Gamma$ are not necessarily so. In fact $\varepsilon_1\Gamma = \varepsilon_2\Gamma$ precisely when $\varepsilon_2 = \varepsilon_1\beta$ for some autometry β of Γ. Let $N(\Gamma)$ be the number of lattices Θ with

$$a\Lambda \subset \Theta \subset \Lambda \tag{6.55}$$

which are isometric† to Γ. Then we have shown that

$$N(\Gamma) = 2^\alpha o(\Lambda)/o(\Gamma). \tag{6.56}$$

Let M denote the number of lattices Θ satisfying (6.55) on which ψ induces a form of the genus $\mathscr{G}$. Then

$$M = \Sigma N(\Gamma), \tag{6.57}$$

† That is, such that the quadratic form induced by ψ on $\Theta \subset W$ is equivalent to that induced by ϕ on $\Gamma \subset V$.

‡ Here $o(\Lambda)$ is the order of $O(\Lambda)$.

the sum being over the isometry classes of lattices Γ in the genus $\mathscr{G}$. On comparing (6.56), (6.57) and using $o(\Lambda) = 2o^+(\Lambda)$, $o(\Gamma) = 2\rho^+(\Gamma)$ (cf. beginning of proof) we have

$$W(\mathscr{G}) = \sum \frac{1}{o^+(\Gamma)} = \frac{M}{2^\alpha o^+(\Lambda)}. \tag{6.58}$$

Since $o^+(\Lambda) = 24$, it remains only to find M.

Any lattice Θ of index a^2 in Λ which satisfies (6.55) determines and is determined by the cyclic subgroup Θ (mod $a\Lambda$) in Λ(mod $a\Lambda$) [recall that a is supposed squarefree]. It is easy to check that such a Θ is in the genus $\mathscr{G}$ precisely when $\phi(\mathbf{b}) \equiv 0 \pmod{a}$ for all $\mathbf{b} \in \Theta$. It is now left to the reader to check that the number M of such Θ is

$$M = \prod_{p|a} (p + 1). \tag{6.59}$$

[By the Chinese Remainder Theorem one may reduce the problem to the case $a = p$.] The truth of the Lemma now follows at once from (6.58) and (6.59) on recalling that α is the number of $p|a$.

COROLLARY . *The number of classes in a genus may be arbitrarily large.*

Proof. For the number of classes in $\mathscr{G}$ is at least equal to $W(\mathscr{G})$ by the definition (6.9) of the weight of a genus. On the other hand, the right-hand side of (6.47) tends to infinity as $a \to \infty$.

We conclude this section by noting that there is a corresponding theory about the representations of forms of lower dimension by forms. When the forms are positive definite the theory is no deeper but distinctly more complicated. A good reference is Pall (1959). When the forms are indefinite there are further complications since the groups of automorphisms will in general be infinite and one has to deal in group indices. Here the modern treatment in terms of "adeles" (which we will discuss briefly in Appendix B) gives perhaps a better formulation, for which see e.g. Weil (1962).

7. SEMI-EQUIVALENCE

Let P be a finite set of primes $p \neq \infty$. We denote by $\mathbf{Z}^{(P)}$ the set of rational numbers which are integral for all $p \in P$, so $\mathbf{Z}^{(P)}$ is a ring. In this section we prove the following Theorem which clearly implies Theorem 1.4.

THEOREM 7.1. *Let f, g be regular forms with rational coefficients. Suppose that they are equivalent over $\mathbf{Q}$ and over $\mathbf{Z}_p$ for each $p \in P$. Then they are equivalent over $\mathbf{Z}^{(P)}$.*

We shall deduce this theorem from a theorem about approximation to elements of the proper automorphism group $O^+(f)$. This is the simplest application of an approach which we shall study in much greater depth in Chapter 11. At the end of the section we indicate briefly an alternative approach.

We must first introduce some notation. Let f be a regular quadratic form whose coefficients we momentarily suppose to lie in some general field k (of characteristic $\neq 2$). Let A be a ring and suppose that there is some field K which contains both k and A. Then we shall denote by $O_A(f)$ the group of A-automorphs of f and by $O_A^+(f)$ the proper automorphs. The elements of $O_A^+(f)$ are matrices with elements in A. Consequently any topology on A (such as the p-adic topology on $\mathbf{Q}_p$ or $\mathbf{Z}_p$) will induce a topology on $O_A^+(f)$. We can now enunciate:

THEOREM 7.2 ("Weak Approximation Theorem"). *Let f be a regular quadratic form with rational coefficients and let P be a finite set of primes $p \neq \infty$. For each $p \in P$ let $\mathbf{T}_p \in O_{\mathbf{Q}_p}^+(f)$ be given. Then there exists a $\mathbf{T} \in O_{\mathbf{Q}}^+(f)$ which is arbitrarily close to each $\mathbf{T}_p$ in the appropriate p-adic topology.*

Note. We could permit $\infty \in P$ with a similar enunciation and proof but will not need this.

Proof of Theorem 7.2. Let n, as usual, be the dimension of f. By Lemma 5.4 of Chapter 2, each $\mathbf{T}_p$ can be written as the product of symmetries, which we shall denote by $\tau(\mathbf{u})$ instead of $\tau_{\mathbf{u}}$ for notational convenience. Here $\mathbf{u}$ may be any element of $\mathbf{Q}_p^n$ for which $f(\mathbf{u}) \neq 0$. Let r_p be the number of symmetries in the expression for $\mathbf{T}_p$. Then r_p is even since $\mathbf{T}_p \in O_{\mathbf{Q}_p}^+(f)$ and $\det(\tau(\mathbf{u})) = -1$. By adding "dummies" of the shape $\tau(\mathbf{v})\tau(\mathbf{v})$ $(=1)$ to the expressions for $\mathbf{T}_p$ we may thus suppose that all the expressions are of the same length, r, say. That is

$$\mathbf{T}_p = \tau(\mathbf{u}_{1p}) \dots \tau(\mathbf{u}_{rp}),$$

where

$$\mathbf{u}_{jp} \in \mathbf{Q}_p^n \qquad (1 \leqslant j \leqslant r)$$

and

$$f(\mathbf{u}_{jp}) \neq 0.$$

We shall approximate to these $\mathbf{T}_p$ by

$$\mathbf{T} = \tau(\mathbf{u}_1) \dots \tau(\mathbf{u}_r)$$

where $\mathbf{u}_j \in \mathbf{Q}^n$ $(1 \leqslant j \leqslant r)$. By the Weak Approximation Theorem for elements of $\mathbf{Q}$ (Lemma 3.2 of Chapter 3) we can choose the $\mathbf{u}_j$ to be arbitrarily close to the $\mathbf{u}_{jp}$ simultaneously for all $p \in P$. This does what is required, since formula (4.3) of Chapter 2 shows that $\tau(\mathbf{v}) \in O_{\mathbf{Q}_p}^+(f)$ is a continuous function of $\mathbf{v} \in \mathbf{Q}_p^n$ for $f(\mathbf{v}) \neq 0$.

Proof of Theorem 7.1. By hypothesis there is an invertible matrix $\mathbf{S}_0$ with rational coefficients and matrices $\mathbf{S}_p$ $(p \in P)$ with elements in $\mathbf{Z}_p$ and $|\det \mathbf{S}_p|_p = 1$ such that

$$g(\mathbf{S}_0\mathbf{x}) = f(\mathbf{x}) \tag{7.1}$$

$$g(\mathbf{S}_p\mathbf{x}) = f(\mathbf{x}). \tag{7.2}$$

Hence $\mathbf{T}_p$ (say) $= \mathbf{S}_p^{-1}\mathbf{S}_0 \in O_{\mathbf{Q}_p}(f)$. Further, by Lemma 3.2 of Chapter 8 the form f has an improper $\mathbf{Z}_p$-automorph $\mathbf{R}_p$ (say) and on taking $\mathbf{R}_p\mathbf{S}_p$ instead of $\mathbf{S}_p$ if necessary we may suppose that each $\mathbf{T}_p$ is proper:

$$\mathbf{T}_p \in O^+_{\mathbf{Q}_p}(f).$$

Let $\mathbf{T}$ be the rational automorph constructed by Theorem 7.2 and put

$$\mathbf{S} = \mathbf{S}_0\mathbf{T}^{-1}. \tag{7.3}$$

Then $\mathbf{S}$ has rational elements and is arbitrarily close simultaneously to the $\mathbf{S}_p$. In particular we can ensure that $\mathbf{S}$ and $\mathbf{S}^{-1}$ have elements in $\mathbf{Z}^{(P)}$, as re-required.

In conclusion we note a rather different approach to Theorem 7.1 due to Siegel (1941), which uses a formula of Cayley.

Let $\mathbf{F}$ be the square symmetric matrix belonging to a form f. Then $\mathbf{T}$ is an automorph when

$$\mathbf{T}'\mathbf{F}\mathbf{T} = \mathbf{F} \tag{7.4}$$

where $\mathbf{T}'$ is the transpose of $\mathbf{T}$. Suppose that

$$\det(\mathbf{I} + \mathbf{T}) \neq 0, \tag{7.5}$$

where $\mathbf{I}$ is the unit matrix, and put

$$\mathbf{L} = (\mathbf{T} - \mathbf{I})(\mathbf{I} + \mathbf{T})^{-1} \tag{7.6}$$

so

$$\mathbf{T} = (\mathbf{I} + \mathbf{L})(\mathbf{I} - \mathbf{L})^{-1}. \tag{7.7}$$

Then it is readily verified that (7.4) is equivalent to

$$\mathbf{L}'\mathbf{F} + \mathbf{F}\mathbf{L} = \mathbf{0}. \tag{7.8}$$

The point is that (7.8) is a collection of linear conditions on the coefficients of $\mathbf{L}$.

Now suppose that we are given $\mathbf{T}_p$ as in Theorem 7.2. If

$$\det(\mathbf{I} + \mathbf{T}_p) \neq 0 \tag{7.9}$$

then we can define $\mathbf{L}_p$ by (7.6) and it satisfies (7.8). We can now find a global solution $\mathbf{L}$ of (7.8) and this, in turn, gives a global $\mathbf{T}$ which satisfies the conclusions of Theorem 7.2.

As a proof of Theorem 7.2, this breaks down when (7.9) fails. To obtain a proof of Theorem 7.1 Siegel notes that in (7.1) and (7.2) we may replace $\mathbf{S}_0$ and $\mathbf{S}_p$ by $\mathbf{S}_0\mathbf{V}_0$ and $\mathbf{S}_p\mathbf{V}_p$ where $\mathbf{V}_0, \mathbf{V}_p$ are respectively $\mathbf{Q}$- and $\mathbf{Z}_p$-automorphs of f. He is then able to show, by a simple device which we shall not reproduce, that $\mathbf{V}_0$ and the $\mathbf{V}_p$ can be chosen in such a way that (7.9) holds for the new $\mathbf{T}_p = \mathbf{S}_p^{-1}\mathbf{S}_0$.

8. REPRESENTATION BY INDIVIDUAL FORMS

The results obtained earlier in this Chapter have been about representations by genera and it was pointed out that they did not imply results about representations by individual forms except, of course, when they are in genera of one class. The main objective of this section is to prove Theorem 1.5 which states that for indefinite forms in $n \geqslant 4$ variables the local conditions do suffice to ensure global solvability, and indeed enables us to approximate to local integral solutions by a global solution. The proof is fairly elaborate and will use much of what we have already proved. Before embarking on it we give a simple result of independent interest which applies to all integral forms, whether definite or indefinite, in any number of variables.

THEOREM 8.1. *Let f be a regular integral form and let $M > 1$ be any integer. Then there is an integer m prime to M with the following property:*

Let $a \in \mathbf{Z}$ and suppose that a is represented by f over every $\mathbf{Z}_p$ (including $p = \infty$). Then am^2 is represented by f over $\mathbf{Z}$.

Proof. Let $f_1, \ldots, f_J$ be representatives of all the equivalence classes in the genus of f. By Theorem 1.4 each f_j is $\mathbf{Z}^{(P)}$-equivalent to f, where the set P of primes may be chosen to include all the $p|M$. This implies that there is an m_j prime to M and a matrix $\mathbf{T}_j$ with elements in $\mathbf{Z}$ such that

$$m_j^2 f_j(\mathbf{x}) = f(\mathbf{T}_j\mathbf{x}). \tag{8.1}$$

We put

$$m = m_1 m_2 \ldots m_J. \tag{8.2}$$

By Theorem 1.3 the integer a is represented over $\mathbf{Z}$ by some form of the genus of f and so

$$a = f_j(\mathbf{b}) \tag{8.3}$$

for some j and some $\mathbf{b} \in \mathbf{Z}^n$. Then

$$m_j^2 a = f(\mathbf{T}_j\mathbf{b}) \tag{8.4}$$

by (8.1): so $m_j^2 a$ and *a fortiori* $m^2 a$ is represented by f, as required.

Proof of Theorem 1.5. We recall that we are given an indefinite integral form f in $n \geqslant 4$ variables, an integer $a \neq 0$ and, for every p, a vector

$$\mathbf{b}_p \in \mathbf{Z}_p^n \tag{8.5}$$

such that

$$f(\mathbf{b}_p) = a. \tag{8.6}$$

We have to find a vector $\mathbf{b} \in \mathbf{Z}^n$ such that

$$f(\mathbf{b}) = a \tag{8.7}$$

and which is arbitrarily close p-adically to $\mathbf{b}_p$ for all $p \neq \infty$ in a given finite set P.

We may suppose, by considering a/c^2 with suitable integer c if necessary instead of a, that all the $\mathbf{b}_p$ are primitive. Further by increasing the original P if necessary, we may suppose that it includes every prime dividing $2ad(f)$.

Let $\varepsilon > 0$ be given. By Lemma 9.1 of Chapter 2, there is an $\mathbf{r} \in \mathbf{Q}^n$ such that

$$f(\mathbf{r}) = a$$

and

$$\|\mathbf{r} - \mathbf{b}_p\|_p < \varepsilon \qquad (\text{all } p \in P).$$

Here, as usual, we have written

$$\|\mathbf{r}\|_p = \max |r_j|_p.$$

There are coprime integers l, c such that

$$l\mathbf{r} = c\mathbf{s}$$

where $\mathbf{s} \in \mathbf{Z}^n$ is primitive and, by the primitivity of the $\mathbf{b}_p$, we have

$$|c|_p = |l|_p = 1 \qquad (\text{all } p \in P).$$

Further

$$f(\mathbf{s}) = (l^2/c^2) f(\mathbf{r}) = l^2 a/c^2,$$

and so $c^2 = 1$ because P contains all the $p|a$.

After an integral unimodular transformation we may suppose that

$$c\mathbf{s} = \pm\mathbf{s} = \mathbf{e}_1 = (1, 0, \ldots, 0).$$

In this new situation we have

$$f(1, 0, \ldots, 0) = al^2 \tag{8.8}$$

where

$$|l|_p = 1 \qquad (\text{all } p \in P) \tag{8.9}$$

and we have to find an integral $\mathbf{b}$ with

$$f(\mathbf{b}) = a \tag{8.10}$$

and

$$\|\mathbf{b} - l^{-1}\mathbf{e}_1\|_p < \varepsilon \qquad (\text{all } p \in P). \tag{8.11}$$

After a further integral unimodular transformation of the variables $x_2, \ldots, x_n$, keeping x_1 fixed, we may suppose that

$$f_{1j} = 0 \qquad (j > 2)$$

and so

$$f(x_1, \ldots, x_n) = al^2x_1^2 + 2f_{12}x_1x_2 + f(0, x_2, \ldots, x_n). \tag{8.12}$$

We shall consider only those representations of a for which x_2 is divisible by al^2 and write

$$x_2 = al^2y, \tag{8.13}$$

$$z = x_1 + f_{12}y. \tag{8.14}$$

Then

$$f(x_1, al^2y, x_3, \ldots, x_n) = al^2z^2 + g(y, x_3, \ldots, x_n) \tag{8.15}$$

for some integral form g with

$$d(g) = al^2\, d(f). \tag{8.16}$$

We now restrict the choice of variables further. Let

$$M = \prod_{p \in P} p^N \tag{8.17}$$

where the positive integer N is chosen so that

$$|M|_p < \varepsilon \qquad (\text{all } p \in P). \tag{8.18}$$

We put

$$y = Mu_2 \tag{8.19}$$

$$x_j = Mu_j \qquad (3 \leqslant j \leqslant n) \tag{8.20}$$

for integers $u_2, \ldots, u_n$. This automatically ensures that

$$|x_j|_p < \varepsilon \qquad (\text{all } p \in P, 2 \leqslant j \leqslant n). \tag{8.21}$$

Hence we have "only" got to find integers $z, u_2, \ldots, u_n$ such that

$$|lz - 1|_p < \varepsilon \qquad (\text{all } p \in P) \tag{8.22}$$

and

$$a(1 - l^2z^2) = M^2g(u_2, \ldots, u_n). \tag{8.23}$$

We now apply Theorem 8.1. This asserts that there is an integer m prime to any given integer, in particular to lM, such that m^2v is represented by g over $\mathbf{Z}$ whenever the integer v is represented over every $\mathbf{Z}_p$ (including $p = \infty$). It will thus be enough to find an integral z such that (8.21) holds and

$$a(1 - l^2z^2) = m^2M^2v, \tag{8.24}$$

where v is represented by g over every $\mathbf{Z}_p$. Since g is a form in $n - 1 \geqslant 3$ variables we need consider only $p = \infty$ and the primes p dividing

$$2d(g) = 2al^2\, d(f). \tag{8.25}$$

We must first ensure that the left-hand side of (8.24) is divisible by m^2M^2 and we do this simply by requiring that

$$lz \equiv 1 \pmod{m^2M^2}. \tag{8.26}$$

This is possible since l was constructed prime to M and m was constructed prime to l. We put

$$1 - lz = m^2M^2w \tag{8.27}$$

so

$$v = a(1 + lz)w \tag{8.28}$$

We now have to impose conditions on z to ensure that v is representable over $\mathbf{Z}_p$ for $p = \infty$ and the p dividing $2d(g)$. By (8.25) there are three cases, which we consider in turn.

(i) $p = \infty$. This is automatic. By (8.24) the signs of a and v are opposite. Since f is indefinite, by hypothesis, equation (8.15) implies that g represents v over $\mathbf{Z}_\infty = \mathbf{R}$.

(ii) $p|2ad(f)$, so $p \in P$ and hence $p^N|M$. It follows from (8.27) that

$$1 + lz \equiv 2 \pmod{p^{2N}}$$

and so

$$1 + lz = 2t_p^2$$

for some p-adic unit r_p. [For $p = 2$ this requires $N \geqslant 2$, which can we suppose]. Hence by (8.28) we must ensure that

$$2aw$$

is represented by g over $\mathbf{Z}_p$. This can be done by imposing the condition that z (and hence w) lies in an arithmetic progression to an appropriate power of p.

(iii) $p|l$. Then $p \nmid 2ad(f)$ and $p \nmid mM$. By (8.24) we thus have to show that a is represented by g over $\mathbf{Z}_p$. In fact a is represented by the form

$$\begin{aligned} h(u_3, \ldots, u_n)\,(\text{say}) &= g(0, u_3, \ldots, u_n) \\ &= f(0, 0, u_3, \ldots, u_n), \end{aligned}$$

as we shall show. By (8.12) we have

$$d(f) \equiv -f_{12}^2\, d(h) \pmod{l},$$

and so $p \nmid 2d(f)$ implies $p \nmid 2d(h)$. Since the form h has dimension $n-2 \geqslant 2$ and $|a|_p = 1$, the required representability of a follows.

There are infinitely many integers z satisfying (8.26) and the conditions imposed on z in (ii) above. Any one of these z will do all that is required. This completes the proof.

NOTES

The results of this chapter generalize to algebraic number fields with comparatively little trouble, cf. O'Meara (1963).

Section 3. The finiteness of the class-number was asserted by Hermite but his proof appears to have been valid only for anisotropic forms, see note on Chapter 13, Section 11. For indefinite ternaries a proof had already been given by Gauss (1801), Sections 274–276. A proof valid in algebraic number fields is in Siegel (1937) Hilfssatz 40. Lemma 3.2 is due to Blaney (1948), who proves it for all regular real forms.

Section 8. A quantitative form of Theorem 1.5 in a rather different shape was obtained by Siegel (1951) using analytic methods. The formulation of the text was discovered independently by Watson (1955) and we follow his proof. Compare also Kneser (1961), which implies Siegel's result. In O'Meara (1963) Theorem 104:3 there is a generalization to algebraic number fields in which indefiniteness may be replaced by isotropy in any localization [see also Chapter 11, Section 8].

Theorem 1.5 breaks down for $n = 3$. There are indefinite ternaries which do not represent every integer which is permitted by congruence conditions [cf. example 23].

Theorem 1.5 will be an essential tool in proving the strong approximation theorem [Chapter 10, Theorem 7.1] and so in showing that for $n \geqslant 3$ an indefinite spinor genus consists of a single class [Theorem 1.4 of Chapter 11]. Conversely, in view of Theorem 7.1 of Chapter 11, Theorem 1.5 follows readily from the theory of spinor genera.

EXAMPLES

1. Find a form $ax^2 + 2bxy + cy^2$ which is properly integrally equivalent to $3x^2 + 2xy + 12y^3$ and for which $b \equiv 0 \pmod{81}$.

2. Show that the integer-valued forms

$$x^2 + xy + 6y^2$$
$$2x^2 + xy + 3y^2$$
$$2x^2 - xy + 3y^2$$

are in the same genus but that no two of them are properly integrally equivalent. Show, further, that every positive-definite integer-valued form with the same determinant as these forms is properly integrally equivalent to one of them.

3. (i) Show that the forms

$$g(x, y) = 3x^2 + 2xy + 23y^2$$
$$h(x, y) = 7x^2 + 6xy + 11y^2$$

are in the same genus but are not equivalent.

(ii) Show that the ternary forms $z^2 - g(x, y)$ and $z^2 - h(x, y)$ are properly integrally equivalent.

[*Hint*. $5^2 - h(1, 1) = 1$. Compare remarks at end of Section 1.]

4. Determine which pairs of the following forms are properly integrally equivalent:

$$x^2 + y^2 - 10z^2$$
$$x^2 - y^2 + 10z^2$$
$$x^2 + 2y^2 - 5z^2$$
$$x^2 - 2y^2 + 5z^2$$
$$-x^2 + 2y^2 + 5z^2.$$

5. Show that $x^2 + y^2 + cz^2$ is in a genus of 1 class for $c = \pm 1, \pm 2, \pm 3$. Deduce necessary and sufficient conditions for an integer m to be primitively represented by each of these forms.

6. Let

$$f_2 = 12x^2 - y^2 + 2xz + 3z^2$$
$$f_3 = -x^2 - y^2 - z^2$$
$$f_5 = 3x^2 - y^2 + 10z^2$$
$$f_7 = 2x^2 - 2xy + 2y^2 + 4z^2.$$

Find a classically integral form $g(x, y, z)$ of determinant 5 which is equivalent to f_p over $\mathbf{Z}_p$ for $p = 2, 3, 5, 7$.

7. (i) Find a classically integral ternary form f which is real-equivalent to $x^2 + y^2 - z^2$ and for which

$$\left.\begin{array}{ll} c_2(f) = c_3(f) = -1 & \\ c_p(f) = +1 & (p \neq 2, 3) \end{array}\right\}*$$

How small can the determinant of f be made?

(ii) Does there exist a classically integral ternary form f which is real-equivalent to $x^2 - y^2 - z^2$ and satisfies (*)?

8. Let $f(\mathbf{x})$ be an integral quadratic form in n variables and let $\mathbf{T}$ be an integral $n \times n$ matrix, $\det(\mathbf{T}) = -1$. Show that $f(\mathbf{x})$ and $f(\mathbf{Tx})$ are in the same genus. Deduce that if f is in a genus consisting of a single proper equivalence class then f has an integral automorph of determinant -1.

9. Let $f(\mathbf{x})$ be an isotropic classically integral form in n variables. Show that f is properly integrally equivalent to a form

$$g(\mathbf{x}) = \Sigma g_{ij} x_i x_j$$

with $g_{1j} = 0$ $(j \neq 2)$. Deduce that $g_{12}^2 | d(f)$.

[*Note.* This can be used to reduce the proof of Theorem 1.1 by induction on n to the consideration of anisotropic forms. For anisotropic forms one may replace the use of Lemma 3.1 by the simpler Lemma 3.2, Corollary 1. For the details see Kneser (1974).]

10. Fill in the details of the following argument to show that there exists a positive definite improperly primitive integral form f with $d(f) = 1$ in n variables precisely when $8|n$.

(i) $c_p(f) = 1$ for $p \neq 2, \infty$ and all integral forms f with $d(f) = 1$.

(ii) $c_\infty(f) = 1$ for all positive definite forms

(iii) Hence $c_2(f) = 1$ for all positive definite integral forms f with $d(f) = 1$.

(iv) A necessary and sufficient condition for the existence of an improperly primitive integral $\mathbf{Z}_2$-form g_2 with $c_2(g_2) = 1, d(g_2) = 1$ is that $8|n$. [*Hint.* See Chapter 8, Example 4.]

(v) If $8|n$, then by Theorem 1.2 there is a global form f with $d(f) = 1$ which is $\mathbf{Z}_2$-equivalent to g_2 and $\mathbf{Z}_p$-equivalent to $x_1^2 + \ldots + x_n^2$ for $p \neq 2$ (including $p = \infty$).

[For existence it is, of course, enough to write down explicitly a form for $n = 8$, cf. Theorem 8.4, Corollary 2, of Chapter 11, but there is a form in 16 variables with the required properties which is not the direct sum of two copies of that in 8 variables. See also Serre (1970) and Appendix A, Example 1.]

11. In this example f, g, h with or without suffices denote classically integral forms of determinant ± 1. The dimensions of f, g, h are respectively $n, n-2$ and 8.

(i) Let $f(\mathbf{x})$ be indefinite and properly primitive of dimension n. Show that it is equivalent to $y_1^2 - y_2^2 + g(\mathbf{z})$ for some g. [*Hint.* cf. example 9.]

(ii) Let f_1, f_2 both be indefinite and properly primitive. Show that if they are **R**-equivalent then they are **Z**-equivalent.

(iii) Let $f(\mathbf{x})$ be indefinite and improperly primitive. Show that it is equivalent to $2y_1y_2 + g(\mathbf{z})$, for some g.

(iv) Let $g_1(\mathbf{z})$ and $g_2(\mathbf{z})$ be improperly primitive. If $y_1^2 - y_2^2 + g_1(\mathbf{z})$ and $y_1^2 - y_2^2 + g_2(\mathbf{z})$ are equivalent, show that $2y_1y_2 + g_1(\mathbf{z})$ and $2y_1y_2 + g_2(\mathbf{z})$ are equivalent.

(v) Let f_1, f_2 both be indefinite and improperly primitive. Show that if they are **R**-equivalent then they are **Z**-equivalent.

(vi) Let f be indefinite and improperly primitive. Show that either $+f$ or $-f$ is equivalent to a sum of copies of the forms $2x_1x_2$ and $h(x_1, \ldots, x_8)$, where h is positive definite [cf. example 10].

[Serre (1964) or see Serre (1970). The proof of (iv) is rather difficult. It permits the reduction of (v) to (ii)].

12. Let p be an odd prime and let V, ϕ be a 2-dimensional regular quadratic space.

(i) Let $\Lambda \subset V$ be a lattice such that the quadratic form induced by ϕ on Λ is primitive integral and has determinant dp^2 for some integer $d \neq 0$. Show that there is a unique lattice $\Gamma \supset \Lambda$ such that ϕ induces on Γ a primitive integral form of determinant d.

(ii) Let Γ be a lattice such that ϕ induces on Γ a primitive integral form of determinant d where d is some integer. Let N be the number of lattices $\Lambda \subset \Gamma$ on which ϕ induces a primitive integral form of determinant dp^2. Show that

$$N = p - (-d/p)$$

where $(-d/p)$ is the quadratic residue symbol [$=0$ if $p \mid d$, $= +1$ if $-d \equiv b^2 \pmod{p}$ for some $b \not\equiv 0$ and $= -1$ otherwise].

(iii) Suppose $d > 3$. Let $h(d), h(dp^2)$ be respectively the number of classes of primitive integral binary forms of determinant d and dp^2.

Show that

$$h(dp^2) = Nh(d),$$

where N is given above.

[*Hint*. A primitive integral form of determinant $d > 3$ has only the automorphs $\pm\mathbf{I}$.]

(iv) Making use of the theory of the automorphs of binary forms [Chapter 13, Section 3] find and prove the appropriate generalization of (iii) to the cases $d = 3, d = 1$ and $d < 0$. [*Note*. The case when $-d$ is a square requires special treatment.]

13. (i) Let $d \in \mathbf{Z}, d \equiv 3 \pmod 4$ and let $h(d), h'(d)$ be respectively the number of classes (for proper integral equivalence) of properly and of improperly primitive integral forms of determinant d. Show that $h(d) = h'(d)$ except when $d \equiv 3 \pmod 8$ and one of the two following conditions holds:

(a) $d > 3$,

(b) $d < 0$ and all integral solutions (t, u) of $t^2 + du^2 = 4$ have t even.

Show that $h(d) = 3h'(d)$ in cases (a), (b).

(ii) Find and prove the appropriate generalization of the previous question to $p = 2$, taking care of the distinction between properly and improperly primitive forms.

[This again presupposes the theory of automorphs of binary forms. For crib to (i) see B. W. Jones (1950), Theorem 82.]

14. Let $G = O_{\mathbf{Z}}^{+}(f)$ be the group of integral automorphs with determinant $+1$ of the positive definite integral binary quadratic form $f(x_1, x_2)$. Show that G consists only of $\pm\mathbf{I}$ except when f is integrally equivalent to a multiple of either $f_4 = x_1^2 + x_2^2$ or $f_6 = 2(x_1^2 + x_1x_2 + x_2^2)$. Show further that the order of G for f_4 and f_6 is respectively 4 and 6.

[*Hint:* consider the set of integral vectors $\mathbf{b} \neq \mathbf{0}$ such that $f(\mathbf{b})$ takes its minimum value.]

15. Let $f(x, y, z) \in \mathbf{Q}[x, y, z]$ be a positive definite quadratic form.

(i) Show that a proper integral automorph $\mathbf{T}$ has order 1, 2, 3, 4 or 6. [*Hint*: Chapter 2, example 9.]

(ii) If f has a proper integral automorph of order 3 show that f is properly integrally equivalent to a form

$$lx^2 + m(y^2 + yz + z^2)$$

or

$$l(3x + y - z)^2 + m(y^2 + yz + z^2)$$

for some $l > 0, m > 0$. Deduce that f also has a proper integral automorph of order 2.

[*Hint:* without loss of generality $\mathbf{Te}_1 = \mathbf{e}_1$. Complete the square].

(iii) If f has a proper integral automorph other than the identity show that it is properly integrally equivalent to a form

$$ax^2 + by^2 + cz^2 + lyz$$

or

$$ax^2 + by^2 + cz^2 + axy + lyz.$$

[*Hint:* see previous hint].

(iv) Show that the order of the group of proper integral automorphs of f is one of 1, 2, 4, 6, 8, 12, 24 and that all of these orders can occur for suitable forms f [cf. Chapter 6, example 7. For a fuller discussion see Mennicke (1967)].

16. Let $\mathbf{x} = (x, y, z)$ and put

$$f(\mathbf{x}) = yz + zx + xy.$$

(i) Let $\sigma_1, \sigma_2, \sigma_3$ be the symmetries in $(0, 1, 1), (1, 0, 1)$ and $(1, 1, 0)$ respectively. Show that

$$\sigma_j \in O_{\mathbf{Z}}(f) \qquad (j = 1, 2, 3).$$

(ii) Let $\mathbf{a} \in \mathbf{R}^3$ satisfy $f(\mathbf{a}) > 0$ and suppose that at least one of a_1, a_2, a_3 is strictly positive and at least one is strictly negative. Show that there is a $j = 1, 2, 3$ such that

$$\mathbf{b} = \sigma_j \mathbf{a}$$

satisfies

$$\max_i |b_i| < \max_i |a_i| \qquad (i = 1, 2, 3).$$

(iii) Let G_1 be the group generated by $\sigma_1, \sigma_2, \sigma_3$ and $\pm \mathbf{I}$. Let $\mathbf{a} \in \mathbf{Z}^3$, $f(\mathbf{a}) > 0$. Show that there is a $\sigma \in G_1$ such that

$$\mathbf{c} = \sigma \mathbf{a}$$

satisfies

$$c_i \geqslant 0 \qquad (i = 1, 2, 3).$$

(iv) Let G_2 be the group generated by the elements of G_1 together with the permutations of x, y, z. Suppose that $\mathbf{a} \in \mathbf{Z}^3$, $f(\mathbf{a}) = 1$. Show that $\mathbf{a} = \sigma(1, 1, 0)$ for some $\sigma \in G_2$.

(v) If $\mathbf{x} = (x, y, z)$ is normal to $(1, 1, 0)$ show that $f(\mathbf{x}) = -(y + z)^2 - z^2$. Deduce that the group G_3 consisting of the $\sigma \in O_{\mathbf{Z}}(f)$ with $\sigma(1, 1, 0) = (1, 1, 0)$ is finite and contained in G_2.

(vi) Deduce that $O_{\mathbf{Z}}(f) = G_2$.

(vii) Find a (possibly redundant) set of generators for $O^+_{\mathbf{Z}}(f)$.

[*Note.* Compare Chapter 13, Sections 6–8.]

17. Let

$$f(\mathbf{x}) = x_0^2 - x_1^2 - x_2^2 - x_3^2$$

and let τ be the symmetry in $(1, 1, 1, 1)$.

(i) If $f(\mathbf{a}) > 0$ where $\mathbf{a} = (a_0, a_1, a_2, a_3)$ and $a_j \geqslant 0$ $(0 \leqslant j \leqslant 3)$ show that $\mathbf{b} = \tau\mathbf{a}$ has $|b_0| < |a_0|$.

(ii) Deduce that $O_{\mathbf{Z}}(f)$ is generated by τ together with the $x_j \to \pm x_j$ and the permutations of x_1, x_2, x_3.

[*Hint*. Example 15. *Crib:* Wall (1964).]

18. Let $3 < m \leqslant 9$ and put

$$f(\mathbf{x}) = x_0^2 - x_1^2 - \ldots - x_m^2.$$

Denote by τ the symmetry in $(1, 1, 1, 1, 0, \ldots, 0)$. Show that $O_{\mathbf{Z}}(f)$ is generated by τ together with the $x_j \to \pm x_j$ and the permutations of $x_1, \ldots, x_m$.

[*Hint*. If $b_0, b_1, \ldots, b_m \in \mathbf{Z}$ with

$$b_1 \geqslant b_2 \geqslant \ldots \geqslant b_m \geqslant 0, b_1^2 + \ldots + b_m^2 = b_0^2 + 1,$$

show that

$$b_0 < b_1 + b_2 + b_3 < 2b_0.$$

Now follow Example 16. *Crib*: Wall (1964). See also Vinberg (1972, 1972a) and Meyer (1977).]

19. Determine $O_{\mathbf{Z}}(f)$ when (i) $f(\mathbf{x}) = x_0^2 - x_1^2 - x_2^2$ and (ii) $f(\mathbf{x}) = x_1^2 + x_2^2 - x_3^2 - x_4^2$.

[*Crib*: Wall (1964).]

20. Let Λ be the lattice with basis $\mathbf{e}_1, \mathbf{e}_2, \mathbf{e}_3, \mathbf{e}_4$ and let $\Gamma_1, \Gamma_2, \Gamma_3, \Delta_1, \Delta_2, \Delta_3$ be the lattices generated by the points of Λ and the pairs of points specified below:

$$\Gamma_1 : \tfrac{1}{2}\mathbf{e}_1, \tfrac{1}{2}\mathbf{e}_2$$
$$\Gamma_2 : \tfrac{1}{2}\mathbf{e}_3, \tfrac{1}{2}\mathbf{e}_4$$
$$\Gamma_3 : \tfrac{1}{2}(\mathbf{e}_1 + \mathbf{e}_3), \tfrac{1}{2}(\mathbf{e}_2 + \mathbf{e}_4)$$
$$\Delta_1 : \tfrac{1}{2}\mathbf{e}_1, \tfrac{1}{2}\mathbf{e}_3$$
$$\Delta_2 : \tfrac{1}{2}\mathbf{e}_2, \tfrac{1}{2}\mathbf{e}_4$$
$$\Delta_3 : \tfrac{1}{2}(\mathbf{e}_1 + \mathbf{e}_2), \tfrac{1}{2}(\mathbf{e}_3 + \mathbf{e}_4).$$

Show that

$$\sum_j M(\mathbf{x}, \Gamma_j) = \sum_j M(\mathbf{x}, \Delta_j)$$

where

$$M(\mathbf{x}, \Lambda) = \begin{cases} 1 & \text{if } \mathbf{x} \in \Lambda \\ 0 & \text{otherwise.} \end{cases}$$

[*Note*: easy but needed below.]

21. Let

$$\begin{aligned} f_1(\mathbf{x}) &= x_1^2 + 2x_2^2 + 4x_3^2 + 8x_4^2 + x_1x_4 - x_2x_3 \\ f_2(\mathbf{x}) &= x_1^2 + x_2^2 + 8x_3^2 + 8x_4^2 + x_1x_4 - x_2x_3 \\ f_3(\mathbf{x}) &= 3(x_1^2 + x_2^2 + x_3^2 + x_4^2) - 2x_1x_3 - 2x_2x_4 + x_1x_4 + x_2x_3 \end{aligned}$$

and denote by $N_j(a)$ the number of integer representations of a by f_j. Show for all a that

$$2N_1(a) = N_2(a) + N_3(a).$$

[*Hint*. In the notation of the previous example define a function ϕ by

$$\phi(y_1\mathbf{e}_1 + \ldots + y_4\mathbf{e}_4) = 4y_1^2 + 4y_2^2 + 8y_3^2 + 8y_4^2 + 2y_1y_4 - 2y_2y_3$$

and consider the quadratic forms induced on the Γ_j and Δ_j. Use basis $\frac{1}{2}(\mathbf{e}_1 \pm \mathbf{e}_3)$, $\frac{1}{2}(\mathbf{e}_2 \pm \mathbf{e}_4)$ for Γ_3 and similarly for Δ_3.]

22. Let $N_j(a)$ denote the number of integral representations of a by f_j where

$$\begin{aligned} f_1 &= x_1^2 + x_1x_2 + 3x_2^2 + x_3^2 + x_3x_4 + 3x_4^2 \\ f_2 &= 2(x_1^2 + x_2^2 + x_3^2 + x_4^2) + 2x_1x_3 + x_1x_4 + x_2x_3 - 2x_2x_4 \\ f_3 &= x_1^2 + 4(x_2^2 + x_3^2 + x_4^2) + x_1x_3 + 4x_2x_3 + 3x_2x_4 + 7x_3x_4. \end{aligned}$$

Show for all a that

$$3N_1(a) = N_2(a) + 2N_3(a).$$

[*Hint*. Apply example 20, cf. example 21.]

[*Note*. For the general theory of such identities in the context of modular forms see Appendix B and for these identities in particular see Hecke (1940) Section 11, Beispiel 2 and Beispiel 4. For a much more elaborate and deep application of the above method, see Kneser (1967a).]

23. Let $m \equiv \pm 3 \pmod 8$. Show that m^2 is represented primitively by $x^2 - 2y^2 + 64z^2$ over every $\mathbf{Z}_p$ but not over $\mathbf{Z}$.

[*Hint*. For $x, y \in \mathbf{Z}$, every $p \equiv \pm 3 \pmod 8$ divides $x^2 - 2y^2$ to an even power. Consider $x^2 - 2y^2 = (m + 8z)(m - 8z)$.]

[*Note*. Cf. Chapter 11, Lemma 7.1 and Examples 7, 8.]

CHAPTER 10

The Spin and Orthogonal Groups

1. INTRODUCTION

The object of this Chapter is primarily utilitarian. In Chapter 9 we proved a "Weak Approximation Theorem" (Theorem 7.2). This deals with the rational automorphs of a regular quadratic form f and asserts that if $\mathbf{T}_p$ are any given proper automorphs of f defined over $\mathbf{Q}_p$, where p runs through a finite set P, then there is a proper rational automorph $\mathbf{T}$ of f which simultaneously approximates to each of the $\mathbf{T}_p$ in the appropriate p-adic sense. The theorem says nothing about the p-adic behaviour at the $p \notin P$. In the next Chapter we shall require a "strong theorem" which, under appropriate conditions, guarantees the existence of a $\mathbf{T}$ which not merely approximates to the $\mathbf{T}_p$ but is p-adically integral at all the $p \notin P$. The relationship between the "Weak Approximation Theorem" which we have already and the "Strong Approximation Theorem" which we are aiming at is similar to that between the two approximation theorems discussed in Section 3 of Chapter 3.

It turns out, as we shall see, that there are very good reasons why there can be no strong approximation theorem of the type suggested without supplementary conditions on the $\mathbf{T}_p$. This is bound up with the fact that the proper orthogonal group O^+ is not simply-connected (in the topological sense if the field we are using is $\mathbf{R}$, otherwise in an appropriate algebraic generalization which we do not need to discuss). When the dimension n is greater than 2, however, there is a simply connected group, the "*spin group*" Spin, which gives a "double covering" of O^+. There is a simple strong approximation theorem in the spin group. In this chapter we shall introduce Spin, prove the strong approximation theorem on it and investigate the consequences for O^+.

As motivation, we shall now describe the spin group for the classical case of the ordinary orthogonal group in 3-dimensional real space $\mathbf{R}^3$. The idea goes back to William Hamilton and uses his famous invention of *quaternions*. This is a noncommutative associative algebra over the reals with basis $1, i, j, k$ where

$$i^2 = j^2 = k^2 = -1$$

$$ij = k, \qquad jk = i, \qquad ki = j.$$

The general quaternion is

$$\mathbf{u} = u_0 + u_1 i + u_2 j + u_3 k \qquad (u_0, u_1, u_2, u_3 \in \mathbf{R}).$$

The "*conjugate*" of $\mathbf{u}$ is

$$\mathbf{u}' = u_0 - u_1 i - u_2 j - u_3 k,$$

so

$$(\mathbf{uv})' = \mathbf{v}'\mathbf{u}'$$

for any two quaternions $\mathbf{u}, \mathbf{v}$. Further,

$$N(\mathbf{u}) \text{ (say)} = \mathbf{u}'\mathbf{u} = u_0^2 + u_1^2 + u_2^2 + u_3^2 \in \mathbf{R}.$$

We call this the "*norm*" of $\mathbf{u}$. Clearly

$$N(\mathbf{uv}) = N(\mathbf{u})N(\mathbf{v}).$$

We identify $\mathbf{R}^3$ with the "pure" quaternions

$$\mathbf{x} = x_1 i + x_2 j + x_3 k:$$

they are characterized by the equation $\mathbf{x}' = -\mathbf{x}$. If $\mathbf{x}$ is pure and $\mathbf{u}$ is any quaternion, then it is easily verified that

$$\mathbf{y} \text{ (say)} = \mathbf{uxu}'$$

is pure and that

$$N(\mathbf{y}) = N(\mathbf{x})\{N(\mathbf{u})\}^2.$$

In particular, if

$$N(\mathbf{u}) = 1 \tag{1.1}$$

then the transformation $\mathbf{x} \to \mathbf{y}$, which we denote by $\sigma(\mathbf{u})$, is in the orthogonal group. It can be shown to be proper†. Further,

$$\sigma(\mathbf{uv}) = \sigma(\mathbf{u})\sigma(\mathbf{v}).$$

Define Spin to be the group (under multiplication) of the $\mathbf{u}$ with $N(\mathbf{u}) = 1$. We therefore have a homomorphism from Spin to O^+.

It can be shown that the homomorphism from Spin to O^+ is onto. The kernel consists only of $\mathbf{u} = \pm 1$, so topologically we have a double cover. Finally Spin, considered as a topological manifold, is just a 3-sphere, so is simply connected. This shows (though we shall not use the fact) that there if no group G related to Spin as Spin is related to O^+. More precisely, let G

† For $\det(\sigma(\mathbf{u}))$ is a continuous function of $\mathbf{u}$ on (1.1).

be a group which is also a connected real manifold. Then any surjective group homomorphism $G \to \text{Spin}$ with finite kernel which is continuous in the real topology is necessarily an isomorphism.

2. THE CLIFFORD ALGEBRA

By an *algebra* A over a field k (a *k-algebra*) we shall mean a finite-dimensional k-vector space endowed with a binary operation ("product") which determines an element $\mathbf{uv} \in A$ for any ordered pair $\mathbf{u}, \mathbf{v}$ of elements of A. This product is required to be associative:

$$(\mathbf{uv})\mathbf{w} = \mathbf{u}(\mathbf{vw}) \qquad (=\mathbf{uvw}), \tag{2.1}$$

and k-linear:

$$(l_1\mathbf{u}_1 + l_2\mathbf{u}_2)\mathbf{v} = l_1\mathbf{u}_1\mathbf{v} + l_2\mathbf{u}_2\mathbf{v} \qquad (l_1, l_2 \in k, \qquad \mathbf{u}_1, \mathbf{u}_2, \mathbf{v} \in A) \tag{2.2}$$

$$\mathbf{u}(l_1\mathbf{v}_1 + l_2\mathbf{v}_2) = l_1\mathbf{uv}_1 + l_2\mathbf{uv}_2 \qquad (l_1, l_2 \in k, \qquad \mathbf{u}, \mathbf{v}_1, \mathbf{v}_2 \in A). \tag{2.3}$$

We also suppose that there is a unit $\mathbf{e} \in A$ such that

$$\mathbf{eu} = \mathbf{ue} = \mathbf{u} \qquad (\text{all } \mathbf{u} \in A) \tag{2.4}$$

Then the set of $l\mathbf{e}$ $(l \in k)$ form a subalgebra of A which is naturally isomorphic to k. We shall normally make this identification and write l instead of $l\mathbf{e}$ (so 1 instead of $\mathbf{e}$).

We shall require a general result about the existences of inverses in algebras.

LEMMA 2.1. *Let* $\mathbf{u}$ *be an element of a k-algebra A. Suppose that there is either a solution* $\mathbf{v}$ *of*

$$\mathbf{uv} = 1 \tag{2.5}$$

or a solution $\mathbf{w}$ *of*

$$\mathbf{wu} = 1. \tag{2.6}$$

Then there is a solution of each equation and

$$\mathbf{v} = \mathbf{w}. \tag{2.7}$$

Further, the solutions $\mathbf{v}, \mathbf{w}$ *are unique.*

Note. When it exists, we shall write $\mathbf{u}^{-1}$ for the common value of $\mathbf{v}$ and $\mathbf{w}$ and call it the *inverse* of $\mathbf{u}$.

Proof. Suppose that $\mathbf{v}$ exists. The correspondence

$$\mathbf{x} \to \mathbf{xu} \qquad (\mathbf{x} \in A) \tag{2.8}$$

is a linear map of A (considered as a k-vector space) into itself. Suppose that $\mathbf{y}$ is in the kernel. Then

$$0 = (\mathbf{yu})\mathbf{v} = \mathbf{y}(\mathbf{uv}) = \mathbf{y}.$$

Hence (2.8) is non-singular and so, since A is finite-dimensional by hypothesis, there is a $\mathbf{w} \in A$ satisfying (2.6). Similarly, if $\mathbf{w}$ exists, then $\mathbf{v}$ exists.

Further,

$$\mathbf{w} = \mathbf{w}(\mathbf{u}\mathbf{v}) = (\mathbf{w}\mathbf{u})\mathbf{v} = \mathbf{v}.$$

Hence any value of $\mathbf{v}$ is equal to any value of $\mathbf{w}$, so both $\mathbf{v}$ and $\mathbf{w}$ must be unique. This concludes the proof.

We now set ourselves in the situation discussed in Chapter 2. It was shown by Clifford that the rôle of the quaternion algebra with respect to the ordinary orthogonal group which was sketched in Section 1 can be generalized to any quadratic space over a field k of characteristic $\neq 2$. We enunciate its principal properties in

THEOREM 2.1. *Let V, ϕ be a quadratic space of dimension n over a field k of characteristic $\neq 2$. Then there is an algebra $C(V)$ (the* "Clifford algebra *of V, ϕ") over k which contains V as a linear subspace and such that*

(i) *$C(V)$ has dimension 2^n.*

(ii) *$C(V)$ is generated by V. More precisely, it is spanned (as a k-vector space) by 1 and the products*

$$\mathbf{x}_1\mathbf{x}_2 \ldots \mathbf{x}_r \qquad (r > 0, \mathbf{x}_j \in V \quad (l \leqslant j \leqslant r)).$$

(iii)
$$\mathbf{x}\mathbf{x} = \phi(\mathbf{x}) \qquad (\text{all } \mathbf{x} \in V). \tag{2.9}$$

Further, these properties determine $C(V)$ uniquely. More precisely, if $C'(V)$ is another algebra containing V which satisfies (i), (ii), (iii) *then there is an isomorphism between $C(V)$ and $C'(V)$ which respects the k-algebra structure and leaves the elements of V fixed.*

Note. We do not require ϕ to be regular, though that is the case of interest.

Proof. We first find some properties which $C(V)$ must have if it exists and show that they determine its structure completely. We then verify tediously that an algebra with this structure exists.

First, if $\mathbf{x}, \mathbf{y}$ are any elements of V, property (iii) implies that

$$\begin{aligned} \mathbf{x}\mathbf{y} + \mathbf{y}\mathbf{x} &= (\mathbf{x} + \mathbf{y})(\mathbf{x} + \mathbf{y}) - \mathbf{x}\mathbf{x} - \mathbf{y}\mathbf{y} \\ &= \phi(\mathbf{x} + \mathbf{y}) - \phi(\mathbf{x}) - \phi(\mathbf{y}) \\ &= 2\phi(\mathbf{x}, \mathbf{y}), \end{aligned} \tag{2.10}$$

where $\phi(\mathbf{x}, \mathbf{y})$ is the bilinear form associated with ϕ.

Next, let $\mathbf{e}_1, \ldots, \mathbf{e}_n$ be any normal basis for V, that is

$$\phi(\mathbf{e}_i, \mathbf{e}_j) = 0 \qquad (i \neq j).$$

By (2.10) we must have

$$\mathbf{e}_i\mathbf{e}_j + \mathbf{e}_j\mathbf{e}_i = 0 \qquad (i \neq j). \tag{2.11}$$

Also, by (2.9)

$$\mathbf{e}_i\mathbf{e}_i = \phi(\mathbf{e}_i) \in k. \tag{2.12}$$

Let J be any subset of $\{1, 2, \ldots, n\}$ arranged in ascending order, say

$$j_1 < j_2 < \ldots < j_r,$$

where $r \leqslant n$. Put

$$\mathbf{e}(J) = \mathbf{e}(j_1)\mathbf{e}(j_2)\ldots\mathbf{e}(j_r), \tag{2.13}$$

where here, and occasionally later, we write

$$\mathbf{e}(j) = \mathbf{e}_j \tag*{(2.13 bis)}$$

for notational convenience. We supplement (2.13) by

$$\mathbf{e}(E) = 1, \tag{2.14}$$

where E (for this proof only) denotes the empty set. It follows from (ii) and (2.12) that the $\mathbf{e}(J)$ span $C(V)$, considered as a k-vector space. Hence by (i) they are linearly independent and form a basis.

Further, if I, J are two subsets of $\{1, \ldots, n\}$ then (2.11), (2.12) imply that

$$\mathbf{e}(I)\mathbf{e}(J) = l(I, J)\mathbf{e}(K), \tag{2.15}$$

where K is the set of indices which occur in precisely one of I, J and where

$$l(I, J) = \{\prod_{\substack{i\in I\\ j\in J\\ i>j}} (-1)\}\{\prod_{i\in I\cap J} \phi(\mathbf{e}_i)\}. \tag{2.16}$$

The precise value of $l(I, J)$ is unimportant: what is important is that if J_1, J_2, J_3 are any three subsets of $\{1, \ldots, n\}$ then

$$\{\mathbf{e}(J_1)\,\mathbf{e}(J_2)\}\,\mathbf{e}(J_3) = \mathbf{e}(J_1)\,\{\mathbf{e}(J_2)\,\mathbf{e}(J_3)\} \tag{2.17}$$

when the products are computed according to the rules (2.15) and (2.16). The verification is left to the reader.

We are now in a position to construct $C(V)$. We take a vector space of dimension 2^n with a basis which we shall denote by $\mathbf{e}(J)$, where J runs through the subsets of $\{1, \ldots, n\}$. We now *define* a product on $C(V)$ by (2.15) and (2.16). By k-linearity the definition extends to the whole of $C(V)$. This product is associative since, as already remarked, (2.17) is a consequence of (2.11) and (2.12). Next, we identify V with a subspace of $C(V)$ by putting

$$\mathbf{e}_i = \mathbf{e}(\{i\}) \qquad (1 \leqslant i \leqslant n)$$

where $\{i\}$ is the set whose only member is i. It then follows from (2.15) and (2.16) that

$$\mathbf{e}_i\mathbf{e}_i = \phi(\mathbf{e}_i) \qquad \mathbf{e}_i\mathbf{e}_j + \mathbf{e}_j\mathbf{e}_i = 0 \qquad (i \neq j). \tag{2.18}$$

Finally, if $\mathbf{x} = \Sigma x_i\mathbf{e}_i$ is any element of V, it follows from (2.14) and linearity that

$$\mathbf{x}\mathbf{x} = \Sigma x_i^2\phi(\mathbf{e}_i) = \phi(\mathbf{x}).$$

The algebra $C(V)$ just constructed thus enjoys the properties (i), (ii), (iii). The truth of the final sentence of the enunciation is also clear, so this completes the proof of the Theorem.

The autometry

$$\mathbf{x} \to -\mathbf{x} \tag{2.19}$$

of V induces an isomorphism of order 2 of $C(V)$ with itself. We denote by $C_0(V)$ the elements $\mathbf{u}$ of $C(V)$ which are fixed under this automorphism and by $C_1(V)$ those $\mathbf{u}$ which are mapped into $-\mathbf{u}$. It follows from Lemma 4.1 that $C_0(V)$, $C_1(V)$ have as bases the $\mathbf{e}(J)$ for which the number of elements of J is respectively even and odd. Clearly $C(V)$ is the direct sum of $C_0(V)$ and $C_1(V)$ and each has dimension 2^{n-1}.

LEMMA 2.2.

$$C_i(V)\, C_j(V) \subset C_m(V)$$

where i, j, $m = 0$ *or* 1 *and* $m \equiv i + j \pmod 2$. *Further, if* $\mathbf{u} \in C_i(V)$ *and* $\mathbf{u}^{-1}$ *exists then* $\mathbf{u}^{-1} \in C_i(V)$.

Proof. By the left-hand side is meant of course the set of $\mathbf{uv}$ with $\mathbf{u} \in C_i(V)$, $\mathbf{v} \in C_j(V)$. The first sentence now follows from consideration of the action of (2.19). The last sentence follows similarly on noting that $\mathbf{u}^{-1}$ if it exists is unique (Lemma 4.1). [In technical parlance, $C(V)$ has a graded algebra structure.]

COROLLARY. $C_0(V)$ *is a subalgebra of* $C(V)$.

Note. We shall call it the "*even Clifford algebra*".

In the next section we shall need

LEMMA 2.3. *Suppose that* V, ϕ *is regular. An element of* $C_0(V)$ *which commutes with every element of* V *must lie in* k.

Proof. If the number of elements of J is even, it is readily verified that

$$\mathbf{e}_i\mathbf{e}(J) = \begin{cases} +\mathbf{e}(J)\mathbf{e}_i & \text{if } i \notin J \\ -\mathbf{e}(J)\mathbf{e}_i & \text{if } i \in J. \end{cases} \tag{2.20}$$

For fixed i the $\mathbf{e}_i\mathbf{e}(J)$ are linearly independent over k because the $\mathbf{e}(J)$ are independent and $\mathbf{e}_i$ has an inverse in $C(V)$ $[\mathbf{e}_i\mathbf{e}_i = \phi(\mathbf{e}_i) \neq 0]$. It follows that if

$$\Sigma l(J)\, \mathbf{e}(J) \qquad (l(J) \in k)$$

commutes with $\mathbf{e}_i$, then the $l(J)$ with $i \in J$ mush vanish. Since this is true for every i, the truth of the lemma follows.

LEMMA 2.4. *There is a k-linear map* $C(V) \to C(V)$ *denoted by* $\mathbf{u} \to \mathbf{u}'$ *with the following properties:*

(i) $\mathbf{u}' = \mathbf{u}$ *if* $\mathbf{u} \in k$ *or* $\mathbf{u} \in V$;

(ii) $(\mathbf{uv})' = \mathbf{v}'\mathbf{u}'$ *for any* $\mathbf{u}, \mathbf{v} \in C(V)$;

(iii) $(\mathbf{u}')' = \mathbf{u}$ *for all* $\mathbf{u} \in C(V)$.

Note. We shall call ′ the *involution* (or the canonical involution) on $C(V)$.

Proof. It is enough to define ′ on the basis $\mathbf{e}(J)$ of $C(V)$. If $\mathbf{e}(J)$ is given by (2.13) and (2.13 *bis*), then we put

$$(\mathbf{e}(J))' = \mathbf{e}(j_r)\mathbf{e}(j_{r-1})\ldots\mathbf{e}(j_1). \tag{2.21}$$

The required properties are readily verified.

LEMMA 2.5. *Let* $\mathbf{u} \in C(V)$ *be such that*

$$\mathbf{uu}' \in k^*. \tag{2.22}$$

Then

$$\mathbf{u}'\mathbf{u} = \mathbf{uu}' \tag{2.23}$$

and an inverse

$$\mathbf{u}^{-1} = (\mathbf{uu}')^{-1}\mathbf{u}' \tag{2.24}$$

exists.

Proof. Put

$$\mathbf{v} = (\mathbf{uu}')^{-1}\mathbf{u}'. \tag{2.25}$$

Then $\mathbf{v}$ satisfies (2.5). The rest follows from Lemma 2.1.

3. THE SPINOR NORM AND THE SPIN GROUP

From now on, we suppose that V, ϕ is regular. The relevance of the Clifford algebra to the orthogonal group is apparent from the following

LEMMA 3.1. *Suppose that* $\mathbf{u} \in C(V)$ *has an inverse* $\mathbf{u}^{-1}$ *in the sense of the definition after Lemma* 2.1. *Suppose, further, that*

$$\mathbf{uxu}^{-1} \in V \qquad (\textit{all } \mathbf{x} \in V).$$

Then the linear map

$$\mathbf{T_u}: \mathbf{x} \to \mathbf{uxu}^{-1} \tag{3.1}$$

is an autometry of V.

Proof.

$$\begin{aligned}\phi(\mathbf{uxu}^{-1}) &= (\mathbf{uxu}^{-1})(\mathbf{uxu}^{-1})\\ &= \mathbf{u}(\mathbf{xx})\mathbf{u}^{-1}\\ &= (\mathbf{xx})(\mathbf{uu}^{-1})\\ &= \mathbf{xx},\end{aligned}$$

since $\mathbf{xx} = \phi(\mathbf{x}) \in k$ is in the centre of $C(V)$.

A particular case is when $\mathbf{u} = \mathbf{y} \in V$ with $\phi(\mathbf{y}) \neq 0$. Then

$$\mathbf{y}^{-1} = (\phi(\mathbf{y}))^{-1}\mathbf{y}.$$

A straightforward calculation using (2.9), (2.10) gives

$$\begin{aligned}\mathbf{yxy} &= (\mathbf{yx} + \mathbf{xy})\mathbf{y} - \mathbf{xyy} \\ &= 2\phi(\mathbf{x}, \mathbf{y})\mathbf{y} - \phi(\mathbf{y})\mathbf{x}\end{aligned}$$

and so

$$\mathbf{yxy}^{-1} = -\tau_{\mathbf{y}}\mathbf{x}. \tag{3.2}$$

Here $\tau_{\mathbf{y}}$ is the symmetry in $\mathbf{y}$ in the sense of Chapter 2, Section 4. Such symmetries will play a vital part in all that follows: for notational convenience we shall sometimes write

$$\tau(\mathbf{y}) = \tau_{\mathbf{y}}. \tag{3.3}$$

It turns out to be convenient in the application of Lemma 3.1 to restrict attention to the even Clifford algebra $C_0(V)$. We denote by $M_0(V)$ the set of $\mathbf{u} \in C_0(V)$ such that

(i) $\mathbf{u}^{-1}$ exists;

(ii) $\mathbf{uxu}^{-1} \in V$ for all $\mathbf{x} \in V$.

Clearly $M_0(V)$ is a group under multiplication.

THEOREM 3.1. *The map* $\mathbf{u} \to \mathbf{T_u}$ *given by* (3.1) *establishes an isomorphism between* $M_0(V)/k^*$ *and* O^+.

Proof. Clearly $\mathbf{u} \to \mathbf{T_u}$ is a group homomorphism. It follows at once from Lemma 2.3 that the kernel of this homomorphism is k^*. We have to show that the image is precisely O^+.

Suppose first that $\sigma \in O^+(V)$. Then by Section 4 of Chapter 2 we have

$$\sigma = \tau(\mathbf{a}_1)\tau(\mathbf{a}_2)\ldots\tau(\mathbf{a}_r) \tag{3.4}$$

in the notation (3.3) for some even r and $\mathbf{a}_j \in V$ with $\phi(\mathbf{a}_j) \neq 0$. Put

$$\mathbf{u} = \mathbf{a}_1 \ldots \mathbf{a}_r. \tag{3.5}$$

Then

$$\mathbf{u}' = \mathbf{a}_r \ldots \mathbf{a}_1$$

and

$$\mathbf{uu}' = \prod_{1 \leqslant j \leqslant r} \phi(\mathbf{a}_j) \in k^*.$$

It follows that $\mathbf{T_u x} = \sigma\mathbf{x}$ for all $\mathbf{x} \in V$ by repeated application of (3.2). Hence the image of $\mathbf{u} \to \mathbf{T_u}$ certainly contains O^+.

Next suppose that there is a $\mathbf{u} \in M_0(V)$ for which $\mathbf{T_u} \notin O^+$. Then $\mathbf{T_u} \in O^-$ by Lemma 3.1 and so

$$\mathbf{T_u} = \tau(\mathbf{a}_1)\ldots\tau(\mathbf{a}_r)$$

with $\mathbf{a}_j \in V$, $\phi(\mathbf{a}_j) \in k^*$, where r is odd. Put

$$\mathbf{v} = \mathbf{a}_1 \ldots \mathbf{a}_r.$$

By (3.2) we now have

$$\mathbf{vxv}^{-1} = -\mathbf{T_u x} \qquad (\text{all } \mathbf{x} \in V).$$

Then by Lemma 2.2 $\mathbf{w} = \mathbf{u}^{-1}\mathbf{v}$ satisfies

$$\mathbf{w} \in C_1(V); \quad \mathbf{wx} = -\mathbf{xw} \qquad (\text{all } \mathbf{x} \in V). \tag{3.6}$$

It is now easy to show along the lines of the proof of Lemma 2.3 that (3.6) implies $\mathbf{w} = \mathbf{0}$. This contradiction shows that $\mathbf{T_u} \in O^+$ and so completes the proof.

Corollary 1. *If* $\mathbf{u} \in M_0(V)$ *then*

$$\mathbf{u} = \mathbf{a}_1 \dots \mathbf{a}_r$$

with $\mathbf{a}_j \in V$ *and some even r. Further,*

$$\mathbf{uu}' \in k^*.$$

Proof. We have shown that $\sigma = \mathbf{T_u}$ is of the shape (3.4) and so by Lemma 2.3 $\mathbf{u} = l\mathbf{a}_1 \dots \mathbf{a}_r$ with $l \in k^*$. We can take $l\mathbf{a}_1$ instead of $\mathbf{a}_1$.

The case when $n = 3$, where n (as always) is the dimension of V, is of special interest and, in any case, we shall need the details later.

Corollary 2. *Suppose that* $n = 3$. *Then*

$$\mathbf{uu}' \in k$$

for all $\mathbf{u} \in C_0(V)$. *Hence*† $M_0(V)$ *consists precisely of the* $\mathbf{u} \in C_0(V)$ *for which* $\mathbf{uu}' \neq 0$.

Note. The case $f(\mathbf{x}) = x_1^2 + x_2^2 + x_3^2$, $k = \mathbf{R}$ was discussed in Section 1.

Proof. Let $\mathbf{e}_1, \mathbf{e}_2, \mathbf{e}_3$ be a normal basis and put

$$f_j = \phi(\mathbf{e}_j).$$

A k-basis for $C_0(V)$ is

$$1, \mathbf{E}_1, \mathbf{E}_2, \mathbf{E}_3,$$

where

$$\mathbf{E}_1 = \mathbf{e}_2\mathbf{e}_3, \qquad \mathbf{E}_2 = \mathbf{e}_3\mathbf{e}_1, \qquad \mathbf{E}_3 = \mathbf{e}_1\mathbf{e}_2.$$

Then

$$\mathbf{E}_j' = -\mathbf{E}_j \qquad (j = 1, 2, 3),$$

$$\mathbf{E}_i\mathbf{E}_j + \mathbf{E}_j\mathbf{E}_i = 0 \qquad (i \neq j)$$

and

$$\mathbf{E}_i\mathbf{E}_i' = -\mathbf{E}_i\mathbf{E}_i = F_i,$$

where

$$f_i F_i = f_1 f_2 f_3,$$

† It is easy to check that $\mathbf{uxu}' \in V$ (all $\mathbf{x} \in V$), e.g. by noting that this follows from (3.2) whenever $\mathbf{u} = \mathbf{yz}\,(\mathbf{y}, \mathbf{z} \in V)$.

as can be verified without trouble. On putting

$$\mathbf{u} = u_0 + u_1\mathbf{E}_1 + u_2\mathbf{E}_2 + u_3\mathbf{E}_3$$

we have

$$\begin{aligned}\mathbf{uu}' &= u_0^2 + F_1u_1^2 + F_2u_2^2 + F_3u_3^2 \\ &= u_0^2 + f_2f_3u_1^2 + f_3f_1u_2^2 + f_1f_2u_3^2 \\ &\in k,\end{aligned}$$

as required. We note for further use that the right-hand side of the last equation is a regular quadratic form in u_0, u_1, u_2, u_3 whose determinant is a perfect square.

We now revert to general n:

COROLLARY 3. *There is a homomorphism* $\theta: \sigma \to \theta(\sigma)$ *of* O^+ *into* $k^*/(k^*)^2$ *defined as follows. Let* $\sigma = \mathbf{T}_{\mathbf{u}}$. *Then*

$$\theta(\sigma) = (\mathbf{uu}')(k^*)^2. \tag{3.7}$$

This follows immediately from the Theorem. We call $\theta(\sigma)$ the *spinor norm* of σ. The kernel of θ will be denoted by Θ; that is

$$\Theta = \{\mathbf{T}_{\mathbf{u}}: \quad \mathbf{uu}' = 1\}. \tag{3.7 bis}$$

Many authors, including O'Meara, write O' for Θ but this might cause confusion with the commutator subgroup (=derived group) which also plays a rôle as must be explained immediately.

We first recall the group-theoretical definition. The *commutator subgroup* G' of a group G is the subgroup generated by the *commutators*

$$[a, b] = aba^{-1}b^{-1} \qquad (u, b \in G). \tag{3.8}$$

It is a normal subgroup since

$$c[a, b]c^{-1} = [cac^{-1}, cbc^{-1}]. \tag{3.9}$$

Then G/G' is an abelian group. If H is a normal subgroup such that G/H is abelian, then $G' \subset H$. All this is familiar. We shall require

LEMMA 3.2. *Let B be a set of generators for a group G [i.e. there is no proper subgroup of G which contains B]. Then the commutator subgroup G' is generated by the commutators*

$$[cb_1c^{-1}, cb_2c^{-1}] \qquad (c \in G, b_1, b_2 \in B). \tag{3.10}$$

Proof. Let H be the subgroup generated by the (3.10). Then H is normal by (3.9) and clearly $H \subset G'$. On the other hand, the group G/H is abelian since the images $\bar{b}$ in it of the $b \in B$ commute and they generate G/H. Hence $G' \subset H$ and so $G' = H$. We shall need only

LEMMA 3.3. *The commutator subgroup Ω of $O(V)$ is generated by the set of*

$$[\tau_{\mathbf{a}}, \tau_{\mathbf{b}}] = \tau_{\mathbf{a}}\tau_{\mathbf{b}}\tau_{\mathbf{a}}\tau_{\mathbf{b}} \qquad (\mathbf{a}, \mathbf{b} \in V, \quad \phi(\mathbf{a}) \neq 0, \quad \phi(\mathbf{b}) \neq 0).$$

Note. Ω is the commutator subgroup of $O(V)$, NOT of $O^+(V)$.

Proof. The expression for the commutator is a consequence of $\tau_{\mathbf{a}}^{-1} = \tau_{\mathbf{a}}$.
By Lemma 4.3. of Chapter 2 we can apply the previous lemma with $G = O(V)$ and $B =$ the set of $\tau_{\mathbf{a}}$. Hence $\Omega(V)$ is generated by the

$$[\sigma\tau_{\mathbf{a}}\sigma^{-1}, \sigma\tau_{\mathbf{b}}\sigma^{-1}] \qquad (\sigma \in O(V); \mathbf{a}, \mathbf{b} \in V). \tag{3.11}$$

The formula

$$\sigma\tau_{\mathbf{a}}\sigma^{-1} = \tau_{\sigma\mathbf{a}} \tag{3.12}$$

follows at once from the definition of the symmetries $\tau_{\mathbf{a}}$. Hence (3.11) is one of the set specified in the Lemma.

COROLLARY 1. *Ω is also generated by the set of*

$$\tau_{\mathbf{a}}\tau_{\mathbf{c}} \qquad (\mathbf{a}, \mathbf{c} \in V, \phi(\mathbf{a}) = \phi(\mathbf{c}) \neq 0).$$

Proof. By Lemma 4.2, Corollary, of Chapter 2 if $\phi(\mathbf{a}) = \phi(\mathbf{c}) \neq 0$ there is a $\sigma \in O(V)$ such that $\mathbf{c} = \sigma\mathbf{a}$ and then

$$\begin{aligned} \tau_{\mathbf{a}}\tau_{\mathbf{c}} &= \tau_{\mathbf{a}}\sigma\tau_{\mathbf{a}}\sigma^{-1} \\ &= \tau_{\mathbf{a}}\sigma\tau_{\mathbf{a}}^{-1}\sigma^{-1} \in \Omega. \end{aligned}$$

Further, the set in the Corollary is contained in that in the Lemma because

$$\tau_{\mathbf{a}}\tau_{\mathbf{b}}\tau_{\mathbf{a}}\tau_{\mathbf{b}} = \tau_{\mathbf{a}}\tau_{\mathbf{c}}$$

with $\mathbf{c} = \tau_{\mathbf{b}}\mathbf{a}$. This concludes the proof.

THEOREM 3.2. $\Omega \subset \Theta$. *Further,* $\Omega = \Theta$ *whenever V is isotropic.*

Note. There are other general cases when $\Omega = \Theta$, for which see O'Meara (1963).

Proof. The generators for Ω given by Lemma 3.3 clearly lie in Θ and so $\Omega \subset \Theta$. It remains to show equality when V is isotropic.

Suppose, first, that $V = H$ is a hyperbolic plane. Then we can verify the Theorem by explicit calculations. There is a basis $\mathbf{e}_1, \mathbf{e}_2$ of H such that

$$\phi(\mathbf{e}_1) = \phi(\mathbf{e}_2) = 0, \qquad \phi(\mathbf{e}_1, \mathbf{e}_2) = 1.$$

The group $O^+(V)$ consists precisely of the

$$\sigma_l^+ : \mathbf{e}_1 \to l\mathbf{e}_1, \quad \mathbf{e}_2 \to l^{-1}\mathbf{e}_2 \qquad (l \in k^*)$$

and there is the identity

$$\sigma_l^+ \sigma_m^+ = \sigma_{lm}^+ \qquad (l, m \in k^*).$$

Further, O^- consists of the

$$\sigma_r^- : \mathbf{e}_1 \to r\mathbf{e}_2, \quad \mathbf{e}_2 \to r^{-1}\mathbf{e}_1 \qquad (r \in k^*)$$

and we have

$$\sigma_r^- \sigma_s^- = \sigma_l^+, \qquad l = r^{-1}s.$$

Further,

$$\tau_{\mathbf{a}} = \sigma_r^-, \qquad \mathbf{a} = (a_1, a_2), \qquad r = -a_1^{-1}a_2.$$

It follows that the spinor norm is given by

$$\theta(\sigma_l^+) = l(k^*)^2.$$

In particular $\sigma_l^+ \in \Theta$ precisely when $l \in (k^*)^2$. If this is so, then

$$\sigma_l^+ = \sigma_r^- \sigma_s^- \sigma_r^- \sigma_s^-$$

for any r, s such that $l = (r^{-1}s)^2$. Hence $\sigma_l^+ \in \Omega$: that is $\Theta \subset \Omega$. This concludes the case $V = H$.

Now let V be any isotropic space. It contains a hyperbolic plane H and there is an orthogonal decomposition

$$V = H \oplus H^{\perp}.$$

We extend the action of $O(H)$ to V by making it leave $H^{\perp}$ elementwise invariant. We shall show first that if $\sigma \in O(V)$ then there is a $\sigma_1 \in O(H)$ such that

$$\sigma\sigma_1^{-1} \in \Omega(V). \tag{3.13}$$

Since Ω is a normal subgroup it is enough to do this when σ runs through a set of generators $\tau_{\mathbf{a}}$ of $O(V)$. Since H is isotropic, there is a $\mathbf{b} \in H$ with $\phi(\mathbf{b}) = \phi(\mathbf{a})$ and then $\mathbf{b} = \rho\mathbf{a}$ for some $\rho \in O(V)$ by Lemma 4.2, Corollary of Chapter 2. Put $\sigma = \tau_{\mathbf{a}}$ and $\sigma_1 = \tau_{\mathbf{b}}$. Then

$$\sigma\sigma_1^{-1} = \tau_{\mathbf{a}}\tau_{\mathbf{b}}^{-1} = \tau_{\mathbf{a}}(\rho\tau_{\mathbf{a}}\rho^{-1})^{-1} \in \Omega(V),$$

as required.

Now suppose that $\sigma \in \Theta(V)$. Then $\sigma_1 \in \Theta(V)$ since $\Omega(V) \subset \Theta(V)$ and so

$$\begin{aligned} \sigma_1 &\in \Theta(V) \cap O(H) \\ &= \Theta(H) \end{aligned}$$

(the last equality because it clearly does not matter whether we compute the spinor norm of an element of $O^+(H)$ in H or in V). We have already seen that $\Theta(H) = \Omega(H)$. Hence

$$\sigma_1 \in \Omega(H) \subset \Omega(V)$$

and $\sigma \in \Omega(V)$ by (3.13). This concludes the proof.

In the applications of this theory we shall primarily be concerned with the spinor norm and the groups Θ, Ω rather than with the spin group itself but we must introduce formally the deity who lurks behind the scenes:

The *spin group* Spin(V) is defined to be the set of $\mathbf{u} \in M_0(V)$ for which $\mathbf{uu}' = 1$, the group law being multiplication in $C(V)$. In other words Spin(V) consists of the $\mathbf{u}$ such that

(i) $\mathbf{u} \in C_0(V)$;
(ii) $\mathbf{uu}' = 1$;
(iii) $\mathbf{uxu}' \in V$ for all $\mathbf{x} \in V$.

THEOREM 3.3. *The map* $\mathbf{u} \to \mathbf{T_u}$ *defined in* (3.1) *gives a homomorphism of* Spin(V) *onto* $\Theta(V)$. *The kernel is the group* $\{1, -1\}$ *of order* 2.

Proof. Follows immediately from Theorem 3.1.

We shall require later

LEMMA 3.4. *Let* $n > 1$. *Suppose that* $\mathbf{T_u} \in \Omega$, *where* $\mathbf{u} \in \mathrm{Spin}(V)$. *Then* $\mathbf{u}$ *is the product of* "commutators" *of the shape*

$$\begin{aligned} \mathbf{w} &= \mathbf{aba}^{-1}\mathbf{b}^{-1} \\ &= \{\phi(\mathbf{a})\}^{-1}\{\phi(\mathbf{b})\}^{-1}\mathbf{abab} \end{aligned}$$

with

$$\mathbf{a}, \mathbf{b} \in V.$$

Proof. Clearly $\mathbf{w} \in \mathrm{Spin}(V)$. Suppose that $\mathbf{u}$ satisfies the conditions of the Lemma. Then

$$\mathbf{u} = l\mathbf{w}_1 \ldots \mathbf{w}_r$$

for some $l \in k^*$ and some commutators $\mathbf{w}_1, \ldots, \mathbf{w}_r$. We must have $l \in \mathrm{Spin}(V)$ since $\mathbf{u}, \mathbf{w}_1, \ldots, \mathbf{w}_r \in \mathrm{Spin}(V)$ and so

$$l = \pm 1.$$

Hence it will be enough to show that -1 is a commutator. Since $n > 1$ we can find $\mathbf{a}_0, \mathbf{b}_0$ which are anisotropic and orthogonal. Then

$$\begin{aligned} \mathbf{w}_0 \text{ (say)} &= \{\phi(\mathbf{a}_0)\}^{-1}\{\phi(\mathbf{b}_0)\}^{-1}\mathbf{a}_0\mathbf{b}_0\mathbf{a}_0\mathbf{b}_0 \\ &= -\{\phi(\mathbf{a}_0)\}^{-1}\{\phi(\mathbf{b}_0)\}^{-1}\mathbf{a}_0\mathbf{a}_0\mathbf{b}_0\mathbf{b}_0 \\ &= -1, \end{aligned}$$

as required.

4. LATTICES OVER INTEGRAL DOMAINS

In this section we discuss how the results of Section 3 generalize to integral domains and give some special results for $\mathbf{Z}_p$.

Let I be an integral domain (which in practice will be $\mathbf{Z}$ or $\mathbf{Z}_p$) and let k be a field containing I. We recall that a lattice Λ in a quadratic space V, ϕ is the set of $a_1\mathbf{e}_1 + \ldots + a_n\mathbf{e}_n$ where the a_j run through I and $\mathbf{e}_1, \ldots, \mathbf{e}_n$ is a k-basis for V. We define $O(\Lambda)$ to be the set of $\sigma \in O(V)$ such that

$$\sigma\Lambda = \Lambda \tag{4.1}$$

and put

$$O^+(\Lambda) = O(\Lambda) \cap O^+(V) \tag{4.2}$$

$$\Omega(\Lambda) = O(\Lambda) \cap \Omega(V) \tag{4.3}$$

$$\Theta(\Lambda) = O(\Lambda) \cap \Theta(V). \tag{4.4}$$

If $O(\Lambda)$ is represented with respect to a basis $\mathbf{e}_1, \ldots, \mathbf{e}_n$ of Λ, then the corresponding matrices have entries in I. Clearly $O(\Lambda)$ does not really depend on the quadratic space V, ϕ but only on the quadratic form induced on Λ by ϕ: and similarly for $O^+(\Lambda)$. On the other hand $\Omega(\Lambda)$ and $\Theta(\Lambda)$ do depend (at least *prima facie*) on the choice of field k containing I. Unless the contrary is stated, this will be taken to be the quotient field of I. Note also that $\Omega(\Lambda)$ need not be the commutator group of $O(\Lambda)$: it may be larger.

Now suppose, further, that

$$f(\mathbf{x}) \in I \qquad (\text{all } \mathbf{x} \in \Lambda) \tag{4.5}$$

(that is, the form induced on Λ by ϕ is integer-valued). We denote by $C(\Lambda)$ the subring of $C(V)$ generated by 1 and the $\mathbf{x} \in \Lambda$. If $\mathbf{e}_1, \ldots, \mathbf{e}_n$ is a basis for Λ, not necessarily normal,† then clearly $C(\Lambda)$ is generated as an I-module by 1 and the expressions

$$\mathbf{e}(J) = \mathbf{e}(j_1) \ldots \mathbf{e}(j_r). \tag{4.6}$$

with

$$j_1 < j_2 < \ldots < j_r \tag{4.7}$$

† Recall that a normal basis is one in which the basis elements are normal (= orthogonal) to each other.

which were used in Section 2. We put

$$C_0(\Lambda) = C(\Lambda) \cap C_0(V) \tag{4.8}$$

$$\mathrm{Spin}(\Lambda) = C(\Lambda) \cap \mathrm{Spin}(V). \tag{4.9}$$

Clearly $C_0(\Lambda)$ is a sub-I-algebra of $C(\Lambda)$: also $\mathrm{Spin}(\Lambda)$ is a subgroup of $\mathrm{Spin}(V)$. Equally clearly, $C(\Lambda), C_0(\Lambda)$ and $\mathrm{Spin}(\Lambda)$ are independent of the choice of the field k and could have been defined intrinsically in terms of Λ and the quadratic form on it.

The case when $n = 3$ is perhaps worth mentioning explicitly. Let $\mathbf{e}_1, \mathbf{e}_2, \mathbf{e}_3$ be a basis for a lattice Λ for which (4.5) holds. Then the discussion of the ternary case shows that $C_0(\Lambda)$ consists of the

$$\mathbf{u} = u_0 + u_1\mathbf{e}_2\mathbf{e}_3 + u_2\mathbf{e}_3\mathbf{e}_1 + u_3\mathbf{e}_1\mathbf{e}_2 \qquad (u_0, u_1, u_2, u_3 \in I)$$

and that then

$$\begin{aligned}\mathbf{u}\mathbf{u}' &= g(u_0, \ldots, u_3)\\ &\in I[u_0, \ldots, u_3].\end{aligned}$$

The quadratic form g need no longer be diagonal as, in general, we cannot choose $\mathbf{e}_1, \mathbf{e}_2, \mathbf{e}_3$ to be normal. The elements of $\mathrm{Spin}(\Lambda)$ correspond precisely to the solutions of

$$g(u_0, u_1, u_2, u_3) = 1 \qquad (u_j \in I).$$

Now let us revert to general n. Theorem 3.3 gives a group homomorphism

$$\mathrm{Spin}(\Lambda) \to \Theta(\Lambda) \tag{4.10}$$

with kernel $\{1, -1\}$. The image need not be the whole of $\Theta(\Lambda)$. We have, however, the following

LEMMA 4.1. *Let $I = \mathbf{Z}_p$ $(p \neq 2, \infty)$ and $n > 1$. Suppose that* (4.5) *holds, and that $d \in U_p$, where d is the determinant of the form induced on Λ and where U_p is the set of p-adic units. Then* (4.10) *is onto. Further, the image $\theta(\Lambda)$ (say) of $O^+(\Lambda)$ under the spinor-norm map satisfies*

$$\theta(\Lambda) = U_p(\mathbf{Q}_p^*)^2. \tag{4.11}$$

Proof. By Lemma 3.3, Corollary 2, of Chapter 8, $O^+(\Lambda)$ consists precisely of the

$$\sigma = \tau(\mathbf{b}_1) \ldots \tau(\mathbf{b}_r), \tag{4.12}$$

where

$$\mathbf{b}_j \in \Lambda, \qquad |\phi(\mathbf{b}_j)|_p = 1 \tag{4.13}$$

and r is even [of course the representation of σ in the shape (4.12) is not unique]. Then

$$\sigma = \mathbf{T}_\mathbf{u}$$

where

$$\mathbf{u} = \mathbf{b}_1 \ldots \mathbf{b}_r,$$

so

$$\mathbf{u}\mathbf{u}' = \phi(\mathbf{b}_1) \ldots \phi(\mathbf{b}_r) \in U_p.$$

Further, since we are supposing $n > 1$, each $\phi(\mathbf{b}_j)$ can take any value in U_p (e.g. by Lemma 3.4 of Chapter 8). This proves (4.11).

Now suppose that $\sigma \in \Theta(\Lambda)$, so

$$\mathbf{u}\mathbf{u}' = t^2, \qquad t \in U_p.$$

Then

$$\mathbf{v} \text{ (say)} = t^{-1}\mathbf{u} \in \mathrm{Spin}(\Lambda)$$

and $\sigma = \mathbf{T}_{\mathbf{v}}$. This concludes the proof.

5. TOPOLOGICAL CONSIDERATIONS

In this section we suppose that k is a topological field. This means that there is a topology on k (considered as a set) under which the field operations (sum, product and taking the inverse) are continuous maps. In practice k will be $\mathbf{Q}_p$ or $\mathbf{R}$ with the usual topology. By definition $C(V)$ is a finite vector space over k and so has a naturally induced topology. It is clear that addition, multiplication and the involution $\mathbf{u} \to \mathbf{u}'$ are continuous on $C(V)$. Further, reciprocation $\mathbf{u} \to \mathbf{u}^{-1}$ is continuous on $M_0(V)$ since $\mathbf{u}^{-1} = (\mathbf{u}\mathbf{u}')^{-1}\mathbf{u}$ by Theorem 3.1, Corollary 1. Hence $M_0(V)$ is a topological group (i.e. product and inverse are both continuous).

The group $O(V)$ also has a natural topology induced by that of k. Indeed if we fix a basis $\mathbf{e}_1, \ldots, \mathbf{e}_n$ of V, the elements of $O(V)$ are represented by $n \times n$ matrices and the topology is the topology on the matrices considered as a subset of an n^2-dimensional vector space. This topology is clearly independent of the basis chosen.

The homomorphism $M_0(V) \to O^+(V)$ given by $\mathbf{u} \to \mathbf{T}_{\mathbf{u}}$ (cf. Theorem 3.1) is clearly continuous. Slightly less trivial is

LEMMA 5.1. *The map* $\mathbf{u} \to \mathbf{T}_{\mathbf{u}}$ *of* $M_0(V)$ *onto* O^+ *has locally a continuous cross-section. That is if* $\sigma_0 = \mathbf{T}_{\mathbf{w}}$ *for some* $\mathbf{w} \in M_0(V)$, *then there is a neighbourhood* $\mathcal{N}$ *of* σ_0 *and a continuous map*

$$\mathbf{v}\colon \sigma \to \mathbf{v}(\sigma)$$

from $\mathcal{N}$ *to* $M_0(V)$ *such that*

$$\mathbf{v}(\sigma_0) = \mathbf{w} \qquad \mathbf{T}_{\mathbf{v}(\sigma)} = \sigma \qquad (\text{all } \sigma \in \mathcal{N}).$$

Proof. It is enough to show that $\mathbf{a}_1, \ldots, \mathbf{a}_r$ in (3.4) can be chosen so that locally they depend continuously on σ. And this is clear from an inspection of the proofs in Section 4 of Chapter 2. This concludes the proof.

COROLLARY 1. *The spinor norm map θ of O^+ to $k^*/(k^*)^2$ is continuous.*

Note. Here $k^*/(k^*)^2$ has the usual quotient topology.†

Proof. The map $M_0(V) \to k^*/(k^*)^2$ given by $\mathbf{u} \to \mathbf{uu}'$ is clearly continuous. The spinor norm is given locally by

$$\sigma \to \mathbf{v}(\sigma) \to \mathbf{v}(\sigma)\{\mathbf{v}(\sigma)\}',$$

where $\sigma \to \mathbf{v}(\sigma)$ is the local cross-section given by the Lemma.

COROLLARY 2. *The map* $\mathrm{Spin}(V) \to O^+(V)$ *has locally a continuous cross-section provided that the map* $k^* \xrightarrow{2} k^*$ *given by* $b \to b^2$ $(b \in k^*)$ *has locally a continuous cross-section.*

Note. The condition about $k^* \xrightarrow{2} k^*$ is clearly satisfied in the fields $k = \mathbf{Q}_p$ and $k = \mathbf{R}\,(=\mathbf{Q}_\infty)$ which we have to deal with: and indeed in any field k which is complete with respect to a valuation, the topology being given by the valuation.

Proof. Clear from the definition of Spin(V).

6. CHANGE OF FIELDS AND RINGS

Let K be any field containing k and let J be a ring with $K \supset J \supset I$. Then, as we have remarked several times before, any quadratic space V, ϕ over k gives rise in a natural way to a quadratic space V_K, ϕ_K over K (the "*tensor product*" with K) and, similarly, any I-lattice Λ in V gives a J-lattice Λ_J in V_K with the "same" basis. There are then natural homomorphisms

$$O(V) \to O(V_K)$$
$$\Theta(V) \to \Theta(V_K)$$
$$C(V) \to C(V_K)$$
$$\mathrm{Spin}(V) \to \mathrm{Spin}(V_K)$$

and (subject to (4.5) to ensure the existence of the objects in question)

$$C(\Lambda) \to C(\Lambda_J)$$
$$\mathrm{Spin}(\Lambda) \to \mathrm{Spin}(\Lambda_J)$$

and so on, with obvious properties. Any algebraist who is worth his salt would make this remark profound by a skillful use of tensor-products and injection maps and a careful (or sloppy) choice of inappropriate notation. In what follows we shall treat it as the triviality it is.

† The canonical topology on k^* is that induced by the embedding $a \to (a, a^{-1})$ of k^* into k^2. When the topology on k is given by a valuation, however, as with $\mathbf{Q}_p$, $\mathbf{R}$, then this is identical with the topology induced in k^* as a subset of k. A subset of $k^*/(k^*)^2$ is open in the quotient topology, by definition, when the corresponding subset of k^* is open.

7. THE STRONG APPROXIMATION THEOREM

In this section we make the restriction

$$p \neq \infty. \tag{7.1}$$

First we apply the general discussion in Section 4 of topological fields to the case $k = \mathbf{Q}_p$. Let Λ_p be a $\mathbf{Z}_p$-lattice in a quadratic space V_p, ϕ over $\mathbf{Q}_p$. Clearly Λ_p is an open and closed subset of V_p and $O(\Lambda_p)$ is an open and closed subgroup of $O(V_p)$.

If we suppose, further, that

$$\phi(\mathbf{x}) \in \mathbf{Z}_p \qquad (\text{all } \mathbf{x} \in \Lambda_p), \tag{7.2}$$

so $C(\Lambda_p)$ is defined, it is clear that it is an open and closed subset (sub-$\mathbf{Z}_p$-algebra) of $C(V_p)$. It follows that $\text{Spin}(\Lambda_p)$ is well-defined and is an open and closed subgroup of $\text{Spin}(V_p)$.

Now let Λ be a $\mathbf{Z}$-lattice in the regular $\mathbf{Q}$-quadratic space V, ϕ. We denote by Λ_p and V_p, ϕ respectively the result of extending the ground-ring from $\mathbf{Z}$ to $\mathbf{Z}_p$ and the ground-field from $\mathbf{Q}$ to $\mathbf{Q}_p$. (A purist would insist on writing V_p, ϕ_p but we are above—or below—such niceties.)

By "*almost all p*" we mean, as always, all p except, possibly, finitely many. With this convention we can enunciate the main subject of this section:

THEOREM 7.1 (Strong Approximation Theorem for the Spin Group). *Let $n \geqslant 3$ and suppose that the $\mathbf{Q}$-quadratic space V, ϕ is indefinite (i.e. isotropic in $\mathbf{R} = \mathbf{Q}_\infty$). Let Λ be any $\mathbf{Z}$-lattice in V, ϕ. With the above notation denote for all (finite) p by $\mathcal{U}_p$ a non-empty open subset of* $\text{Spin}(V_p)$ *and suppose that*

$$\mathcal{U}_p = \text{Spin}(\Lambda_p) \qquad (\textit{almost all } p). \tag{7.3}$$

Then there is a $\mathbf{u} \in \text{Spin}(V)$ *such that*

$$\mathbf{u} \in \mathcal{U}_p \qquad (\textit{all } p \neq \infty). \tag{7.4}$$

Before embarking on the proof we make some comments. In the first place, the condition (7.2) holds for almost all p (consider a basis for Λ), so $\text{Spin}\Lambda_p$ exists for almost all p, and the condition (7.3) makes sense. Secondly, if Γ is any other $\mathbf{Z}$-lattice in V, ϕ, we have $\Gamma_p = \Lambda_p$ for almost all p, as will be explained more minutely in the next Chapter. Hence the condition (7.3) does not actually depend on the choice of the lattice Λ. Thirdly, the statement (7.4) presupposes the natural homomorphism (actually, an embedding) $\text{Spin}(V) \rightarrow \text{Spin}(V_p)$ discussed in Section 6.

What we shall actually use is not Theorem 7.1 itself but the

COROLLARY. *Suppose that the conditions of Theorem 7.1 are satisfied. For all p denote by* $\mathscr{V}_p$ *a non-empty open subset of* $\Theta(V_p)$ *and suppose that*

$$\mathscr{V}_p = \Theta(\Lambda_p) \qquad (\textit{almost all } p). \tag{7.5}$$

Then there is a $\sigma \in \Theta(V)$ *such that*

$$\sigma \in \mathscr{V}_p \qquad (\textit{all } p). \tag{7.6}$$

To deduce the Corollary from the Theorem it is enough to observe that the homomorphism $\mathrm{Spin}(V_p) \to \Theta(V_p)$ given by Theorem 3.2 is clearly continuous.

We should give the reasons here why the statements for $O^+(V)$ analogous to the above Corollary for $\Theta(V)$ must be false. By Lemma 5.1 there is a finite set P of primes such that $\theta(\sigma_p) \in U_p(\mathbf{Q}_p^*)^2$ whenever $p \notin P$ and $\sigma_p \in O^+(\Lambda_p)$. Suppose that $\sigma_p \in O^+(V_p)$ is given for $p \in P$. If the hypothetical result were true, we would be able (by Lemma 5.1 Corollary) to find a $\sigma \in O^+(V)$ such that

$$\theta(\sigma) \in \theta(\sigma_p)(\mathbf{Q}_p^*)^2 \qquad (p \in P)$$
$$\theta(\sigma) \in U_p(\mathbf{Q}_p^*)^2 \qquad (p \notin P).$$

However it may happen that there is no element of $\mathbf{Q}^*$ which satisfies these conditions on $\theta(\sigma)$. It is easy to construct a concrete example. Take

$$f(\mathbf{x}) = x_1^2 + x_2^2 - x_3^2$$

and $P = \{2\}$. Put

$$\sigma_2 = \tau_{\mathbf{a}}\tau_{\mathbf{b}},$$

where

$$\mathbf{a} = (1, 0, 0), \qquad \mathbf{b} = (2, 1, 0).$$

Then

$$\theta(\sigma_2) = 5(\mathbf{Q}_2^*)^2.$$

It is easy to see that there is no $v \in \mathbf{Q}^*$ such that

$$v \in 5(\mathbf{Q}_2^*)^2$$
$$v \in U_p(\mathbf{Q}_p^*)^2 \qquad (p \neq 2).$$

A fortiori there is no σ such that $v = \theta(\sigma)$ has these properties.

For the proof of the Theorem we require the following.

LEMMA 7.1. *Let* $g(\mathbf{x}) \in \mathbf{Q}[\mathbf{x}]$ *be a regular indefinite quadratic form of determinant d in* $n \geqslant 4$ *variables and let* $a \in \mathbf{Q}^*$. *Let P be a finite set of primes and suppose that*

$$2 \in P \tag{7.7}$$

and

$$a \in \mathbf{Z}_p, \qquad g(\mathbf{x}) \in \mathbf{Z}_p[\mathbf{x}] \qquad (\textit{all } p \notin P), \tag{7.8}$$

$$|d|_p = 1 \qquad (\textit{all } p \notin P). \tag{7.9}$$

For $p \in P$ *let* $\mathbf{b}_p \in \mathbf{Q}_p^n$ *be a solution of* $g(\mathbf{b}_p) = a$. *Then there is a* $\mathbf{b} \in \mathbf{Q}^n$ such *that*

$$f(\mathbf{b}) = a, \qquad \mathbf{b} \in \mathbf{Z}_p^n \qquad (\textit{all } p \notin P) \tag{7.10}$$

and which is arbitrarily close p-adically to each of the $\mathbf{b}_p$.

This is just a reformulation of Theorem 1.5 of Chapter 9. For $p \notin P$ the condition (7.8) ensures that there is certainly a $\mathbf{b}_p \in \mathbf{Z}_p^n$ with $f(\mathbf{b}_p) = a$. On multiplying $g(\mathbf{x})$, $\mathbf{b}_p$ $(p \in P)$ and a respectively by m, m and m^3 where

$$m = \prod_{p \in P} p^N$$

and N is sufficiently large, we may suppose without loss of generality that $g(\mathbf{x})$ and the $\mathbf{b}_p$ $(p \in P)$ are integral: and then we have the conditions of the theorem quoted.

We now prove Theorem 7.1. The cases $n = 3$ and $n > 3$ are dealt with differently.

Proof of Theorem 7.1, $n = 3$. Let $\mathbf{e}_1, \mathbf{e}_2, \mathbf{e}_3$ be a basis for Λ. Then $C_0(V)$ consists of the

$$\mathbf{u} = u_0 + u_1\mathbf{e}_2\mathbf{e}_3 + u_2\mathbf{e}_3\mathbf{e}_1 + u_3\mathbf{e}_1\mathbf{e}_2 \qquad (u_j \in \mathbf{Q}) \tag{7.11}$$

and

$$\mathbf{u}\mathbf{u}' = g(u_0, u_1, u_2, u_3) \text{ (say)} \in \mathbf{Q}[u_0, u_1, u_2, u_3].$$

is a quadratic form. Further, Spin(V) consists of the $\mathbf{u}$ with

$$g(u_0, u_1, u_2, u_3) = 1.$$

Let P be any finite set of primes p such that

(i) $2 \in P$
(ii) $\mathscr{U}_p = \text{Spin}(\Lambda_p)$ (all $p \notin P$)
(iii) $g(u_0, \ldots, u_3) \in \mathbf{Z}_p[u_0, \ldots, u_3]$ (all $p \notin P$).
(iv) $|d|_p = 1$ (all $p \notin P$),

where d is the determinant of g. Since V, ϕ is indefinite, the form $g(u_0, \ldots, u_4)$ is also indefinite (as is easily checked, particularly when $\mathbf{e}_1, \mathbf{e}_2, \mathbf{e}_3$ are normal).

Now let

$$\mathbf{u}^{(p)} \in \mathscr{U}_p \qquad (p \in P),$$

say with coordinates $(u_0^{(p)}, \ldots, u_3^{(p)})$. By Lemma 7.1 there are $(u_0, \ldots, u_3) \in \mathbf{Q}^4$ with $g(u_0, \ldots, u_3) = 1$ such that the u_j are arbitrarily close p-adically to the $u_j^{(p)}$ $(0 \leqslant j \leqslant 3,\ p \in P)$ and the u_j are p-adic integers for $p \notin P$. Then $\mathbf{u}$ given by (7.11) has the properties required by the Theorem.

Proof of Theorem 7.1, $n \geqslant 4$. We denote by f the form induced by ϕ on Λ and its determinant will be d. We denote by P_1 a finite set of primes such that

(i) $2 \in P_1$

(ii) $\mathscr{U}_p = \text{Spin}(\Lambda_p)$ for all $p \notin P_1$

(iii) f has coefficients in $\mathbf{Z}_p$ for all $p \notin P_1$

(iv) $|d|_p = 1$ for all $p \notin P_1$.

For $p \in P_1$ select $\mathbf{u}^{(p)} \in \mathscr{U}_p$. Then (as in the proof of Theorem 3.1, Corollary 1) we have

$$\mathbf{u}^{(p)} = l^{(p)}\mathbf{a}_1^{(p)} \ldots \mathbf{a}_r^{(p)} \qquad (p \in P_1)$$

where

$$l^{(p)} \in \mathbf{Q}_p^*, \qquad \mathbf{a}_j^{(p)} \in V_p.$$

Here r is even and *prima facie* may depend on p but by appending terms of the type $\mathbf{aa} = \phi(\mathbf{a})$ we may suppose without loss of generality that r is the same for all $p \in P_1$. Then by taking $l^{(p)}\mathbf{a}_1^{(p)}$ instead of $\mathbf{a}_1^{(p)}$ we may suppose that

$$l^{(p)} = 1 \qquad (\text{all } p \in P_1).$$

We have

$$\phi(\mathbf{a}_p^{(1)}) \ldots \phi(\mathbf{a}_r^{(p)}) = 1 \qquad (\text{all } p \in P_1)$$

since $\mathbf{u}^{(p)} \in \text{Spin}(V_p)$.

The next stage is to find $\tilde{\mathbf{a}}_j \in V$ such that

$$\phi(\tilde{\mathbf{a}}_1) \ldots \phi(\tilde{\mathbf{a}}_r) = 1$$

and the $\tilde{\mathbf{a}}_j$ are arbitrarily close to the $\mathbf{a}_j^{(p)}$. Note that we are NOT requiring that $\tilde{\mathbf{a}}_j \in \Lambda_p$ for $p \notin P_1$. For $j \geqslant 2$ we just take any $\tilde{\mathbf{a}}_j \in V$ which is close to the $\mathbf{a}_j^{(p)}$. Define $m \in \mathbf{Q}^*$ by

$$m\phi(\tilde{\mathbf{a}}_2) \ldots \phi(\tilde{\mathbf{a}}_r) = 1.$$

There are then $\mathbf{a}_1'^{(p)} \in V_p$ close to the $\mathbf{a}_1^{(p)}$ such that

$$\phi(\mathbf{a}_1'^{(p)}) = m \qquad (\text{all } p \in P_1).$$

Since V, ϕ is indefinite in $n \geqslant 4$ variables it is universal by the Strong Hasse Principle (Theorem 1.5 of Chapter 6 and its two corollaries) and so by Lemma 9.1, Corollary, of the same Chapter, there is an $\tilde{\mathbf{a}}_1 \in V$ with $\phi(\tilde{\mathbf{a}}_1) = m$ which is close to the $\mathbf{a}_1'^{(p)}$.

Put

$$\tilde{\mathbf{u}} = \tilde{\mathbf{a}}_1 \ldots \tilde{\mathbf{a}}_r \in \text{Spin}(V).$$

Now let P_2 be the finite set of those $p \notin P_1$ for which $\tilde{\mathbf{u}} \notin \text{Spin}(\Lambda_p)$. If P_2 is empty, we are done. Otherwise define open sets $\mathscr{W}_p \subset \text{Spin}(V_p)$ by

$$\mathscr{W}_p = \tilde{\mathbf{u}}^{-1}\mathscr{U}_p,$$

so

$$\mathscr{W}_p = \text{Spin}(\Lambda_p) \quad (p \notin P_1 \cup P_2).$$

Put

$$\mathbf{v}_p = 1 \qquad (p \in P_1)$$
$$\mathbf{v}_p = \tilde{\mathbf{u}}^{-1} \qquad (p \in P_2),$$

so

$$\mathbf{v}_p \in \mathscr{W}_p \qquad (p \in P_1 \cup P_2).$$

The space V_p for $p \in P_2$ is isotropic by conditions (i), (iii), (iv) in the definition of P_1 (which is disjoint from P_2) and since $n \geqslant 4$. Hence $\Theta(V_p) = \Omega(V_p)$ for $p \in P_2$ by theorem 3.2. It follows that $\mathbf{T}_{\mathbf{v}_p} \in \Omega(V_p)$ for $p \in P_2$: and this is trivially the case for $p \in P_1$.

By Lemma 3.4 each $\mathbf{v}_p$ ($p \in P_1 \cup P_2$) is a product of "commutators", say

$$\mathbf{v}_p = \mathbf{w}_1^{(p)} \ldots \mathbf{w}_s^{(p)} \qquad (p \in P_1 \cup P_2).$$

where each $\mathbf{w}_j^{(p)}$ is of the type

$$\mathbf{w}^{(p)} = \{\phi(\mathbf{a}_p)\}^{-1}\{\phi(\mathbf{b}_p)\}^{-1}\mathbf{a}_p\mathbf{b}_p\mathbf{a}_p\mathbf{b}_p$$

with

$$\mathbf{a}_p, \mathbf{b}_p \in V_p.$$

Prima facie the number s of terms depends on p but we can make it independent of p by appending "trivial" commutators with $\mathbf{a}_p = \mathbf{b}_p$, which are equal to 1.

To conclude the proof of Theorem 7.1 it will thus be enough to prove

LEMMA 7.2. *Let $n \geqslant 4$ and suppose that V, ϕ, Λ are as in the Theorem. Let P be a set of p such that*

(i) $2 \in P$;

(ii) *the form f induced by ϕ on Λ has coefficients in $\mathbf{Z}_p$ for $p \notin P$;*

(iii) *the determinant d of f satisfies $|d|_p = 1$ for $p \notin P$.*

For $p \in P$ let

$$\mathbf{w}_p = \{\phi(\mathbf{a}_p)\}^{-1}\{\phi(\mathbf{b}_p)\}^{-1}\mathbf{a}_p\mathbf{b}_p\mathbf{a}_p\mathbf{b}_p$$

be a commutator in Spin(V_p). *Then there is a* $\mathbf{u} \in \text{Spin}(V)$ *which is arbitrarily close to each of the* $\mathbf{w}_p$ *and such that*

$$\mathbf{u} \in \text{Spin}(\Lambda_p) \qquad (\textit{all } p \notin P).$$

Proof. Choose first $\mathbf{a}, \mathbf{b} \in V$ which are arbitrarily close to $\mathbf{a}_p, \mathbf{b}_p$ respectively for $p \in P$: we do not make any conditions for $p \notin P$. Then

$$\mathbf{bab} = \phi(\mathbf{b})\mathbf{c},$$

where

$$\mathbf{c} = -\tau_{\mathbf{a}}\mathbf{b}$$

by (3.1). In particular,

$$\phi(\mathbf{c}) = \phi(\mathbf{a}) = m \text{ (say)}. \tag{7.12}$$

Let P_3 be a finite set of primes disjoint from P such that

$$\mathbf{a}, \mathbf{c} \in \Lambda_p \qquad (p \notin P \cup P_3) \tag{7.13}$$

$$|m|_p = 1 \qquad (p \notin P \cup P_3). \tag{7.14}$$

By Lemma 7.1 there is a $\mathbf{d} \in V$ such that

$$\phi(\mathbf{d}) = m, \tag{7.15}$$

$$\mathbf{d} \text{ is near } \mathbf{c} \text{ } p\text{-adically for } p \in P, \tag{7.16}$$

$$\mathbf{d} \text{ is near } \mathbf{a} \text{ } p\text{-adically for } p \in P_3 \tag{7.17}$$

and

$$\mathbf{d} \in \Lambda_p \, (p \notin P \cup P_3). \tag{7.18}$$

Then

$$\mathbf{u} = m^{-1}\mathbf{ad} \in \mathrm{Spin}(V)$$

by (7.12), (7.15). Further, $\mathbf{u}$ is near to $\mathbf{w}_p$ for $p \in P$ and $\mathbf{u}$ is near 1 for $p \in P_3$, so we may certainly suppose that

$$\mathbf{u} \in \mathrm{Spin}(\Lambda_p) \qquad (p \in P_3).$$

Finally

$$\mathbf{u} \in \mathrm{Spin}(\Lambda_p) \qquad (p \notin P \cup P_3)$$

by (7.13), (7.14), and (7.18).

NOTES

Most of the work of this Chapter generalizes to algebraic number fields, cf. O'Meara (1963), where there is a further study of Clifford algebras. For Clifford algebras, see also Lam (1973) and Scharlau (1969). For the wider context of linear algebraic groups see Artin (1957) and Dieudonné (1955).

The arithmetic of the quaternions introduced in Section 1 has been much investigated as a natural generalization of the arithmetic of $\mathbf{Z}$ and $\mathbf{Z}[i]$ and the resultant theory has been used to study representation of integers as sums of squares. Lipschitz (1886) defined an integral quaternion as one for which $u_0, u_1, u_2, u_3 \in \mathbf{Z}$ (in the notation of Section 1) but Hurwitz (1896, 1919) used the definition $2u_j \in \mathbf{Z}$, $2u_0 \equiv 2u_1 \equiv 2u_2 \equiv 2u_3 \pmod 2$. There is an account in Hardy and Wright (1938) [the first edition follows Lipschitz, the later ones Hurwitz]. Venkov (1922, 1928, 1929, 1931) applies quaternions to various questions on the representation of integers and binary forms as sums of squares: he also obtains an elementary proof of the class-number of definite binary forms of given determinant in a large number of cases [cf. Appendix B].

For an application of quaternions over an algebraic number field, see Kirmse (1924).

If f is any ternary form over $\mathbf{Q}$, the even Clifford algebra $C_0(f)$ is a natural generalization of the quaternions and is referred to as a generalized quaternion algebra. It is a division algebra provided that f is anisotropic [cf. Chapter 2, example 4]. Generalized quaternions may be used to study integral representations by ternary forms. They were extensively investigated by Brandt [see his survey Brandt (1943)] and by Linnik under the name of "hermitions" [Linnik (1939, 1940, 1949), Linnik and Malyšev (1953). There is an account in Malyšev (1962), Chapter 4]. For the arithmetic of quaternions in a general context see Deuring (1935), especially pp. 135–137 where there are also references to a number of earlier papers dealing with special problems. See also Jones and Pall (1939) [cf. Chapter 11, example 4], Pall (1946a) and Peters (1969).

The use of generalized quaternion algebras for ternary forms may be regarded as a very special case of the theory of the next chapter and some of the special results in the papers mentioned above now fit neatly into the more general framework.

Section 3. Lipschitz (1886, 1959) has given an explicit formula for the spinor norm in the ordinary orthogonal group.

Section 10. There is a general theory of strong approximation theorems over linear algebraic groups. In particular there can be such a theorem only when the group is simply connected. See Kneser (1965, 1966).

EXAMPLES

1. Let Λ_p be a $\mathbf{Z}_p$-lattice in the 3-dimensional regular quadratic space V_p, ϕ where $p \neq 2$. Suppose that ϕ is integer-valued on Λ_p and that the determinant of the quadratic form induced by ϕ on Λ_p is divisible by p but not by p^2. Show that $\theta(\alpha)$ runs through all cosets of $\mathbf{Q}_p^*$ modulo $(\mathbf{Q}_p^*)^2$ as α runs through $O^+(\Lambda_p)$.

2. Notations and hypotheses as in 1. Suppose also that V_p, ϕ is anisotropic. Let $\mathbf{b}$ be any element of Λ_p with $|\phi(\mathbf{b})|_p = p^{-1}$ and let $\alpha \in O^+(\Lambda_p)$. Show that

$$\alpha\mathbf{b} \equiv \varepsilon\mathbf{b} \pmod{p\Lambda_p}$$

where

$$\varepsilon = \begin{cases} +1 & \text{if } \theta(\alpha) \subset U_p(\mathbf{Q}_p^*)^2 \\ -1 & \text{if } \theta(\alpha) \subset pU_p(\mathbf{Q}_p^*)^2. \end{cases}$$

3. Let Λ be a $\mathbf{Z}$-lattice in a 3-dimensional regular quadratic space V, ϕ. Suppose that

(i) ϕ is indefinite:

(ii) ϕ is integer-valued on Λ;

(iii) For every prime p the set U_p of p-adic units is contained in the union of the $\theta(\alpha)$, where α runs through $O^+(\Lambda_p)$ and Λ_p is the localization of Λ.

Let P^* be a set of primes p such that Λ_p satisfies the hypotheses of examples 1, 2 above and let $\eta_p = \pm 1$ be given arbitrarily for $p \in P^*$.
Let $\mathbf{b} \in \Lambda$ satisfy

$$|\phi(\mathbf{b})|_p = p^{-1} \qquad (\text{all } p \in P^*).$$

Show that there is a $\beta \in O^+(\Lambda)$ such that

$$\beta\mathbf{b} \equiv \eta_p \mathbf{b} \qquad (\text{mod } p\Lambda)$$

for all $p \in P^*$.

4. Let V, ϕ be a 3-dimensional regular quadratic space over a field k and let its determinant be $d(k^*)^2$.

(i) Show that the centre of the full Clifford algebra $C(V)$ has dimension 2 over k and that it has a basis $1, \mathbf{c}$ with $\mathbf{c}^2 \in -d(k^*)^2$.

(ii) Suppose that V, ϕ is isotropic. Show that there is an isomorphism

$$\lambda\colon C_0(V) \to A$$

of the even Clifford algebra with the algebra A of 2×2 matrices with entries in k. Show that λ can be chosen so that

$$(\lambda(\mathbf{u}))' = \lambda(\mathbf{u}') \qquad (\mathbf{u} \in C_0(V))$$

where the (′) on the left-hand side denotes the transpose of a matrix and where the (′) on the right-hand side is the canonical involution on $C_0(V)$.

(iii) Hypotheses as in (ii) but suppose also that $-1 \in d(k^*)^2$. Show that the isomorphism λ can be extended in two ways to give an algebra-homomorphism

$$\mu_j\colon C(V) \to A \qquad (j = 1, 2).$$

(iv) Hypotheses as in (iii). For each j show that A is spanned as a k-vector space by $\mu_j(1)$ and the $\mu_j(\mathbf{v})$, $\mathbf{v} \in V$. Show, further, that the $\mu_j(\mathbf{v})$ are precisely the elements of A whose trace is 0.

5. Let Λ be a $\mathbf{Z}_p$-lattice ($p \neq 2$) in a 3-dimensional quadratic space V, ϕ with $k = \mathbf{Q}_p$. Suppose that ϕ takes integral values on Λ and that the determinant of the quadratic form induced on Λ is a p-adic unit. Let $\mathbf{u} \in M_0(V)$ be such that $\mathbf{T}_\mathbf{u} \in O^+(\Lambda)$ in the notation of Theorem 3.1. Show that $\mathbf{u} = l\mathbf{u}_0$ where $l \in \mathbf{Q}_p^*$, $\mathbf{u}_0 \in M_0(\Lambda)$ and $\mathbf{u}_0'\mathbf{u}_0$ is a unit of $\mathbf{Z}_p$.

[*Hint*. On replacing ϕ by $t\phi$, where $t \in \mathbf{Q}_p^*$ is a unit, we may suppose that all the hypotheses of the preceding example are satisfied. Show that μ_1 can be chosen so that the image of $M_0(\Lambda)$ consists of the group A_0 of 2×2 matrices with entries in $\mathbf{Z}_p$ whose determinant is a unit. Now use the fact that every 2×2 non-singular matrix $\mathbf{B}$ with entries in $\mathbf{Q}_p$ is of the shape $\mathbf{B} = \mathbf{B}_1\mathbf{B}_2\mathbf{B}_3$ where $\mathbf{B}_1, \mathbf{B}_3 \in A_0$ and $\mathbf{B}_2$ is diagonal.]

[*Note*. Special case of example 7.]

6. Let Λ be a $\mathbf{Z}$-lattice in a 3-dimensional quadratic space V, ϕ with $k = \mathbf{Q}$. Suppose that ϕ induces an integral form on Λ of determinant d. Let $\mathbf{u} \in M_0(V)$ be such that $\mathbf{T}_{\mathbf{u}} \in O^+(\Lambda)$. Show that $\mathbf{u} = l\mathbf{u}_0$ where $l \in \mathbf{Q}^*$, $\mathbf{u}_0 \in M_0(\Lambda)$ and $|\mathbf{u}_0'\mathbf{u}_0|_p = 1$ for all $p \nmid 2d$.

[*Hint*. Localize and apply previous example.]

[*Note*. Special case of example 7.]

7. Let I be an integral domain with quotient field k and let Λ be an I-lattice in the regular 3-dimensional k-quadratic space V, ϕ. Let $\mathbf{e}_1, \mathbf{e}_2, \mathbf{e}_3$ be an I-basis of Λ and suppose that

$$\phi(\Sigma x_i\mathbf{e}_i) = \sum_{i,j} f_{ij}x_ix_j$$

with

$$f_{ij} \in I.$$

(i) Show that $C_0(\Lambda)$ has the I-basis

$$1,\ \boldsymbol{\varepsilon}_1, \boldsymbol{\varepsilon}_2, \boldsymbol{\varepsilon}_3$$

where

$$\boldsymbol{\varepsilon}_1 = \mathbf{e}_2\mathbf{e}_3 - f_{23},$$
$$\boldsymbol{\varepsilon}_2 = \mathbf{e}_3\mathbf{e}_1 - f_{31},$$
$$\boldsymbol{\varepsilon}_3 = \mathbf{e}_1\mathbf{e}_2 - f_{12}.$$

(ii) Show that $\boldsymbol{\varepsilon}_j' = -\boldsymbol{\varepsilon}_j$ $(j = 1, 2, 3)$ and that

$$(y_1\boldsymbol{\varepsilon}_1 + y_2\boldsymbol{\varepsilon}_2 + y_3\boldsymbol{\varepsilon}_3)^2 = -\sum_{i,j} F_{ij}y_iy_j,$$

where (F_{ij}) is the adjoint matrix to (f_{ij}).

(iii) Let $\mathbf{e}_i^* = \sum_j t_{ij}\mathbf{e}_j$ $(1 \leqslant i \leqslant 3)$ be another basis of Λ, where $t_{ij} \in I$, $\det(t_{ij}) = 1$ and let $\boldsymbol{\varepsilon}_i^*$ be defined in terms of the $\mathbf{e}_i^*$ in the same way that the $\boldsymbol{\varepsilon}_i$ are defined in terms of the $\mathbf{e}_i$. Show that $\boldsymbol{\varepsilon}_i^* = \sum_j s_{ij}\boldsymbol{\varepsilon}_j$ where $s_{ij} \in I$, $\det(s_{ij}) = 1$. Show, further, that the matrix (s_{ij}) is the adjoint of (t_{ij}).

(iv) Let I be a principal ideal domain and let $\mathbf{u}$ be any element of $C_0(\Lambda)$. Show that a basis $\mathbf{e}_1$, $\mathbf{e}_1$, $\mathbf{e}_3$ of Λ can be chosen in such a way that $\mathbf{u} = l + m\boldsymbol{\varepsilon}_1$ for some $l, m \in I$.

(v) Let I be a principal ideal domain. We say that $\mathbf{u} \in C_0(\Lambda)$ is primitive if it is a primitive element of $C_0(\Lambda)$ considered as a 4-dimensional I-lattice. If $\mathbf{u}$ is primitive and $\mathbf{T}_\mathbf{u} \in O^+(\Lambda)$ [in the notation of Theorem 3.1], show that

$$\mathbf{u}'\mathbf{u} \mid 4d,$$

where $d = \det(f_{ij})$.

[*Hint*. Use (iv). If $\mathbf{u}'\mathbf{e}_i\mathbf{u} = \sum_j r_{ij}\mathbf{e}_j$ $(1 \leqslant i \leqslant 3)$ we have $\mathbf{u}'\mathbf{u} \mid r_{ij}$ for all i, j and the result follows after some manipulation. For a proof not using (iv) see Bachmann (1898), Chapter 4, Section 1.]

[*Note*. This generalizes examples 5, 6.]

CHAPTER 11

Spinor Genera

1. INTRODUCTION

In this Chapter we introduce a new subdivision of the quadratic forms over **Z** which is intermediate between the proper equivalence class and the genus, namely the spinor genus. We can always decide effectively whether or not two forms are in the same spinor genus: it turns out that under quite general conditions there is only one spinor genus in a genus. On the other hand, for indefinite forms in three or more variables, it will be a consequence of the results of the preceding chapter that any two forms in the same spinor genus are properly equivalent. Hence the results of this chapter give an effective procedure for deciding whether or not two given indefinite forms in $n \geqslant 3$ variables are properly equivalent. The corresponding problem for definite forms is trivial (at least in principle) and for indefinite forms in two variables can be dealt with by the methods of Section 3 of Chapter 13. Hence we will have filled the logical lacuna mentioned at the end of Section 1 of Chapter 9.

The theory in this chapter will be mainly developed in terms of lattices in a quadratic space V and we start by supplementing what was said in Chapter 7 and in the same general context. Let I be a principal ideal domain contained in the field k (of characteristic $\neq 2$) and let V, ϕ be a regular k-quadratic space. We shall say that two I-lattices Λ, Γ in V are *equivalent* if there is an $\alpha \in O(V)$ such that $\Gamma = \alpha\Lambda$. We shall say that the equivalence is *proper* or *improper* according as $\alpha \in O^+(V)$ or $\alpha \in O^-(V)$. If we wish to emphasize that an equivalence may be either proper or improper, then we shall speak of equivalence *in the wide sense*.

If $\mathbf{a}_1, \ldots, \mathbf{a}_n$ is any I-basis of Λ, then

$$f(x_1, \ldots, x_n) = \phi(x_1\mathbf{a}_1 + \ldots + x_n\mathbf{a}_n) \tag{1.1}$$

is a quadratic form taking values in k. Different choices of the bases $\mathbf{a}_1, \ldots, \mathbf{a}_n$ gives precisely all the forms which are I-equivalent to f (in the wide sense).

LEMMA 1.1. *The process just described sets up a 1–1 correspondence between the wide equivalence classes of I-lattices and the wide I-equivalence classes of quadratic forms which are k-equivalent to f.*

Proof. Suppose, first, that Λ, Γ are equivalent lattices, say $\Gamma = \alpha\Lambda$ and that $\mathbf{a}_1, \ldots, \mathbf{a}_n$ is a basis of Λ. Then $\alpha\mathbf{a}_1, \ldots, \alpha\mathbf{a}_n$ is a basis of Γ. Since α is an autometry we have

$$\begin{aligned}\phi(x_1\mathbf{a}_1 + \ldots + x_n\mathbf{a}_n) &= \phi(\alpha(x_1\mathbf{a}_1 + \ldots + x_n\mathbf{a}_n)) \\ &= \phi(x_1\alpha\mathbf{a}_1 + \ldots + x_n\alpha\mathbf{a}_n)\end{aligned}$$

Hence Λ, Γ correspond to the same equivalence class of quadratic forms.

Conversely, suppose that two lattices Λ, Γ correspond to the same equivalence class of quadratic forms. Then we can choose bases $\mathbf{a}_1, \ldots, \mathbf{a}_n$ of Λ and $\mathbf{b}_1, \ldots, \mathbf{b}_n$ of Γ such that identically

$$\phi(x_1\mathbf{b}_1 + \ldots + x_n\mathbf{b}_n) = \phi(x_1\mathbf{a}_1 + \ldots + x_n\mathbf{a}_n).$$

Then there is a unique linear transformation α of V into itself for which

$$\mathbf{b}_j = \alpha\mathbf{a}_j \qquad (1 \leqslant j \leqslant n).$$

Clearly α is an autometry and $\alpha\Lambda = \Gamma$, as required.

We shall actually be concerned primarily with proper equivalence and note the

COROLLARY 1. *Suppose that the class of the lattice Λ and the class of the form f correspond. Then Λ is improperly equivalent to itself precisely when f is improperly equivalent to itself.*

Proof. Clear.

When n is odd, both Λ and f have the trivial improper automorph $\mathbf{x} \to -\mathbf{x}$: so the Corollary has real content only when n is even.

Similarly when $I = \mathbf{Z}_p$, $k = \mathbf{Q}_p$, we need not distinguish between proper and improper equivalence since by Lemma 3.2, Corollary, a $\mathbf{Q}_p$-valued quadratic form is always improperly $\mathbf{Z}_p$-equivalent to itself. When $I = \mathbf{Z}$, $k = \mathbf{Q}$ we can make a more precise statement.

COROLLARY 2. *There is a 1–1 correspondence between proper equivalence classes of* $\mathbf{Z}$*-lattices in the* $\mathbf{Q}$*-quadratic space V and the proper* $\mathbf{Z}$*-equivalence classes of quadratic forms in some fixed proper* $\mathbf{Q}$*-equivalence class of forms.*

Proof. Let $\mathbf{e}_1, \ldots, \mathbf{e}_n$ be some $\mathbf{Q}$-basis for V fixed in all that follows. We shall say that a set of n linearly independent elements $\mathbf{a}_1, \ldots, \mathbf{a}_n$ is *positively oriented* if $\det(a_{ij}) > 0$ where

$$\mathbf{a}_i = \sum_j a_{ij}\mathbf{e}_j.$$

If $\mathbf{a}_1, \ldots, \mathbf{a}_n$ and $\mathbf{b}_1, \ldots, \mathbf{b}_n$ are two positively oriented bases of the same lattice Λ then the two forms $\phi(x_1\mathbf{a}_1 + \ldots + x_n\mathbf{a}_n)$ and $\phi(x_1\mathbf{b}_1 + \ldots + x_n\mathbf{b}_n)$ are properly equivalent. We map Λ into this proper equivalence class. If $\alpha \in O^+(V)$ and $\mathbf{a}_1, \ldots, \mathbf{a}_n$ is a positively oriented basis of Λ then $\alpha\mathbf{a}_1, \ldots, \alpha\mathbf{a}_n$ is a positively oriented basis of $\alpha\Lambda$. The rest is obvious.

Note. The correspondence is not "canonical" since it depends on the arbitrary choice of the basis $\mathbf{e}_1, \ldots, \mathbf{e}_n$.

After these preliminaries we return to the main subject-matter of this Chapter. Any two quadratic forms in the same genus are $\mathbf{Q}$-equivalent by Theorem 1.4 of Chapter 9, and so can be made to correspond to $\mathbf{Z}$-lattices in the *same* $\mathbf{Q}$-quadratic space. We can therefore investigate forms in a genus by considering the set of lattices in a fixed $\mathbf{Q}$-quadratic space V and their behaviour under "localization" with respect to rational primes p.

In what follows p will always denote a rational prime, that is

$$p \neq \infty. \tag{1.2}$$

The symbols Λ, Γ will be reserved for $\mathbf{Z}$-lattices in V. The symbols V_p, Λ_p, Γ_p will denote localization in the sense of Sections 6 and 7 of the preceding Chapter, so Λ_p is a $\mathbf{Z}_p$-lattice in V_p. On the other hand, $\Lambda^{(p)}, \Gamma^{(p)}$ will be used for $\mathbf{Z}_p$-lattices in V_p which are not, in general, related to any specific $\mathbf{Z}$-lattices. The group of proper autometries of V and V_p are denoted by $O^+(V)$ and $O^+(V_p)$ respectively: it is not necessary to indicate the field of definition in the notation since it is clear that it must be $\mathbf{Q}$ for $O^+(V)$ but $\mathbf{Q}_p$ for $O^+(V_p)$. We regard $O^+(V)$ as a subgroup of $O^+(V_p)$ and so use the same symbol, α (say), both for an element of $O^+(V)$ and for its extension (= localization) as an autometry of V_p. Elements of $O^+(V_p)$ will be denoted by symbols such as $\alpha_p, \beta_p, \ldots$. Note that we are being inconsistent here: α_p is *not* the localization of some α in general. Finally, if $\Lambda, \Lambda^{(p)}$ are respectively a $\mathbf{Z}$-lattice in V and a $\mathbf{Z}_p$-lattice in V_p then $O^+(\Lambda), O^+(\Lambda^{(p)})$ are respectively the subgroups of $O^+(V), O^+(V_p)$ which leave $\Lambda, \Lambda^{(p)}$ invariant (as sets).

There is a surprisingly simple criterion for deciding whether or not a set of local lattices can be the localizations of a global lattice.

THEOREM 1.1. *Let $\Gamma^{(p)}$ for all primes p be a $\mathbf{Z}_p$-lattice in V_p and let Λ be a $\mathbf{Z}$-lattice. A necessary and sufficient condition that there exists a $\mathbf{Z}$-lattice Γ such that*

$$\Gamma_p = \Gamma^{(p)} \qquad (\textit{all } p) \tag{1.3}$$

is that

$$\Gamma^{(p)} = \Lambda_p \qquad (\textit{almost all } p). \tag{1.4}$$

Further Γ, if it exists, is uniquely determined by (1.3).

The proof will be given in Section 2.

It is natural to say that two **Z**-lattices Λ, Γ are in the same *genus* if they are properly equivalent everywhere locally for every p: that is

$$\Gamma_p = \beta_p \Lambda_p \qquad \text{(all } p) \tag{1.5}$$

for some

$$\beta_p \in O^+(V_p) \qquad \text{(all } p). \tag{1.6}$$

We could have written $O(V_p)$ instead of $O^+(V_p)$ in (1.6) since every $\mathbf{Z}_p$-lattice is improperly equivalent to itself by Lemma 3.2 Corollary of Chapter 8, but in this Chapter the emphasis is on the proper orthogonal group. Clearly a genus of lattices corresponds to a genus of quadratic forms in the sense of Lemma 1.1.

By Theorem 1.1 if (1.5) holds, we have $\Gamma_p = \Lambda_p$ for almost all p and so (1.5) implies that

$$\beta_p \in O^+(\Lambda_p) \qquad \text{(almost all } p). \tag{1.7}$$

Conversely if β_p is given satisfying (1.6) and (1.7) then by Theorem 1.1 there is a unique **Z**-lattice Γ satisfying (1.5). More generally let $\{\beta'_p\}$ and $\{\beta''_p\}$ be two sets satisfying (1.6), (1.7) and let Γ', Γ'' be the corresponding lattices, so

$$\Gamma'_p = \beta'_p \Lambda_p \tag{1.8}$$
$$\Gamma''_p = \beta''_p \Lambda_p. \tag{1.9}$$

Then we have

LEMMA 1.2 (i) $\Gamma' = \Gamma''$ *if and only if there exist*

$$\gamma_p \in O^+(\Lambda_p) \tag{1.10}$$

such that

$$\beta''_p = \beta'_p \gamma_p \qquad \textit{(all } p). \tag{1.11}$$

(ii) *The lattices* Γ', Γ'' *are properly equivalent, say*

$$\Gamma'' = \alpha \Gamma' \tag{1.12}$$

with

$$\alpha \in O^+(V), \tag{1.13}$$

precisely when there exist γ_p *satisfying* (1.10) *such that*

$$\beta''_p = \alpha \beta'_p \gamma_p \qquad \textit{(all } p). \tag{1.14}$$

Proof. (i) By theorem 1.1 we have $\Gamma' = \Gamma''$ when

$$\beta''_p \Lambda_p = \beta'_p \Lambda_p \qquad \text{(all } p) \tag{1.15}$$

and this is equivalent to (1.11)

(ii) Similar.

The problem of determining the proper equivalence classes in a genus has thus been translated into one about the proper orthogonal group and its localizations, albeit a not obviously tractable one. We now explain how the problem can be broken down into two simpler ones by the application of the spinor norm.

We recall that the spinor norm θ gives maps

$$O^{+}(V) \to \mathbf{Q}^{*}/(\mathbf{Q}^{*})^{2} \tag{1.16}$$

and

$$O^{+}(V_p) \to \mathbf{Q}_p^{*}/(\mathbf{Q}_p^{*})^{2} \tag{1.17}$$

for every p. We have agreed to consider $O^{+}(V)$ as a subgroup of $O^{+}(V_p)$ and this injection corresponds to the natural homomorphism of $\mathbf{Q}^{*}/(\mathbf{Q}^{*})^{2}$ into $\mathbf{Q}_p^{*}/(\mathbf{Q}_p^{*})^{2}$. The kernels of (1.16) and (1.17) are denoted by $\Theta(V), \Theta(V_p)$ respectively and $\Theta(\Lambda) = \Theta(V) \cap O^{+}(\Lambda)$; $\Theta(\Lambda_p) = \Theta(V_p) \cap O^{+}(\Lambda_p)$. It is convenient to introduce a notation for the images of the spinor norm map. We put

$$\theta(V) = \{\theta(\alpha): \quad \alpha \in O^{+}(V)\} \tag{1.18}$$

and, similarly,

$$\theta(V_p) = \{\theta(\alpha): \quad \alpha \in O^{+}(V_p)\} \tag{1.19}$$

$$\theta(\Lambda) = \{\theta(\alpha): \quad \alpha \in O^{+}(\Lambda)\} \tag{1.20}$$

$$\theta(\Lambda_p) = \{\theta(\alpha): \quad \alpha \in O^{+}(\Lambda_p)\}. \tag{1.21}$$

We shall need

Lemma 1.3. (i) *If Δ, Γ are equivalent lattices then*

$$\theta(\Lambda) = \theta(\Gamma) \tag{1.22}$$

(ii) *If the lattices Δ, Γ are in the same genus, then*

$$\theta(\Lambda_p) = \theta(\Gamma_p) \tag{1.23}$$

for all p.

Proof. (i) If $\Gamma = \gamma\Lambda$ then $O^{+}(\Gamma)$ consists of the $\gamma\alpha\gamma^{-1}$, $\alpha \in O^{+}(\Lambda)$ and

$$\theta(\gamma\alpha\gamma^{-1}) = \theta(\gamma)\,\theta(\alpha)\,\{\theta(\gamma)\}^{-1} = \theta(\alpha). \tag{1.24}$$

Alternatively, old Nick Bourbaki would say that (1.22) holds "by transport of structure".

(ii) This follows from the local version of (i) since Γ_p is equivalent to Λ_p by the definition of genus.

Let us say temporarily that $S(\Lambda, \Gamma)$ holds for the ordered pair of lattices Λ, Γ if there exist

$$\gamma \in O^{+}(V) \tag{1.25}$$

and

$$\delta_p \in \Theta(V_p) \tag{1.26}$$

such that

$$\Gamma_p = \gamma\delta_p\Lambda_p \qquad \text{(all } p\text{)}. \tag{1.27}$$

LEMMA 1.4. (i) *The relation S just described is an equivalence relation (that is: reflexive, symmetric and transitive).*

(ii) *If* Λ, Γ *are properly equivalent then* $S(\Lambda, \Gamma)$ *holds.*

(iii) *If* $S(\Lambda, \Gamma)$ *holds then* Λ, Γ *are in the same genus.*

(iv) *If* $\Gamma', \Gamma'', \beta'_p, \beta''_p$ *(all p) are as in (1.8), (1.9) then* $S(\Gamma', \Gamma'')$ *holds precisely when there exists a* $\gamma \in O^+(V)$ *such that*

$$\theta(\beta'_p)\,\theta(\beta''_p)\,\theta(\gamma) \in \theta(\Lambda_p) \qquad \textit{(all } p\textit{)}. \tag{1.28}$$

Note. Once we have proved (i) we shall speak of an equivalence class for the relation *S* as a *spinor genus*. It then follows from (ii) and (iii) that spinor genera are intermediate between proper equivalence classes and genera. By Lemma 1.1, Corollary 2, we may also speak of a *spinor genus* of quadratic forms.

Proof. (ii) and (iii). Obvious from (1.27).

(iv) Suppose that $S(\Gamma', \Gamma'')$ holds, so there exist $\gamma \in O^+(V)$ and $\delta_p \in \Theta(V_p)$ such that

$$\beta''_p\Lambda_p = \gamma\delta_p\beta'_p\Lambda_p \qquad \text{(all } p\text{)}. \tag{1.29}$$

Hence

$$\beta''_p = \gamma\delta_p\beta'_p\varepsilon_p \tag{1.30}$$

for some

$$\varepsilon_p \in O^+(\Lambda_p). \tag{1.31}$$

On applying the spinor norm θ to (1.30) and recalling that $\theta(\delta_p) = 1$ by definition, we have

$$\theta(\beta''_p) = \theta(\gamma)\,\theta(\beta'_p)\,\theta(\varepsilon_p).$$

This implies (1.28) since θ takes its values in the commutative group $\mathbf{Q}_p^*/(\mathbf{Q}_p^*)^2$ of exponent 2 and since $\theta(\varepsilon_p) \in \theta(\Lambda_p)$.

Now suppose conversely that there exists a $\gamma \in O^+(V)$ such that (1.28) holds. By the definition of $\theta(\Lambda_p)$ there is an $\varepsilon_p \in O^+(\Lambda_p)$ such that

$$\theta(\beta'_p)\,\theta(\beta''_p)\,\theta(\gamma) = \theta(\varepsilon_p). \tag{1.32}$$

Then (1.30) defines δ_p as an element of $O^+(V_p)$. It follows from (1.30) and (1.32) that $\theta(\delta_p) = 1$ and so $\delta_p \in \Theta(V_p)$, as required.

(i) By (iv) the relation $S(\Gamma', \Gamma'')$ for two lattices in the same genus is clearly an equivalence relation. Hence (i) follows from (iii).

This concludes the proof. We note in passing that the right-hand side of (1.28) apparently depends on the choice of the lattice Λ in its genus. It is clear from the theorem that this cannot be so, and this is confirmed by Lemma 1.3 (ii).

We can now enunciate the theorems which will be the main objective of this section.

THEOREM 1.2. *The number of spinor genera in a genus is always finite and is a power of 2. The number of spinor genera in a given genus can always be determined effectively.*

The finiteness follows at once from the finiteness of the number of equivalence classes in a genus (i.e. from Theorem 1.1 of Chapter 9). That the number of spinor genera is a power of 2 is an almost immediate consequence of the fact that the spinor norm takes its values in a group of exponent 2. Hence the sting of the theorem is in the last sentence, and this will be proved in Theorem 3.1.

There is very often only one spinor genus in a genus. A simple criterion is given by

THEOREM 1.3. *Let V, ϕ be a quadratic space of dimension n. Suppose that ϕ takes integral values on the lattice Λ and denote by d the determinant of the corresponding quadratic forms. Suppose that the genus of Λ contains more than one spinor genus. Then at least one of the following must occur:*

(i) *there is an odd prime p such that*

$$p^{n(n-1)/2} \mid d.$$

(ii) *the quadratic forms induced on Λ are integral in the classical sense and*

$$2^{n(n-3)/2+[(n+1)/2]} \mid d.$$

Here the [ ] denotes "integral part". Note that the conditions (i), (ii) are not in general sufficient for the genus to contain more than one spinor genus.

The proof will be given in Section 3.

THEOREM 1.4. *Let V, ϕ be an indefinite quadratic space of dimension $n \geqslant 3$. Then every spinor genus contains precisely one proper equivalence class.*

Proof. The hard work has all been done in the previous Chapter so we can give the complete proof here. We have to show that if Γ, Λ are lattices and

$$\Gamma_p = \beta_p \Lambda_p, \qquad \beta_p \in \Theta(V_p) \qquad \text{(all } p\text{)},$$

then Γ, Λ are properly equivalent. We have already seen in (1.7) that

$$\beta_p \in \Theta(\Lambda_p) \qquad \text{(almost all } p\text{)}.$$

By Theorem 7.1, Corollary, of Chapter 10 there is an $\alpha \in \Theta(V)$ such that

$$\alpha \in \beta_p \Theta(\Lambda_p) \qquad \text{(all } p\text{)}.$$

[Take $\mathscr{V}_p = \beta_p \Theta(\Lambda_p)$.] Then $\Gamma_p = \alpha \Lambda_p$ for all p and so $\Gamma = \alpha \Lambda$, as required. This concludes the proof.

Nothing similar to Theorem 1.4 can hold for definite forms. In fact we have

LEMMA 1.5. *For every $n \geqslant 3$ there are lattices whose spinor genus contains more than one class.*

Proof.† Let Λ be the lattice with basis $\mathbf{b}_1, \dots, \mathbf{b}_n$ and let ϕ be defined by

$$\phi(\mathbf{b}_i, \mathbf{b}_j) = 0$$
$$\phi(\mathbf{b}_i) = q_i$$

where q_i is the ith odd prime (so $q_1 = 3$, $q_2 = 5, \dots$). Then the genus of Λ contains only 1 spinor genus by Theorem 1.3. On the other hand it is easy to see that 1 is represented by ϕ on Λ_p for every p (including $p = \infty$). Hence 1 is represented by ϕ on some lattice Γ in the genus of Λ by Theorem 1.3 of Chapter 9. But 1 is not represented by ϕ on Λ since

$$\phi(x_1\mathbf{b}_1 + \dots + x_n\mathbf{b}_n) = q_1x_1^2 + \dots + q_nx_n^2 > 1.$$

for $x_1, \dots, x_n \in \mathbf{Z}$, not all zero. Hence Λ, Γ are in the same spinor genus but are not equivalent.

In view of its importance we enunciate explicitly a consequence of Theorems 1.3 and 1.4 in the language of quadratic forms.

THEOREM 1.5. *Let $f(x_1, \dots, x_n)$ be an indefinite form in $n \geqslant 3$ variables which is integer-valued and has determinant $d \neq 0$. Suppose that*

(i) $p^{n(n-1)/2} \nmid d$ *for every odd prime p.*

(ii) either *f is not integral in the classical sense* or $2^{n(n-3)/2+[(n+1)/2]} \nmid d$.

Then every form in the genus of f is properly equivalent to f. Further, f is improperly equivalent to itself.

Proof. The form f corresponds to a lattice Λ in some $\mathbf{Q}$-quadratic space V, ϕ. By Theorems 1.3 and 1.4 every lattice in the genus of Λ is properly equivalent to Λ. The last sentence of the enunciation follows because two forms which are improperly equivalent are in the same genus: or, alternatively, by Lemma 1.1, Corollary 1.

The discussion of spinor genera in this chapter is largely in terms of lattices. In Sections 4–6 we shall indicate how this discussion can be interpreted in terms of quadratic forms. Let f, g be integral forms of the same genus in $n \geqslant 3$ variables. By Theorem 1.4 of Chapter 9, the forms are equivalent over the ring $\mathbf{Z}^{(P)}$ where, as usual, P is a finite set of primes p and $\mathbf{Z}^{(P)}$ consists of the rationals which are p-adic integers for all primes $p \in P$.

† Lemma 6.6 of Chapter 9 and its Corollary show that there are spinor genera with $n = 3$ whose class number is arbitrarily large. The genus $\mathscr{G}$ considered there consists of a single spinor genus by Theorem 1.3. See also Appendix A.

We may suppose that $2 \in P$ and that the determinant d of the genus is a unit for all primes $q \notin P$. One particular form of such an equivalence is the following. Let

$$f(\mathbf{x}) = \Sigma f_{ij} x_i x_j$$

and let $r \in \mathbf{Z}$ be a p-adic unit for all $p \in P$. The form

$$h(\mathbf{x}) \text{ (say)} = f(r^{-1}x_1, rx_2, x_3, \ldots, x_n) \tag{1.33}$$

will be integral provided that f is integral and

$$r^2 | f_{11}, \quad r | f_{1j} \qquad (j > 2).$$

Clearly then h is $\mathbf{Z}^{(P)}$-equivalent to f. It turns out that every $\mathbf{Z}^{(P)}$-equivalence can be expressed as the composition of a sequence of $\mathbf{Z}$-equivalences and of equivalences of the special type (1.33). Further, the spinor genus of h depends only on r and on the spinor genus of f.

The results described above give a very satisfactory understanding of the forms in a genus of indefinite forms, at least when the number of variables is at least three. They shed rather less light on the theory of definite forms. It has already been remarked more than once that there is no difficulty in principle in deciding whether two definite forms over $\mathbf{Z}$ are equivalent but the process does not give much insight. In Sections 7, 8 we consider some of the implications of the theory for definite forms and we conclude in Section 9 by proving

THEOREM 1.6. *Let $f(\mathbf{x})$ be a positive definite integral form in $n \geqslant 4$ variables. Then there is an integer N with the following property:*

Let $a \geqslant N$ be an integer which is primitively represented by $f(\mathbf{x})$ over $\mathbf{Z}_p$ for all primes p. Then a is primitively represented by f over $\mathbf{Z}$.

We shall show by examples that the conclusion of the theorem no longer holds true for $n = 3$ or if the word "primitively" is omitted.

2. LOCALIZATION OF LATTICES

In this section we prove Theorem 1.1.

Suppose, first, that Λ, Γ are two $\mathbf{Z}$-lattices in the same $\mathbf{Q}$-vector space V, say with respective bases $\mathbf{a}_1, \ldots, \mathbf{a}_n$ and $\mathbf{b}_1, \ldots, \mathbf{b}_n$. Then there are $s_{ij}, t_{ij} \in \mathbf{Q}$ such that

$$\mathbf{a}_i = \sum_{j=1}^{n} s_{ij} \mathbf{b}_j \qquad (1 \leqslant i \leqslant n) \tag{2.1}$$

and

$$\mathbf{b}_i = \sum_{j=1}^{n} t_{ij} \mathbf{a}_j \qquad (1 \leqslant i \leqslant n). \tag{2.2}$$

Then

$$s_{ij} \in \mathbf{Z}_p, \qquad t_{ij} \in \mathbf{Z}_p \qquad (1 \leqslant i \leqslant n, 1 \leqslant j \leqslant n) \tag{2.3}$$

for almost all p. Clearly (2.3) implies

$$\Gamma_p = \Lambda_p \tag{2.4}$$

and so (2.4) holds for almost all p.

Now let $\mathbf{Z}_p$-lattices $\Gamma^{(p)}$ be given in V_p for all p and suppose that

$$\Gamma^{(p)} = \Lambda_p \qquad \text{(almost all } p\text{)}. \tag{2.5}$$

We have to show that there is a $\mathbf{Z}$-lattice Γ such that

$$\Gamma_p = \Gamma^{(p)} \qquad \text{(all } p\text{)}. \tag{2.6}$$

Let P be a finite set of primes p such that

$$\Gamma^{(p)} = \Lambda_p \qquad \text{(all } p \notin P\text{)}. \tag{2.7}$$

For each $p \in P$ there is an integer $m(p) \geqslant 0$ such that

$$p^{m(p)}\Gamma^{(p)} \subset \Lambda_p.$$

On considering the lattice

$$\prod_{p \in P} p^{-m(p)}\Lambda$$

instead of Λ we do not disturb (2.7) and may suppose without loss of generality that

$$\Gamma^{(p)} \subset \Lambda_p \qquad \text{(all } p \in P\text{)}. \tag{2.8}$$

Put

$$\Gamma = \bigcap_{p \in P} (\Lambda \cap \Gamma^{(p)}). \tag{2.9}$$

We shall show that Γ is a lattice and that it satisfies (2.6).

For $p \in P$ there are integers $h(p) \geqslant 0$ such that

$$p^{h(p)}\Lambda_p \subset \Gamma^{(p)}. \tag{2.10}$$

Hence

$$\prod_{p \in P} p^{h(p)}\Lambda \subset \Gamma \subset \Lambda. \tag{2.11}$$

Since Γ as defined by (2.9) is clearly a group under addition, it must be a lattice by Lemma 3.3 of Chapter 7.

It is clear from (2.7) and (2.9) that

$$\Gamma_p \subset \Gamma^{(p)} \qquad \text{(all } p\text{)}.$$

To conclude the proof we have thus only to show that $\Gamma^{(p)} \subset \Gamma_p$. Since Γ_p is the closure of Γ in V_p with respect to the p-adic topology it will be enough to show that any point of $\Gamma^{(p)}$ can be arbitrarily closely approximated p-adically by points of Γ. This is a straightforward exercise in the Strong Approximation Theorem for $\mathbf{Z}$ (= Chinese Remainder Theorem), Lemma 3.1 of Chapter 3. More precisely, let p be fixed, and let $\mathbf{c}^{(p)}$ be any element of $\Gamma^{(p)}$. Then

$$\mathbf{c}^{(p)} = c_1^{(p)}\mathbf{a}_1 + \ldots + c_n^{(p)}\mathbf{a}_n \tag{2.12}$$

where $\mathbf{a}_1, \ldots, \mathbf{a}_n$ is a fixed basis of Λ and $c_i^{(p)} \in Z_p$ $(1 \leqslant i \leqslant n)$. Let N be arbitrarily large, in particular

$$N \geqslant h(p) \tag{2.13}$$

if $p \in P$. By the Chinese Remainder Theorem there are $c_1, \ldots, c_n \in \mathbf{Z}$ such that

$$c_i \equiv c_i^{(p)} \pmod{p^N} \qquad (1 \leqslant i \leqslant n) \tag{2.14}$$

and

$$c_i \equiv 0 \pmod{q^{h(q)}} \qquad (1 \leqslant i \leqslant n, q \in P, q \neq p). \tag{2.15}$$

Put

$$\mathbf{c} = c_1\mathbf{a}_1 + \ldots + c_n\mathbf{a}_n.$$

Then

$$\mathbf{c} \in q^{h(q)}\Lambda \subset \Gamma^{(q)} \qquad (q \in P, q \neq p)$$

by (2.15). If $p \in P$ we have

$$\mathbf{c}^{(p)} - \mathbf{c} \in p^{h(p)}\Lambda \subset \Gamma^{(p)}$$

by (2.13), (2.14); and so $\mathbf{c} \in \Gamma^{(p)}$ since $\mathbf{c}^{(p)} \in \Gamma^{(p)}$ by hypothesis. Hence $\mathbf{c} \in \Gamma$ by the definition (2.19). But $\mathbf{c}$ can be made arbitrarily close p-adically to $\mathbf{c}^{(p)}$ by choice of N, and so $\mathbf{c}^{(p)} \in \Gamma_p$. This concludes the proof.

By similar arguments the reader can easily prove

LEMMA 2.1. *Suppose that Λ, Γ are $\mathbf{Z}$-lattices with $\Gamma \subset \Lambda$. Then the index of Γ in Λ is the product of the indices of Γ_p in Λ_p over all primes p.*

We conclude this section with a straightforward translation of Theorem 2.1 of Chapter 9 into our present language. If Λ is an I-lattice, where I is a ring, we denote by $SL(\Lambda) = SL_I(\Lambda)$ the group of automorphs of Λ with determinant $+1$.

THEOREM 2.1. *Let Λ be a $\mathbf{Z}$-lattice and for a finite set P of primes p let $\beta_p \in SL(\Lambda_p)$ be given. Then there is a $\beta \in SL(\Lambda)$ which is arbitrarily close p-adically to each of the β_p.*

3. NUMBER OF SPINOR GENERA

In this section we show that the number of spinor genera in a genus is equal to the order of a certain group which depends, of course, on the genus and which is readily computable. We shall suppose throughout that

$$n \geqslant 3, \tag{3.1}$$

where n is the dimension. There are additional complications when $n = 2$ but if he so wishes the reader should have little difficulty in making the appropriate modification to the arguments and to the enunciations (or see Kneser (1956)). We shall not need the theory of spinor genera for $n = 2$ since then we have the fuller information given by the composition of binary forms (Chapter 14).

LEMMA 3.1. *Suppose that* $n \geqslant 3$. *Then*

$$\theta(V_p) = \mathbf{Q}_p^*. \tag{3.2}$$

Proof. The form ϕ represents all the elements of $\mathbf{Q}_p^*$ on V_p except possibly when $n = 3$. In the exceptional case the elements of $\mathbf{Q}_p^*$ which cannot be represented lie in a single coset $e_p(\mathbf{Q}_p^*)$ (say) of $\mathbf{Q}_p^*$ modulo $(\mathbf{Q}_p^*)^2$ by Lemma 2.5, Corollary, of Chapter 4. Let w be any element of $\mathbf{Q}_p^*$. Let u, v be elements of $\mathbf{Q}_p^*$ with

$$uv = w.$$

In the exceptional case we may suppose, further, that

$$u \notin e_p(\mathbf{Q}_p^*)^2, \qquad v \notin e_p(\mathbf{Q}_p^*)^2$$

since $\mathbf{Q}_p^*/(\mathbf{Q}_p^*)^2$ has order 4 or 8. There are $\mathbf{a}, \mathbf{b} \in V_p$ such that

$$\phi(\mathbf{a}) = u, \qquad \phi(\mathbf{b}) = v.$$

Then

$$\sigma = \tau_{\mathbf{a}}\tau_{\mathbf{b}} \in O^+(V_p)$$

and

$$\theta(\sigma) = \phi(\mathbf{a})\,\phi(\mathbf{b})\,(\mathbf{Q}_p^*)^2 = w(\mathbf{Q}_p^*)^2.$$

This completes the proof.

LEMMA 3.2. *Suppose that* $n \geqslant 3$. *Then*

$$\theta(V) = \mathbf{Q}^* \quad (V, \phi \textit{ indefinite}) \tag{3.3}$$

$$= \mathbf{Q}_+ \quad (V, \phi \textit{ definite}), \tag{3.4}$$

where $\mathbf{Q}_+$ *denotes the group of positive elements of* $\mathbf{Q}^*$.

Proof. We use the Hasse principle (more precisely, Theorem 1.1 Corollary 2 of Chapter 6), together with the arguments in the preceding proof. The form ϕ represents all elements of each $\mathbf{Q}_p^*$ over V_p except that when $n = 3$ there may be a finite set P of primes p such that the elements of a single coset $e_p(\mathbf{Q}_p^*)^2$ are omitted. Let w be any element of $\mathbf{Q}^*$ or $\mathbf{Q}_+$ as the case may be. Then we can pick $u, v \in \mathbf{Q}^*$ such that $uv = w$ and

$$u \notin e_p(\mathbf{Q}_p^*)^2, \quad v \notin e_p(\mathbf{Q}_p^*)^2 \qquad (\text{all } p \in P).$$

If V, ϕ is definite, we may suppose, further, that the sign of u is chosen so that u is represented by ϕ over V_∞; and then v is also so represented. By the Hasse principle, there exist $\mathbf{a}, \mathbf{b} \in V$ such that $\phi(\mathbf{a}) = u, \phi(\mathbf{b}) = v$. The argument is now the same as for the previous lemma.

LEMMA 3.3. *Suppose that $n \geqslant 2$. Let $p \neq 2$ and let $\Lambda^{(p)}$ be a lattice in V_p. Suppose that ϕ takes integral values on $\Lambda^{(p)}$ and that the determinant of the quadratic form induced on $\Lambda^{(p)}$ is a p-adic unit. Then*

$$\theta(\Lambda^{(p)}) = U_p(\mathbf{Q}_p^*)^2, \tag{3.5}$$

where U_p denotes the group of p-adic units.

Proof. Every $u \in U_p$ is represented by ϕ on $\Lambda^{(p)}$ (e.g. by Lemma 3.4 of Chapter 8). As in the proof of Lemma 3.1 it follows that $\theta(\Lambda^{(p)}) \supset U_p(\mathbf{Q}_p^*)^2$.

By Lemma 3.3, Corollary 2, of Chapter 8, the conditions of the enunciation imply that every $\sigma \in O(\Lambda^{(p)})$ can be expressed as the product of a number of $\tau_{\mathbf{b}}$ where $\mathbf{b} \in \Lambda^{(p)}$ and $\phi(\mathbf{b}) \in U_p$. In particular, if $\sigma \in O^+(\Lambda^{(p)})$ we have $\theta(\sigma) \in U_p(\mathbf{Q}_p^*)^2$. This concludes the proof.

We now construct a group whose order will be shown to be the number of spinor genera in the genus of a given lattice Λ. We denote by P any finite set of primes such that

$$2 \in P \tag{3.6}$$

and

$$\theta(\Lambda_p) = U_p(\mathbf{Q}_p^*)^2 \qquad (\text{all } p \notin P). \tag{3.7}$$

Such sets P exist by Lemma 3.3. Note that, by Lemma 1.3, the conditions (3.6), (3.7) depend only on the genus of Λ, not on Λ itself. Let

$$R = R_P = \prod_{p \in P} \mathbf{Q}_p^* \tag{3.8}$$

be the product of the groups $\mathbf{Q}_p^*$ $(p \in P)$: that is, its elements are the sets of J-tuples

$$\{c(p)\}_{p \in P} = \{c(p_1), c(p_2), \ldots, c(p_J)\} \tag{3.9}$$

where $p_1, \ldots, p_J$ are the elements of P in some order, and where the $c(p_j)$ run independently through $\mathbf{Q}_p^*$ $(p = p_j)$: the group laws being defined component-wise. In an obvious sense R contains

$$S = S_P = \prod_{p \in P} \theta(\Lambda_p) \tag{3.10}$$

and

$$R \supset S \supset R^2 \tag{3.11}$$

where R^2 is the set of $r^2, r \in R$.

We denote by $T = T_P$ the set of $t \in \mathbf{Q}^*$ such that

$$|t|_p = 1 \qquad (\text{all } p \notin P) \tag{3.12}$$

and

$$t > 0 \quad \text{if } V, \phi \text{ is definite.} \tag{3.13}$$

We can regard T as a subgroup of R by the "diagonal map", that is identifying t with

$$\{t, t, \ldots, t\} \in R. \tag{3.14}$$

Finally, we define the quotient group

$$G = G_P = R_P/S_P T_P \tag{3.15}$$

where $S_P T_P$ denotes the group consisting of the st, $s \in S$, $t \in T$. By (3.11) we have

$$g^2 = 1 \qquad (\text{all } g \in G). \tag{3.16}$$

It will follow from what is proved below, or can be verified without trouble directly, that the group G_P does not really depend on P, provided (of course) that P satisfies (3.6) and (3.7).

LEMMA 3.4. *Suppose that* $n \geqslant 3$. *Every lattice* Δ *in the genus of* Λ *is properly equivalent to a lattice* Γ *for which*

$$\Gamma_p = \beta_p \Lambda_p, \qquad \beta_p \in O^+(V_p) \tag{3.17}$$

with

$$\theta(\beta_p) \in U_p(\mathbf{Q}_p^*)^2 \qquad (p \in P) \tag{3.18}$$

and

$$\theta(\beta_p) = (\mathbf{Q}_p^*)^2 \qquad (p \notin P). \tag{3.19}$$

Proof. Suppose that

$$\Delta_p = \delta_p \Lambda_p \qquad \delta_p \in O^+(V_p)$$

where, without loss of generality, δ_p is the identity for almost all p. We can then choose a positive rational c such that

$$c\theta(\delta_p) \in U_p(\mathbf{Q}_p^*)^2 \qquad (\text{all } p).$$

By Lemma 3.2 there is a $\gamma \in O^+(V)$ with $\theta(\gamma) = c$. Put

$$\Gamma = \gamma\Delta \tag{3.20}$$

so

$$\Gamma_p = \gamma\delta_p\Lambda_p \qquad (\text{all } p)$$

with

$$\theta(\gamma\delta_p) \in U_p(\mathbf{Q}_p^*)^2.$$

By (3.7) for each $p \notin P$ there is an $\alpha_p \in O^+(\Lambda_p)$ such that

$$\theta(\alpha_p) = \theta(\gamma\delta_p). \tag{3.21}$$

Then

$$\beta_p = \begin{cases} \gamma\delta_p & (p \in P) \\ \gamma\delta_p\alpha_p & (p \notin P) \end{cases}$$

satisfy (3.17), (3.18) and (3.19), as required.

COROLLARY. *Every spinor genus in the genus of* Λ *contains a lattice* Γ *of the form* (3.17), (3.18) *for which*

$$\Gamma_p = \Lambda_p \qquad (p \notin P).$$

Proof. The statement (3.19) is equivalent to $\beta_p \in \Theta(V_p)$ $(p \notin P)$. Define a new lattice Γ' by

$$\Gamma'_p = \Gamma_p = \beta_p\Lambda_p \qquad (p \in P)$$
$$\Gamma'_p = \Lambda_p \qquad (p \notin P).$$

Then Γ' is in the same spinor genus as Γ. Since Γ, Δ are equivalent, the lattice Γ' satisfies the conclusions of the Corollary.

LEMMA 3.5 $(n \geqslant 3)$. *Let* Γ, Δ *be in the genus of* Λ *with*

$$\Gamma_p = \beta_p\Lambda_p, \qquad \Delta_p = \delta_p\Lambda_p, \qquad \beta_p, \delta_p \in O^+(V_p) \tag{3.23}$$

and

$$\beta_p, \delta_p \in \Theta(V_p) \qquad (p \notin P). \tag{3.24}$$

Then a necessary and sufficient condition that Γ, Δ *be in the same spinor genus is that*

$$\{\theta(\beta_p)\}_{p\in P} \text{ and } \{\theta(\delta_p)\}_{p\in P}$$

should be in the same coset of R *modulo* ST.

Note. The notation is as in (3.9) and (3.15). There is an "abuse of notation" in that $\theta(\beta_p)$, $\theta(\delta_p)$ are defined as cosets of $\mathbf{Q}_p^*$ modulo $(\mathbf{Q}_p^*)^2$, not as elements of $\mathbf{Q}_p^*$. But by (3.11) it is immaterial which representatives of $\theta(\beta_p)$, $\theta(\delta_p)$ in $\mathbf{Q}_p^*$ are chosen.

Proof. By Lemma 1.4(iv) a necessary and sufficient condition that Γ, Δ be in the same spinor genus is that there is a $\gamma \in O^+(V)$ such that

$$\theta(\beta_p)\,\theta(\delta_p)\,\theta(\gamma) \in \theta(\Lambda_p) \qquad (\text{all } p). \tag{3.25}$$

By (3.7) and (3.24) the condition (3.25) for $p \notin P$ is equivalent to

$$\theta(\gamma) \in U_p(\mathbf{Q}_p^*)^2 \qquad (\text{all } p \notin P)$$

and so by (3.12) to

$$\theta(\gamma) \in T(\mathbf{Q}^*)^2. \tag{3.26}$$

On the other hand, the condition (3.25) for all $p \in P$ is equivalent to

$$\{\theta(\beta_p)\,\theta(\delta_p)\,\theta(\gamma)\}_{p\in P} \in S. \tag{3.27}$$

By (3.11) this concludes the proof.

COROLLARY. *Let Γ be a lattice in the genus of Λ with*

$$\Gamma_p = \Lambda_p \qquad (\textit{all } p \notin P) \tag{3.28}$$

and let Δ be the lattice determined by

$$\Delta_p = \alpha_p\Gamma_p \qquad (p \in P) \tag{3.29}$$

$$\Delta_p = \Lambda_p \qquad (p \notin P) \tag{3.30}$$

where

$$\alpha_p \in O^+(V_p). \tag{3.31}$$

Then the spinor genus $\mathscr{S}_2$ (say) of Δ depends only on the spinor genus $\mathscr{S}_1$ of Γ and the image g of $\{\theta(\alpha_p)\}_{p\in P}$ in $G = R/ST$.

Proof. Clear.

In the situation of this Corollary we shall write

$$\mathscr{S}_2 = g\mathscr{S}_1. \tag{3.32}$$

THEOREM 3.1. (i) *The spinor genus $g\mathscr{S}_1$ is defined for all $g \in G$ and every spinor genus $\mathscr{S}_1$ in the genus of Λ.*

(ii) $g_1\mathscr{S}_1 = g_2\mathscr{S}_1$ *only when* $g_1 = g_2$ $(g_1, g_2 \in G)$.

(iii) *if $\mathscr{S}_1, \mathscr{S}_2$ are any two spinor genera in the genus then there is a $g \in G$ such that $\mathscr{S}_2 = g\mathscr{S}_1$.*

(iv) $g_1(g_2\mathscr{S}_1) = (g_1g_2)\mathscr{S}_1$ *for $g_1, g_2 \in G$, where g_1g_2 denotes the group product in G.*

Note 1. This proves Theorem 1.2.

Note 2. In technical parlance Theorem 3.1 states that the set of spinor genera in the genus is a "principal homogeneous space" over G.

Proof. (i) By Lemma 3.4 Corollary every spinor genus $\mathscr{S}_1$ of the genus of Λ contains a lattice Γ satisfying (3.28). Let g be any element of G and let $\mathbf{r} = \{r_p\}_{p\in P}$ be any representative of g in R. By Lemma 3.1 there are $\alpha_p \in O^+(V_p)$ with $\theta(\alpha_p) = r_p(\mathbf{Q}_p^*)^2$. Then, by definition, (3.29) and (3.30) determine a lattice Δ in the spinor genus $g\mathscr{S}_1$.

(ii), (iii) and (iv). These are all immediate consequences of Lemma 3.5.

The definition of the action of G on the set of spinor genera depends *prima facie* on the choice of a special lattice Λ in the genus, since we have used (3.28). In fact we have the

COROLLARY. *The definition of* (3.32) *is independent of the choice of* Λ *in its genus. More precisely: let* Γ *be any lattice in the genus of* Λ *(not necessarily satisfying* (3.28)) *and let* Δ *be any lattice for which*

$$\Delta_p = \beta_p\Gamma_p \qquad \beta_p \in O^+(V_p) \qquad (p\in P)$$
$$\Delta_p = \Gamma_p \qquad (p \notin P).$$

Then $\mathscr{S}_2 = g\mathscr{S}_1$, *where* $\mathscr{S}_1, \mathscr{S}_2$ *are the spinor genera of* Γ, Δ *respectively and where* g *is the image of* $\{\theta(\beta_p)\}_{p\in P}$ *in* G.

Proof. Clear by the proof of Lemma 3.4, Corollary.

In the proof above we have made no use of (3.18). This indicates that we could have used a smaller group instead of R in the description of G. We shall now verify this by purely group-theoretic manipulations. The disadvantage is that the description of the subgroup Y below is more complicated than that of ST.

Put

$$X = \prod_{p\in P} U_p \subset R = \prod_{p\in P} \mathbf{Q}_p^* \tag{3.33}$$

and

$$Y = X \cap ST. \tag{3.34}$$

Clearly $R = XT$ and so

$$G = R/ST = X/Y. \tag{3.35}$$

The complication in the description of Y arises from the possible existence of primes $q \in P$ for which there is an $\alpha_q \in O^+(\Lambda_q)$ with

$$\theta(\alpha_q) \subset qU_q(\mathbf{Q}_q^*)^2.$$

Let P_1 be the (possibly empty) set of such q and for each q select one α_q. Put

$$\mathbf{s}_q = \{s_{qp}\}_{p\in P}$$

where

$$s_{qp} = q \qquad (p \neq q)$$

and

$$s_{qq} \in q\theta(\alpha_q) \cap U_q.$$

Clearly $\mathbf{s}_q \in ST$ and so $\mathbf{s}_q \in Y$.

LEMMA 3.6. *The group Y is generated by the following sets:*

(i) $S \cap X = \prod_{p \in P} (\theta(\Lambda_p) \cap U_p)$;

(ii) $T \cap X = \begin{cases} \pm 1 & \textit{if } V, \phi \textit{ is indefinite}; \\ +1 & \textit{if } V, \phi \textit{ is definite}; \end{cases}$

(iii) *the* $\mathbf{s}_q$ *described above.*

Proof. Clear.

COROLLARY. *Suppose that* $U_p \subset \theta(\Lambda_p)$ *for all p. Then the genus of* Λ *consists of a single spinor genus.*

Proof. For then $S \cap X = X$.

We conclude this section by giving sufficient (but not necessary) conditions for the hypothesis of Lemma 3.6, Corollary, to hold. We shall say that a lattice $\Lambda' \subset \Lambda$ in a subspace V' of V is a *direct orthogonal summand*† of Λ if there is a lattice $\Lambda'' \subset \Lambda$ in the orthogonal complement $(V')^\perp$ of V' such that every $\mathbf{a} \in \Lambda$ can be put in the form $\mathbf{a} = \mathbf{a}' + \mathbf{a}''$ ($\mathbf{a}' \in \Lambda'$, $\mathbf{a}'' \in \Lambda''$). In terms of quadratic forms this means that there is a basis $\mathbf{b}_1, \ldots, \mathbf{b}_n$ of Λ such that $\mathbf{b}_1, \ldots, \mathbf{b}_m$ ($m = \dim \Lambda'$) is a basis for Λ' and

$$\phi(x_1\mathbf{b}_1 + \ldots + x_n\mathbf{b}_n) = \phi(x_1\mathbf{b}_1 + \ldots + x_m\mathbf{b}_m) + \phi(x_{m+1}\mathbf{b}_{m+1} + \ldots + x_n\mathbf{b}_n).$$

When Λ' is a direct orthogonal summand of Λ, every element of $O^+(\Lambda')$ can be extended to an element of $O^+(\Lambda)$ by making it act trivially on Λ''. Hence

$$\theta(\Lambda') \subset \theta(\Lambda).$$

We shall apply this remark to the Λ_p.

LEMMA 3.7. *Let* $p \neq 2$. *Suppose that* $\Lambda^{(p)}$ *is a 2-dimensional* $\mathbf{Z}_p$*-lattice with a basis* $\mathbf{b}_1, \mathbf{b}_2$ *such that*

$$\phi(x_1\mathbf{b}_1 + x_2\mathbf{b}_2) = a_1x_1^2 + a_2x_2^2$$

with

$$|a_1|_p = |a_2|_p \neq 0.$$

Then $U_p \subset \theta(\Lambda^{(p)})$.

† For more on this concept, see Appendix A.

Proof. Follows from Lemma 3.3.

COROLLARY. *Suppose that ϕ takes p-adic integral values on Λ_p and that $U_p \not\subset \theta(\Lambda_p)$. Then the determinant of Λ_p is divisible by $p^{n(n-1)/2}$.*

Proof. By Theorem 3.1 of Chapter 8 there is a $\mathbf{Z}_p$-basis $\mathbf{b}_1, \ldots, \mathbf{b}_n$ of Λ_p such that

$$\phi(x_1\mathbf{b}_1 + \ldots + x_n\mathbf{b}_n) = a_1x_1^2 + a_2x_2^2 + \ldots + a_nx_n^2$$

with $a_j \in \mathbf{Z}_p$. If two of the $|a_j|_p$ were equal, then Λ_p would contain a direct orthogonal summand of the type $\Lambda^{(p)}$ and so we should have $U_p \subset \theta(\Lambda^{(p)}) \subset \theta(\Lambda_p)$, contrary to hypothesis. Hence the $|a_j|_p$ are all different and the power of p dividing $a_1a_2\ldots a_n$ is at least

$$0 + 1 + \ldots + (n-1) = n(n-1)/2.$$

As usual, the 2-adic case is more complicated.

LEMMA 3.8 ($p = 2$). *The following cases imply that $U_2 \subset \theta(\Lambda^{(2)})$.*

(i) *$\Lambda^{(2)}$ is 2-dimensional, with basis $\mathbf{b}_1, \mathbf{b}_2$ and*
$\phi(x_1\mathbf{b}_1 + x_2\mathbf{b}_2) = 2^e x_1x_2$ *or* $= 2^e(x_1^2 + x_1x_2 + x_2^2)$
for some $e \in \mathbf{Z}$.

(ii) *$\Lambda^{(2)}$ is 3-dimensional with basis $\mathbf{b}_1, \mathbf{b}_2, \mathbf{b}_3$,*
$\phi(x_1\mathbf{b}_1 + x_2\mathbf{b}_2 + x_3\mathbf{b}_3) = a_1x_1^2 + a_2x_2^2 + a_3x_3^2$
and

$$|a_1|_2 = |a_2|_2,$$

$$|a_3|_2 = \begin{cases} |a_1|_2 & \textit{or} \\ 2|a_1|_2 & \textit{or} \\ \tfrac{1}{2}|a_1|_2. \end{cases}$$

(iii) *As in* (ii) *but with*

$$|a_2|_2 = 2|a_1|_2, \qquad |a_3|_2 = 4|a_1|_2.$$

Proof. This is left to the reader. For a stronger result see Earnest and Hsia (1975) Section IV.

COROLLARY. *Suppose that ϕ takes 2-adic integral values on Λ_2 and that $U_2 \not\subset \theta(\Lambda_2)$. Then the quadratic form induced on Λ_2 by ϕ is 2-adically integral in the classical sense and its determinant is divisible by 2^T, where $T = \frac{1}{2}n(n-3) + [\frac{1}{2}(n+1)]$.*

Proof. By Lemma 4.1 of Chapter 8, the lattice Λ_2 is the direct orthogonal sum of 1-dimensional lattices and of 2-dimensional lattices of type (i) of the Lemma. By the hypotheses of the Corollary, the latter cannot occur and so there is a $\mathbf{Z}_2$-basis of Λ_2 for which

$$\phi(x_1\mathbf{b}_1 + \ldots + x_n\mathbf{b}_n) = a_1x_1^2 + \ldots + a_nx_n^2$$

with $a_j \in \mathbf{Z}_2$. The condition that Λ_2 can contain no direct summands $\Lambda^{(2)}$ of types (ii) or (iii) shows that the power of 2 dividing $a_1a_2 \ldots a_n$ must be at least

$$\begin{aligned} &0 + 0 + 2 + 2 + 4 + \ldots \\ &\qquad = \tfrac{1}{2}n(n-3) + [\tfrac{1}{2}(n+1)]. \end{aligned}$$

This concludes the proof of the Corollary.

Finally we note that Theorem 1.3 is an immediate consequence of the Corollaries to Lemmas 3.6, 3.7 and 3.8.

4. AN ALTERNATIVE APPROACH

Let Λ, Γ be two $\mathbf{Z}$-lattices in the $\mathbf{Q}$-quadratic space V, ϕ and suppose that they have the *same determinant* (with respect to some fixed basis of V: alternatively one could say that the relative determinant $d(\Lambda/\Gamma)$ of Chapter 7, Section 3 is 1). Then $\Lambda \cap \Gamma$ is also a lattice and it has the same index in Λ and Γ. Put

$$I(\Lambda, \Gamma) = [\Lambda \cap \Gamma : \Lambda] = [\Lambda \cap \Gamma : \Gamma]. \tag{4.1}$$

By Lemma 2.1 the localizations Λ_p, Γ_p satisfy

$$\Lambda_p = \Gamma_p \tag{4.2}$$

if and only if

$$p \nmid I(\Lambda, \Gamma). \tag{4.3}$$

Since Λ, Γ have the same determinant, the quadratic forms induced by ϕ on Λ and Γ also have the same determinant.

LEMMA 4.1. *Suppose that Λ, Γ are as above and that $I(\Lambda, \Gamma)$ is odd. For all $p | I(\Lambda, \Gamma)$ suppose that the quadratic forms induced by ϕ on Λ, Γ are p-adically integral and that their determinants are p-adic units. Then Λ, Γ are in the same genus.*

Proof. The hypotheses imply that for $p | I(\Lambda, \Gamma)$ the quadratic forms induced on Λ_p, Γ_p are $\mathbf{Z}_p$-equivalent (Lemma 3.4 of Chapter 8) and so the lattices Λ_p, Γ_p are equivalent. For $p \nmid I(\Lambda, \Gamma)$ we have (4.2). This proves the Lemma. The next Lemma goes in the opposite direction.

LEMMA 4.2. *Let Λ, Δ be lattices in the same genus and let P_1 be any finite set of primes p. Then there is a lattice Γ in the proper equivalence class of Δ such that $I(\Lambda, \Gamma)$ is not divisible by any $p \in P_1$.*

Note. This is essentially the translation into lattice terms of the theorem that forms in the same genus are "semi-equivalent". (Theorem 1.4 of Chapter 9.)

Proof. Suppose that

$$\Delta_p = \delta_p \Lambda_p \qquad \delta_p \in O^+(V_p)$$

for $p \in P_1$. By the "weak approximation theorem" (Theorem 7.2 of Chapter 9) there is a $\delta \in O^+(V)$ such that

$$\delta \in \delta_p O^+(\Lambda_p) \qquad (\text{all } p \in P_1).$$

Put

$$\Gamma = \delta^{-1}\Delta.$$

Then $\Gamma_p = \Lambda_p$ for all $p \in P_1$. The conclusion of the Lemma now follows from the equivalence of (4.2) and (4.3).

We now discuss the relation between the machinery we have just constructed and that of the preceding section. We consider one fixed genus, say that of the lattice Λ, and denote by P^* any finite set of primes such that

(i) $2 \in P^*$

(ii) For all $p \notin P^*$ the quadratic form induced by ϕ on Λ_p is integral and its determinant is a p-adic unit.

Note that these conditions depend only on the genus of Λ and not on Λ itself. Further, P^* satisfies the conditions (3.6) (3.7) imposed on the set P, but (ii) may be strictly stronger than (3.7). In any case we shall suppose that

$$P \subset P^*. \tag{4.4}$$

Next we must ask the reader to recall the definition of R, S, T, G given in (3.8)–(3.15). The group $\mathbf{Q}^*$ can be identified with a subgroup of $R = \prod_{p \in P} \mathbf{Q}_p^*$ by the "diagonal map" (cf. (3.14)). For $c \in \mathbf{Q}^*$ we shall denote by $g(c) \in G$ the image of c in the composite map

$$\mathbf{Q}^* \to R = \prod_{p \in P} \mathbf{Q}_p^* \to G = R/ST. \tag{4.5}$$

For later use we note at once

LEMMA 4.3. (i) *If* $c_1, c_2 \in \mathbf{Q}^*$ and

$$\left|\frac{c_1}{c_2} - 1\right|_2 \leqslant 2^{-3}$$

$$\left|\frac{c_1}{c_2} - 1\right|_p \leqslant p^{-1} \qquad (p \in P, p \neq 2)$$

then

$$g(c_1) = g(c_2).$$

(ii) *Given* $g^* \in G$ *and any positive integer* M *there is a positive integer* c *prime to* M *such that*

$$g(c) = g^*.$$

Proof. (i) The given conditions imply that

$$c_1/c_2 \in (\mathbf{Q}_p^*)^2 \qquad (\text{all } p \in P)$$

and so

$$c_1/c_2 \in S, \; g(c_1/c_2) = 1.$$

(ii) As was noted in the derivation of (3.25), the image of

$$X = \prod_{p \in P} U_p \subset R = \prod_{p \in P} \mathbf{Q}_p^*$$

in $G = R/ST$ is the whole of G, where U_p is the group of p-adic units. Hence there is a $\{u_p\}_{p \in P} \in X \subset R$ whose image in $R/XY = G$ is g^*. It is then enough to choose the positive integer c such that it satisfies

$$c \equiv u_2 \pmod{2^3}$$
$$c \equiv u_p \pmod{p} \qquad (p \in P, p \neq 2)$$

and is prime to M. This concludes the proof.

We can now enunciate the main result of this section.

THEOREM 4.1. *Suppose that* $n \geqslant 3$. *Let* Γ, Δ *be in the genus of* Λ *and suppose that* $I(\Gamma, \Delta)$ *is not divisible by any* $p \in P^*$. *Then*

$$\mathscr{S}_2 = g_0 \mathscr{S}_1 \tag{4.6}$$

where $\mathscr{S}_1, \mathscr{S}_2$ *are the spinor genera of* Γ, Δ *respectively and where*

$$g_0 = g(I(\Gamma, \Delta)) \tag{4.7}$$

is the image of $I(\Gamma, \Delta)$ *under* (4.5).

Before giving the proof we require a number of lemmas. The first is quite general: we have worded the enunciation so that it includes the case $p = 2$, though we shall not need it.

LEMMA 4.4. *Let* Λ, Γ *be* **Z**-*lattices in* V, ϕ *with the same determinant, and let* p *be a prime dividing* $I(\Lambda, \Gamma)$. *Suppose that the quadratic forms induced by* ϕ *on* Λ_p, Γ_p *are integral in the classical sense. Then there is a lattice* Δ *with the same determinant as* Λ *and* Γ *such that*

$$I(\Lambda, \Delta) = p \tag{4.8}$$

$$I(\Delta, \Gamma) = I(\Lambda, \Delta)/p \tag{4.9}$$

and the form induced by ϕ *on* Δ *is classically integral.*

Proof. By hypothesis there is an

$$\mathbf{a} \in \Gamma$$

such that

$$\mathbf{a} \notin \Lambda, \quad p\mathbf{a} \in \Lambda.$$

Then by the integrality condition on Λ we have

$$p\phi(\mathbf{a}, \mathbf{b}) = \phi(p\mathbf{a}, \mathbf{b}) \in \mathbf{Z}_p \qquad (\text{all } \mathbf{b} \in \Lambda)$$

and by that on Γ we have

$$\phi(\mathbf{a}, \mathbf{b}) \in \mathbf{Z}_p \qquad (\text{all } \mathbf{b} \in \Gamma).$$

Let $\Lambda^{(1)}$ denote the set of $\mathbf{c} \in \Lambda$ such that

$$\phi(p\mathbf{a}, \mathbf{c}) \equiv 0 \pmod{p}.$$

Then

$$\Lambda \cap \Gamma \subset \Lambda^{(1)}$$

and $\Lambda^{(1)}$ either coincides with Λ or is of index p in Λ. In the second case take $\Lambda^{(1)} = \Lambda^{(2)}$ and in the first let $\Lambda^{(2)}$ be any sublattice of Λ of index p such that

$$\Lambda \cap \Gamma \subset \Lambda^{(2)} \subset \Lambda$$

(such a $\Lambda^{(2)}$ is easily seen to exist, e.g. using Lemma 3.2 of Chapter 7). In both cases

$$\phi(\mathbf{a}, \mathbf{c}) \in \mathbf{Z}_p \qquad (\text{all } \mathbf{c} \in \Lambda^{(2)}).$$

Let now Δ be the lattice generated by $\Lambda^{(2)}$ and $\mathbf{a}$. Then $\Lambda^{(2)} = \Delta \cap \Lambda$ and has index p in both Δ and Λ, which gives (4.8). Finally, (4.9) holds since $\Delta \cap \Gamma$ is the lattice generated by $\Lambda \cap \Gamma$ and $\mathbf{a}$. This concludes the proof.

COROLLARY. *Suppose that the quadratic forms induced by ϕ on Λ, Γ are classically p-adically integral for a set $\hat{P}$ of primes which includes all $p \mid I(\Lambda, \Gamma)$. Then there are lattices $\Delta^{(j)}$ $(1 \leqslant j \leqslant J)$ for some J all with the same determinant as Λ, Γ and such that*:

(i) $\Delta^{(1)} = \Lambda, \Delta^{(J)} = \Gamma$;

(ii) $I(\Delta^{(j)}, \Delta^{(j+1)})$ *is prime for* $1 \leqslant j < J$;

(iii) $I(\Lambda, \Gamma) = \prod_{j<J} I(\Delta^{(j)}, \Delta^{(j+1)})$;

(iv) *the quadratic forms induced by ϕ on the $\Delta^{(j)}$ are classically p-adically integral for all the* $p \in \hat{P}$.

Proof. Follows from the Lemma by induction on $I(\Lambda, \Gamma)$ on noting that $\Delta_q = \Lambda_q$ for all $q \in \hat{P}, q \neq p$ by (4.8) and the equivalence of (4.2) and (4.3).

We can define $I(\Lambda_p, \Gamma_p)$ for p-adic lattices Λ_p, Γ_p with the same determinant quite analogously to (4.1). If Λ_p, Γ_p are localizations of global lattices Λ, Γ of the same determinant, then $I(\Lambda_p, \Gamma_p)$ is the power of p dividing $I(\Lambda, \Gamma)$ by Lemma 2.1. With this convention we have:

LEMMA 4.5. *Let $p \neq 2$, let Λ_p, Γ_p be lattices of the same determinant in V_p, ϕ and let*

$$I(\Lambda_p, \Gamma_p) = p. \tag{4.10}$$

Suppose that ϕ induces an integral (p-adic) quadratic form on Λ_p, Γ_p and that the determinant of one (and so of both) of these quadratic forms is a p-adic unit. Then there are $\mathbf{b}_1, \ldots, \mathbf{b}_n \in V_p$ such that

$$\text{(i)} \qquad \phi(\mathbf{b}_1) = \phi(\mathbf{b}_2) = 0, \qquad \phi(\mathbf{b}_1, \mathbf{b}_2) = 1/p \tag{4.11}$$

$$\text{(ii)} \qquad \phi(\mathbf{b}_1, \mathbf{b}_j) = \phi(\mathbf{b}_2, \mathbf{b}_j) = 0 \qquad (j > 2) \tag{4.12}$$

and (iii)

$$p\mathbf{b}_1, \mathbf{b}_2, \ldots, \mathbf{b}_n; \tag{4.13}$$

$$\mathbf{b}_1, p\mathbf{b}_2, \mathbf{b}_3, \ldots, \mathbf{b}_n \tag{4.14}$$

are bases of Λ_p, Γ_p respectively.

Proof. Let $\mathbf{b}$ be a vector which is in Γ_p but not in Λ_p. Then $p\mathbf{b} \in \Lambda_p$ and is clearly a primitive element of Λ_p. Hence by the hypothesis about the determinant of the form induced on Λ_p there is a $\mathbf{c} \in \Lambda_p$ such that

$$|\phi(p\mathbf{b}, \mathbf{c})|_p = 1.$$

Since $\phi(\mathbf{b}) \in \mathbf{Z}_p, \phi(\mathbf{c}) \in \mathbf{Z}_p$ we can find by Hensel's lemma (Chapter 3, Section 4) an $h \in \mathbf{Z}_p$ such that

$$\begin{aligned}\phi(\mathbf{b} + ph\mathbf{c}) &= \phi(\mathbf{b}) + 2h\phi(p\mathbf{b}, \mathbf{c}) + p^2h^2\phi(\mathbf{c}) \\ &= 0.\end{aligned}$$

Then $ph\mathbf{c} \in p\Lambda_p \subset \Gamma_p$ and so $\mathbf{b} + ph\mathbf{c} \in \Gamma_p$ but $\mathbf{b} + ph\mathbf{c} \notin \Lambda_p$. Put $\mathbf{b}_1 = \mathbf{b} + ph\mathbf{c}$ and define $\mathbf{b}_2$ similarly interchanging the rôles of Λ_p and Γ_p.

For $\mathbf{d} \in \Lambda_p \cap \Gamma_p$ we have

$$\phi(p\mathbf{b}_1, \mathbf{d}) = p\phi(\mathbf{b}_1, \mathbf{d}) \in p\mathbf{Z}_p.$$

By the condition on the determinants of the form induced on Λ_p there must be some $\mathbf{e} \in \Lambda_p$ for which $\phi(\mathbf{b}_1, \mathbf{e})$ is a p-adic unit. Since Λ_p is generated (as a group) by $\Lambda_p \cap \Gamma_p$ and $\mathbf{b}_2$ it follows that

$$\phi(p\mathbf{b}_1, \mathbf{b}_2) \in U_p.$$

Hence on multiplying $\mathbf{b}_2$ by an element of U_p we may suppose that

$$p\phi(\mathbf{b}_1, \mathbf{b}_2) = \phi(p\mathbf{b}_1, \mathbf{b}_2) = 1.$$

Finally, if $\mathbf{d} \in \Lambda_p \cap \Gamma_p$, then

$$\mathbf{d}' = \mathbf{d} - a_1 p \mathbf{b}_1 - a_2 p \mathbf{b}_2$$

with

$$a_1 = \phi(\mathbf{b}_2, \mathbf{d}) \in \mathbf{Z}_p, \qquad a_2 = \phi(\mathbf{b}_1, \mathbf{d}) \in \mathbf{Z}_p$$

satisfies

$$\phi(\mathbf{d}', \mathbf{b}_1) = \phi(\mathbf{d}', \mathbf{b}_2) = 0.$$

Hence we can extend $p\mathbf{b}_1, p\mathbf{b}_2$ to a basis $p\mathbf{b}_1, p\mathbf{b}_2, \mathbf{b}_3, \ldots, \mathbf{b}_n$ of $\Lambda_p \cap \Gamma_p$ so that (4.12) holds. This completes the proof.

Proof of Theorem 4.1. By Lemma 4.4, Corollary, and by Theorem 3.1 and its Corollary we need consider only the case when

$$I(\Gamma, \Delta) = q \text{ (say)}$$

is a prime. By hypothesis

$$q \notin P^*.$$

By the equivalence of (4.2) and (4.3) we have

$$\Gamma_p = \Delta_p \qquad (p \neq q).$$

By Lemma 4.5 we have

$$\Gamma_q = \beta_q \Delta_q$$

where β_q is defined by

$$\begin{aligned} \mathbf{b}_1 &\to q\mathbf{b}_1 \\ \mathbf{b}_2 &\to q^{-1}\mathbf{b}_2 \\ \mathbf{b}_j &\to \mathbf{b}_j \qquad (j \geqslant 3) \end{aligned}$$

and $\mathbf{b}_1, \ldots, \mathbf{b}_n$ is a basis of V_p such that

$$\phi(x_1\mathbf{b}_1 + \ldots + x_n\mathbf{b}_n) = q^{-1}x_1x_2 + \phi(x_3\mathbf{b}_3 + \ldots + x_n\mathbf{b}_n).$$

By the argument at the beginning of the proof of Theorem 3.2 of Chapter 10 (with $p = q, \mathbf{e}_1 = \mathbf{b}_1, \mathbf{e}_2 = \mathbf{b}_2$) we have

$$\theta(\beta_q) = q(\mathbf{Q}_q^*)^2.$$

By Lemma 3.2 there is an $\alpha \in O^+(V)$ such that

$$\theta(\alpha) = q(\mathbf{Q}^*)^2.$$

Then $\Gamma' = \alpha\Gamma$ is properly equivalent to Γ and satisfies

$$\Gamma'_p = \gamma_p \Delta_p$$

where

$$\begin{aligned} \gamma_q &= \alpha\beta_q \\ \gamma_p &= \alpha \qquad (\text{all } p \neq q). \end{aligned}$$

Then

$$\theta(\gamma_q) = q^2(\mathbf{Q}_q^*)^2 \subset U_q(\mathbf{Q}_q^*)^2$$

and

$$\theta(\gamma_p) = q(\mathbf{Q}_p^*)^2 \subset U_p(\mathbf{Q}_p^*)^2 \qquad (p \neq q).$$

Hence Γ is in the same spinor genus as Γ'' defined by

$$\begin{aligned}\Gamma_p'' &= \alpha\Delta_p \qquad (p \in P)\\ &= \Delta_p \qquad (p \notin P).\end{aligned}$$

The truth of the Theorem now follows from Theorem 3.1, Corollary.

5. SIMULTANEOUS BASES OF TWO LATTICES

In this section we prove some elementary results which will be needed in the next section. We state and prove them for $\mathbf{Z}$ but with minor modifications the enunciations and the proofs go over to any principal ideal ring.

THEOREM 5.1. *Let Λ, Γ be $\mathbf{Z}$-lattices of dimension n with*

$$\Gamma \subset \Lambda. \tag{5.1}$$

Then there is a basis $\mathbf{b}_1, \ldots, \mathbf{b}_n$ of Λ and integers $s_1, \ldots, s_n$ such that

$$s_1\mathbf{b}_1, \ldots, s_n\mathbf{b}_n \tag{5.2}$$

is a basis of Γ. Further, we may suppose that

$$s_j > 0 \qquad (1 \leqslant j \leqslant n) \tag{5.3}$$

and

$$s_{j+1} \mid s_j \qquad (1 \leqslant j < n). \tag{5.4}$$

Proof. The quotient group Λ/Γ is a finite abelian group and our proof follows the strategy of one of the standard proofs of the structure theorem for such groups.

Let

$$s_1 = \prod_{1 \leqslant u \leqslant U} p_u^{m(u)} \qquad (m(u) > 0) \tag{5.5}$$

be the least common multiple of the orders of the $\mathbf{b} \in \Lambda$ modulo Γ. We first show that there is a $\mathbf{b}_1$ whose order is precisely s_1. Clearly for each u there is a $\mathbf{c}_u \in \Lambda$ whose order modulo Γ is precisely $p_u^{m(u)}$. Put

$$\mathbf{d} = \sum_{1 \leqslant u \leqslant U} \mathbf{c}_u.$$

We shall show now that $\mathbf{d}$ has order s_1. For each u put

$$P_u = \prod_{v \neq u} p_v^{m(v)}.$$

Then

$$P_u \mathbf{d} \equiv P_u \mathbf{c}_u \pmod{\Gamma}$$

and so, since P_u is prime to p_u, the order of $\mathbf{d}$ modulo Γ is divisible by that of $\mathbf{c}_u$, namely $p_u^{m(u)}$. Hence the order of $\mathbf{d}$ is divisible by $\Pi p_u^{m(u)} = s_1$, as asserted.

Then $\mathbf{d}$ is a multiple of a primitive element of Λ, say

$$\mathbf{d} = t\mathbf{b}_1.$$

The order of $\mathbf{b}_1$ modulo Γ must be at least equal to that of $\mathbf{d}$, namely s_1 and so must be exactly s_1 by the definition of s_1. We now show that $s_1\mathbf{b}_1$ is a primitive element of Γ. For if not, then $s_1\mathbf{b}_1 = q\mathbf{e}$, where $\mathbf{e} \in \Gamma$ and q is prime. If $q \mid s_1$, then we have a contradiction with the fact that s_1 is the order of $\mathbf{b}_1$ modulo Γ. If, however, q is prime to s_1 then $\mathbf{b}_1$ must be divisible by q in Λ, again a contradiction.

Now let $\bar{\Lambda}$ be the $(n-1)$-dimensional lattice Λ modulo $\mathbf{b}_1$. Then the image $\bar{\Gamma}$ of Γ in $\bar{\Lambda}$ is clearly just Γ modulo $s_1\mathbf{b}_1$. By induction on the dimension n there exists a basis $\bar{\mathbf{b}}_2, \dots, \bar{\mathbf{b}}_n$ of $\bar{\Lambda}$ and positive integers $s_2, \dots, s_n$ such that (5.4) holds for $j \neq 1$ and such that $s_2\bar{\mathbf{b}}_2, \dots, s_n\bar{\mathbf{b}}_n$ is a basis of $\bar{\Gamma}$.

Let $\mathbf{b}_j\,(j \neq 1)$ be any elements of $\bar{\mathbf{b}}_j$. Then clearly $\mathbf{b}_1, \mathbf{b}_2, \dots, \mathbf{b}_n$ is a basis of Λ. Further, the order of $\bar{\mathbf{b}}_j$ modulo $\bar{\Gamma}$ divides the order of b_j modulo Γ and *a fortiori* divides s_1, say

$$s_1 = s_j t_j. \tag{5.6}$$

This, incidentally, confirms (5.4) also for $j = 1$.

By the definition of the s_j we have

$$s_j\mathbf{b}_j \equiv u_j\mathbf{b}_1 \pmod{\Gamma} \qquad (j \neq 1) \tag{5.7}$$

for some $u_j \in \mathbf{Z}$. Hence

$$\begin{aligned} t_j u_j \mathbf{b}_1 &\equiv t_j s_j \mathbf{b}_j \\ &= s_1 \mathbf{b}_j \\ &\equiv \mathbf{0} \pmod{\Gamma}. \end{aligned}$$

But $\mathbf{b}_1$ has precise order s_1 modulo Γ and so $s_1 \mid t_j u_j$, that is $s_j \mid u_j$ by (5.6). On replacing $\mathbf{b}_j$ by $\mathbf{b}_j - v_j\mathbf{b}_1$, where $s_j v_j = u_j$, we may thus suppose without loss of generality that

$$s_j\mathbf{b}_j \equiv \mathbf{0} \pmod{\Gamma}.$$

Finally, $s_1\mathbf{b}_1, s_2\mathbf{b}_2, \dots, s_n\mathbf{b}_n$ is a basis for Γ because $\bar{\Gamma}$ is Γ modulo $s_1\mathbf{b}_1$ and because $s_2\bar{\mathbf{b}}_2, \dots, s_n\bar{\mathbf{b}}_n$ is a basis of $\bar{\Gamma}$. This concludes the proof.

COROLLARY. *Let Γ, Δ be $\mathbf{Z}$-lattices of the same determinant in the $\mathbf{Q}$-vector space V. Suppose that*

$$I(\Gamma, \Delta) = p \tag{5.8}$$

for some prime p. Then there are $\mathbf{b}_1, \ldots, \mathbf{b}_n \in V$ such that

$$p\mathbf{b}_1, \mathbf{b}_2, \ldots, \mathbf{b}_n \tag{5.9}$$

and

$$\mathbf{b}_1, p\mathbf{b}_2, \mathbf{b}_3, \ldots, \mathbf{b}_n \tag{5.10}$$

are bases of Γ, Δ respectively.

Proof. Follows from the Theorem applied to $p\Gamma \subset \Delta$. It is, of course, also easy to construct proofs from first principles.

The following Lemma will be required only to prove a result of rather minor importance.

LEMMA 5.1. *Let Δ, Γ be $\mathbf{Z}$-lattices of the same determinant in the $\mathbf{Q}$-quadratic space V, ϕ of dimension, n. For every $p \mid I(\Lambda, \Gamma)$ suppose that the quadratic forms induced by ϕ on Λ_p and Γ_p are classically integral and that their determinants are p-adic units. Then there are a basis $\mathbf{b}_1, \ldots, \mathbf{b}_n$ of Λ and $s_1, \ldots, s_n \in \mathbf{Q}$ such that*

$$s_1\mathbf{b}_1, \ldots, s_n\mathbf{b}_n \tag{5.11}$$

is a basis of Γ and such that the following conditions hold:

(i) $s_j > 0$

(ii) $s_j/s_{j+1} \in \mathbf{Z}$ $\quad (1 \leqslant j < n)$

(iii) $s_j s_{n+1-j} = 1$ $\quad (1 \leqslant j \leqslant n)$

(iv) $s_j \in \mathbf{Z}$ $\quad (1 \leqslant j \leqslant [(n+1)/2])$.

Proof. We have $t\Lambda \subset \Gamma$ for some positive integer t. On applying Theorem 5.1 to Γ and $t\Lambda$ one obtains $\mathbf{b}_1, \ldots, \mathbf{b}_n, s_1, \ldots, s_n$ satisfying all the conditions of the Lemma except possibly (iii) and (iv). We shall show that (iii) and (iv) are also satisfied. Since (iv) follows readily from (ii) and (iii), it will be enough to verify (iii).

In this proof denote by P the set of primes p dividing $I(\Lambda, \Gamma)$. Then $\Lambda_p = \Gamma_p$ for $p \notin P$ and so

$$|s_j|_p = 1 \qquad (1 \leqslant j \leqslant n, p \notin P). \tag{5.12}$$

Put

$$a_{ij} = \phi(\mathbf{b}_i, \mathbf{b}_j) \qquad (1 \leqslant i, j \leqslant n). \tag{5.13}$$

By the hypotheses of the theorem we have

$$a_{ij} \in \mathbf{Z}_p \qquad (p \in P) \tag{5.14}$$

and

$$s_i s_j a_{ij} \in \mathbf{Z}_p \qquad (p \in P). \tag{5.15}$$

It follows from (ii) that

$$s_u s_v a_{ij} \in \mathbf{Z}_p \qquad (u \leqslant i, v \leqslant j). \tag{5.16}$$

Now suppose, if possible, that there exist $p \in P$ and u, v with $u + v \leqslant n + 1$ such that $|s_u s_v|_p > 1$. Then (5.16) implies that $a_{ij} \in p\mathbf{Z}_p$ for $i \geqslant u, j \geqslant v$: and then by (5.14) $\det(a_{ij}) \in p\mathbf{Z}_p$, contrary to hypothesis. Hence

$$|s_u s_v|_p \leqslant 1 \qquad (p \in P, u + v \leqslant n + 1). \tag{5.17}$$

Similarly, on interchanging the roles of Λ and Γ and so replacing s_j by $(s_{n+1-j})^{-1}$ we have

$$|s_u s_v|_p \geqslant 1 \qquad (p \in P, u + v \geqslant n + 1). \tag{5.18}$$

Finally, (5.17) and (5.18) imply

$$|s_j s_{n+1-j}|_p = 1 \qquad (p \in P \ 1 \leqslant j \leqslant n),$$

which with (5.12) and (i) gives (iii).

6. THE LANGUAGE OF FORMS

In this section we translate some of the results already obtained into the language of quadratic forms.

THEOREM 6.1. *Let f_1, f_2 be two classically integral quadratic forms over $\mathbf{Z}$ in $n \geqslant 2$ variables and with the same determinant $d \neq 0$. Suppose that they are rationally equivalent. Then there is a sequence of quadratic forms g_v $(1 \leqslant v \leqslant V)$, all classically integral and of determinant d, such that*

$$g_1 = f_1 \qquad g_V = f_2$$

and such that for each $v < V$ one of the two following holds:

(i) *g_{v+1} is properly $\mathbf{Z}$-equivalent to g_v*

(ii) *there is a prime p (depending on v) such that*

$$g_{v+1}(x_1, \ldots, x_n) = g_v(p^{-1}x_1, px_2, x_3, \ldots, x_n).$$

Proof. Every form is improperly $\mathbf{Q}$-equivalent to itself and so f_1, f_2 are properly $\mathbf{Q}$-equivalent. Let† V be an n-dimensional $\mathbf{Q}$-vector space with basis $\mathbf{e}_1, \ldots, \mathbf{e}_n$ and define ϕ on V by

$$\phi(x_1\mathbf{e}_1 + \ldots + x_n\mathbf{e}_n) = f_1(x_1, \ldots, x_n).$$

† The use of V in two senses is unfortunate but should not cause difficulty.

Then the proper equivalence of f_1 and f_2 gives us another $\mathbf{Q}$-basis $\mathbf{e}'_1, \ldots, \mathbf{e}'_n$ of V such that

$$\phi(x_1\mathbf{e}'_1 + \ldots + x_n\mathbf{e}'_n) = f_2(x_1, \ldots, x_n)$$

and such that the determinant of $\mathbf{e}'_1, \ldots, \mathbf{e}'_n$ with respect to $\mathbf{e}_1, \ldots, \mathbf{e}_n$ is equal to $+1$. Then in the sense introduced in the proof of Lemma 1.1, Corollary 2, the sets $\mathbf{e}_1, \ldots, \mathbf{e}_n$ and $\mathbf{e}'_1, \ldots, \mathbf{e}'_n$ have the same orientation, which we shall call "positive". Let Λ, Γ be the $\mathbf{Z}$-lattices with bases $\mathbf{e}_1, \ldots, \mathbf{e}_n$ and $\mathbf{e}'_1, \ldots, \mathbf{e}'_n$ respectively. Then the hypotheses of Lemma 4.3, Corollary, are satisfied and there is a sequence of lattices $\Delta^{(j)}$ $(1 \leqslant j \leqslant J)$ all with the same determinant as Λ and Γ such that

$$\Delta^{(1)} = \Lambda, \qquad \Delta^{(J)} = \Gamma$$

and

$$I(\Delta^{(j)}, \Delta^{(j+1)}) = p_j \qquad (1 \leqslant j < J),$$

where p_j is prime. Further, ϕ induces a classically integral form on every $\Delta^{(j)}$ (take $\bar{P}$ = the set of all primes in the quoted Corollary).

By Theorem 5.1, Corollary, there is for each j a sequence $\mathbf{b}_1, \ldots, \mathbf{b}_n$ of elements of V such that

$$p\mathbf{b}_1, \mathbf{b}_2, \ldots, \mathbf{b}_n \tag{6.1}$$

$$\mathbf{b}_1, p\mathbf{b}_2, \ldots, \mathbf{b}_n \tag{6.2}$$

with $p = p_j$ are respectively bases of $\Delta^{(j)}$ and $\Delta^{(j+1)}$. On taking $-\mathbf{b}_1$ for $\mathbf{b}_1$ if necessary, we may suppose that the sequence $\mathbf{b}_1, \ldots, \mathbf{b}_n$ is positively oriented.

The truth of the Theorem is now apparent. The v of type (ii) correspond to a change from a basis of type (6.1) to that of type (6.2) whereas the v of type (i) correspond to a change from one positively oriented basis of a $\Delta^{(j)}$ to another such basis of the same lattice.

THEOREM 6.2. *Let f_1, f_2 be two classically integral forms of the same genus of determinant $d \neq 0$ in $n \geqslant 2$ variables and let $\bar{P}$ be any finite set of primes. Then the conclusion of Theorem* 6.1 *holds and we can suppose in* (ii) *that $p \notin \bar{P}$.*

Conversely, if f_1, f_2 are classically integral forms of determinant d and there is a sequence of forms g_v such that for each v either (i) *holds or* (ii) *holds with $p \nmid 2d$, then f_1, f_2 are in the same genus.*

Proof. The first paragraph of the enunciation follows from Lemma 4.2 and the above proof of Theorem 6.1. The converse proposition is straightforward since if (ii) holds with $p \nmid 2d$ then g_{v+1} and g_v are $\mathbf{Z}_p$-equivalent for every p.

COROLLARY. *Suppose that* $n \geqslant 3$. *Let* $\mathscr{S}_j$ *be the spinor genus of* f_j $(j = 1, 2)$ *and let W be the product of the primes occurring in the substitutions of type* (ii). *Then*

$$\mathscr{S}_2 = g_0 \mathscr{S}_1$$

in the notation of Theorem 3.1, *where* $g_0 = g(W)$ *is the image of W in G under the map* (4.5).

Proof. Follows at once from Theorem 4.1.

THEOREM 6.3. *Let* f_1, f_2 *be classically integral forms in* $n \geqslant 2$ *variables and of determinant* $d \neq 0$. *Let* $\mathbf{y} = \mathbf{T}\mathbf{x}$:

$$y_i = \sum_j t_{ij} x_j \tag{6.3}$$

be a transformation such that identically

$$f_2(\mathbf{y}) = f_1(\mathbf{x}). \tag{6.4}$$

Suppose further that

$$t_{ij} \in \mathbf{Q}, \quad t_{ij} \in \mathbf{Z}_p \qquad (\text{all } p \mid 2d) \tag{6.5}$$

and

$$\det(t_{ij}) = +1. \tag{6.6}$$

Then there are positive integers

$$s_1, \ldots, s_m \qquad m = [n/2] \tag{6.7}$$

and transformations

$$x'_i = \sum_j u_{ij} x_j \qquad u_{ij} \in \mathbf{Z} \quad \det(u_{ij}) = +1 \tag{6.8}$$

$$y'_i = \sum v_{ij} y_j \qquad v_{ij} \in \mathbf{Z} \quad \det(v_{ij}) = +1 \tag{6.9}$$

such that

$$\left.\begin{aligned} y'_i &= s_i x'_i && (1 \leqslant i \leqslant m) \\ x'_{n+1-i} &= s_i y'_{n+1-i} && (1 \leqslant i \leqslant m) \\ x'_{m+1} &= y'_{m+1} && (n \text{ odd}). \end{aligned}\right\} \tag{6.10}$$

and

We can suppose, further, that

$$s_{i+1} \mid s_i \qquad (1 \leqslant i < m). \tag{6.11}$$

Finally, if $n \geqslant 3$, *the spinor genera* $\mathscr{S}_1, \mathscr{S}_2$ *of* f_1, f_2 *satisfy*

$$\mathscr{S}_2 = g_0 \mathscr{S}_1 \tag{6.12}$$

where

$$g_0 = g(W) \qquad W = s_1 \dots s_m \tag{6.13}$$

in the notations of Theorem 3.1 *and Theorem* 4.1.

Proof. As in the proof of Theorem 6.1 the transformation (6.3) can be realized by two similarly orientated bases of lattices Λ, Γ in the same **Q**-quadratic space V. The condition (6.5) implies that $\Lambda_p = \Gamma_p$ for $p|2d$ and so $I(\Lambda, \Gamma)$ is prime to $2d$ (by the equivalence of (4.2) and (4.3)). Thus the conditions of Lemma 5.1 apply. The transformations (6.8) and (6.9) are those which take the initial bases of Λ, Γ to the special bases furnished by Lemma 5.1. (We can always ensure that the basis $\mathbf{b}_1, \dots, \mathbf{b}_n$ of Lemma 5.1 is positively orientated by changing the sign of one of the $\mathbf{b}_j$ if need be.) On using (iii) of Lemma 5.1 it is easy to see that the transformation (6.10) corresponds to the special bases of Λ, Γ given by that Lemma.

Finally, it is obvious that

$$I(\Lambda, \Gamma) = s_1 \dots s_m$$

and so (6.13) follows from Theorem 4.1.

This concludes the proof. To find W from **T** it is not, in fact, necessary to find the transformations (6.8) and (6.9):

COROLLARY. *W is the least common multiple of the denominators of the $m \times m$ minors of the matrix (t_{ij}), where $m = [n/2]$.*

Proof. Follows from the standard result in matrix theory [the "Binet–Cauchy Identity"] that if

$$\mathbf{A} = \mathbf{BC}$$

where $\mathbf{A}, \mathbf{B}, \mathbf{C}$ are matrices then every $m \times m$ minor of $\mathbf{A}$ is expressible as a sum $\Sigma_I b_I c_I$ where b_I, c_I are $m \times m$ minors of $\mathbf{B}, \mathbf{C}$ respectively. Here m is any integer and the matrices $\mathbf{A}, \mathbf{B}, \mathbf{C}$ need not be square. See any sufficiently old-fashioned school algebra textbook.

7. REPRESENTATION BY SPINOR GENERA

The results of this section hold indifferently for definite and indefinite forms but the main interest is for definite forms.

THEOREM 7.1. *Let f be a classically integral form over $\mathbf{Z}$ in $n \geqslant 4$ variables. Suppose that $a \in \mathbf{Z}$, $a \neq 0$ is representable primitively by f over every $\mathbf{Z}_p$ (including $p = \infty$). Then a is primitively representable over $\mathbf{Z}$ by some form in the spinor genus of f.*

Note 1. For indefinite forms this is weaker than Theorem 1.5 of Chapter 9.

Note 2. The theorem becomes false for $n = 3$. See below.

Proof. By Theorem 1.3 of Chapter 9 the integer a is primitively representable by some form f^* in the genus of f. In the notation of Theorem 3.1 let g^* be the element of G such that

$$\mathscr{S}^* = g^*\mathscr{S}, \tag{7.1}$$

where $\mathscr{S}$, $\mathscr{S}^*$ are the spinor genera of f, f^* respectively. By Lemma 4.3 there is a positive integer c which is prime to $2ad$ (d being the determinant of f) and such that

$$g^* = g(c). \tag{7.2}$$

We can suppose without loss of generality that

$$f^*(1, 0, \ldots, 0) = a$$

and so

$$af^*(x_1, \ldots, x_n) = (ax_1 + b_2x_2 + \ldots + b_nx_n)^2 + h(x_2 \ldots x_n)$$

for some $b_2, \ldots, b_n \in \mathbf{Z}$ and for some classically integral form h in $n - 1$ variables of determinant $a^{n-2}d$. Since $n - 1 \geqslant 3$ and c is prime to $2ad$, the form $h(x_2, \ldots, x_n)$ is isotropic over $\mathbf{Q}_p$ for all $p|c$. Hence and by Lemma 4.2 of Chapter 9, after a suitable integral transformation of $x_2, \ldots, x_n$ with determinant $+1$ we may suppose without loss of generality that the form

$$h^*(x_2, \ldots, x_n) = h(c^{-1}x_2, cx_3, x_4, \ldots, x_n)$$

is classically integral. It is clearly in the genus of h and so by Lemma 4.2 of Chapter 9 again, there is a form h' properly equivalent to h^* and arbitrarily close to h p-adically for all $p|2ad$. In particular we can choose h' so that $f'(x_1, \ldots, x_n)$ defined by

$$af'(x_1, \ldots, x_n) = (ax_1 + b_2x_2 + \ldots + b_nx_n)^2 + h'(x_2, \ldots, x_n)$$

is classically integral and in the same genus as f^* (and so as f). The spinor genera $\mathscr{S}', \mathscr{S}^*$ of f', f^* respectively satisfy

$$\mathscr{S}^* = g(c)\mathscr{S}' \tag{7.3}$$

by Theorem 6.2, Corollary. Since $g(c) = g^*$ we have $\mathscr{S} = \mathscr{S}'$ by Theorem 3.1. This concludes the proof.

Theorem 7.1 fails for $n = 3$. For example, [Watson (1960), p. 115]

$$\left.\begin{aligned} f_1 &= x_1^2 + x_1x_2 + x_2^2 + 9x_3^2 \\ f_2 &= x_1^2 + 3(x_2^2 + x_2x_3 + x_3^2) \end{aligned}\right\} \tag{7.4}$$

are representatives of the two classes in their genus and are in different spinor genera. Clearly f_2 represents 4 primitively but f_1 does not. [See also the examples to this chapter.]

We shall say that an integer a is *exceptional* for a ternary genus if it is represented primitively by at least one but not by all of the spinor genera in the genus. Thus 4 is exceptional for the genus represented by (7.4).

LEMMA 7.1. *Let $p \neq 2$ be a prime which does not divide the determinant of a genus $\mathscr{F}$ of integral ternary forms. Then for any integer $a \neq 0$ if one of a, ap^2 is exceptional, then so is the other.*

Proof. Every form f of the genus $\mathscr{F}$ is equivalent to a form

$$f^* = \Sigma f^*_{ij} x_i x_j \tag{7.5}$$

where

$$\begin{aligned} 2f^*_{13} = 2f^*_{31} &\equiv 1 \qquad (p^2) \\ f^*_{22} &\equiv -4d \qquad (p^2) \\ f^*_{ij} &\equiv 0 \qquad (p^2) \text{ otherwise} \end{aligned}$$

or, say,

$$f^* \equiv x_1 x_3 - 4dx_2^2 \qquad (p^2). \tag{7.6}$$

Then the form

$$f^*_p(x_1, x_2, x_3) = f^*(px_1, x_2, p^{-1}x_3) \tag{7.7}$$

is also in the genus $\mathscr{F}$. Further, by Theorem 3.1, f^*_p runs through representatives of all the spinor genera of $\mathscr{F}$ when f (and so f^*) does so. Suppose now and a is represented primitively by f (and so by f^*), say

$$a = f^*(b^*_1, b^*_2, b^*_3) \qquad \text{g.c.d.}(b^*_1, b^*_2, b^*_3) = 1. \tag{7.8}$$

Then

$$p^2 a = f^*_p(b^*_1, pb^*_2, p^2 b^*_3)$$

and this representation is primitive provided that

$$p \nmid b^*_1. \tag{7.9}$$

Conversely, suppose that ap^2 is represented primitively by f, say

$$ap^2 = f^*(c^*_1, c^*_2, c^*_3) \qquad \text{g.c.d.}(c^*_1, c^*_2, c^*_3) = 1. \tag{7.10}$$

Then

$$a = f^*_p(p^{-2}c^*_1, p^{-1}c^*_2, c^*_3): \tag{7.11}$$

this representation is integral provided that

$$p \mid c^*_1, \ p \mid c^*_2 \tag{7.12}$$

since then $p \nmid c_3$ by primitivity and consequently $p^2 \mid c_1$ by (7.6) and (7.10). To prove the Lemma it is thus enough to ensure that we can choose f^* equivalent to f in such a way that (7.6) holds and so that (7.9) or (7.12) holds as the case may be.

Consider first (7.12). By hypothesis here is a $\mathbf{c} \in \mathbf{Z}^3$ such that

$$f(\mathbf{c}) = ap^2.$$

We shall work first p-adically. Since p does not divide the discriminant of f, there is a $\mathbf{c}' \in \mathbf{Z}_p^3$ such that

$$f(\mathbf{c}, \mathbf{c}') = \tfrac{1}{2}$$

and on taking $\alpha\mathbf{c} + \beta\mathbf{c}'$ instead of $\mathbf{c}'$ $(\alpha, \beta \in \mathbf{Z}_p)$ we may suppose without loss of generality that

$$f(\mathbf{c}') = 0.$$

By a similar argument there is a $\mathbf{c}'' \in \mathbf{Z}_p^3$ such that

$$f(\mathbf{c}, \mathbf{c}'') = f(\mathbf{c}', \mathbf{c}'') = 0$$

and

$$\det(\mathbf{c}'', \mathbf{c}', \mathbf{c}) = 1.$$

By Theorem 2.1 of Chapter 9 there is a basis $\mathbf{e}_1, \mathbf{e}_2, \mathbf{e}_3$ of $\mathbf{Z}^3$ such that $\mathbf{e}_1, \mathbf{e}_2, \mathbf{e}_3$ are arbitrarily close p-adically to $\mathbf{c}'', \mathbf{c}', \mathbf{c}$ respectively. Then clearly $f^*(x_1, x_2, x_3) = f(x_1\mathbf{e}_1 + x_2\mathbf{e}_2 + x_3\mathbf{e}_3)$ does what is required.

One can achieve (7.9) by a similar argument but it is more messy as one has to distinguish the cases $p \nmid a$ and $p \mid a$. Alternatively, one can note that if $p \mid b_1^*$ but $p \nmid b_3^*$ it is enough to make the substitution

$$x_1 \to x_3, x_3 \to x_1$$

while if $p \mid b_1^*$, $p \mid b_3^*$, so $p \nmid b_2^*$ then one can put

$$x_1 \to x_1' + 8dx_2' + 4dx_3'$$

$$x_3 \to x_3'$$

$$x_2 \to x_2' + x_3'$$

for which

$$x_1x_3 - 4dx_2^2 = x_1'x_3' - 4dx_2'^2.$$

8. A GENERALIZED STRONG APPROXIMATION

In this section we sketch a proof of a generalization of the "Strong Approximation Theorem for the Spin Group", Theorem 7.1 of Chapter 10 and then give some applications to definite forms. In Section 9 we shall use the generalization to prove Theorem 1.

We start with some observations of a theological nature. We have already noted in Chapter 3 and subsequently that there is an analogy between the ordinary absolute value $|\ |_\infty$ and the p-adic valuations $|\ |_p$ and we have taken this analogy into account in the notation $\mathbf{R} = \mathbf{Q}_\infty$. Some results (e.g. the "Product Theorem for the Norm Residue Symbol", Lemma 3.4 of Chapter 3) necessarily refer to all the valuations including $|\ |_\infty$. Many of the results of this book, however, refer only to the set of p-adic valuations and $|\ |_\infty$ either is not mentioned at all or appears in a different role from that of the $|\ |_p$. In many cases there is a more general formulation in which some p-adic valuation plays the exceptional role. We did not give these more general formulations earlier because the analogy between $|\ |_\infty$ and the $|\ |_p$ is far from perfect and we should have been involved in additional explanations and complications in the proofs. Now we must mention some of these generalizations. We shall in general give only very rough indications of proof. We shall denote the set of all valuations of $\mathbf{Q}$ by Ω:

$$\Omega = \{v: \quad v = \infty \quad \text{or} \quad v = p = \text{prime}\}. \tag{8.1}$$

LEMMA 8.1. *Let $u \in \Omega$ and for all $v \in \Omega$, $v \neq u$ let $\mathscr{V}_v$ be a non-empty open set of $\mathbf{Q}_v$ (in the v-topology). Suppose that $\mathscr{V}_v = \mathbf{Z}_v$ for almost all v. Then there is an $a \in \mathbf{Q}$ such that*

$$a \in \mathscr{V}_v \ (\text{all } v \neq u). \tag{8.2}$$

Note. As usual, "almost all" means "all except at most finitely many". We have used the convention that $\mathbf{Z}_\infty = \mathbf{R}$ but it is clear that the significance of the Lemma is independent of this convention.

Proof. When $u = \infty$ this is essentially another formulation of the "Strong Approximation Theorem", Lemma 3.1 of Chapter 3. When $u = q = $ prime we can use Lemma 3.1 to find a $b \in \mathbf{Q}$ such that

$$b \in \mathscr{V}_p \qquad (p \neq q, p \neq \infty).$$

We can now solve (8.2) by putting

$$a = b + rs/q^N \qquad r, s \in \mathbf{Z}$$

where successively (i) s is chosen so that for the finitely many p for which $\mathscr{V}_p \neq \mathbf{Z}_p$ it is divisible by a sufficiently high power of p to ensure that $a \in \mathscr{V}_p$ whatever the values of r and N, (ii) N is chosen so that s/q^N is arbitrarily small in absolute value and (iii) r is chosen so that $a \in \mathscr{V}_\infty$.

THEOREM 8.1. *Let f be a regular form over $\mathbf{Z}$ in $n \geqslant 4$ variables and suppose that f is isotropic in $\mathbf{Q}_u$ where $u \in \Omega$ is fixed. Suppose that $a \in \mathbf{Z}$ is represented by f over $\mathbf{Z}_v$ for all $v \in \Omega$. Let $\mathbf{b}_v \in \mathbf{Z}_v^n$ be given for $v \in \Omega_1$, where Ω_1 is a* *finite subset of* Ω *not containing* u. *Then there is a* $\mathbf{b} \in \mathbf{Q}^n$ *such that*

(i) $f(\mathbf{b}) = a$,

(ii) $\mathbf{b} \in \mathbf{Z}_v^n$ *for all* $v \neq u$.

(iii) $\mathbf{b}$ *is arbitrarily close to* $\mathbf{b}_v$ *in the* v-*topology for all* $v \in \Omega_1$.

Proof. For $u = \infty$ this is Theorem 1.5 of Chapter 9. The proof for $u \neq \infty$ is left to the reader. There is a completely detailed proof of a generalization of Theorem 8.1 to algebraic number-fields in O'Meara (1963), Theorem 104.3.

There is a similar generalization of the "Strong Approximation Theorem for the Spingroup", Theorem 7.1 of Chapter 10 and of its Corollary. We content ourselves with giving the generalization of the Corollary as

THEOREM 8.2. *Let* V, ϕ *be a regular* $\mathbf{Q}$-*quadratic space of dimension* $n \geqslant 3$ *which is isotropic over* $\mathbf{Q}_u$ *for some fixed* $u \in \Omega$. *For all* $v \in \Omega$, $v \neq u$ *denote by* $\mathscr{V}_v$ *a non-empty open subset of* $\Theta(V_p)$ *and suppose that*

$$\mathscr{V}_v = \Theta(\Lambda_v)$$

for almost all v, *where* Λ *is some* $\mathbf{Z}$-*lattice in* V, ϕ. *Then there is a* $\sigma \in \Theta(V)$ *such that*

$$\sigma \in \mathscr{V}_v \ (\textit{all } v \neq u).$$

Proof. Left to the reader, or see O'Meara (1963), Theorem 104.4.

We now come to a stage where the reformulation of results with some valuation other than ∞ in a special role is no longer quite so straightforward. The reason is that $\mathbf{Q}_\infty = \mathbf{Z}_\infty$ and so we did not need, indeed could not, distinguish between rational and integral equivalence at ∞. We shall not attempt to formulate definitions or theorems about "u-genera" or "u-spinor genera" but indicate the consequences of Theorem 8.2 in a context where we retain the ordinary definition of genus and spinor genus. If p is a prime we denote by $\mathbf{Z}^{[p]}$ the ring of rationals whose denominators are powers of p. In other words

$$\mathbf{Z}^{[p]} = \{a\colon |a|_q \leqslant 1 \quad \text{all } q \neq p, \infty\}.$$

THEOREM 8.3. *Let* f, g *be two integral forms in the same spinor genus and in* $n \geqslant 3$ *variables. Suppose that they are isotropic over* $\mathbf{Q}_q$ *for some prime* q. *Then they are properly equivalent over* $\mathbf{Z}^{[q]}$.

Proof. As for Theorem 1.4 but using Theorem 8.2 above instead of Theorem 7.1, Corollary, of Chapter 10. In the notation of Theorem 1.4 we have to find an $\alpha \in \Theta(V)$ which satisfies $\alpha \in \beta_p \Theta(\Lambda_p)$ for all primes $p \neq q$: there is no condition at ∞.

COROLLARY. *Let f be an integral form in n variables which is isotropic over $\mathbf{Q}_q$ for some q. Suppose either that $n \geqslant 4$ or that $n = 3$ and that the genus of f contains only one spinor genus. Then there is an integer $m \geqslant 0$ with the following property: suppose that $c \in \mathbf{Z}$, $c \neq 0$ is divisible by q^{2m} and that c is representable by f over every $\mathbf{Z}_p$ and over $\mathbf{R} = \mathbf{Z}_\infty$. Then c is representable by f over $\mathbf{Z}$, say $c = f(\mathbf{c})$. Further, we may suppose that $p \nmid \mathbf{c}$ for all $p \neq q$ for which c is primitively represented over $\mathbf{Z}_p$.*

Proof. For suppose that $b \in \mathbf{Z}$, $b \neq 0$, is representable by f over every $\mathbf{Z}_p$ (including $p = \infty$). Then b is representable by some form g in the spinor genus of f by Theorem 7.1 if $n \geqslant 4$ and by Theorem 1.3 of Chapter 9 if $n = 3$. By Theorem 8.3 it follows that $c = q^{2m}b$ is representable by f over $\mathbf{Z}$ for some $m \geqslant 0$ which can be chosen independent of b since there are only finitely many equivalence classes in the spinor genus. This concludes the proof.

Theorem 8.3 does not, however, exploit the full potentialities of Theorem 8.2 which even gives connections between forms in different genera. We give only a particular case due to Kneser (1957).

THEOREM 8.4. *Let f_1, f_2 be two classically integral forms of the same determinant d in $n \geqslant 3$ variables, both isotropic over $\mathbf{Q}_2$. Suppose that they are equivalent over $\mathbf{R} = \mathbf{Z}_\infty$ and over $\mathbf{Z}_p$ for all $p \neq 2$. Suppose, further, that $p^{n(n-1)} \nmid d$ for all $p \neq 2$. Then f, g are properly equivalent over $\mathbf{Z}^{[2]}$.*

Proof. Since f_1, f_2 are equivalent over $\mathbf{Q}_\infty$ and over $\mathbf{Q}_p$ for $p \neq 2$, they must be equivalent over $\mathbf{Q}$. They can therefore be represented by lattices Λ, Γ in the same $\mathbf{Q}$-vector space. Further, Λ_p and Γ_p are equivalent for $p \neq 2$ and the condition $p^{n(n-1)/2} \nmid d$ implies that $U_p(\mathbf{Q}_p^*)^2 \subset \theta(\Lambda_p)$ for $p \neq 2$ by Lemma 3.7, Corollary. The rest of the proof can be left with confidence to the reader (cf. Lemma 3.6, Corollary).

COROLLARY 1. *The forms f_1, f_2 can be connected with a sequence of forms g_v as described in Theorem 6.1 such that always $p = 2$ in* (iii).

As a very special case we have

COROLLARY 2. *There is precisely one proper class of classically integral forms of determinant $d = 1$ in n variables for $n \leqslant 7$. For $n = 8$ there are precisely two, one properly and one improperly primitive.*

Proof (sketch). For $n < 5$ the result has already been proved in Lemma 3.2, Corollary 2, of Chapter 9. We may therefore suppose that $n \geqslant 5$. Any two such forms are $\mathbf{Z}_p$-equivalent for $p \neq 2$, since $p \nmid d$. Further, since $n \geqslant 5$ any such form is isotropic over $\mathbf{Q}_2$. Hence the hypotheses of the Theorem apply. One form satisfying the hypotheses is $x_1^2 + \ldots + x_n^2$. We want to show that it gives the only class for $n = 5, 6, 7$ and that there is one more class for $n = 8$. It is convenient to use the language of lattices. Let Λ be the lattice with basis $\mathbf{e}_1, \ldots, \mathbf{e}_n$ and define ϕ by

$$\phi(x_1\mathbf{e}_1 + \ldots + x_n\mathbf{e}_n) = x_1^2 + \ldots + x_n^2.$$

By the theorem there is a lattice Γ corresponding to any other form g satisfying the conditions of the Corollary and such that $I(\Lambda, \Gamma) = 2^t$ for some t. We can suppose that t is minimal (for fixed g). We know that there is a sequence of lattices

$$\Lambda = \Delta_1, \Delta_2, \ldots, \Delta_J = \Gamma$$

all corresponding to integral forms of determinant 1 and such that

$$I(\Delta_j, \Delta_{j+1}) = 2.$$

Let $\mathbf{b} \in \Delta_1, \mathbf{b} \notin \Lambda$. Then $2\mathbf{b} \in \Lambda$. Since $d = 1$ is odd, the lattice Λ' consisting of the $\mathbf{c} \in \Lambda$ such that

$$\phi(2\mathbf{b}, \mathbf{c}) \equiv 0 \pmod 2$$

must be of index 2 in Λ: and Δ_1 is the lattice generated by Λ' and $\mathbf{b}$. The lattice Δ_1 is thus entirely determined by $\mathbf{b}$. Further, $\mathbf{b}$ and $\mathbf{b} + \mathbf{c}$ with $\mathbf{c} \in \Lambda$ both give rise to the same Δ_1 whenever $\phi(2\mathbf{b}, \mathbf{c})$ is even.

We have

$$2\mathbf{b} = b_1\mathbf{e}_1 + \ldots + b_n\mathbf{e}_n$$

for some $b_j \in \mathbf{Z}$. At least one of the b_j is odd since $\mathbf{b} \notin \Lambda$. Since $\phi(\mathbf{b}) \in \mathbf{Z}$, the number N of odd b_j must be divisible by 4.

Suppose, first, that $5 \leqslant n \leqslant 7$. Then the only possibility is $N = 4$. We can suppose, by symmetry, that $b_1 \equiv b_2 \equiv b_3 \equiv b_4 \equiv 1 \pmod 2$ and $b_5, \ldots, b_n$ are even. By replacing $\mathbf{b}$ by $\mathbf{b} + \mathbf{c}$, $\mathbf{c} \in \Lambda$ as described above, it is easy to see that we need consider only

$$2\mathbf{b} = \pm\mathbf{e}_1 \pm \mathbf{e}_2 \pm \mathbf{e}_3 \pm \mathbf{e}_4$$

(independent signs) and by changing the signs of $\mathbf{e}_1, \ldots, \mathbf{e}_4$ as necessary that

$$2\mathbf{b} = \mathbf{e}_1 + \ldots + \mathbf{e}_4.$$

It is then easy to verify that Δ_1 is properly equivalent to Λ.

For $n = 8$ there are the two possibilities $N = 4$ and $N = 8$. When $N = 8$ the lattice Δ_1 is generated by $\mathbf{b}$, where

$$2\mathbf{b} = \mathbf{e}_1 + \ldots + \mathbf{e}_8$$

and Λ' consists of the $\mathbf{c} = c_1\mathbf{e}_1 + \ldots + c_8\mathbf{e}_8 \in \Lambda$ with

$$c_1 + \ldots + c_8 \equiv 0 \pmod 2.$$

It is easily checked that $\phi(\mathbf{d}) \equiv 0 \pmod 2$ for all $\mathbf{d} \in \Delta_1$, and so Δ_1 corresponds to an improperly primitive form. It can further be verified that Δ_2 is now properly equivalent either to $\Lambda = \Delta_1$ or to Δ_2. Hence there are only two equivalence classes.

For further applications of this technique see Kneser (1957) and Example 6.

9. REPRESENTATION BY DEFINITE FORMS

The object of this section is to prove Theorem 1.6, which asserts that if $f(\mathbf{x})$ is a regular positive definite integral form in $n \geqslant 4$ variables then there is a constant $N = N(f)$ with the following property: if $a \geqslant N$ is an integer which is represented primitively by f over all $\mathbf{Z}_p$ then a is represented primitively by f over $\mathbf{Z}$. It is clear that it does not matter which of the two definitions of integrality we use ("classically integral" or "integer-valued") since we can take $N(cf) = cN(f)$ for any integer $c > 0$. Indeed we could, if we wished, have assumed only that f has rational coefficients.

The theorem would become false if we permitted $n = 3$ as follows at once from Lemma 7.1 and Note 2 to Theorem 7.1. The theorem would also become false for $n = 4$ if the word "primitively" were omitted. Consider, for example, the form

$$f_0(\mathbf{x}) = x_1^2 + x_2^2 + 5(x_3^3 + x_4^2). \tag{9.1}$$

For any $m \geqslant 0$ this represents 3.2^{2m} over every $\mathbf{Z}_p$ (it is enough to check only $p = 2, 5$) but it does not represent 3.2^{2m} over $\mathbf{Z}$. For suppose $a_1^2 + a_2^2 + 5(a_3^2 + a_4^2) = 3.2^{2m}$ for some integer $m > 0$ and some $a_1, \ldots, a_4$. Then $a_1, \ldots, a_4$ must be all even and so there is a similar representation of $3.2^{2(m-1)}$. Hence by induction there would be a representation with $m = 0$: which is clearly impossible.

The counter-example above works because f_0 is anisotropic over $\mathbf{Q}_2$. If a quadratic form f is isotropic over $\mathbf{Q}_p$, then there is some $M = M(f, p)$ with the following property: if a is represented by f over $\mathbf{Z}_p$ then there is some $m = m(a, p, f) \leqslant M$ such that $p^{-2m}a$ is primitively represented over $\mathbf{Z}_p$. The verification of this statement is left to the reader. It follows that when f is isotropic over all $\mathbf{Q}_p$, in particular whenever $n \geqslant 5$, then the assertion of Theorem 1.6 remains true if the word "primitively" is omitted in the two places where it occurs.

As the proof of Theorem 1.6 is somewhat complicated we make no attempt to obtain good values of the constant $N = N(f)$. In particular we shall adopt the simplest methods to deal with any difficulties even though their effect on $N(f)$ is extravagant and could have been mitigated by a more complicated choice of tactic. The proof for $n \geqslant 5$ is on rather different lines from that for $n = 4$ and, as it is distinctly simpler, will be given first. For both proofs we shall denote by P a finite set of primes p such that

(i) $2 \in P$

(ii) If p divides the determinant of f, then $p \in P$.

We note that any $a \in \mathbf{Z}$ is represented primitively over $\mathbf{Z}_p$ for all $p \notin P$ and so the hypotheses of Theorem 1.6 are really only making assertions about the $p \in P$. As the counter-example above suggests, we shall have particular difficulty in dealing with the a which are divisible by a high power of some prime p in P.

Proof of Theorem 1.6 ($n \geqslant 5$). The key step is to find a finite set Ψ of substitutions

$$\mathbf{x} = \psi(\mathbf{y}) \tag{9.2}$$

given by

$$\psi:\ x_i = s_{i1}y_1 + \ldots + s_{in}y_n \qquad (1 \leqslant i \leqslant n) \tag{9.3}$$

with

$$s_{ij} \in \mathbf{Z}, \qquad \det(s_{ij}) \neq 0, \tag{9.4}$$

such that

$$f(\psi(\mathbf{y})) = g(y_1, y_2) + h(y_3, \ldots, y_n), \tag{9.5}$$

where the integral quadratic forms $g = g^{\psi}$ and $h = h^{\psi}$ have a number of pleasant properties. In particular we shall require

(α) If $a \in \mathbf{Z}$ is representable primitively by f over all $\mathbf{Z}_p$ ($p \in P$) then there is a $\psi \in \Psi$ such that a is similarly represented by $g = g^{\psi}$.

(β) If $b_1, \ldots, b_n \in \mathbf{Z}$ and g.c.d.$(b_1, b_2) = 1$, then $\psi(\mathbf{b}) \in \mathbf{Z}^n$ is primitive.

(γ) There is a prime $q \notin P$ and an $m \geqslant 0$ such that if $c \in \mathbf{Z}$ is positive, divisible by q^{2m}, and representable by h over every† $\mathbf{Z}_p$ then c is representable by h over $\mathbf{Z}$.

(δ) For each ψ there is a finite set $B = B(\psi)$ of coprime pairs b_1, b_2 of integers such that if $a \in \mathbf{Z}$ is representable primitively by $g = g^{\psi}$ over every $\mathbf{Z}_p$ ($p \in P$) then there is a pair $(b_1, b_2) \in B$ such that

$$c = a - g(b_1, b_2) \tag{9.6}$$

satisfies all the hypotheses of (γ) except, possibly, the hypothesis that c is positive.

† Note that we do not here make the restriction $p \in P$. Also that (γ) says "representable", not "primitively representable".

Once we have established (α), (β), (γ), (δ) we are done, as we show at once. For let

$$N = \sup_{\psi} \sup_{(b_1, b_2) \in B(\psi)} g^{\psi}(b_1, b_2) \tag{9.7}$$

and suppose that $a > N$ is primitively representable by f over every $\mathbf{Z}_p$ $(p \in P)$. Then a is so represented by some g^{ψ} and (9.7) implies that the c given by (9.6) is positive. Hence (γ) applies and

$$a - g(b_1, b_2) = c = h(b_3, \ldots, b_n)$$

is for some $b_3, \ldots, b_n$. Then $a = f(\psi(\mathbf{b}))$ by (9.5) where $\mathbf{b} = (b_1, \ldots, b_n)$ and $\mathbf{b}$ is primitive by (β).

We now embark on the details and must start with a local discussion.

Let $p \in P$ be fixed temporarily. Since $n \geqslant 5$, the form $f(\mathbf{x})$ is isotropic over $\mathbf{Q}_p$ and so there is a primitive $\mathbf{u}_0 = \mathbf{u}_0(p) \in \mathbf{Z}_p^n$ such that $f(\mathbf{u}_0) = 0$. Since f is regular, there is a $\mathbf{v}_0 = \mathbf{v}_0(p) \in \mathbf{Z}_p^n$ such that $f(\mathbf{u}_0, \mathbf{v}_0) \neq 0$ and we can suppose without loss of generality that $\mathbf{u}_0, \mathbf{v}_0$ is a primitive pair of vectors (that is, by definition, that there is a basis $\mathbf{u}_0, \mathbf{v}_0, \mathbf{w}_3, \ldots, \mathbf{w}_n$ of $\mathbf{Z}_p^n$). Then

$$\begin{aligned} l_p^{(0)}(\mathbf{z}) = l_p^{(0)}(z_1, z_2) &= f(z_1\mathbf{u}_0 + z_2\mathbf{v}_0) \\ &= 2z_1z_2 f(\mathbf{u}_0, \mathbf{v}_0) + z_2^2 f(\mathbf{v}_0) \end{aligned} \tag{9.8}$$

is a regular form. It represents primitively all sufficiently small $a_p \in \mathbf{Z}_p$, indeed with $z_1 = 1$.

If $f(\mathbf{z})$ represents some element a_p (say) of $\mathbf{Z}_p$ primitively, then clearly it represents primitively all elements of $a_p U_p^2$ (where U_p, as usual, is the group of p-adic units). Hence there is only a finite number $J = J(p)$ (say) of classes of $\mathbf{Z}_p/U_p^2$ which are represented primitively by f but not by $l_p^{(0)}$. Let $\mathbf{u}_j = \mathbf{u}_j(p)$ be primitive elements of $\mathbf{Z}_p^n$ such that $f(\mathbf{u}_j)$ $(1 \leqslant j \leqslant J = J(p))$ runs through the classes of $\mathbf{Z}_p/U_p^2$ which are represented by f but not by $l_p^{(0)}$. For each j let $\mathbf{v}_j \in \mathbf{Z}_p^n$ be chosen so that $\mathbf{u}_j$, $\mathbf{v}_j$ is a primitive pair and put

$$l_p^{(j)}(\mathbf{z}) = f(z_1\mathbf{u}_j + z_2\mathbf{v}_j). \tag{9.9}$$

Then every $a_p \in \mathbf{Z}_p$ which is represented primitively by f is also represented primitively by one of the $l_p^{(j)}$ $(0 \leqslant j \leqslant J(p))$.

We now construct a global form. Let $k = k(p)$ be given with $0 \leqslant k(p) \leqslant J(p)$ for each $p \in P$. By Theorem 2.1 of Chapter 9 we can find a primitive pair of vectors $\mathbf{u}, \mathbf{v} \in \mathbf{Z}^n$ which are arbitrarily close p-adically to $\mathbf{u}_k(p), \mathbf{v}_k(p)$ $\big(k = k(p)\big)$ for each $p \in P$. We shall suppose, in particular, that $\mathbf{u}, \mathbf{v}$ are chosen so that

$$g(z_1, z_2) = f(z_1\mathbf{u} + z_2\mathbf{v}) \text{ (say)}$$

is $\mathbf{Z}_p$-equivalent to $l_p^{(k)}(\mathbf{z})$ with $k = k(p)$ for each $p \in P$. We may suppose, further, that the coefficients of $g(z_1, z_2)$ are not all divisible by some $p \notin P$:

that this can be done follows from the details of the proof of Theorem 2.1 of Chapter 9 (just quoted) since $\mathbf{u}, \mathbf{v}$ are constructed in succession and we can ensure first that $f(\mathbf{u}) \neq 0$ and then that $f(\mathbf{v})$ is not divisible by any prime $p \notin P$ which divides $f(\mathbf{u})$. It is clear that any $a \in \mathbf{Z}$ which is represented primitively by f over all the $\mathbf{Z}_p$ $(p \in P)$ is also so represented by $g(\mathbf{z})$ for suitable choice of the $k(p)$. This is (α) of the sketch of the proof above.

We now turn to the construction of the polynomial h which appears in (9.5). Let $V^{(1)}$ be the 2-dimensional $\mathbf{Q}$-vector space spanned by $\mathbf{u}, \mathbf{v}$ and let $\Delta \subset V^{(1)}$ be the $\mathbf{Z}$-lattice generated by $\mathbf{u}$, $\mathbf{v}$. Denote by $V^{(2)}$ the $(n-2)$-dimensional subspace of V normal to $V^{(1)}$, so V is the direct orthogonal sum of $V^{(1)}$ and $V^{(2)}$. We shall define h in terms of a lattice Γ in $V^{(2)} \cap \mathbf{Z}^n$: and Γ itself will be determined by its localizations Γ_p. In order to obtain (β) of the sketch-proof we must ensure that if $\mathbf{e}^{(1)}$ is a primitive element of Δ and $\mathbf{e}^{(2)}$ is any element of Γ then $\mathbf{e}^{(1)} + \mathbf{e}^{(2)}$ is a primitive element of $\mathbf{Z}^n$. For this it will be clearly sufficient to ensure that

$$\mathbf{e}^{(1)} + \mathbf{e}^{(2)} \in p\mathbf{Z}_p^n, \quad \mathbf{e}^{(1)} \in \Delta_p, \qquad \mathbf{e}^{(2)} \in \Gamma_p \tag{9.10$_1$}$$

implies

$$\mathbf{e}^{(1)} \in p\Delta_p \tag{9.10$_2$}$$

for every prime p. We shall also require that

$$\theta(\Gamma_p) \supset U_p^2 \tag{9.11}$$

for all p, since this by Lemma 3.6, Corollary, will ensure that the genus of Γ consists of a single spinor genus. We distinguish three kinds of prime p.

(I) $p \in P$. First we require that $\Gamma_p \subset p\mathbf{Z}_p^n$. Then (9.10$_1$) certainly implies (9.10$_2$). Second we require (9.11). This certainly holds if Γ_p has basis $\mathbf{w}_1, \ldots, \mathbf{w}_{n-2} \in \mathbf{Z}_p^n \cap V_p^{(2)}$ such that $f(\mathbf{w}_i, \mathbf{w}_j) = 0$ $(i \neq j)$ and $|f(\mathbf{w}_j)|_p$ takes only the values p^{-l} and p^{-l-1} for some $l \in \mathbf{Z}$. We choose for Γ_p any lattice in $V_p^{(2)}$ satisfying the two conditions just mentioned.

(II) $p \notin P$ *but* $p | d(g)$, *where* $d(g)$ *is the determinant of* $g(\mathbf{z})$. We have ensured that p does not divide all the coefficients of g and so when it is diagonalized over $\mathbf{Z}_p$ one coefficient is a unit and one is not. In terms of (9.10) this means that there are $\mathbf{r}, \mathbf{s} \in \mathbf{Z}_p^n$ which generate the same 2-dimensional $\mathbf{Z}_p$-lattice Δ_p as do $\mathbf{u}, \mathbf{v}$ and for which

$$|f(\mathbf{r})|_p = 1, \qquad f(\mathbf{r}, \mathbf{s}) = 0, \qquad |f(\mathbf{s})|_p < 1. \tag{9.12}$$

Since $p \notin P$, the determinant of f is a p-adic unit and so there is a $\mathbf{t} \in \mathbf{Z}_p^n$ such that

$$f(\mathbf{r}, \mathbf{t}) = 0, \qquad |f(\mathbf{s}, \mathbf{t})|_p = 1. \tag{9.13}$$

Since $p \notin P$ also implies $p \neq 2$, the binary form induced on the space spanned by $\mathbf{s}, \mathbf{t}$ is a hyperbolic plane. In particular, there are $\mathbf{r}_2, \mathbf{r}_3 \in \mathbf{Z}_p^n$ such that

$$f(\mathbf{r}_2) = -f(\mathbf{r}_3) = 1 \qquad f(\mathbf{r}_2, \mathbf{r}_3) = 0 \tag{9.14}$$

and

$$\mathbf{s} = l\mathbf{r}_2 + m\mathbf{r}_3, \quad l, m \in \mathbf{Z}_p,$$

where

$$|l|_p = |m|_p = 1, \qquad f(\mathbf{s}) = l^2 - m^2.$$

Hence there are $\mathbf{r}_4, \mathbf{r}_5, \ldots, \mathbf{r}_n \in \mathbf{Z}_p^n$ such that

$$\mathbf{r} = \mathbf{r}_1, \mathbf{r}_2, \mathbf{r}_3, \ldots, \mathbf{r}_n$$

is a basis for $\mathbf{Z}_p^n$ with

$$|f(\mathbf{r}_j)|_p = 1 \quad f(\mathbf{r}_i, \mathbf{r}_j) = 0 \qquad (1 \leqslant i, j \leqslant n, i \neq j). \tag{9.15}$$

Further, $\mathbf{Z}_p^n \cap V_p^{(2)}$ has basis

$$\mathbf{w}, \mathbf{r}_4, \ldots, \mathbf{r}_n, \tag{9.16}$$

where

$$\mathbf{w} = m\mathbf{r}_2 + l\mathbf{r}_3,$$

so

$$|f(\mathbf{w})|_p = |f(\mathbf{s})|_p < 1.$$

We take for Γ_p the lattice with basis

$$p\mathbf{w}, \mathbf{r}_4, \ldots, \mathbf{r}_n.$$

Then clearly (9.10_1) implies (9.10_2), as required. Further, (9.11) holds by (9.15) and since $p \neq 2$ (because $p \notin P$). We note that f represents the whole of U_p on Γ_p (and indeed does so, by (9.15), on the two-dimensional lattice spanned by $\mathbf{r}_4, \mathbf{r}_5$).

(III) $p \notin P$ *and* $p \nmid d(g)$. By an argument similar to that in case (II) there is a basis $\mathbf{r}_1, \ldots, \mathbf{r}_n$ of $\mathbf{Z}_p^n$ such that (9.15) holds and $\mathbf{r}_1, \mathbf{r}_2$ are a basis for Δ_p.

We take for Γ_p the $\mathbf{Z}_p$-lattice generated by

$$\mathbf{r}_3, \ldots, \mathbf{r}_n$$

so

$$\Gamma_p = \mathbf{Z}_p^n \cap V_p^{(2)}. \tag{9.17}$$

Clearly (9.10_1) implies (9.10_2) and (9.11) holds, as required. We note that f represents the whole of $\mathbf{Z}_p$ on Γ_p by (9.15).

Since (9.17) holds for almost all primes p, there certainly is by Theorem 1.1 a global lattice Γ with the specified localizations Γ_p. Let $\mathbf{w}_3, \ldots, \mathbf{w}_n$ be a $\mathbf{Z}$-basis of Γ. For $\mathbf{y} \in \mathbf{Z}^n$ we define

$$\psi(\mathbf{y}) = y_1\mathbf{u} + y_2\mathbf{v} + y_3\mathbf{w}_3 + \ldots + y_n\mathbf{w}_n \tag{9.18}$$

and then we have (9.5), where h is the form determined by the basis $\mathbf{w}_3, \ldots, \mathbf{w}_n$ of Γ. We note that we have achieved (β) of the sketch proof by ensuring that (9.10_1) implies (9.10_2) for every prime p. For later use we note that we have also proved

(ε) If $p \notin P$, $p \mid d(g)$ then h represents all of U_p over $\mathbf{Z}_p$. If $p \notin P$, $p \nmid d(g)$ then h represents all of $\mathbf{Z}_p$ over $\mathbf{Z}_p$.

We now choose a prime $q = q(\psi)$ which will play a special role in the argument. It must satisfy

$$q \notin P \qquad q \nmid d(h) \tag{9.19}$$

and

$$-d(g) \in U_q^2 \tag{9.20}$$

but otherwise can be chosen arbitrarily†. It follows from (9.20) that $g(y_1, y_2)$ is equivalent over $\mathbf{Z}_q$ to $2y_1y_2$ and so represents primitively every element of $\mathbf{Z}_q$. Further, as already remarked, $h(y_3, \ldots, y_n)$ is in a genus of one spinor genus (by (9.11) and Lemma 3.6, Corollary) and it is isotropic over $\mathbf{Q}_q$ by (9.19) and since it is of dimension $n - 2 \geqslant 3$. Hence by Theorem 8.3 there is an integer $m > 0$ with the following property: if $c \in \mathbf{Z}, c \neq 0$ is divisible by q^{2m} and it representable by h over every $\mathbf{Z}_p$ (including $\mathbf{Z}_\infty = \mathbf{R}$) then c is representable by h over $\mathbf{Z}$. We thus have (γ) of the sketch-proof.

We can now commence the final stage of the proof. Let $\psi \in \Psi$ and $p \in P$ be fixed, and let $\mathscr{A}_p = \mathscr{A}_p(\psi)$ denote the set of elements a_p of $\mathbf{Z}_p$ which can be represented primitively by $g = g^\psi$, say

$$a_p = g(b'_p, b''_p) \qquad \max\{|b'_p|_p, |b''_p|_p\} = 1.$$

Then if c'_p, c''_p run through neighbourhoods of b'_p, b''_p (in the p-adic sense) it is clear that $a_p - g(c'_p, c''_p)$ runs through a neighbourhood of the origin and so we can find c'_p, c''_p such that

$$a_p - g(c'_p, c''_p) = d_p \text{ (say)} \tag{9.21}$$

is non-zero and is represented by h over $\mathbf{Z}_p$. There is then a neighbourhood D_p of d_p, consisting of non-zero elements of $\mathbf{Z}_p$, all of whose elements are represented by h over $\mathbf{Z}_p$ and so in an obvious notation

$$D_p + g(c'_p, c''_p) \subset \mathscr{A}_p. \tag{9.22}$$

The set $\mathscr{A}_p \subset \mathbf{Z}_p$ is clearly closed in the p-adic topology and so $\mathscr{A}_p$ is covered by a finite collection of the open sets (9.22). [The form of argument was chosen so as to cover the case when g is isotropic over $\mathbf{Z}_p$. Otherwise it is enough to note that if g represents a_p primitively, then it represents the whole of $a_pU_p^2$.]

Now suppose that $a \in \mathbf{Z}$ is represented by f primitively over every $\mathbf{Z}_p$. Then by construction there is at least one ψ such that a is represented by $g = g^\psi$ primitively over every $\mathbf{Z}_p$. We now choose $b_1, b_2 \in \mathbf{Z}$ to satisfy the following conditions

† Such q exist. For $x^2 + d(g)y^2$ represents an integer G which is prime to $d(g)d(h)$ and not divisible by any $p \in P$. Then any prime $q \mid G$ will do.

C_1: g.c.d.$(b_1, b_2) = 1$.

C_2: Let $p \in P$. Then $a \in \mathscr{A}_p$ and so a is in one of the sets (9.22). We require that b_1, b_2 are sufficiently close p-adically to c'_p, c''_p that $a - g(b_1, b_2) \in D_p$.

C_3: Let $p \notin P$ but $p \mid d(g)$. We require that $a - g(b_1, b_2)$ is a p-adic unit. This is always possible since we have arranged that not all the coefficients of g are divisible by p (case II) above).

C_4: We require that $a - g(b_1, b_2)$ is divisible by q^{2m}. This is possible since, as already remarked, g is equivalent over $\mathbf{Z}_q$ to $2y_1y_2$.

Clearly these conditions can be satisfied simultaneously. Indeed, there is a finite set $B = B(\psi)$ of pairs $\mathbf{b} = (b_1, b_2)$ of integers such that for any given $a \in \mathbf{Z}$ the conditions are satisfied for at least one $\mathbf{b} \in B$. This follows since there are only a finite number of primes p to be considered in C_3 and C_4 and since for given p there are only a finite number of sets (9.22). Further, $a - g(b_1, b_2)$ is representable by h over all $\mathbf{Z}_p$: for $p \in P$ by C_2 and for $p \notin P$ by C_3 and (ε). We have thus achieved (δ) of the sketch proof. Since we already have (α), (β), (γ) this completes the proof of the Theorem for $n \geqslant 5$.

Proof of Theorem 1.6 ($n = 4$). This will be along the same general lines as the proof for $n > 4$ except that the form g will now be a form in a single variable. This leads to additional complications. For an integral quadratic form f and for $a \in \mathbf{Z}$ it is convenient to introduce $\mathscr{R}(f, a)$ for the proposition (which may be true or false):

$\mathscr{R}(f, a)$: If a is primitively represented by f over all $\mathbf{Z}_p$, then a is primitively represented by a over f.

With this convention we shall require

LEMMA 9.1. *Let f be a positive definite integral form in $n = 4$ variables of determinant $d(f) = d \neq 0$ and let $a > 0$ be an integer. Let $q \neq 2$ be a prime such that $a \in U_q$ and $d \in U_q^2$. Suppose that $\mathscr{R}(g, a)$ is true for the given a and for all positive definite integral forms g in 4 variables with $d(g) = d$. Then $\mathscr{R}(f, q^m a)$ is true for all integers $m \geqslant 0$.*

Proof. The statement is trivial for $m = 0$, so let $m > 0$ be fixed. There is nothing to prove unless $q^m a$ is primitively represented by f over all $\mathbf{Z}_p$, so we suppose this.

Since $d \in U_q^2$ and $q \neq 2$, the form f is $\mathbf{Z}_q$-equivalent to $2x_1x_3 + 2\delta x_2x_4$ with $\delta^2 = d$. Hence on substituting a $\mathbf{Z}$-equivalent form for f we may suppose without loss of generality that

$$f(\mathbf{x}) = \sum_{i,j} f_{ij}x_ix_j \tag{9.23}$$

where

$$\left.\begin{aligned} f_{13} = f_{31} &\equiv 1 \quad (q^{2m+1}) \\ f_{24} = f_{42} &\equiv \delta \quad (q^{2m+1}) \\ f_{ij} &\equiv 0 \quad (q^{2m+1}) \quad \text{otherwise} \end{aligned}\right\} \tag{9.24}$$

or, say,

$$f(\mathbf{x}) \equiv 2x_1x_3 + 2\delta x_2x_4 \qquad (q^{2m+1}). \tag{9.25}$$

Hence the form

$$g(\mathbf{x}) = q^{-m}f(x_1, x_2, q^m x_3, q^m x_4) \tag{9.26}$$

is integral and satisfies

$$g(\mathbf{x}) \equiv 2x_1x_3 + 2\delta x_2x_4 \qquad (q). \tag{9.27}$$

Further,

$$f(\mathbf{x}) = q^{-m}g(q^m x_1, q^m x_2, x_3, x_4). \tag{9.28}$$

Clearly $d(g) = d$ and a is represented primitively by g over every $\mathbf{Z}_p$. [This is trivial for $p \neq q$ and automatic for $p = q$ since $q \nmid 2d$.] Hence by the hypotheses of the Lemma there is a primitive $\mathbf{b} \in \mathbf{Z}^4$ such that

$$a = g(\mathbf{b}).$$

By (9.27) and since $q \nmid a$, at least one of b_1, b_2 is not divisible by q. Hence the representation

$$q^m a = f(b_1, b_2, q^m b_3, q^m b_4)$$

is primitive. This proves the Lemma.

After Lemma 9.1 it will be enough to prove

LEMMA 9.2. *Let f be a positive definite integral form in 4 variables of determinant d and let $q \neq 2$ be a prime such that $d \in U_q^2$. Then there is a constant $N_1 = N_1(f, q)$ such that $\mathscr{R}(f, a)$ holds for all integers a with $a > N_1$ and $q \nmid a$.*

Before proving Lemma 9.2 we show that it does in fact imply Theorem 1.6 for $n = 4$. Let f, d, q be as in Lemmas 9.1 and 9.2. Put

$$N_2 = N_2(d, q) = \max_g N_1(g, q) \tag{9.29}$$

where the maximum is over the equivalence classes of integral forms g in 4 variables and of determinant d. Then Lemmas 9.1 and 9.2 imply that

$$\mathscr{R}(f, q^m a_0) \tag{9.30}$$

holds whenever

$$m \geqslant 0, \qquad q \nmid a_0, \qquad a_0 > N_2(d, q). \tag{9.31}$$

Now let $q(1), q(2)$ be two distinct primes such that† $d \in U_{q(1)}^2$, $d \in U_{q(2)}^2$. We show that

$$N(f) = N_2(d, q(1))\, N_2(d, q(2)) \tag{9.32}$$

† Such primes exist. Cf. footnote to p. 240.

has the properties required by Theorem 1.6. For suppose that

$$a > N(f). \tag{9.33}$$

Then

$$a = a_3(q(1))^{m(1)}(q(2))^{m(2)} \tag{9.34}$$

for some $m(1) \geqslant 0, m(2) \geqslant 0$ and a_3 prime to both $q(1)$ and $q(2)$. By (9.32), (9.33) we have at least one of the inequalities

$$a_1 = a_3(q(2))^{m(2)} > N_2(d, q(1)) \tag{9.35}$$

or

$$a_2 = a_3(q(1))^{m(1)} > N_2(d, q(2)). \tag{9.36}$$

Hence the conditions (9.31) are satisfied either with $q = q(1), a_0 = a_1$, $m = m(1)$ or with $q = q(2), a_0 = a_2, m = m(2)$. In either case (9.30) is now the assertion $\mathscr{R}(f, a)$. Hence we have shown that (9.33) implies $\mathscr{R}(f, a)$. This is just the assertion of Theorem 1.6.

Proof of Lemma 9.2. We shall require

LEMMA 9.3. *Let f be a regular quaternary isotropic form over $\mathbf{Z}_p^4$. Then there is a $\mathbf{t} = \mathbf{t}_p \in \mathbf{Z}_p^4$ with $f(\mathbf{t}) \neq 0$ and a lattice $\Delta = \Delta_p$ in the 3-dimensional space normal to $\mathbf{t}$ with the following properties:*

(i) *$-f(\mathbf{t})$ is represented by f on Δ_p.*

(ii) *If $|f(\mathbf{t} + \mathbf{c})|_p < |f(\mathbf{t})|_p$, $\mathbf{c} \in \Delta_p$, then $\mathbf{t} + \mathbf{c}$ is a primitive element of $\mathbf{Z}_p^4$.*

(iii) *$\theta(\Delta_p) \supset U_p$.*

Proof. By hypothesis there is a primitive $\mathbf{e}_1 \in \mathbf{Z}_p^4$ such that

$$f(\mathbf{e}_1) = 0.$$

Let $\mathbf{e}_1, \mathbf{e}_2, \mathbf{e}_3, \mathbf{e}_4$ be a basis for $\mathbf{Z}_p^4$. Then at least one of $f(\mathbf{e}_1, \mathbf{e}_j)$ is non-zero and we may suppose that $|f(\mathbf{e}_1, \mathbf{e}_2)| \geqslant |f(\mathbf{e}_1, \mathbf{e}_j)| \quad (j = 3, 4)$.

On replacing $\mathbf{e}_j$ by $\mathbf{e}_j - l_j \mathbf{e}_2$ $(j = 3, 4)$ with $l_j \in \mathbf{Z}_p$ we have without loss of generality

$$f(\mathbf{e}_1, \mathbf{e}_2) \neq 0, \qquad f(\mathbf{e}_1, \mathbf{e}_3) = f(\mathbf{e}_1, \mathbf{e}_4) = 0.$$

Since f is regular, $f(x_3\mathbf{e}_3 + x_4\mathbf{e}_4)$ does not vanish identically and so without loss of generality

$$f(\mathbf{e}_3) \neq 0.$$

Put

$$f(\mathbf{e}_1, \mathbf{e}_2) = p^\beta b, \qquad b \in U_p$$
$$f(\mathbf{e}_3) = p^\gamma c, \qquad c \in U_p$$

and

$$\lambda = 1 \quad \text{if} \quad p = 2$$
$$= 0 \text{ otherwise.}$$

Let s, t be positive integers and define r, μ by

$$r + \beta + \lambda = 2s + \gamma = \mu.$$

Put

$$\mathbf{r}_1 = \mathbf{e}_1, \qquad \mathbf{r}_2 = p^r\mathbf{e}_2, \qquad \mathbf{r}_3 = p^s\mathbf{e}_3, \qquad \mathbf{r}_4 = p^t\mathbf{e}_4.$$

Then

$$f(\mathbf{r}_1) = f(\mathbf{r}_1, \mathbf{r}_3) = f(\mathbf{r}_1, \mathbf{r}_4) = 0$$

and

$$2f(\mathbf{r}_1, \mathbf{r}_2) \in p^\mu U_p.$$
$$f(\mathbf{r}_3) \in p^\mu U_p.$$

We can clearly choose s, t so large that

$$f(\mathbf{r}_i, \mathbf{r}_j) \in p^{\mu+1}\mathbf{Z}_p \qquad ((i, j) \neq (1, 2), (2, 1), (3, 3)).$$

Hence we can find $k_2, k_4 \in p\mathbf{Z}_p$ such that

$$f(\mathbf{r}_j - k_j\mathbf{r}_3, \mathbf{r}_3) = 0 \qquad (j = 2, 4).$$

Put $\mathbf{t} = \mathbf{t}_p = \mathbf{r}_3$ and let Δ_p be the lattice with basis

$$\mathbf{v}_1 = \mathbf{r}_1 = \mathbf{e}_1$$
$$\mathbf{v}_2 = \mathbf{r}_2 - k_2\mathbf{r}_3$$
$$\mathbf{v}_3 = \mathbf{r}_4 - k_4\mathbf{r}_3.$$

Then

$$f(\mathbf{t}) \in p^\mu U_p.$$

Further,

$$\begin{aligned} f(y_1\mathbf{v}_1 + y_2\mathbf{v}_2 + y_3\mathbf{v}_3) &= 2f(\mathbf{v}_1, \mathbf{v}_2)y_1y_2 \\ &\quad + \text{terms in } y_2, y_3 \\ &\equiv p^\mu b y_1 y_2 \pmod{p^{\mu+1}} \end{aligned}$$

for $y_1, y_2, y_3 \in \mathbf{Z}_p$. In particular, f represents all of $p^\mu U_p$ primitively on Δ_p, indeed with $y_2 = 1, y_3 = 0$. This implies (i) and (iii) of the enunciation.

Finally, suppose that

$$|f(\mathbf{t} + \mathbf{c})|_p < p^{-\mu}$$

where

$$\mathbf{c} = y_1\mathbf{v}_1 + y_2\mathbf{v}_2 + y_3\mathbf{v}_3, \qquad y_j \in \mathbf{Z}_j.$$

Then

$$c + by_1y_2 \equiv 0 \qquad (p)$$

where $c = p^{-\mu}\phi(\mathbf{t}) \in U_p$ and $b = 2p^{-\mu}\phi(\mathbf{v}_1, \mathbf{v}_2) \in U_p$. Hence

$$y_1 \in U_p.$$

On expressing $\mathbf{t} + \mathbf{c}$ in terms of the basis $\mathbf{e}_1, \ldots, \mathbf{e}_4$ of $\mathbf{Z}_p^4$ we have

$$\mathbf{t} + \mathbf{c} = x_1\mathbf{e}_1 + \ldots + x_4\mathbf{e}_4$$

where

$$x_1 \equiv y_1 \qquad (p),$$

so $x_1 \in U_p$. It follows that $\mathbf{t} + \mathbf{c}$ is primitive in $\mathbf{Z}_p^4$. This proves (ii) and so concludes the proof of the Lemma.

COROLLARY. *For every $z \in \mathbf{Z}_p$ with $|z|_p < |f(\mathbf{t})|_p$ there is a $\mathbf{c} \in \Delta_p$ such that $f(\mathbf{t} + \mathbf{c}) = z$.*

Proof: Clear.

After this preliminary, we proceed with the proof of Lemma 9.2 on the same lines as the proof of Theorem 1.6 for $n \geqslant 5$ except that now the form g will be 1-dimensional. We recall that P is any finite set of primes such that $2 \in P$ and such that $d(f) \in U_p$ for all $p \notin P$. We suppose, as we may, that

$$q \notin P. \tag{9.37}$$

First, the local discussion. Let $p \in P$ and suppose first that f is isotropic at $\mathbf{Q}_p$. We define $\mathbf{u}_0 = \mathbf{u}_0(p) \in \mathbf{Z}_p^4$ to be the $\mathbf{t}$ given by Lemma 9.3. There are only a finite number of classes of $z \in \mathbf{Z}_p$ modulo† U_p^2 for which $|z|_p \geqslant |f(\mathbf{u}_0)|_p$. We pick a finite set $\mathbf{u}_j = \mathbf{u}_j(p)$ $(1 \leqslant j \leqslant J(p))$ of primitive elements of $\mathbf{Z}_p^4$ such that the $f(\mathbf{u}_j)U_p^2$ contain all the $z \in \mathbf{Z}_p$ with $|z|_p \geqslant |f(\mathbf{u}_0)|_p$ which are primitively represented by f over $\mathbf{Z}_p^4$.

If $p \in P$ but f is anisotropic over $\mathbf{Q}_p$, then f represents primitively only a finite number of classes of $\mathbf{Z}_p$ modulo U_p^2. We choose primitive $\mathbf{u}_j = \mathbf{u}_j(p) \in \mathbf{Z}_p^4 (1 \leqslant j \leqslant J(p))$ such that $\bigcup_j f(\mathbf{u}_j)\, U_p^2$ contains all the $z \in \mathbf{Z}_p$ which are primitively represented by f.

Finally, for $p = q$ we select two primitive vectors $\mathbf{u}_1 = \mathbf{u}_1(q), \mathbf{u}_2 = \mathbf{u}_2(q)$ in $\mathbf{Z}_q^4$ such that $f(\mathbf{u}_1) \in U_q^2$ and $f(\mathbf{u}_2) \in U_q, \notin U_q^2$.

We now come to the global construction and require a further auxiliary prime π, where ‡

$$d(f) \in U_\pi^2$$

and

$$\pi \notin P, \pi \neq q. \tag{9.38}$$

[The fact that we have used a Greek letter is merely an indication that we have exhausted the resources of the Latin alphabet and has no other significance.]

Now let $k(p)$ be given for $p \in P$ and $p = q$ where $k(q) = 1$ or 2 and $0 \leqslant k(p) \leqslant J(p)$ or $1 \leqslant k(p) \leqslant J(p)$ according as f is isotropic or anisotropic

† In the multiplicative sense.

‡ Cf. footnote to p. 240.

over $\mathbf{Q}_p\,(p \in P)$. By Dirichlet's theorem on primes in arithmetic progressions† we can find an integer $g_1 > 0$ such that

(i) $$g_1 \in f(\mathbf{u}_{k(p)}(p))U_p^2 \qquad (p \in P \cup \{q\}), \tag{9.39$_1$}$$

(ii) $$g_1 \in U_\pi, \tag{9.39$_2$}$$

and

(iii) $$g_1 = g_2 p(0), \tag{9.40}$$

where $p(0) \notin P \cup \{q, \pi\}$ is prime and all the prime divisors of g_2 are in P. Then g_1 is representable by f over every $\mathbf{Z}_p$, and so by Theorem 8.1 there is a primitive $\mathbf{u} \in \mathbf{Z}^4$ such that

$$f(\mathbf{u}) = g_1 \pi^{2l} = g \text{ (say)} \tag{9.41}$$

for some $l \geqslant 0$, and such that $\pi^{-l}\mathbf{u}$ is arbitrarily close to $\mathbf{u}_{k(p)}(p)$ for $p \in P \cup \{q\}$.

We now construct a lattice Γ on the 3-dimensional space $V^{(2)}$ consisting of the vectors normal to $\mathbf{u}$. As for $n \geqslant 5$ we define Γ in terms of its local completions Γ_p and so have to consider a number of cases.

(a) $p \in P, k(p) = 0$. In this case $\mathbf{u}$ is close to $\pi^l\mathbf{t}$, where $\mathbf{t}$ is the vector given by Lemma 9.3. Since π^l is a p-adic unit, there is thus an element σ of $O(\mathbf{Z}_p^4)$ [that is, an automorph of f over $\mathbf{Z}_p$] such that $\mathbf{u} = u\sigma\mathbf{t}$ for some $u \in U_p$. We take for Γ_p the 3-dimensional lattice $\Gamma_p = \sigma\Delta_p$. Then $\mathbf{u}, \Gamma_p$ have all the properties stated in Lemma 9.3 and its Corollary for $\mathbf{t}, \Delta_p$.

(b) $p \in P, k(p) \neq 0$. As in case (I) for $n \geqslant 5$, we pick any

$$\Gamma_p \subset p\mathbf{Z}_p^4 \cap V_p^{(2)} \tag{9.42}$$

with

$$\theta(\Gamma_p) \supset U_p^2. \tag{9.43}$$

Then, as before,

$$y\mathbf{u} + \mathbf{c} \notin p\mathbf{Z}_p^4 \qquad (y \in U_p, \mathbf{c} \in \Gamma_p). \tag{9.44}$$

(c) $p \notin P$, $p \mid g$. This case comprises only the two primes $p = \pi$ and $p = p(0)$.

As in the case (II) of $n \geqslant 5$, there are $\mathbf{r}_1, \mathbf{r}_2, \mathbf{r}_3, \mathbf{r}_4 \in \mathbf{Z}_p^4$ such that

$$f(\mathbf{r}_i, \mathbf{r}_j) = 0 \qquad (i \neq j) \tag{9.45}$$

$$f(\mathbf{r}_1) = -f(\mathbf{r}_2) = 1, \tag{9.46}$$

$$|f(\mathbf{r}_3)|_p = |f(\mathbf{r}_4)|_p = 1, \tag{9.47}$$

and

$$\mathbf{u} = \lambda\mathbf{r}_1 + \mu\mathbf{r}_2 \tag{9.48}$$

† The examples at the end of the chapter indicate how an appeal to Dirichlet's theorem can be avoided.

with

$$\lambda, \mu \in U_p, \qquad \lambda^2 - \mu^2 = g. \tag{9.49}$$

A basis for $\mathbf{Z}_p^4 \cap V_p^{(2)}$ is given by

$$\mathbf{w}, \mathbf{r}_3, \mathbf{r}_4, \tag{9.50}$$

where

$$\mathbf{w} = \mu\mathbf{r}_1 + \lambda\mathbf{r}_2, \tag{9.51}$$

so

$$f(\mathbf{w}) = -g. \tag{9.52}$$

Since

$$\lambda^2 - \mu^2 = g \equiv 0 \qquad (p) \tag{9.53}$$

we may suppose without loss of generality that

$$\lambda \equiv \mu \qquad (p). \tag{9.54}$$

Then a necessary and sufficient condition that

$$y_1\mathbf{u} + y_2\mathbf{w} + y_3\mathbf{r}_3 + y_4\mathbf{r}_4 \in p\mathbf{Z}_p^4 \qquad (y_i \in \mathbf{Z}_p) \tag{9.55}$$

is that

$$y_1 + y_2 \equiv y_3 \equiv y_4 \equiv 0 \qquad (p). \tag{9.56}$$

We now have to split cases.

(c_1) $p = \pi$. Since $d(f) \in U_\pi^2$, we may suppose further that

$$f(\mathbf{r}_3) = -f(\mathbf{r}_4) = 1. \tag{9.57}$$

We take for Γ_π the lattice with basis

$$\pi\mathbf{w}, \quad \mathbf{r}_3, \quad \mathbf{r}_4. \tag{9.58}$$

Then (9.44) holds for $p = \pi$, as follows at once from the equivalence of (9.55) and (9.56). We note also, by (9.57), that f represents primitively the whole of $\mathbf{Z}_\pi$ on Γ_π.

(c_2) $p = p(0)$. We put

$$\Gamma_p = \mathbf{Z}_p^4 \cap V_p^{(2)}. \tag{9.59}$$

We note that this does not necessarily imply (9.44) and we will have to take special precautions later. We note also that f represents primitively on Γ_p at least the whole of U_p and of $-gU_p^2$.

(d) $p \notin P, p \nmid g$. This case includes $p = q$. Again we put (9.59). Now there is a normal basis

$$\mathbf{u} = \mathbf{r}_1, \mathbf{r}_2, \mathbf{r}_3, \mathbf{r}_4$$

of $\mathbf{Z}_p^4$ such that

$$f(\mathbf{r}_j) \in U_p \qquad (1 \leqslant j \leqslant 4).$$

and $\mathbf{r}_2, \mathbf{r}_3, \mathbf{r}_4$ is a basis for Γ_p. Clearly (9.44) holds and f represents the whole of $\mathbf{Z}_p$ primitively on Γ_p.

This concludes the determination of the Γ_p and so of Γ. We note that

$$\theta(\Gamma_p) \supset U_p \tag{9.60}$$

for all p (in case (a), by Lemma 9.3). Hence the genus of Γ contains only one spinor genus. It follows, as for the $n \geqslant 5$ case, that there is an integer $m \geqslant 0$ such that if $c \in \mathbf{Z}$ is positive, divisible by q^{2m}, and representable on Γ_p for all p, then c is representable on Γ.

Now let $a \in \mathbf{Z}$ be given. We choose the $k(p)$ $(p \in P)$ appropriately as for $n \geqslant 5$. By the hypotheses of Lemma 9.2 (which is what we are engaged in proving) we have $q \nmid a$ and so may choose $k(q) = 1$ or 2 so that

$$a \in gU_q^2.$$

Let us denote by P^* the set of all primes which have been considered so far, that is

$$P^* = P \cup \{q, \pi, p(0)\}.$$

Then there is a $b \in \mathbf{Z}$ such that

$$b \in U_p \qquad (p \in P^*),$$
$$c \text{ (say)} = a - gb^2 \equiv 0 \qquad (q^{2m})$$

and c is representable by f on all Γ_p $(p \in P^*)$. The verification is left to the reader [the only case requiring thought is $p = p(0)$]. The representation may be chosen to be primitive, but we do not need this.

Suppose also that

$$c > 0.$$

Then by the definition of m there is a $\mathbf{c} \in \Gamma$ such that

$$c = f(\mathbf{c})$$

and so

$$a = f(b\mathbf{u} + \mathbf{c}).$$

We want to show that we can choose $\mathbf{c}$ in such a way that $b\mathbf{u} + \mathbf{c}$ is a primitive element of $\mathbf{Z}^4$. If not, there is a prime p such that $p \mid (b\mathbf{u} + \mathbf{c})$.

We discuss various cases.

$p \notin P^*$. This is included in case (d) above. By Theorem 8.3, Corollary, we may choose $\mathbf{c} \in \Gamma$ not to be of the form $p\mathbf{d}, \mathbf{d} \in \Gamma$ and then $p \nmid (b\mathbf{u} + \mathbf{c})$ by (9.44).

$p = p(0)$. This is case (c_2). In the notation of (9.55) we have

$$y_1 = b \in U_p$$

$+ y_3\mathbf{r}_3 + y_4\mathbf{r}_4 \qquad (y_2, y_3, y_4 \in \mathbf{Z}_p).$

$-\mathbf{c}$ if need be we can ensure that (9.56) does not hold; and then $p \nmid (b\mathbf{u} + \mathbf{c})$, as required.

$p \in P^*, p \neq p(0)$. In case (a) this follows from Lemma 9.3 and in (b), (c_1), (d) by (9.44) since in both cases $b \not\equiv 0\,(p)$.

As in the case $n \geqslant 5$, one then completes the proof of Lemma 9.2 by showing that a finite set of b will cover all values of a. We omit the details and remark only that this completes the proof of Theorem 1.6.

NOTES

Much of this Chapter generalizes to algebraic number fields, cf. O'Meara (1963).

Meyer (1891) was the first to prove results about the class-number of indefinite forms in $n \geqslant 3$ variables. Eisenstein (1851) had already noted from his tabulation of indefinite ternary forms that the class-number seemed always to be 1.

The notion of spinor genus was introduced by Eichler (1952a) but his concept is different from that which has subsequently established itself and which is used by O'Meara (1963). Watson (1960) introduced a different definition. Although it appears to have been common knowledge that the definitions of O'Meara and Watson are equivalent, Section 4 of this chapter appears to be the first proof of the equivalence in print. For another related concept see Jones (1977).

Section 2. Theorem 1.1 generalizes to Dedekind domains (e.g. the ring of integers of an algebraic number-field) if lattice is defined in sense (ii) of the Notes to Chapter 7.

Section 7. Kneser (1961) proved the more precise result that over any algebraic number field for $n \geqslant 4$ the weight of the representations of a given $a \neq 0$ by any two spinor genera in the same genus is the same. For indefinite ternaries he shows that either the weight of representations by all classes (=spinor genera) is the same or there are precisely two weights, each taken by half the classes: see also Jones and Watson (1955). There are also results for representation of forms by forms, see Weil (1962).

Section 9. The arguments of this section are taken largely from Kneser (19..
For a generalization to the representation of forms by forms and to algebraic number-fields, see Hsia, Kitaoka and Kneser (1978).

A form of Theorem 1.6 was proved analytically by Tartakovskiĭ (1927) after the diagonal case had been treated by Kloosterman (1926) and an alternative treatment was given by Ross and Pall (1946). Watson (1960a) considers the more general equation

$$\Sigma a_{ij}x_ix_j + \Sigma b_ix_i + c = 0$$

with a linear term and obtains estimates for the number of representations. Malyšev [see account in Malyšev (1962), Chapter 3] and Pommerenke (1959) obtain estimates analytically for the distribution of integral points **b** on the surface $f(\mathbf{x}) = b$, where f is a definite integral form in $n \geqslant 4$ variables and b is a large positive integer.

The situation with regard to representation by definite ternaries is much less clear-cut. Some definite ternaries fail to represent infinitely many integers which are permitted by congruence conditions [Jones and Pall (1939) cf. example 6]. On the other hand, Linnik devised a method using generalized quaternions and techniques from probability theory which gives results in the opposite direction. He gives conditions on the genus of a definite ternary form f which ensure that it represents all sufficiently large integers b which satisfy a further auxiliary condition

$$\left(\frac{b}{q}\right) = \text{prescribed}. \qquad (*)$$

Here q may be any prime from a certain infinite set depending on f. The condition (*) could be omitted if an appropriate generalization of the Riemann hypothesis is true. An extension of the method shows that the representations of b tend to equidistribution on the ellipsoid $f(\mathbf{x}) = b$ as $b \to \infty$ subject to (*). [There is a considerable volume of work by Linnik and Malyšev which is described in the monograph Malyšev (1962). Pall (1949) indicates a lacuna in the original paper Linnik (1940).] Martin Kneser notes that Linnik's conditions imply that his forms are all in genera containing only one spinor genus.

Linnik (1956) also applied his method to the indefinite form $x_1x_3 - x_2^2$. The result can be interpreted as a statement about the distribution of the sets of coefficients (a, b, c) of integral forms $ax^2 + 2bxy + cy^2$ with given determinant $d = ac - b^2$. For an unconditional result without a condition (*), see Patterson (1975) and for related results see Fricker (1971).

Watson (1976) has determined all definite ternaries which are in genera of more than one class but which represent all integers permitted by genus considerations. He has also [Watson (1954)] constructed definite ternaries with large sets of primitive exceptional values m [m is exceptional if it is permitted by genus considerations but not represented. It is primitive exceptional if it is not of the shape $m = m_0 m_1^2$ where m_0 is exceptional]. It is not apparently known whether a positive ternary can have infinitely many primitive exceptional values. For representation by ternaries, see also Peters (1978) and Lomadze (1978). For definite quaternaries representing all integers see Willerding (1948).

There are two monographs on representation by positive definite quadratic forms. Malyšev (1962) is largely an exposition of the work of himself and of Linnik which has been mentioned above. Kogan (1971) is primarily concerned with results about specific forms.

Liouville (1858–65) has given explicit formulae for the number of representations of integers by a substantial number of special definite forms as consequences of some rather elaborate general identities. The identities involve general functions of several variables and the formulae are obtained by specializing these, with, perhaps, subsequent manipulation. The proofs of the identities are combinatorial and are absolutely elementary, but the original motivation appears to have come from identities for modular forms [cf. Appendix B]. See also Dickson (1919), Vol **2**, Chapter 9, Bachmann (1910), Vol. **2**, pp. 365–433, Uspensky and Heaslet (1939), Chapter 13 and Kogan (1971).

EXAMPLES

1. Use the theory of spinor genera to show that the genus of $f = x^2 + 5y^2 - 7z^2$ contains only one proper equivalence class. Deduce necessary and sufficient conditions for an integer a to be primitively represented by f.

2. Show that the forms

$$x^2 - 3y^2 - 2yz - 23z^2$$
$$x^2 - 7y^2 - 6yz - 11z^2$$

are integrally equivalent. [cf. end of Section 1 of Chapter 9].

3. Show that

$$f = x^2 + 5y^2 - 25z^2$$

is in a genus of 1 class. Deduce that it is integrally equivalent to

$$g = -x^2 + 5y^2 + 25z^2.$$

Find an equivalence.

4. Let $p > 2$ be prime and let Λ_p be an n-dimensional $\mathbf{Z}_p$-lattice. Suppose that Λ_p has a normal basis $\mathbf{b}_j (1 \leqslant j \leqslant n)$ such that the $|\phi(\mathbf{b}_j)|_p$ are distinct. Show that $\theta(\Lambda_p)$ is generated as a group by $(\mathbf{Q}_p^*)^2$ and the $\phi(\mathbf{b}_1)\phi(\mathbf{b}_j)$ $(1 < j \leqslant n)$.

[*Hint*: Consider the proof of Lemma 3.3, Corollary 1, of Chapter 8].

5. Let $p > 2$ be prime. Show that

$$f = x^2 + py^2 - p^2z^2$$

is in a genus of 2 classes if $p \equiv 1 \pmod 8$, but otherwise in a genus of 1 class.

6. (i) Let

$$f(x, y, z) = x^2 + y^2 + 16z^2.$$

Show that the set P occurring in the definition (3.15) of the group G can be taken to consist of $p = 2$ alone and that G has order 2. Deduce that the genus of f consists of two spinor genera namely that of f and that of

$$g = (x - y)^2 + (x + y + z)^2 + 4z^2.$$

(ii) In the notation of Theorem 4.1 show that $g(w) \in G$ for the odd positive integer w is the identity of G precisely when $w \equiv 1 \pmod 4$.

(iii) Show that the two spinor genera of the genus of f have precisely one class.

[*Hint*: The heroic may use the methods of Chapter 9 to enumerate all the classes of definite integral forms of determinant 16. Alternatively one can use the technique of Theorem 8.4, Corollary 2, but using sequences of lattices Δ_j with $I(\Delta_j, \Delta_{j+1}) = 3$].

(iv) Show that every odd square m^2 is primitively representable by precisely one of f, g. Indeed it is representable by f or g according as $m \equiv 1$ or $m \equiv 3 \pmod 4$, where $m > 0$. [cf. Jones and Pall (1939), who use other techniques.]

7. (i) Show that

$$f = x^2 + xy + y^2 + 9z^2$$
$$g = x^2 + 3(y^2 + yz + z^2)$$

represents the two spinor genera of a genus and that each spinor genus consists of a single class.

(ii) Show that $4m^2$ for integer $m > 0, 2 \nmid m$ is representable primitively by f if $m \equiv 2 \pmod 3$ and by g if $m \equiv 1 \pmod 3$ but never by both f and g.

(iii) Prove the assertions of (ii) without using the theory of spinor genera.

[*Hint*: If $4m^2$ is representable by f, then

$$x^2 + xy + y^2 = (2m + 3z)(2m - 3z).$$

Any prime $p \neq 3$ which divides the left-hand side to an odd power must have $p \equiv 1 \pmod 3$. Compare Watson (1960), p. 115.]

8. Apply the methods of the two preceding examples to the squares which are represented primitively integrally by the two forms

$$x^2 - 2y^2 + 64z^2$$

$$(2x + z)^2 - 2y^2 + 16z^2.$$

[*Note.* Cf Chapter 9, example 23 and Siegel (1951), Section 1, para 7].

9. Show that the genus of

$$f(x, y, z, t) = x^2 + xy + 7y^2 + 3(z^2 + zt + t^2)$$

contains more than one spinor genus. Determine all the classes in the genus. [Watson (1960), p. 114.]

10. Prove the following statement whose proof was left to the reader at the beginning of Section 9:

Let f be an isotropic form in $n \geqslant 2$ variables taking values in $\mathbf{Z}_p$. Then there is an $M = M(f, p)$ with the following property. If a is represented by f over $\mathbf{Z}_p$ then there is some $m = m(a, p, f)$ with $0 \leqslant m \leqslant M$ such that $p^{-2m}a$ is primitively represented over $\mathbf{Z}_p$.

11. Let Λ be a lattice in the regular $\mathbf{Q}$-quadratic space V, ϕ of dimension 3. Suppose that ϕ is $\mathbf{Z}$-valued on Λ. Let q be a prime such that ϕ is isotropic over $\mathbf{Q}_q$.

Let P^* be a set of primes $p \neq 2$ such that

(i) ϕ is anisotropic over $\mathbf{Q}_p$ $(p \in P^*)$.

(ii) the determinant of the form induced by ϕ on Λ is divisible by p but not by p^2 $(p \in P^*)$.

Further, let $\mathbf{b} \in \Lambda$ be such that

$$|\phi(\mathbf{b})|_p = p^{-1} \qquad (p \in P^*),$$

and let $\eta_p = \pm 1$ be given arbitrarily for $p \in P^*$. Show that there is then an $\alpha \in O^+(V)$ such that

$$\alpha\mathbf{b} \equiv \eta_p \mathbf{b} \pmod{p\Lambda} \qquad (\text{all } p \in P^*)$$

and

$$\alpha \in O^+(\Lambda_p) \qquad (\text{all } p \neq q).$$

[*Hint.* This is just the analogue, in the sense of Section 8, of example 3 of Chapter 10.]

12. Show that the proof of Theorem 1.5 in Section 9 can be freed of the

reference to Dirichlet's theorem about the existence of primes in arithmetic progressions.

[*Hint*: Dirichlet's theorem is required to show the existence of a g_1 satisfying (9.39) and (9.40). We can in any case find an integral $g_1 > 0$ which satisfies (9.39) and which is not divisible by the square of a $p \notin P$. Let g be given by (9.41). If $p \notin P$, $p|g$ and $V_p^{(2)}$ is isotropic we argue as in the given proof. For the set P_A of $p \notin P, p|g$ for which $V_p^{(2)}$ is anisotropic, we may by the preceding example find an $\alpha \in O^+(V^{(2)})$ such that

$$\alpha \in O^+(\Lambda_p^{(2)}) \qquad (\text{all } p \neq q)$$

and such that

$$\alpha\mathbf{c} \equiv \eta_p \mathbf{c} \pmod{p\Lambda_p^{(2)}} \qquad (\text{all } p \in P_A),$$

where the $\eta_p = \pm 1$ are arbitrary. We may now choose the η_p so that $b\mathbf{u} + \alpha\mathbf{c}$ is not divisible by p, precisely as in the treatment of $p(0)$ (penultimate paragraph of Section 9)].

13. In Lemma 9.3 if $p \neq 2$ show that $\mathbf{t}$ can always be chosen to be a primitive element of $\mathbf{Z}_p^4$.

[*Hint*. The form f is $\mathbf{Z}_p$-equivalent to a diagonal form $\Sigma a_j x_j^2$.]

CHAPTER 12

The Reduction of Positive Definite Quadratic Forms

1. INTRODUCTION

The theory of reduction of definite quadratic forms is a purely real-valued theory so we consider forms

$$f(\mathbf{x}) = \sum_{i,j} f_{ij} x_i x_j \qquad (f_{ij} \in \mathbf{R}) \tag{1.1}$$

which are (strictly) positive definite. The object is to select from the infinitely many forms which are integrally equivalent to f one which is characterized in some intrinsic way. The process which turns out to be most suitable was defined by Minkowski.

Definition. The strictly positive definite quadratic form (1.1) is said to be *reduced* (in the sense of Minkowski) if for each j

$$f(\mathbf{e}_j^*) \geqslant f(\mathbf{e}_j) \tag{1.2}$$

where $\mathbf{e}_j = (0, \ldots, 0, 1, 0, \ldots, 0)$ and where $\mathbf{e}_j^*$ runs through all the integral vectors which with $\mathbf{e}_1, \ldots, \mathbf{e}_{j-1}$ can be extended to give a basis $\mathbf{e}_1, \ldots, \mathbf{e}_{j-1}$, $\mathbf{e}_j^*, \ldots$ of the lattice of integral vectors. (For $j = 1$ this means only that $\mathbf{e}_1^*$ runs through all the primitive vectors.)

Alternatively, condition (1.2) can be written

$$f(0, \ldots, 0, 1, 0, \ldots, 0) \leqslant f(b_1, \ldots, b_n) \tag{1.3}$$

where $b_1, \ldots, b_n$ run through all sets of integers such that

$$\text{g.c.d.}(b_j, b_{j+1}, \ldots, b_n) = 1. \tag{1.4}$$

As we use them continually, we single out two special cases of (1.4).

LEMMA 1.1. *If f is reduced, then*

$$0 < f_{11} \leqslant f_{22} \leqslant \ldots \leqslant f_{nn} \tag{1.5}$$

and

$$|2f_{ij}| \leqslant f_{ii} \qquad (1 \leqslant i < j \leqslant n). \tag{1.6}$$

Proof. For if $i < j$ we can take $\mathbf{e}_i^* = \mathbf{e}_j$ and $\mathbf{e}_j^* = \mathbf{e}_j \pm \mathbf{e}_i$.

We have at once

THEOREM 1.1. *Every positive strictly definite form is equivalent to at least one and at most finitely many reduced forms.*

Proof. We first note that for any $M > 0$ the set of $\mathbf{x} \in \mathbf{R}^n$ with

$$f(\mathbf{x}) \leqslant M$$

is bounded, and so in particular there are only finitely many integral vectors $\mathbf{m}$ such that

$$f(\mathbf{m}) \leqslant M.$$

It is now obvious that we can choose inductively a basis $\mathbf{b}_1, \ldots, \mathbf{b}_n$ of $\mathbf{Z}^n$ such that

$$f(\mathbf{b}_j) = \inf_{\mathbf{b}_j^*} f(\mathbf{b}_j^*), \tag{1.7}$$

where the infimum is over all vectors $\mathbf{b}_j^*$ which are such that $\mathbf{b}_1, \ldots, \mathbf{b}_{j-1}, \mathbf{b}_j^*$ can be extended to a basis of $\mathbf{Z}^n$. Then clearly the form

$$g(y_1, \ldots, y_n) = f(y_1\mathbf{b}_1 + \ldots + y_n\mathbf{b}_n)$$

is reduced and equivalent to f. In general the vectors $\mathbf{b}_1, \ldots, \mathbf{b}_n$ will be uniquely determined by f: but in particular cases there will be several vectors $\mathbf{b}_j$ which satisfy the condition (1.7).

The key facts about reduction are given by Theorem 1.1 above and the two following theorems which will be proved in subsequent sections of this chapter.

THEOREM 1.2. *Suppose that the forms*

$$f(\mathbf{x}), \qquad f(\mathbf{T}\mathbf{x})$$

are both reduced, where

$$\mathbf{T} = (t_{ij})$$

is an integral unimodular matrix. Then

$$|t_{ij}| \leqslant C_1$$

where C_1 is a constant depending only on the number n of variables.

THEOREM 1.3. *There is a* finite *subset $\mathscr{C}$ of the conditions* (1.3) *with the following property:*

Let $f(x_1, \ldots, x_n)$ *be a (not necessarily definite) form which satisfies the conditions* $\mathscr{C}$ *together with*

$$f(1, 0, \ldots, 0) > 0.$$

Then f *is strictly positive definite and reduced.*

Theorem 1.3, which will be proved in Section 5, has a geometrical description. We can represent a form $f(\mathbf{x}) = \Sigma f_{ij} x_i x_j$ in n variables by the point with coordinates f_{ij} $(i \leqslant j)$ in $\mathbf{R}^N$, where $N = n(n + 1)/2$. Theorem 1.3 then says that the set of reduced positive definite forms is represented in $\mathbf{R}^N$ by a cone $\mathscr{R}$ bounded by a finite number of hyperplanes. The group $SL_n^{\pm}$ of integral $n \times n$ matrices $\mathbf{T}$ with $\det \mathbf{T} = \pm 1$ acts on $\mathbf{R}^N$ if we define $\mathbf{T}f$ to be the form $f(\mathbf{Tx})$. Since every positive definite form is equivalent to a reduced form, the union $\bigcup_{\mathbf{T}} \mathbf{T}\mathscr{R}$ $(\mathbf{T} \in SL_n^{\pm})$ represents precisely the set of all positive definite forms. We shall study this geometrical configuration in more detail in Sections 5–7.

The shape of $\mathscr{R}$ is known explicitly only for $n \leqslant 7$. For $n = 5, 6$ it was stated without proof by Minkowski [(1886), (1887) reproduced in (1911), Band I, pp. 154, 218] and proofs have been given recently by Ryškov (1971) and Tammela (1973). The case $n = 7$ is in Tammela (1977). For $n \leqslant 4$, Minkowski's proof is so simple that we give it here.

LEMMA 1.2. *Let* $n \leqslant 4$. *A necessary and sufficient condition that* $f(\mathbf{x})$ *be a reduced positive definite form is that*

(i) $0 < f_{11} \leqslant f_{22} \leqslant \ldots \leqslant f_{nn}$

(ii) $f(\mathbf{s}) \geqslant f_{JJ}$ *for* $1 \leqslant J \leqslant n$ *and for all* $\mathbf{s}$ *with*

$$s_j = 0 \quad \text{or } \pm 1 \qquad (j < J)$$
$$s_J = 1$$
$$s_j = 0 \qquad (j > J).$$

Note 1. For $n = 2$ the conditions (i) and (ii) are just the familiar

$$0 < f_{11} \leqslant f_{22}, \qquad 2|f_{12}| \leqslant f_{11}.$$

Note 2. The set of $\mathbf{s}$ in (ii) cannot be replaced by a smaller set. This can be seen by considering forms in the neighbourhood of

$$f_n(\mathbf{x}) = x_1^2 + \ldots + x_n^2 + (x_1 + \ldots + x_n)^2 \qquad (2 \leqslant n \leqslant 4)$$

and

$$g_4(\mathbf{x}) = 2x_1^2 + 2x_2^2 + 2x_3^2 + 2x_4^2 + 2x_4(x_1 + x_2 + x_3)$$

and the forms obtainable from them by substitutions $x_j \to \pm x_j$ and also permuting the variables in g_4.

Proof. The conditions (i) and (ii) are clearly necessary. To prove them sufficient it will clearly be enough to show that

$$f(\mathbf{a}) \geqslant f_{JJ}$$

for any integral vector **a** with

$$a_J \neq 0, \qquad a_j = 0 \qquad (j > J).$$

If a form f satisfies (i) and (ii) then so does the form in $n - m$ variables obtained from it by equating m of the variables to 0. Hence we may suppose without loss of generality that

$$a_j \neq 0 \qquad (1 \leqslant j \leqslant n)$$

and have to prove that

$$f(\mathbf{a}) \geqslant f_{nn}. \tag{1.7 bis}$$

After a substitution $x_j \to \pm x_j$ we may suppose that

$$a_j > 0 \qquad (1 \leqslant j \leqslant n). \tag{1.7 ter}$$

Put $A = \max a_j$. When $A = 1$ the required inequality is just one of the given inequalities (ii). We may therefore suppose that $A > 1$ and will use induction on $\sum_j a_j$ and n. Let $B = \min a_j > 0$. If $a_n = B$ put $k = n$, otherwise let k be any index for which $a_k = B$. Put

$$b_j = a_j - B \qquad (j \neq k)$$
$$b_k = a_k = B,$$

so that

$$0 \leqslant b_j \leqslant a_j, \qquad b_n \neq 0.$$

A simple calculation shows that

$$f(\mathbf{a}) - f(\mathbf{b}) = B^2\{f(1, \ldots, 1) - f_{nn}\} + 2 \sum_{i \neq k} B\, b_i \sum_{j \neq k} f_{ij}.$$

If we can show that $f(\mathbf{a}) \geqslant f(\mathbf{b})$ we are done, since $f(\mathbf{b}) \geqslant f_{nn}$ by the induction hypothesis. But

$$f(1, \ldots, 1) - f_{nn} \geqslant 0$$

by (ii) and

$$\sum_{j \neq k} f_{ij} = (2 - \tfrac{1}{2}n) f_{ii} + \tfrac{1}{2} \sum_{\substack{j \neq k \\ j \neq i}} (f_{ii} + 2f_{ij})$$
$$\geqslant 0$$

by (1.6), which is clearly a consequence of (i) and (ii). This concludes the proof.

Siegel has shown that Theorem 1.2 is a special case of a property of what are now called Siegel domains which are useful not only for definite forms but also in the study of indefinite forms and in the general theory of linear algebraic groups.

Definition. Let $\delta > 0, \eta > 0$. The *Siegel domain* $\mathscr{S}_n(\delta, \eta) = \mathscr{S}(\delta, \eta)$ is the set of quadratic forms

$$f(x_1, \ldots, x_n) = h_1(x_1 + c_{12}x_2 + \ldots + c_{1n}x_n)^2$$
$$+ h_1(x_2\, c_{23}x_3 + \ldots + c_{2n}x_n)^2 + \ldots + h_n x_n^2$$

with

$$0 < h_j \leqslant \delta h_{j+1} \qquad (1 \leqslant j < n)$$

and

$$|c_{ij}| \leqslant \eta \qquad (1 \leqslant i < j \leqslant n).$$

In Sections 3 and 4 we shall prove

LEMMA 1.3. *There exist δ_n, η_n depending only on n such that every reduced form is in $\mathscr{S}_n(\delta_n, \eta_n)$.*

THEOREM 1.4. *Let $f(\mathbf{x})$ and $f(\mathbf{Tx})$ be in the same Siegel domain $\mathscr{S}_n(\delta, \eta)$ where $\mathbf{T}$ is an integral unimodular matrix. Then there is a $C = C(n, \delta, \eta)$ such that the elements t_{ij} of $\mathbf{T}$ satisfy*

$$|t_{ij}| \leqslant C(n, \delta, \eta).$$

Clearly Theorem 1.2 follows from Lemma 1.3 and Theorem 1.4.

Minkowski's definition of reduction is not the original one, which was introduced by Hermite. Hermite† defined a reduced strictly definite form f recursively as follows:

(i) $f(1, 0, \ldots, 0) = \inf_{\mathbf{m} \text{ integral}, \neq \mathbf{0}} f(\mathbf{m})$

(ii) $|2f_{1j}| \leqslant f_{11}$ $(j > 1)$ and $g(x_2, \ldots, x_n)$ is reduced as a function of $(n-1)$ variables, where

$$f(x_1, \ldots, x_n) = f_{11}\left(x_1 + \frac{f_{12}}{f_{11}}x_2 + \ldots + \frac{f_{1n}}{f_{11}}x_n\right)^2 + g(x_2, \ldots, x_n). \tag{1.8}$$

It is easy to see that a Hermite-reduced form is in $\mathscr{S}(\frac{4}{3}, \frac{1}{2})$. Hence the analogue of Theorem 1.1 holds for Hermite-reduction. On the other hand there is no simple characterization of Hermite-reduction corresponding to Theorem 1.3 above in terms of the coefficients f_{ij}.

Before embarking on the main discussion we recall some simple facts which will be useful.

† There are two other definitions of reduction which go by the name of Hermite. One, also for definite forms, is mentioned in the Notes to this Chapter. The other is for indefinite forms and will be a major topic in Chapter 13.

A positive definite form $f(\mathbf{x})$ in n variables is equivalent over the reals $\mathbf{R}$ to a sum of squares $g(\mathbf{y}) = \sum_{i=1}^{m} y_i^2$ for $m \leqslant n$. If $m = n$ then f is strictly definite but if $m < n$ then f is only semi-definite. It follows that a positive definite f is semi-definite if and only if it is singular.

For any two vectors $\mathbf{x}_1, \mathbf{x}_2$ we have

$$\sqrt{f(\mathbf{x}_1 + \mathbf{x}_2)} \leqslant \sqrt{f(\mathbf{x}_1)} + \sqrt{f(\mathbf{x}_2)};$$

since the corresponding inequality for $g(\mathbf{y})$ is just the triangle inequality for distances in m-dimensional space. It follows that

$$\sqrt{f\left(\sum_j \lambda_j \mathbf{x}_j\right)} \leqslant \sum_j |\lambda_j| \sqrt{f(\mathbf{x}_j)} \tag{1.9}$$

for any real vectors $\mathbf{x}_j$ and real numbers λ_j.

Finally, for any M the set of $\mathbf{x} \in \mathbf{R}^n$ for which $f(\mathbf{x}) \leqslant M$ is symmetric and convex: it is bounded if and only if f is strictly definite. For the corresponding statements are clearly true in terms of $g(\mathbf{y})$.

2. SUCCESSIVE MINIMA

We denote by

$$f(\mathbf{x}) = \sum_{i,j=1}^{n} f_{ij} x_i x_j$$

a strictly definite quadratic form with determinant

$$D = d(f) = \det(f_{ij}) > 0.$$

THEOREM 2.1. *There is a constant γ_n such that*

$$M_1 \text{ (say)} = \inf_{\mathbf{m} \neq \mathbf{0} \text{ integral}} f(\mathbf{m}) \leqslant \gamma_n D^{1/n}.$$

This is a special case of Lemma 3.2 of Chapter 9, but we give two self-contained proofs.

(i) (Hermite). Suppose that f is Hermite-reduced in the sense discussed at the end of the last paragraph, so that

$$f_{11} = M_1 \tag{2.1}$$

and

$$\det g = D/f_{11}.$$

Suppose that γ_{n-1} has already been found. Then there exist integers $m_2, \ldots, m_n$ not all 0 such that

$$g(m_2, \ldots, m_n) \leqslant \gamma_{n-1}(D/f_{11})^{1/(n-1)}.$$

Now pick the integer m_1, as is clearly possible, so that

$$\left| m_1 + \frac{f_{12}}{f_{11}} m_2 + \dots + \frac{f_{1n}}{f_{11}} m_n \right| \leqslant \tfrac{1}{2}.$$

Then

$$f(m_1, \dots, m_n) \leqslant f_{11}/4 + g(m_2, \dots, m_n)$$
$$\leqslant f_{11}/4 + \gamma_{n-1}(D/f_{11})^{1/n-1}.$$

By (2.1) we thus have

$$f_{11} \leqslant \tfrac{1}{4} f_{11} + \gamma_{n-1}(D/f_{11})^{1/n-1}:$$

that is

$$f_{11}^n \leqslant (\tfrac{4}{3})^{n-1} \gamma_{n-1}^{n-1} D.$$

Since we can clearly take $\gamma_1 = 1$ we deduce by induction that we can take

$$\gamma_n = (\tfrac{4}{3})^{(n-1)/2}.$$

(ii) (Minkowski). For any M the set of $\mathbf{x} \in \mathbf{R}^n$ satisfying

$$f(\mathbf{x}) \leqslant M$$

is convex and symmetric (being affinely equivalent to a sphere) and has volume

$$K_n M^{n/2} D^{-\frac{1}{2}} \tag{2.2}$$

where K_n is the volume of the unit n-dimensional sphere. By Minkowski's convex body theorem† this sphere contains an integral point other than the origin provided that (2.2) is $\geqslant 2^n$. This shows that we can take

$$\gamma_n = 4K_n^{-2/n}.$$

In fact the best possible values of γ_n are known for $n \leqslant 8$ (see the Appendix of Cassels (1959) for values and further references). We content ourselves with showing that the best value for $n = 2$ is $2/\sqrt{3}$. We may suppose that f is (Minkowski) reduced, so

$$2|f_{12}| \leqslant f_{11} \leqslant f_{22}.$$

Then

$$D = f_{11} f_{22} - f_{12}^2$$
$$\geqslant f_{11}^2 - (f_{11}/2)^2$$
$$= (\tfrac{3}{4}) f_{11}^2,$$

that is

$$f_{11} \leqslant (2/\sqrt{3}) D^{\frac{1}{2}}.$$

The sign of equality is required for $x_1^2 + x_1 x_2 + x_2^2$.

† See *Note* after proof of Theorem 2.4 of Chapter 5 (p. 71).

We now introduce the *successive minima* of a strictly definite form:

Definition. We define the jth minimum M_j of f to be the positive number such that

(i) the set of integral $\mathbf{m}$ with $f(\mathbf{m}) \leqslant M_j$ spans a subspace of dimension $\geqslant j$;

(ii) the set of integral $\mathbf{m}$ with $f(\mathbf{m}) < M_j$ spans a subspace of dimension $<j$.

It is clear that $M_1, \ldots, M_n$ are well defined and satisfy

$$M_1 \leqslant M_2 \leqslant \ldots \leqslant M_n \tag{2.3}$$

and that there is a set of linearly independent integral vectors $\mathbf{m}_j$ with

$$f(\mathbf{m}_j) = M_j. \tag{2.4}$$

The M_j are uniquely determined by the form f but in general there may be several sets of independent $\mathbf{m}_j$ satisfying (2.4). For any set of $\mathbf{m}_j$ we have however the obvious

LEMMA 2.1. *Let* $\mathbf{c}$ *be integral and suppose that* $f(\mathbf{c}) < M_j$. *Then* $\mathbf{c}$ *is linearly dependent on* $\mathbf{m}_1, \ldots, \mathbf{m}_{j-1}$.

The following estimate, due to Minkowski, is the key to much that follows.

THEOREM 2.2

$$M_1 \ldots M_n \leqslant \gamma_n^n D \tag{2.5}$$

where γ_n *is any number satisfying the conditions of Theorem* 2.1.

Proof. We introduce new variables $\mathbf{y} = (y_1, \ldots, y_n)$ by

$$\mathbf{x} = y_1\mathbf{m}_1 + \ldots + y_n\mathbf{m}_n; \tag{2.6}$$

so

$$\mathbf{y} = \mathbf{T}\mathbf{x} \tag{2.7}$$

for some linear transformation $\mathbf{T}$. On successively completing the square with respect to the variables $\mathbf{y}$ we have

$$f(\mathbf{x}) = L_1^2 + L_2^2 + \ldots + L_n^2 \tag{2.8}$$

where

$$L_j = L_j(\mathbf{y}) = L_j(y_j, \ldots, y_n) \tag{2.9}$$

is a real linear form in $y_j, \ldots, y_n$. Now define a new quadratic form $h(\mathbf{x})$ by

$$h(\mathbf{x}) = L_1^2/M_1 + \ldots + L_n^2/M_n, \tag{2.10}$$

so the determinant $d(h)$ of h is

$$d(h) = D/M_1 \ldots M_n. \tag{2.11}$$

We shall show that

$$h(\mathbf{b}) \geqslant 1 \tag{2.12}$$

for all $\mathbf{b} \in \mathbf{Z}^n$ other than $\mathbf{b} = \mathbf{0}$. For let $\mathbf{Tb} = (t_1, \ldots, t_n)$ and let J be the largest index j such that $t_J \neq 0$. Then $\mathbf{b}$ is not linearly dependent on $\mathbf{m}_1, \ldots, \mathbf{m}_{J-1}$ and so

$$f(\mathbf{b}) \geqslant M_J \tag{2.13}$$

by Lemma 2.1. On the other hand, (2.9) implies that

$$L_j(\mathbf{t}) = 0, \qquad (j > J) \tag{2.14}$$

and so

$$\begin{aligned} h(\mathbf{b}) &= \sum_{j \leqslant J} L_j^2(\mathbf{t})/M_j \\ &\geqslant M_J^{-1} \sum_{j \leqslant J} L_j^2(\mathbf{t}) \\ &= M_J^{-1} f(\mathbf{b}) \\ &\geqslant 1 \end{aligned} \tag{2.15}$$

by (2.4) and (2.13). Hence by Theorem 2.1 applied to the form h we have

$$1 \leqslant \gamma_n (d(h))^{1/n}. \tag{2.16}$$

The theorem now follows at once from (2.11) and (2.16).

3. REDUCED FORMS AND SIEGEL DOMAINS

Our main objective is

THEOREM 3.1. *Let*

$$\begin{aligned} f(\mathbf{x}) &= \Sigma f_{ij} x_i x_j \\ &= h_1(x_1 + c_{12}x_2 + \ldots + c_{1n}x_n)^2 \\ &\quad + h_2(x_2 + c_{23}x_3 + \ldots + c_{2n}x_n)^2 \\ &\quad + \ldots + h_n x_n^2 \end{aligned} \tag{3.1}$$

be Minkowski-reduced with successive minima $M_1, \ldots, M_n$. Then for each $j (1 \leqslant j \leqslant n)$ the ratio of any two of h_j, f_{jj}, M_j is bounded by a constant depending only on n. More precisely, there are constants $C_4(j)$ and $C_5(n), C_6(n)$ depending only on j, n respectively such that

(i) $h_j \leqslant f_{jj}$,
(ii) $f_{jj} \leqslant C_4(j)\, M_j$,
(iii) $M_j \leqslant C_5(n)\, h_j$,
(iv) $f_{jj} \leqslant C_6(n)\, h_j$,
(v) $M_j \leqslant f_{jj}$.

Proofs. (i) We have

$$\begin{aligned} f_{jj} &= h_j + \sum_{i<j} h_i c_{ij}^2 \\ &\geqslant h_j. \end{aligned} \tag{3.2}$$

(ii) Let $\mathbf{m}_j$, as before, be a set of linearly independent integral vectors with $f(\mathbf{m}_j) = M_j$. For fixed j at least one of $\mathbf{m}_1, \ldots, \mathbf{m}_j$ must be linearly independent of $\mathbf{e}_1, \ldots, \mathbf{e}_{j-1}$ (where $\mathbf{e}_i$ is the ith unit vector). Suppose that it is $\mathbf{m}_k$. Then there is a vector $\mathbf{e}_j^*$ which together with $\mathbf{e}_1, \ldots, \mathbf{e}_{j-1}$ can be extended to a basis and such that

$$\mathbf{m}_k = t_1\mathbf{e}_1 + \ldots + t_{j-1}\mathbf{e}_{j-1} + s\mathbf{e}_j^*,$$

where $s > 0$ and $t_1, \ldots, t_{j-1}$ are integers.

On replacing $\mathbf{e}_j^*$ by $\mathbf{e}_j^* + r_1\mathbf{e}_1 + \ldots + r_{j-1}\mathbf{e}_{j-1}$ with suitable integral $r_1, \ldots, r_{j-1}$ we may suppose without loss of generality that

$$|t_i| \leqslant s/2 \qquad (1 \leqslant i < j).$$

Now

$$\mathbf{e}_j^* = (1/s)\,\mathbf{m}_k - (t_1/s)\,\mathbf{e}_1 - \ldots - (t_{j-1}/s)\,\mathbf{e}_{j-1}$$

and so by (1.9) [the convexity of $\sqrt{f(\mathbf{x})}$] we have

$$\begin{aligned} (f(\mathbf{e}_j^*))^{\frac{1}{2}} &\leqslant (1/s)(f(\mathbf{m}_k))^{\frac{1}{2}} + |t_1/s|(f(\mathbf{e}_1))^{\frac{1}{2}} \\ &\quad + \ldots + |t_{j-1}/s|\,(f(\mathbf{e}_{j-1}))^{\frac{1}{2}} \\ &\leqslant (f(\mathbf{m}_k))^{\frac{1}{2}} + \tfrac{1}{2}\sum_{i<j}(f(\mathbf{e}_i))^{\frac{1}{2}}. \end{aligned} \tag{3.3}$$

By induction we may suppose that (ii) holds for $i < j$, and so

$$\begin{aligned} f(\mathbf{e}_i) &= f_{ii} \\ &\leqslant C_4(i)\,M_i \\ &\leqslant C_4(i)\,M_j. \end{aligned}$$

Since $k \leqslant j$ we have

$$\begin{aligned} f(\mathbf{m}_k) &= M_k \\ &\leqslant M_j. \end{aligned}$$

Hence (3.3) implies that

$$f(\mathbf{e}_j^*) \leqslant C_4(j)\,M_j$$

where

$$(C_4(j))^{\frac{1}{2}} = 1 + \tfrac{1}{2}\sum_{i<j}(C_4(i))^{\frac{1}{2}}.$$

But $\mathbf{e}_1, \ldots, \mathbf{e}_{j-1}, \mathbf{e}_j^*$ can be extended to a basis and so, by the definition of reduction,

$$f_{jj} \leqslant f(\mathbf{e}_j^*).$$

Hence we have proved (ii) by induction.

Note 1. We have estimated $C_4(j)$ crudely. For a discussion of better estimates see van der Waerden (1956) Section 6.

Note 2. For a related result due to Mahler see Cassels (1959) Lemma 8 on p. 135.

(iii) On comparing determinants in (3.1) we have

$$h_1 \dots h_n = d(f) = D.$$

By Theorem 5 we have

$$M_1 \dots M_n \leqslant \gamma_n^n D.$$

Hence

$$\begin{aligned}\gamma_n^n &\geqslant (M_1/h_1)\dots(M_n/h_n)\\ &\geqslant (M_j/h_j)\prod_{i \neq j} C_4(i)\end{aligned}$$

by (ii), which we have just proved. Hence

$$M_j/h_j \leqslant C_5(n),$$

for some $C_5(n)$ depending only on n, as was required.

Finally, (iv) follows immediately from (ii) and (iii); and (v) is trivial.

Proof of Lemma 1.3. This asserts that a reduced form is in the Siegel domain $\mathscr{S}_n(\delta_n, \eta_n)$ for some δ_n, η_n depending only on n.

By the definition of reduction and the previous theorem we have

$$\begin{aligned}h_j &\leqslant f_{jj}\\ &\leqslant f_{j+1,j+1}\\ &\leqslant C_6(n)\, h_{j+1}.\end{aligned} \tag{3.4}$$

We may thus take $\delta_n = C_6(n)$.

Further, for $i < j$ we have

$$f_{ij} = h_i c_{ij} + \sum_{k<j} h_k c_{ki} c_{kj}$$

and so

$$|c_{ij}| \leqslant \frac{|f_{ij}|}{h_i} + \sum_{k<j} \frac{h_k}{h_i} |c_{ki} c_{kj}|. \tag{3.5}$$

Then by the condition of reduction

$$\begin{aligned}|f_{ij}| &\leqslant \tfrac{1}{2} f_{ii}\\ &\leqslant \tfrac{1}{2} C_6(n)\, h_i\end{aligned}$$

and by (3.4) we have

$$\frac{h_k}{h_i} \leqslant (C_6(n))^{i-k} \qquad (k < i).$$

Writing temporarily $H_i = \max_{j>i} |c_{ij}|$ it thus follows from (3.5) that

$$H_i \leqslant \tfrac{1}{2} C_6 + \sum_{k<j} C_6^{i-k} H_k^2. \tag{3.6}$$

But since f is reduced we have

$$|c_{1j}| = \left|\frac{f_{1j}}{f_{11}}\right| \leqslant \tfrac{1}{2};$$

that is

$$H_1 \leqslant \tfrac{1}{2}.$$

It then follows from (3.6) by induction that $H_2, H_3, \dots, H_n$ are all bounded by a constant η_n (say) which depends only on n. Then f is in $\mathscr{S}_n(\delta_n, \eta_n)$, as required.

4. SIEGEL DOMAINS

In this section

$$\begin{aligned} f(\mathbf{x}) &= \sum_{i,j=1}^{n} f_{ij} x_i x_j \\ &= \sum_{i=1}^{n} h_i (x_i + c_{i,i+1} x_{i+1} + \dots + c_{in} x_n)^2 \end{aligned} \tag{4.1}$$

is a positive definite form in some Siegel domain $\mathscr{S}_n(\delta, \eta)$ with $\delta > 1$ but is not necessarily reduced. We denote by M_j the successive minima of f and by

$$\mathbf{m}_j = (m_{j1}, \dots, m_{jn}) \qquad (1 \leqslant j \leqslant n) \tag{4.2}$$

any set of linearly independent integral vectors for which

$$f(\mathbf{m}_j) = M_j \qquad (1 \leqslant j \leqslant n). \tag{4.3}$$

We first prove the analogue of Theorem 3.1 for Siegel domains.

LEMMA 4.1. *Suppose that the form* (4.1) *lies in* $\mathscr{S}_n(\delta, \eta)$. *Then for each* $j\,(1 \leqslant j \leqslant n)$ *the ratio of any two of* h_j, f_{jj}, M_j *is bounded by a constant depending only on* n, δ, η.

Proof.

$$\begin{aligned} h_j \leqslant f_{jj} &= h_j + \sum_{i<j} h_i c_{ij}^2 \\ &\leqslant h_j + \sum_{i<j} h_i \eta^2 \\ &\leqslant h_j + h_j \Sigma \delta^{j-i} \eta^2 \\ &\leqslant C_{10} h_j, \end{aligned}$$

where

$$C_{10} = C_{10}(n, \delta, \eta) = 1 + \eta^2 \sum_{1 \leq j < n} \delta^j.$$

The vectors $\mathbf{e}_1, \ldots, \mathbf{e}_j$ are linearly independent, and so

$$\begin{aligned} M_J &\leq \max_{1 \leq i \leq J} f(\mathbf{e}_i) \\ &\leq C_{10} \max_{1 \leq i \leq J} h_i \\ &\leq C_{11} h_J, \end{aligned}$$

where

$$C_{11} = C_{11}(n, \delta, \eta) = \delta^{n-1} C_{10}.$$

Finally, let $\mathbf{a} = (a_1, \ldots, a_n)$ be an integral vector which is not linearly dependent on $\mathbf{e}_1, \ldots, \mathbf{e}_{J-1}$. Then there is a $k \geq j$ such that $a_k \neq 0$, $a_i = 0$ $(i > k)$. It follows immediately from (4.1) that $f(\mathbf{a}) \geq h_k$. Hence

$$\begin{aligned} M_j &\geq \min(h_J, h_{J+1}, \ldots, h_n) \\ &\geq C_{12} h_J, \end{aligned}$$

where

$$C_{12} = \delta^{1-n}.$$

This completes the proof.

Lemma 4.2. *For any fixed J with the notation* (4.1), (4.2) *write*

$$\mu_i = m_{Ji} + c_{i,i+1} m_{J,i+1} + \ldots + c_{i,n} m_{Jn} \quad (1 \leq i \leq n); \qquad (4.3\ bis)$$

so that

$$\begin{aligned} M_J &= f(\mathbf{m}_J) \\ &= \Sigma h_i \mu_i^2. \end{aligned} \qquad (4.4)$$

Then

$$|\mu_i| \leq C_{13} = C_{13}(n, \delta, \eta) \qquad (1 \leq i \leq n).$$

Proof. We fix i and distinguish two cases dealing first with the case when i is comparatively small.

(i) Suppose, first, that

$$f(\mathbf{e}_u) < M_J \qquad (1 \leq u \leq i). \qquad (4.5)$$

In particular, $i < J$. Put

$$\mathbf{m}_J^* = \mathbf{m}_J + t_1 \mathbf{e}_1 + \ldots + t_i \mathbf{e}_i$$

where $t_1, \ldots, t_i$ are integers. We claim that

$$f(\mathbf{m}_J^*) \geq M_J = f(\mathbf{m}_J). \qquad (4.6)$$

For otherwise $\mathbf{m}_J$ would be a linear combination of the vectors $\mathbf{m}_J^*, \mathbf{e}_1, \ldots, \mathbf{e}_i$ all lying in $f(\mathbf{x}) < M_J$, contrary to the definition of the successive minima.

We pick the integers $t_i, t_{i-1}, t_{i-2}, \ldots, t_1$ in order, as we clearly may, so that

$$|\mu_u^*| \leqslant \tfrac{1}{2} \qquad (1 \leqslant u \leqslant i)$$

where $\mu_1^*, \ldots, \mu_n^*$ are the analogues of (4.3 *bis*) for $\mathbf{m}_J^*$. Clearly

$$\mu_u^* = \mu_u \qquad (u > i).$$

Hence (4.6) implies that

$$\begin{aligned}\sum_{u\leqslant i} h_u\mu_u^2 &\leqslant \sum_{u\leqslant i} h_u\mu_u^{*2}\\ &\leqslant \tfrac{1}{4}\sum_{u\leqslant i} h_u\\ &\leqslant \tfrac{1}{4}(\delta^{i-1} + \ldots + 1)\,h_i.\end{aligned}$$

In particular,

$$\begin{aligned}\mu_i^2 &\leqslant \tfrac{1}{4}(\delta^{i-1} + \ldots + 1)\\ &\leqslant \tfrac{1}{4}(\delta^{n-1} + \ldots + 1);\end{aligned}$$

which proves the Lemma in this case.

(ii) Suppose now that (4.5) is false, so that $M_J \leqslant f_{uu} = f(\mathbf{e}_u)$ for some $u \leqslant i$. Then by Lemma 4.1 and using the notation of its proof we have

$$M_J \leqslant f_{uu} \leqslant C_{10}h_u \leqslant \delta^{i-u}C_{10}h_i \leqslant \delta^{n-1}C_{10}h_i.$$

Then (4.4) gives

$$\mu_i^2 \leqslant \delta^{n-1}C_{10};$$

and the Lemma is proved also in this case.

COROLLARY.

$$|m_{Ji}| \leqslant C_{14}(n, \delta, \eta) \qquad (1 \leqslant J \leqslant n, 1 \leqslant i \leqslant n).$$

Proof. For we can determine $m_{Jn}, m_{J,n-1}, \ldots, m_{J1}$ in order from

$$\begin{aligned}\mu_n &= m_{Jn}\\ \mu_{n-1} &= m_{J,n-1} + c_{n-1,n}m_{Jn}\\ &\;\;\vdots\\ \mu_1 &= m_{J1} + \ldots + c_{1n}m_{Jn}\end{aligned}$$

where

$$|\mu_i| \leqslant C_{13} \qquad \text{and} \qquad |c_{ij}| \leqslant \eta.$$

Proof of Theorem 1.4. This asserts that if $f(\mathbf{x})$ and $f(\mathbf{Tx})$ are both in $\mathscr{S}_n(\delta, \eta)$ then the elements t_{ij} of the integral unimodular matrix $\mathbf{T}$ satisfy

$$|t_{ij}| \leqslant C(n, \delta, \eta)$$

for some $C(n, \delta, \eta)$.

For now we have two bases $\mathbf{e}_1, \ldots, \mathbf{e}_n$ and $\mathbf{e}_1^*, \ldots, \mathbf{e}_n^*$ (say) such that both

$$f(x_1\mathbf{e}_1 + \ldots + x_n\mathbf{e}_n)$$

and

$$f(x_1\mathbf{e}_1^* + \ldots + x_n\mathbf{e}_n^*)$$

are in $\mathscr{S}_n(\delta, \eta)$. The previous Corollary asserts that

$$\mathbf{m}_j = \Sigma m_{ji}\mathbf{e}_i$$

with

$$|m_{ji}| \leqslant C_{14}.$$

Since the situation is symmetric between the $\mathbf{e}_i$ and the $\mathbf{e}_i^*$ we have

$$\mathbf{m}_j = \Sigma m_{ji}^*\mathbf{e}_i^*$$

for some $m_{ji}^* \in \mathbf{Z}$ with

$$|m_{ji}^*| \leqslant C_{14}.$$

On eliminating the n linearly independent vectors $\mathbf{m}_j$ we deduce that

$$\mathbf{e}_j = \Sigma t_{ij}\mathbf{e}_i^*$$

where the $t_{ij} \in \mathbf{Q}$ are bounded by some $C(n, \delta, \eta)$. Since $\{\mathbf{e}_i\}$ and $\{\mathbf{e}_i^*\}$ are both bases we must have $t_{ij} \in \mathbf{Z}$. This completes the proof.

5. GEOMETRY OF DEFINITE AND REDUCED FORMS

In this section we represent the real quadratic form

$$f(\mathbf{x}) = \sum_{i,j=1}^{n} f_{ij}x_ix_j \qquad (f_{ij} = f_{ji})$$

in n variables by the point with coordinates f_{ij} $(i \leqslant j)$ in $n(n+1)/2$-dimensional space and consider the geometric and topological nature of the subsets of $\mathbf{R}^{n(n+1)/2}$ corresponding to the positive definite and to the reduced forms. We denote the ordinary euclidean length in both $\mathbf{R}^{n(n+1)/2}$ and in $\mathbf{R}^n$

by $| \ |$ so that

$$|f|^2 = \sum_{1 \leqslant i \leqslant j \leqslant n} f_{ij}^2$$

and

$$|\mathbf{x}|^2 = \sum_{1 \leqslant i \leqslant n} x_i^2$$

for a real variable $\mathbf{x}$. We shall require the two following simple lemmas.

LEMMA 5.1.

$$|f(\mathbf{x})| \leqslant 2|f| \ |\mathbf{x}|^2$$

for any quadratic form and real vector $\mathbf{x}$.

Proof. For by Cauchy–Bunjakovskij–Schwarz we have

$$\begin{aligned} |f(\mathbf{x})|^2 &= (\textstyle\sum f_{ii} x_i^2 + \sum_{i<j} 2 f_{ij} x_i x_j)^2 \\ &\leqslant (\textstyle\sum_{i \leqslant j} f_{ij}^2)(\sum_i x_i^4 + \sum_{i<j} 4 x_i^2 x_j^2) \\ &\leqslant 4|f|^2 \, |\mathbf{x}|^4. \end{aligned}$$

LEMMA 5.2. *Suppose that* f *is strictly positive definite. Then there is a* $K = K_f > 0$ *such that*

$$f(\mathbf{x}) \geqslant K|\mathbf{x}|^2$$

for all real vectors $\mathbf{x}$.

Proof. The function $f(\mathbf{x})$ of the variable $\mathbf{x}$ is continuous on the compact set $|\mathbf{x}| = 1$ and so attains its lower bound K (say). Then $K > 0$ since f is strictly definite. The required inequality now follows by homogeneity in $\mathbf{x}$.

THEOREM 5.1. (i) *The set* $\mathscr{P}^\circ$ *of all strictly positive definite forms* f *is an open convex subset of* $\mathbf{R}^{n(n+1)/2}$.

(ii) *The closure of* $\mathscr{P}^\circ$ *is the set* $\mathscr{P}$ *of all positive definite or semi-definite forms.*

Proof. (i) If f_0, f_1 are strictly definite then so is

$$f_\lambda = (1 - \lambda) f_0 + \lambda f_1 \tag{5.1}$$

for all real λ in $0 \leqslant \lambda \leqslant 1$: which shows that $\mathscr{P}^\circ$ is convex. Now let $f_0(\mathbf{x})$ be any strictly definite form and let K be defined by Lemma 5.2. Let f be any form with

$$|f - f_0| < K/4. \tag{5.2}$$

Then by Lemma 5.1 in an obvious notation for any $\mathbf{x}$ we have

$$|(f - f_0)(\mathbf{x})| \leqslant K|\mathbf{x}|^2/2$$

and so

$$f(\mathbf{x}) \geqslant f_0(\mathbf{x}) - |(f - f_0)(\mathbf{x})| \\ \geqslant K|\mathbf{x}|^2/2. \tag{5.3}$$

Hence f is strictly definite: which shows that $\mathscr{P}^\circ$ is open.

(ii) We first show that $\mathscr{P}$ is closed. For suppose that $f_0 \notin \mathscr{P}$. Then there is a real vector $\mathbf{x}_0$ such that $f_0(\mathbf{x}_0) < 0$. Clearly $f(\mathbf{x}_0) < 0$ for all f in some neighbourhood of f_0: which shows that the complement of $\mathscr{P}$ is open.

Now let f_0 be any point of $\mathscr{P}$ and let f_1 be any point of $\mathscr{P}^\circ$. Then clearly f_λ defined by 5.1 is in $\mathscr{P}^\circ$ for $0 < \lambda \leqslant 1$. Hence f_0 is in the closure of $\mathscr{P}^\circ$ as required.

Note. We have $d(f) > 0$ on $\mathscr{P}^\circ$ and $d(f) = 0$ on $\mathscr{P} - \mathscr{P}^\circ$. Hence $\mathscr{P}^\circ$ is one of the components of the complement of the algebraic hypersurface $d(f) = 0$.

We now require the

Definition. A form $f(\mathbf{x})$ is *strictly reduced* if it is reduced and if the only integral unimodular transformations $\mathbf{T}$ such that $f(\mathbf{Tx})$ is also reduced are the diagonal transformations with elements ± 1. In other words if $\mathbf{e}_1, \ldots, \mathbf{e}_{j-1}, \mathbf{e}_j^*$ is extendible to a basis and $\mathbf{e}_j^* \neq \pm \mathbf{e}_j$, then

$$f(\mathbf{e}_j^*) > f(\mathbf{e}_j). \tag{5.4}$$

LEMMA 5.3. (i) *the set $\mathscr{R}^\circ$ of strictly reduced forms is convex and open.*

(ii) *the set $\mathscr{R}$ of all reduced forms is the relative closure of $\mathscr{R}^\circ$ in $\mathscr{P}^\circ$.*

This Lemma will be superseded by Theorem 5.2 below. The proof follows the paradigm of that of Theorem 5.1.

Proof. (i) Convexity of $\mathscr{R}^\circ$ is proved exactly as for $\mathscr{P}^\circ$ in Theorem 5.1. Now let $f_0 \in \mathscr{R}^\circ$ and let $\mathscr{N}_0$ be the set of f with (5.2), so (5.3) holds. In particular there is a constant X such that

$$\{f \in \mathscr{N}_0; |\mathbf{x}| \geqslant X\} \Rightarrow f(\mathbf{x}) > 1 + \max_{1 \leqslant i \leqslant n} f_0(\mathbf{e}_j). \tag{5.5}$$

Now let $\mathscr{N}_1$ be the set of $f \in \mathscr{N}_0$ for which

$$f(\mathbf{e}_j) < f_0(\mathbf{e}_j) + 1. \tag{5.6}$$

Then $\mathscr{N}_1$ is open. Further, (5.5) and (5.6) imply that (5.4) holds for all those $\mathbf{e}_j^*$ for which $|\mathbf{e}_j^*| \geqslant X$. Let $\mathscr{N}_2$ be the subset of $\mathscr{N}_1$ which also satisfies the remaining inequalities (5.4), that is those for which $|\mathbf{e}_j^*| < X$ $(1 \leqslant j \leqslant n)$. Then $\mathscr{N}_2$ is open and $f_0 \in \mathscr{N}_2 \subset \mathscr{R}^\circ$.

(ii) We have $\mathscr{R} \subset \mathscr{P}^\circ$ by definition. Let $f_0 \in \mathscr{P}^\circ, f_0 \notin \mathscr{R}$. Then there is some j and an integral vector $\mathbf{e}_j^*$ such that $\mathbf{e}_1, \ldots, \mathbf{e}_{j-1}, \mathbf{e}_j^*$ extends to a basis and such that $f_0(\mathbf{e}_j^*) < f_0(\mathbf{e}_j)$. Then there is a neighbourhood $\mathcal{N}$ of f_0 in which $f(\mathbf{e}_j^*) < f(\mathbf{e}_j)$ and so no point of $\mathcal{N}$ is in $\mathcal{R}$. This shows that $\mathcal{R}$ is relatively closed. The proof that $\mathcal{R}$ is the relative closure of $\mathcal{R}^\circ$ follows exactly as for Theorem 5.1.

Definition. For $1 \leqslant j \leqslant n$ we denote by W_j the set of integral vectors $\mathbf{e}_j^*$ with the following properties: (i) There are integral vectors $\mathbf{e}_{j+1}^*, \ldots, \mathbf{e}_n^*$ such that $\mathbf{e}_1, \ldots, \mathbf{e}_{j-1}, \mathbf{e}_j^*, \mathbf{e}_{j+1}^*, \ldots, \mathbf{e}_n^*$ is a basis. (ii) There is a quadratic form $f(\mathbf{x})$ such that both

$$f(\mathbf{x}) = f(x_1\mathbf{e}_1 + \ldots + x_n\mathbf{e}_n)$$

and

$$f^*(\mathbf{x}) = f(x_1\mathbf{e}_1 + \ldots + x_{j-1}\mathbf{e}_{j-1} + x_j\mathbf{e}_j^* + \ldots + x_n\mathbf{e}_n^*)$$

are reduced. By Theorem 1.2 the set W_j is finite $(1 \leqslant j \leqslant n)$.

THEOREM 5.2. *The set $\mathcal{R}^\circ$ of strictly reduced forms is defined by the finite set of linear inequalities:*

(α) $f(\mathbf{e}_1) > 0$

(β) $f(\mathbf{e}_j^*) > f(\mathbf{e}_j) \qquad (1 \leqslant j \leqslant n, \mathbf{e}_j^* \in W_j)$.

Note. These inequalities are linear and homogeneous in the f_{ij}.

Proof. Every $f \in \mathcal{R}^\circ$ satisfies (α) and (β). We want to show that an $f \notin \mathcal{R}^\circ$ fails to satisfy either (α) or one of the (β). Suppose that $f_0 \notin \mathcal{R}^\circ$ and $f_1 \in \mathcal{R}^\circ$ and define

$$f_\lambda = (1-\lambda)f_0 + \lambda f_1 \qquad (0 \leqslant \lambda \leqslant 1). \tag{5.7}$$

Let μ be the upper bound of the λ such that $f_\lambda \notin \mathcal{R}^\circ$. Then f_μ is on the frontier of $\mathcal{R}^\circ$. We distinguish two cases:

(i) f_μ is strictly definite. Then f_μ is reduced but not strictly reduced and so there is some j and some $\mathbf{e}_j^* \in W_j$ such that

$$f_\mu(\mathbf{e}_j^*) = f_\mu(\mathbf{e}_j).$$

Since f_1 is strictly reduced we have

$$f_1(\mathbf{e}_j^*) > f_1(\mathbf{e}_j)$$

and so

$$f_0(\mathbf{e}_j^*) \leqslant f_0(\mathbf{e}_j)$$

by (5.7). Hence f_0 does not satisfy (β).

(ii) f_μ is not strictly definite. Then f_λ is strictly reduced for $\lambda > \mu$. In particular

$$f_\lambda(\mathbf{b}) \geqslant f_\lambda(\mathbf{e}_1)$$

for all $\lambda > \mu$ and all integral $\mathbf{b} \neq \mathbf{0}$. By continuity

$$f_\mu(\mathbf{b}) \geqslant f_\mu(\mathbf{e}_1) \qquad \text{(all integral } \mathbf{b} \neq \mathbf{0}).$$

Since f_μ is semi-definite but not definite it follows from Lemma 5.4, which will be enunciated below, that $f_\mu(\mathbf{e}_1) = 0$. As in (i) we deduce that $f_0(\mathbf{e}_1) \leqslant 0$. Hence f_1 does not satisfy (α). This concludes the proof of the theorem.

COROLLARY. *The set* $\mathcal{R}$ *of reduced forms is determined by* (α) *and the inequalities* ($\beta^{\geqslant}$) *obtained by substituting* $\geqslant$ *for* $>$ *in* (β).

Proof. follows from the Theorem and Lemma 5.3.

To complete the proof of Theorem 5.2 we need thus only prove

LEMMA 5.4. *Suppose that* f *is positive semidefinite but not definite. Then the infimum of* $f(\mathbf{b})$ *for integral* $\mathbf{b} \neq \mathbf{0}$ *is* 0.

Proof. Let $\varepsilon > 0$. The set of $\mathbf{x}$ for which

$$f(\mathbf{x}) < \varepsilon$$

is convex, and symmetric about the origin. It has infinite volume and so contains an integral point $\mathbf{b} \neq \mathbf{0}$ by Minkowski's convex body theorem.

6. GEOMETRY OF THE BINARY CASE

Before continuing the study of the general case we examine the reduction of binary forms. We recall that

$$f(\mathbf{x}) = f_{11}x_1^2 + 2f_{12}x_1x_2 + f_{22}x_2^2 \tag{6.1}$$

is strictly definite when

$$f_{11} > 0, \qquad f_{11}f_{22} - f_{12}^2 > 0, \tag{6.2}$$

and that the conditions for reduction are

$$\left.\begin{aligned} f_{11} + 2f_{12} &\geqslant 0 \\ f_{11} - 2f_{12} &\geqslant 0 \\ f_{22} - f_{11} &\geqslant 0. \end{aligned}\right\} \tag{6.3}$$

Both the set $\mathcal{P}^\circ$ of strictly positive forms and the set $\mathcal{R}$ of reduced forms are cones in $\mathbf{R}^3$ with vertices at the origin; and this is reflected by the fact that (6.2) and (6.3) are both homogeneous. It is convenient to put

$$\frac{2f_{12}}{f_{22} + f_{11}} = \xi, \qquad \frac{f_{22} - f_{11}}{f_{22} + f_{11}} = \eta. \tag{6.4}$$

The condition (6.2) for strict definiteness becomes

$$\xi^2 + \eta^2 < 1, \tag{6.5}$$

and similarly (6.3) gives

$$\left.\begin{aligned} \eta + 2\xi &\leqslant 1 \\ \eta - 2\xi &\leqslant 1 \\ \eta &\geqslant 0 \end{aligned}\right\}. \tag{6.6}$$

Note that given values of ξ and η correspond both to the set of forms

$$\lambda f(\mathbf{x}) \qquad \lambda > 0$$

and to

$$-\lambda f(\mathbf{x}) \qquad \lambda > 0$$

and so the condition $f_{11} > 0$ disappears from (6.5).

FIGURE 12.1

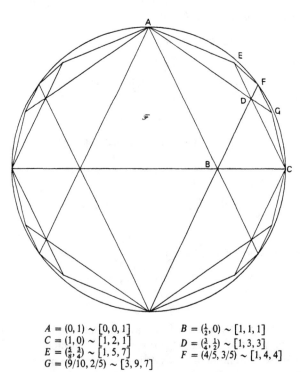

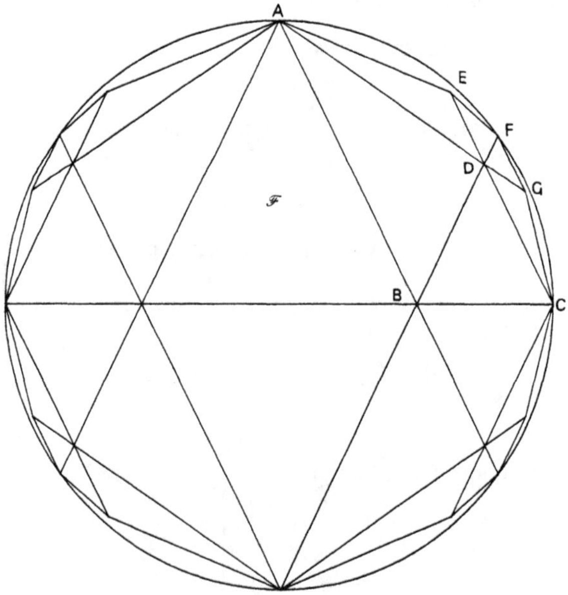

$A = (0, 1) \sim [0, 0, 1]$ $\quad$ $B = (\frac{1}{2}, 0) \sim [1, 1, 1]$
$C = (1, 0) \sim [1, 2, 1]$ $\quad$ $D = (\frac{3}{4}, \frac{1}{2}) \sim [1, 3, 3]$
$E = (\frac{5}{8}, \frac{3}{4}) \sim [1, 5, 7]$ $\quad$ $F = (4/5, 3/5) \sim [1, 4, 4]$
$G = (9/10, 2/5) \sim [3, 9, 7]$

The circle has unit radius. The coordinates are (ξ, η) and we have written $(\xi, \eta) \sim [f_{11}, 2f_{12}, f_{22}]$ in the notation of (6.4). The figure shows the fundamental domain $\mathscr{F}$ and some of the transforms $\mathbf{T}\mathscr{F}$. Each $\mathbf{T}\mathscr{F}$ has one vertex on the circumference and two in the interior of the circle. There are infinitely many $\mathbf{T}\mathscr{F}$ and they completely cover the interior of the circle without overlapping; they accumulate only towards the circumference. Six triangles $\mathbf{T}\mathscr{F}$ meet at every interior vertex. The set of vertices on the circumference is everywhere dense and a vertex on the circumference belongs to infinitely many triangles $\mathbf{T}\mathscr{F}$.

Thus the strictly definite forms correspond to the points (ξ, η) in the interior of the unit circle. The reduced forms correspond to the points of the triangle $\mathscr{F}$ with vertices $(0, 1)$, $(\pm\frac{1}{2}, 0)$ which correspond respectively to the semi-definite form x_2^2 and to $x_1^2 \pm x_1x_2 + x_2^2$. The interior points of $\mathscr{F}$ correspond to strictly reduced forms and the boundary points of $\mathscr{F}$ other than $(0, 1)$ to forms which are reduced but not strictly reduced.

We now consider the effect of an integral unimodular transformation $\mathbf{T} \in SL^{\pm}$ of the variables:

$$\mathbf{T}(x_1, x_2) = (ax_1 + bx_2, cx_1 + dx_2).$$

This takes a form f into the equivalent form $\mathbf{T}f$ defined by $f(\mathbf{T}(x_1 \,.\, x_2)) = (\mathbf{T}f)(x_1 \,.\, x_2)$. We shall not actually need the explicit expression for $\mathbf{T}f$ but give it for completeness:

$$\mathbf{T}(f_{11}, f_{12}, f_{22}) = (a^2f_{11} + 2acf_{12} + c^2f_{22}, abf_{11} + (ad + bc)f_{12} + cdf_{22}, \\ b^2f_{11} + 2bdf_{12} + d^2f_{22})$$

or, in terms of ξ and η: $\mathbf{T}(\xi, \eta) = (2\rho/\tau, \sigma/\tau)$ where

$$\begin{aligned} \rho &= (ab + cd) + (ad + bc)\xi + (cd - ab)\eta \\ \sigma &= (b^2 + d^2 - a^2 - c^2) + 2(bd - ac)\xi + (a^2 + d^2 - b^2 - c^2)\eta \\ \tau &= (a^2 + b^2 + c^2 + d^2) + 2(ac + bd)\xi + (c^2 + d^2 - a^2 - b^2)\eta. \end{aligned}$$

The transformation $\mathbf{T}$ takes strictly definite forms into strictly definite forms and so maps the interior of the unit circle one–one onto itself.

Since $\mathbf{T}$ acts linearly on the coefficients of the forms f, the transform $\mathbf{T}\mathscr{F}$ of the triangle $\mathscr{F}$ in the (ξ, η)-plane is another triangle. We now are in a position to prove the following facts where we recall that $\mathscr{F}$ corresponds to the reduced forms and so contains all its boundary points except $(0, 1)$:

(i) The $\mathbf{T}\mathscr{F}$ completely cover the open disc $\xi^2 + \eta^2 < 1$. For any strictly definite form f can be written $f = \mathbf{T}f_0$ for some $\mathbf{T}$ and some reduced form f_0.

(ii) If an interior point of $\mathbf{T}_1\mathscr{F}$ belongs to $\mathbf{T}_2\mathscr{F}$ then $\mathbf{T}_1\mathscr{F}$ and $\mathbf{T}_2\mathscr{F}$ coincide. For suppose that $\mathbf{T}_1(\xi_1, \eta_1) = \mathbf{T}_2(\xi_2, \eta_2)$ where (ξ_1, η_1) is an interior point of $\mathscr{F}$ and (ξ_2, η_2) is in $\mathscr{F}$ but not necessarily interior. Then $\mathbf{T}_2^{-1}\mathbf{T}_1(\xi_1, \eta_1) = (\xi_2, \eta_2)$. Thus in an obvious notation $\mathbf{T}_2^{-1}\mathbf{T}_1$ takes the strictly reduced form f_1 into the reduced form f_2, But this implies, by the very definition of a reduced form, that $\mathbf{T}_2^{-1}\mathbf{T}_1$ must be one of the diagonal transformations

$$\begin{pmatrix} \pm 1 & 0 \\ 0 & \pm 1 \end{pmatrix}$$

which take $\mathscr{F}$ into itself.

(iii) Every side of a $\mathbf{T}\mathscr{F}$ belongs to precisely one other $\mathbf{T}\mathscr{F}$. One vertex of $\mathbf{T}\mathscr{F}$ lies on the unit circle $\xi^2 + \eta^2 = 1$ and the other two belong to precisely 5 other $\mathbf{T}\mathscr{F}$. For this is true of $\mathscr{F}$ itself and $\mathbf{T}$ maps the whole configuration of triangles onto itself.

Before leaving the binary case we note that our diagram can be used to prove facts about the unimodular group. It is easy to verify that the three triangles having a side in common with $\mathscr{F}$ are $\mathbf{S}_j\mathscr{F}$ where

$$\mathbf{S}_1(x_1, x_2) = (-x_2, x_1)$$

$$\mathbf{S}_2(x_1, x_2) = (x_1 + x_2, x_2)$$

$$\mathbf{S}_3(x_1, x_2) = (x_1 - x_2, x_2).$$

Hence the three triangles having a side in common with $\mathbf{T}\mathscr{F}$ are precisely $\mathbf{T}\mathbf{S}_j\mathscr{F}$ $(j = 1, 2, 3)$. Now let $\mathbf{T}^*$ be any unimodular transformation. Then it is easy to see that we can find a sequence of triangles

$$\mathscr{F}, \mathbf{T}_1\mathscr{F}, \mathbf{T}_2\mathscr{F}, \ldots, \mathbf{T}_N\mathscr{F}$$

such that $\mathbf{T}_N\mathscr{F} = \mathbf{T}^*\mathscr{F}$ and any two successive triangles have a side in common (for a formal proof see the next section). Hence we can find $\mathbf{S}^{(m)} \in \{\mathbf{S}_1, \mathbf{S}_2, \mathbf{S}_3\}$ for $1 \leqslant m \leqslant N$ such that

$$\mathbf{T}_1\mathscr{F} = \mathbf{S}^{(1)}\mathscr{F}$$

$$\mathbf{T}_2\mathscr{F} = \mathbf{S}^{(1)}\mathbf{S}^{(2)}\mathscr{F}$$

$$\cdots\cdots$$

$$\mathbf{T}^*\mathscr{F} = \mathbf{T}_N\mathscr{F} = \mathbf{S}^{(1)}\mathbf{S}^{(2)}\ldots\mathbf{S}^{(N)}\mathscr{F}.$$

Hence

$$\mathbf{T}^* = \mathbf{S}^{(1)}\mathbf{S}^{(2)}\ldots\mathbf{S}^{(N)}\mathbf{R}.$$

where $\mathbf{R}$ maps $\mathscr{F}$ onto itself and so is of the shape

$$\begin{pmatrix} \pm 1 & 0 \\ 0 & \pm 1 \end{pmatrix}.$$

This shows that the unimodular group is generated by $\mathbf{S}_1, \mathbf{S}_2, \mathbf{S}_3$ and the diagonal matrices. We actually have $\mathbf{S}_3 = \mathbf{S}_2^{-1}$, so we may omit $\mathbf{S}_3$.

In conclusion we note that our diagram is equivalent to the more traditional diagram associated with the unimodular group. For the details see Chapter 13, Section 8.

7. GEOMETRY OF THE GENERAL CASE

In this section we show that most of the results proved in the previous section generalize but some of the proofs are more sophisticated.

THEOREM 7.1.

$$\mathscr{P}^\circ = \bigcup_{\mathbf{T}} \mathbf{T}\mathscr{R}$$

where **T** runs through all unimodular transformations.

Proof. This is just another way of saying that every strictly definite form f_0 is of the shape $f_0 = \mathbf{T}f_1$ for some **T** and some reduced form f_1.

THEOREM 7.2. *If* $\mathbf{T}\mathscr{R} = \mathscr{R}$ *then* **T** *is of the shape*

$$\mathbf{T}x_j = \pm x_j \qquad (1 \leqslant j \leqslant n). \tag{7.1}$$

Proof. For let f_0 be a strictly reduced form. By hypothesis $\mathbf{T}f_0$ is also reduced and (7.1) now follows by the definition of strict reduction.

THEOREM 7.3. *For any integral unimodular* **T** *the set* $\mathbf{T}\mathscr{R}$ *is determined by*

$$\mathbf{T}\mathscr{R} \subset \mathscr{P}^\circ \tag{7.2}$$

and a finite number of conditions

$$L_s^{(\mathbf{T})}(f) \geqslant 0, \tag{7.3}$$

where the $L_s^{(\mathbf{T})}$ *are homogeneous linear forms in the* f_{ij}. *Alternatively the condition* (7.2) *can be replaced by*

$$L_0^{(\mathbf{T})}(f) > 0, \tag{7.4}$$

where $L_0^{(\mathbf{T})}$ *is a further homogeneous linear form.*

Proof. For the above statements are true of $\mathscr{R}$ by Theorem 5.2, Corollary. They hold also for $\mathbf{T}\mathscr{R}$ because **T** acts linearly on the space of forms and is an isomorphism of $\mathscr{P}^\circ$.

THEOREM 7.4. *Suppose that* $\mathbf{T}_1\mathscr{R}$ *and* $\mathbf{T}_2\mathscr{R}$ *have a form f in common and that it is an interior point of at least one of them. Then* $\mathbf{T}_1\mathscr{R} = \mathbf{T}_2\mathscr{R}$.

Proof. For suppose that $f_0 \in \mathbf{T}_1\mathscr{R} \cap \mathbf{T}_2\mathscr{R}$ and is an inner point of $\mathbf{T}_1\mathscr{R}$. Then the forms $\mathbf{T}_1^{-1}f_0$ and $\mathbf{T}_2^{-1}f_0$ are equivalent and they are both reduced, the first strictly. The result now follows from the definition of strict reduction.

THEOREM 7.5. *There is a finite set* $\Sigma = \{\mathbf{S}_1, \mathbf{S}_2, \ldots, \mathbf{S}_t\}$ *of unimolar transformations such that* $\mathbf{T}_1\mathscr{R} \cap \mathbf{T}_2\mathscr{R}$ *is non-empty if and only if* $\mathbf{T}_2 = \mathbf{T}_1\mathbf{S}$ *for some* $\mathbf{S} \in \Sigma$.

Proof. We take for Σ the set of $\mathbf{S}$ such that $\mathscr{R} \cap \mathbf{S}\mathscr{R}$ is not empty. If $f_0 \in \mathscr{R} \cap \mathbf{S}\mathscr{R}$ then both f_0 and $\mathbf{S}^{-1}f_0$ are reduced so the finiteness of Σ follows from Theorem 1.1. The general result now follows since

$$\mathbf{T}_1\mathscr{R} \cap \mathbf{T}_2\mathscr{R} = \mathbf{T}_1(\mathscr{R} \cap \mathbf{T}_1^{-1}\mathbf{T}_2\mathscr{R}).$$

COROLLARY. *Any form f belongs to at most t sets $\mathbf{T}\mathscr{R}$, where t is the cardinal of Σ.*

So far the arguments have followed those for $n = 2$ with little modification. The following theorem, however, is rather more tricky for general n.

THEOREM 7.6. *A compact subset $\mathscr{C}$ of $\mathscr{P}^\circ$ meets only finitely many sets $\mathbf{T}\mathscr{R}$.*

Note. This property of the cover of $\mathscr{P}^\circ$ by the $\mathbf{T}\mathscr{R}$ is called "*local finiteness*". It excludes many kinds of pathological behaviour.

Proof. By Lemma 1.3 $\mathscr{R}$ is contained in some Siegel domain $\mathscr{S}(\delta, \eta)$ and so in the interior $\mathscr{S}^\circ = \mathscr{S}^\circ(\delta', \eta')$ of any Siegel domain $\mathscr{S}(\delta', \eta')$ with $\delta' > \delta, \eta' > \eta$.
By Theorem 7.1 the open sets $\mathbf{T}\mathscr{S}^\circ$ cover $\mathscr{C}$ and so by compactness $\mathscr{C}$ is covered by a finite collection $\mathbf{T}_j\mathscr{S}^\circ$ $(1 \leqslant j \leqslant J)$, say. But each $\mathbf{T}_j\mathscr{S}^\circ$, meets only a finite number of $\mathbf{T}\mathscr{S}^\circ$ by Theorem 1.4. Since $\mathbf{T}\mathscr{R} \subset \mathbf{T}\mathscr{S}^\circ$, this completes the proof.

COROLLARY. *Let f_0 be any element of $\mathscr{P}^\circ$. Then there is some neighbourhood $\mathscr{N}$ of f_0 which is contained in the union of the $\mathbf{T}\mathscr{R}$ which contain f_0.*

Proof. Since $\mathscr{P}^\circ$ is open, there is an open set $\mathscr{N}_1$ and a compact set $\mathscr{C}$ such that

$$f_0 \in \mathscr{N}_1 \subset \mathscr{C} \subset \mathscr{P}^\circ.$$

By the Theorem the $\mathbf{T}\mathscr{R}$ which meet $\mathscr{C}$ but do not contain f_0 are finite in number, say $\mathbf{T}_k\mathscr{R}$ $(1 \leqslant k \leqslant K)$. By Lemma 5.3 (ii), the complement $\mathscr{P}^\circ - \mathbf{T}_k\mathscr{R}$ is open. Then

$$\mathscr{N} = \mathscr{N}_1 \cap \{ \bigcap_{1 \leqslant k \leqslant K} (\mathscr{P}^\circ - \mathbf{T}_k\mathscr{R})\}$$

does what is required.

THEOREM 7.7. *Let Σ be the set of unimodular integral transformations $\mathbf{S}$ such that $\mathscr{R}$ meets $\mathbf{S}\mathscr{R}$. Then Σ is finite and every integral unimodular transformation $\mathbf{T}$ is the product of elements of Σ.*

Proof. The finiteness of Σ is given by Theorem 1.2. Clearly $\mathbf{T}_1\mathscr{R}$ and $\mathbf{T}_2\mathscr{R}$ intersect if and only if $\mathbf{T}_2 = \mathbf{T}_1\mathbf{S}$ for some $\mathbf{S} \in \Sigma$. Now let $\mathbf{T}$ be any integral unimodular transformation and let f_0, f_1 be any points in $\mathscr{R}$ and $\mathbf{T}\mathscr{R}$ respectively. The line segment

$$\mathscr{C}: (1 - \lambda)f_0 + \lambda f_1 \qquad (0 \leqslant \lambda \leqslant 1)$$

joining f_0 and f_1 is compact and so meets only a finite number of transforms $\mathbf{T}_k\mathscr{R}$ $(1 \leqslant k \leqslant K)$ of $\mathscr{R}$ by Theorem 7.6. The $\mathbf{T}_k\mathscr{R}$ are convex and so meet $\mathscr{C}$ either in a single point or in a closed segment. Hence we can suppose that $\mathbf{T}_1$ is the identity, that $\mathbf{T}_K = \mathbf{T}$ and that $\mathbf{T}_k\mathscr{R}$ meets $\mathbf{T}_{k+1}\mathscr{R}$ for $1 \leqslant k < K$. It follows that $\mathbf{T}_{k+1} = \mathbf{T}_k\mathbf{S}_k$ for some $\mathbf{S}_k \in \Sigma$ $(1 < k < K)$ and $\mathbf{T} = \mathbf{T}_K = \mathbf{S}_2\mathbf{S}_3 \ldots \mathbf{S}_{K-1}$. as required.

We can improve Theorem 7.7 slightly. For any $\mathbf{T}_1, \mathbf{T}_2$ the sets $\mathbf{T}_1\mathscr{R}, \mathbf{T}_2\mathscr{R}$ are convex and hence so is their intersection $\mathbf{T}_1\mathscr{R} \cap \mathbf{T}_2\mathscr{R}$: we say that $\mathbf{T}_1\mathscr{R}$ and $\mathbf{T}_2\mathscr{R}$ are *neighbours* if the intersection has codimension 1. Clearly $\mathbf{T}_1\mathscr{R}, \mathbf{T}_2\mathscr{R}$ are neighbours precisely when $\mathbf{T}_2 = \mathbf{T}_1\mathbf{S}$ for some $\mathbf{S}$ such that $\mathscr{R}$ and $\mathbf{S}\mathscr{R}$ are neighbours. We denote the set of $\mathbf{S}$ with the latter property by Σ_1, so $\Sigma_1 \subset \Sigma$.

COROLLARY. $\mathbf{T}$ *can be expressed as the product of elements of* Σ_1 *and a matrix having* ± 1 *along the diagonal and zeros elsewhere.*

Proof. Let f_0 be an inner point of $\mathscr{R}$ and let $\mathscr{A}$ be a closed ball contained in $\mathbf{T}\mathscr{R}$. Then the set

$$\mathscr{C}: (1 - \lambda)f_0 + \lambda f_1 \qquad (0 \leqslant \lambda \leqslant 1, f_1 \in \mathscr{A})$$

is compact and so meets only finitely many $\mathbf{L}\mathscr{R}$ ($\mathbf{L}$ integral, unimodular). Denote by $\mathscr{D}$ the union of all the $\mathbf{L}_1\mathscr{R} \cap \mathbf{L}_2\mathscr{R}$ where $\mathbf{L}_1\mathscr{R} \cap \mathbf{L}_2\mathscr{R}$ has codimension at least 2 and $\mathbf{L}_1\mathscr{R}, \mathbf{L}_2\mathscr{R}$ both meet $\mathscr{C}$. Then $\mathscr{D}$ is the union of a finite number or sets of codimension at least 2 and so it is possible to find $f_1 \in \mathscr{A}$ such that the line segment from f_0 to f_1 does not meet $\mathscr{D}$ (cf. Fig. 12.2). We can now argue as in the proof of the Theorem.

In conclusion we note one fact about the 2-dimensional case which we have not proved to generalize to n dimensions. In 2-dimensions if $\mathbf{T}_1\mathscr{R}$ and $\mathbf{T}_2\mathscr{R}$ have an intersection of codimension 1, then they have a complete face in common. It is not, apparently, known whether this is true for all n.

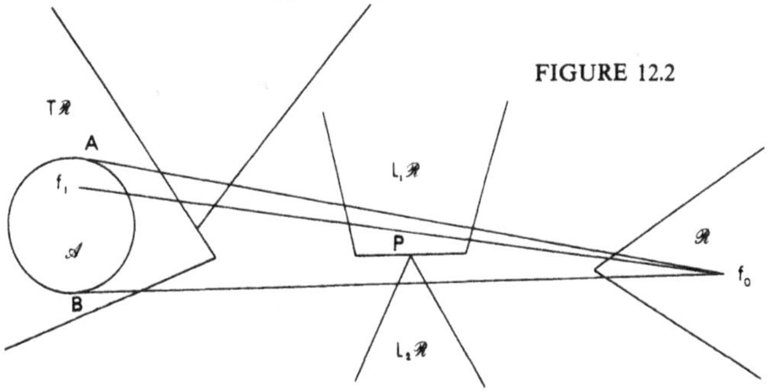

FIGURE 12.2

This Figure illustrates the proof of Theorem 7.1, Corollary in 2 dimensions. The set $\mathscr{D}$ consists of points such as P which are common to two $\mathbf{L}\mathscr{R}$ which are not neighbours (i.e do not have a side in common). There are only finitely many such points P in the triangle ABf_0, and so it is possible to choose $f_1 \in \mathscr{A}$ so that the line-segment f_0f_1 does not pass through any of them.

NOTES

Minkowski's major paper on reduction is Minkowski (1905). Siegel's approach is developed in numerous papers: in particular Siegel (1940) contains many of the key ideas of this chapter and the next. The largely expository paper of van der Waerden (1956) contains historical material and references. For generalization to algebraic number fields see Humbert (1940, 1949). For Siegel domains in the context of linear algebraic groups, see Borel (1969). For a survey, with special reference to the work of Soviet mathematicians, see Ryškov (1974).

Conditions for reduction when $n = 5, 6$ were stated without proof by Minkowski (1886, 1887) and proved by Ryškov (1971, 1973) and Tammela (1973, 1975). The conditions for $n = 7$ have been found by Tammela (1977).

We have not attempted to obtain good estimates for the various constants occurring in the statements of the theorems, since this would have complicated the arguments. For such estimates see Remak (1938), van der Waerden (1956), Siegel (1972) and the references given there.

There are a number of tables of reduced integral ternary and quaternary quadratic forms using various definitions of reduction and of integrality. For ternaries see Eisenstein (1851), Dickson (1930), Jones (1935), Brandt and Intrau (1948); and for quaternaries, S. B. Townes (1940), K. German (1963).

Minkowski's method of reduction of definite forms is the usual one but Venkov (1940) has introduced another approach which we shall briefly describe. It depends on the fact that the set of all real quadratic forms of dimension n has a natural structure as a quadratic space of dimension $N = \frac{1}{2}n(n+1)$ with respect to which the set $\mathscr{P}$ of positive definite or semi-definite forms of dimension n has a special rôle. For $f(\mathbf{x}) = \Sigma f_{ij}x_i x_j$ and $h(\mathbf{x}) = \Sigma h_{ij}x_i x_j$ of dimension n one considers the bilinear form

$$\{f, h\} = \sum_{i,j} f_{ij}h_{ij} = \text{Trace } \mathbf{FH},$$

where $\mathbf{F}, \mathbf{H}$ are the corresponding matrices. For a special form

$$h_0(\mathbf{x}) = (l_1x_1 + \ldots + l_nx_n)^2 \qquad (*)$$

we have

$$\{f, h_0\} = f(l_1, \ldots, l_n).$$

It readily follows by bilinearity that

$$\{f, h\} \geqslant 0 \qquad (\text{all } f, h \in \mathscr{P}).$$

[Indeed a necessary and sufficient condition that $f \in \mathscr{P}$ is that $\{f, h\} \geqslant 0$ for all $h \in \mathscr{P}$.]

Now let h_1 be some fixed strictly positive definite form and let $V = V(h_1)$ be the set of strictly definite forms f such that

$$\{f, h_1\} = \inf_{\mathbf{S}} \{f, \mathbf{S}h_1\}$$

where $\mathbf{S}$ runs through SL^+ and where $\mathbf{S}h_1(\mathbf{x}) = h_1(\mathbf{Sx})$ by definition. Venkov shows that $V(h_1)$ is bounded by a finite number of hyperplanes. It is a fundamental domain for the action of SL^+ on $\mathscr{P}$ provided that h_1 has no proper integral automorphs. If h_1 has a non-trivial group G of proper integral automorphs, then $V(h_1)$ is the union of g fundamental domains, where g is the order of G. For small values of n one obtains the Minkowski set of reduced forms on putting $h_1(\mathbf{x}) = \sum_j x_j^2$.

Venkov's method is closely related to the work of Voronoï (1908) on "perfect" quadratic forms. Consider the set M of quadratic forms $f = f_{\mathbf{m}}(\mathbf{x}) = (m_1 x_1 + \dots + m_n x_n)^2$ where $\mathbf{m}$ runs through all non-zero integral vectors. A form h is *perfect* (of minimum 1) if the hyperplane

$$\{f, h\} = 1$$

is a face of the convex cover of M. This last statement is equivalent to the two following conditions:

(i) $\{f_{\mathbf{m}}, h\} \geqslant 1$ (all $\mathbf{m} \in M$).

(ii) if $M_0(h)$ denotes the set of $\mathbf{m} \in M$ for which there is equality in (i), then the only quadratic form f for which

$$\{f_{\mathbf{m}}, f\} = 1 \qquad (\text{all } \mathbf{m} \in M_0(h))$$

is $f = h$.

Voronoï assigns to each perfect form h the set $W(h)$ of forms f for which

$$\{f, h\} = \inf_g \{f, g\} \qquad (g \text{ perfect}).$$

He shows that each $W(h)$ is bounded by a finite number of hyperplanes and so obtains an algorithm for determining all the perfect forms (up to integral equivalence). Perfect forms may be used in determining the best value of the constant γ_n in Theorem 2.1. A form h is said to be *extreme* (of minimum 1) if

(i) $$\inf_{\mathbf{m} \neq \mathbf{0}} h(\mathbf{m}) = 1, \text{ and}$$

(ii) $d(h)$ is a local maximum taken over the forms which satisfy (i).

To obtain the best possible value of γ_n one need consider only extreme forms.

Voronoï shows that the extreme forms are precisely the perfect forms that possess a further property ("eutactic"). The perfect forms are known for $n \leqslant 6$, see Barnes (1957). For $n = 7$, see Stacey (1975, 1976).

It should be noted that some writers, particularly of the Soviet school, use the term "Hermite-reduced" with a different meaning from that introduced in Section 1. A definite form f is reduced in this sense if

$$f_{ii} \leqslant f^*_{ii} \qquad (1 \leqslant i \leqslant n)$$

for all forms f^* integrally equivalent to f. Such a form is clearly Minkowski-reduced, but not necessarily *vice-versa*. There is a discussion of the relation between this concept, Minkowski-reduction and Venkov-reduction in Ryškov (1971, 1972, 1973).

EXAMPLES

1. Verify that the forms f listed below are positive definite and find an integral unimodular transformation $\mathbf{T}$ such that $f(\mathbf{Tx})$ is Minkowski-reduced:

(i) $6x^2 - 14xy + 9y^2$

(ii) $25x^2 + 22xy + 5y^2$

(iii) $x^2 + y^2 + z^2 + 10(x + y + z)^2$.

2. Let $f(x, y) = ax^2 + 2bxy + cy^2$ be Minkowski-reduced with determinant $d = ac - b^2$. Show that

$$ac \leqslant (4/3)d$$

and determine all the cases of equality.

3. Let $f(\mathbf{x}) = \Sigma f_{ij}x_ix_j$ be a Minkowski-reduced ternary form of determinant d. Show that

$$f_{11}f_{22}f_{33} \leqslant 2d.$$

Determine all the cases of equality.

[*Note.* This is much more tedious. One approach due to Gauss is as follows. First show that without loss of generality either (i) $f_{ij} \geqslant 0$ (all $i \neq j$) or (ii) $f_{ij} \leqslant 0$ (all $i \neq j$). In the first case put $g_{ij} = f_{ii} - 2f_{ij}$ and show that

$$2d = f_{11}f_{22}f_{33} + g_{32}g_{21}g_{13} + \Sigma(f_{11}f_{23}g_{23} + f_{23}g_{13}g_{21}),$$

the sum being over cyclical permutations of 1, 2, 3. In the second case put $h_{ij} = f_{ii} + 2f_{ij}$ and $k_i = f(1, 1, 1) - f_{ii}$ and show that

$$6d = 3f_{11}f_{22}f_{33} + h_{23}h_{31}h_{12} + 2h_{32}h_{13}h_{21}$$
$$- \Sigma(f_{11}f_{23}h_{23} + 2f_{11}f_{23}k_1 + f_{23}f_{13}h_{21}).]$$

4. Let $\mathbf{b}_1, \ldots, \mathbf{b}_n$ be vectors in the real positive definite quadratic space V, ϕ. Show that

$$\det_{ij} \phi(\mathbf{b}_i, \mathbf{b}_j) \leqslant \prod_j \phi(\mathbf{b}_j).$$

[*Hint.* (i) trivial if the $\mathbf{b}_j$ are linearly dependent; (ii) also trivial if the $\mathbf{b}_j$ are mutually normal. (iii) Otherwise there is a set of mutually normal vectors $\mathbf{c}_j$ such that

$$\mathbf{b}_j = \sum_{i<j} t_{ij}\mathbf{c}_i + \mathbf{c}_j \qquad (t_{ij} \in \mathbf{R}).$$

Show that $\phi(\mathbf{c}_j) \leqslant \phi(\mathbf{b}_j)$ and reduce the case of the $\mathbf{b}_j$ to that of the $\mathbf{c}_j$.]

[*Note.* This inequality, due to Hadamard, generalizes the assertion that the volume of a parallelepiped is majorized by the product of the lengths of the edges.]

5. Let f be a positive definite form of dimension n with successive minima $M_j = f(\mathbf{m}_j)$.

(i) Show that

$$\prod M_j \geqslant I^2 d(f)$$

where $I = |\det(\mathbf{m}_1, \ldots, \mathbf{m}_n)|$.

(ii) Deduce that I is bounded above by a constant depending only on n.

(iii) Show that $I = 1$ for $n = 2$ and $n = 3$.

CHAPTER 13

Automorphs of Integral Forms

1. INTRODUCTION

In this chapter the group of integral automorphs of a form f will be denoted for convenience by $O(f)$ rather than by $O_{\mathbf{Z}}(f)$. We shall normally regard it as a subgroup of the group $O_{\mathbf{R}}(f)$ of real automorphs. As usual the number of variables is denoted by n.

When f is definite, the group $O(f)$ is clearly finite. The elucidation of the group-theoretic structure of $O(f)$ may have interesting consequences, as Conway has recently shown, but we are not concerned with them here.

For indefinite f the general principle is that, roughly speaking, the group $O(f)$ is as big as the geometry of the situation permits. When f is anisotropic (by which we mean, of course, anisotropic over $\mathbf{Q}$), this principle takes a particularly simple form: for every real automorph $\mathbf{S} \in O_{\mathbf{R}}(f)$ there is an integral automorph $\mathbf{T} \in O(f)$ such that all the elements of $\mathbf{TS}^{-1}$ are bounded by a constant depending only on f (Theorem 2.2). It follows (Theorem 2.2, Corollary) that if $\mathbf{b}$ is integral and $f(\mathbf{b}) = m$ then there is a $\mathbf{T} \in O(f)$ such that the coordinates of $\mathbf{Tb}$ are bounded by a constant depending only on f and m. For anisotropic forms this gives a procedure to decide effectively whether or not a form f represents a given integer m and whether or not two given forms are integrally equivalent. For binary forms this fills the gap left in Chapter 11 where the theory of spinor genera was developed only for $n > 2$ variables, and for other anisotropic forms the procedures of this Chapter are sometimes easier to apply in practice than those of Chapter 11. There is a discussion of effectiveness in Section 12, where we give a general algorithm which does not depend on the theory of spinor genera.

By Meyer's Theorem (Theorem 1.1, Corollary 1, of Chapter 6) an indefinite form can be anisotropic only when $n \leqslant 4$. Hence the results just discussed are inapplicable to "most" forms. For isotropic forms f the formulation of the general principle about the existence of integral automorphs is rather more sophisticated (Theorem 11.2). For binary isotropic forms (which are in any case anomalous) and for isotropic ternaries there is however a simple direct discussion (Sections 3 and 5) and for general n there is a simple way of manufacturing integral automorphs (Theorem 4.2).

The theory of integral automorphs is closely bound up with a theory of reduction of indefinite forms. Such a theory was invented by Hermite† who makes it depend on reduction of definite forms. In contrast to the reduction theory for definite forms, that for indefinite forms is essentially arithmetical: the key theorems depend on the integrality (or at least the rationality) of the forms.

We now review briefly the contents of this chapter. In Section 2 we introduce Hermite reduction and consider in detail the anisotropic case. The more sophisticated isotropic case is deferred to Section 11. In Section 3 we dispose of binary forms: we consider them in considerable detail, partly in preparation for Chapter 14.

For indefinite ternaries the real group $O_{\mathbf{R}}(f)$ is closely related to the geometry of the non-euclidean plane. In §§6–8 we discuss this relation and use it to discuss $O(f)$ for anisotropic indefinite ternaries and also the integral representation of integers by such f. The case of isotropic f, which can be dealt with explicitly, will already have been dealt with in Section 5.

In Section 9 we give a description of the real automorphs of indefinite quaternaries in terms of non-euclidean geometry and in Section 10 we give a more general representation of the real automorphism group valid for all indefinite forms. We do not discuss in detail the implications for the integral automorphism group. Finally, in Section 11 we treat the Hermite reduction of isotropic forms.

2. HERMITE REDUCTION: ANISOTROPIC FORMS

Hermite introduced a definition of reduction of indefinite forms which is closely related to the theory of automorphs. Hermite's own methods work only for anisotropic forms and after the initial definitions we treat only such forms in this section. Siegel showed that some but not all of the results do extend to isotropic forms but the proofs are much more sophisticated: we reserve the discussion of isotropic forms until Section 11.

Hermite's idea is a very natural one. Any real regular form $f(\mathbf{x})$ can be written, in many ways, in the shape

$$f(\mathbf{x}) = L_1(\mathbf{x})^2 + \ldots + L_l(\mathbf{x})^2 - L_{l+1}(\mathbf{x})^2 - \ldots - L_{l+m}(\mathbf{x})^2, \tag{2.1}$$

where $n = l + m$ is the number of variables and $L_j(\mathbf{x})\,(1 \leqslant j \leqslant n)$ are n

† This notion of reduction for indefinite forms should not be confused with the methods of reduction for definite forms, also named after Hermite, which are discussed on p. 259 and p. 282.

linearly independent real linear forms. The real quadratic form

$$g(\mathbf{x}) = L_1(\mathbf{x})^2 + \ldots + L_{l+m}(\mathbf{x})^2 \tag{2.2}$$

is called a *Hermite majorant* of $f(\mathbf{x})$. Clearly $g(\mathbf{x})$ is positive definite and the determinants $d(f)$, $d(g)$ satisfy

$$d(g) = |d(f)|. \tag{2.3}$$

Further,

$$|f(\mathbf{a})| \leqslant g(\mathbf{a}) \tag{2.4}$$

for any $\mathbf{a} \in \mathbf{R}^n$.

We shall say that $f(\mathbf{x})$ is *Hermite-reduced* (or just *reduced*)† if it possesses a Hermite majorant $g(\mathbf{x})$ which is Minkowski-reduced (in the sense of the previous chapter). This definition is slightly unhistorical since Hermite actually worked with a different definition of reduction for definite forms.

LEMMA 2.1. *Any regular form $f(\mathbf{x})$ is integrally equivalent to a reduced form.*

Proof. Let $g(\mathbf{x})$ be any Hermite majorant of $f(\mathbf{x})$. By Theorem 1.1 of Chapter 12 there is an integral unimodular transformation $\mathbf{T}$ such that $g(\mathbf{Tx})$ is Minkowski-reduced. Clearly $g(\mathbf{Tx})$ is a Hermite majorant of $f(\mathbf{Tx})$, so $f(\mathbf{Tx})$ is Hermite reduced.

THEOREM 2.1. *There are only finitely many reduced integral anisotropic forms f of given determinant $d \neq 0$.*

Note. We shall show later (Theorem 11.1) that the enunciation remains true if the word "anisotropic" is omitted.

Proof. Since f is integral and anisotropic we have

$$|f(\mathbf{a})| \geqslant 1 \qquad (\mathbf{a} \in \mathbf{Z}^n, \mathbf{a} \neq \mathbf{0}). \tag{2.5}$$

Hence by (2.4) we have

$$g(\mathbf{a}) \geqslant 1 \qquad (\mathbf{a} \in \mathbf{Z}^n, \mathbf{a} \neq \mathbf{0}) \tag{2.6}$$

for any Hermite majorant g. It follows that the successive minima $M_1, M_2, \ldots, M_n$ of g (in the sense of Chapter 12, Section 2) satisfy

$$1 \leqslant M_1 \leqslant M_2 \leqslant \ldots \leqslant M_n. \tag{2.7}$$

But now

$$M_1 \ldots M_n \leqslant C d(g)$$

by Theorem 2.2 of Chapter 12 where C depends only on n. Further, $d(g) = |d(f)| = |d|$, where d is specified in the enunciation, and so

$$M_j \leqslant C|d| \qquad (1 \leqslant j \leqslant n).$$

By hypothesis, we may suppose that g is Minkowski-reduced and so by

† Cf. footnote on p. 285.

Lemma 1.1 and Theorem 3.1 of Chapter 12 the coefficients g_{ij} of g satisfy

$$|g_{ij}| \leqslant C'|d|$$

where C' depends only on d. Finally, by (2.4), the coefficients f_{ij} of f are given by

$$2f_{ij} = f(\mathbf{e}_i + \mathbf{e}_j) - f(\mathbf{e}_i) - f(\mathbf{e}_j)$$

where $\mathbf{e}_1, \ldots, \mathbf{e}_n$ are the unit vectors, and so

$$\begin{aligned} |2f_{ij}| &\leqslant |f(\mathbf{e}_i + \mathbf{e}_j)| + |f(\mathbf{e}_i)| + |f(\mathbf{e}_j)| \\ &\leqslant g(\mathbf{e}_i + \mathbf{e}_j) + g(\mathbf{e}_i) + g(\mathbf{e}_j) \\ &\leqslant C''|d|, \end{aligned}$$

where C'' depends only on n. But $f_{ij} \in \mathbf{Z}$, so there are only a finite number of possibilities for each f_{ij} and hence also for f.

For later use we record

COROLLARY 1. *The coefficients of both f and g are bounded by $C''|d|$, where C'' depends only on n.*

More important is

COROLLARY 2. *Let $f(\mathbf{x})$ be a regular anisotropic quadratic form with rational coefficients. Then there is at least one and at most a finite number of Hermite-reduced forms which are integrally equivalent to f.*

Proof. (i) "At least one". Follows from Lemma 2.1.

(ii) "At most finitely many". On multiplying f by a suitable integer we may suppose that f is an integral form. The required result follows from Theorem 2.1 since integrally equivalent forms have the same determinant.

THEOREM 2.2. *Suppose that f is integral, regular and anisotropic. Then there is a constant k depending only on f such that for every real automorph $\mathbf{S}$ of f there are integral automorphs $\mathbf{T}, \mathbf{W}$ such that all the elements of $\mathbf{TS}^{-1}$ and $\mathbf{S}^{-1}\mathbf{W}$ are $\leqslant k$ in absolute value.*

Note. The conclusions of the theorem are false for *every* isotropic f. For example x_1x_2 has the real automorphs $\mathbf{S}: x_1 \to sx_1, x_2 \to s^{-1}x_2$ for every $s \in \mathbf{R}^*$ but the group of integral automorphs is finite. There is, however, a weaker statement which is valid also for isotropic forms; see Theorem 11.2.

Proof. We note first that it is enough to prove the existence of $\mathbf{W}$ since $(\mathbf{TS}^{-1})^{-1} = \mathbf{ST}^{-1}$ and so if $\mathbf{W}_1$ is constructed for $\mathbf{S}^{-1}$ instead of $\mathbf{S}$ then it is enough to take $\mathbf{T} = \mathbf{W}_1^{-1}$.

Let $g(\mathbf{x})$ be a Hermite majorant of f. Then $g(\mathbf{Sx})$ is also a Hermite majorant of f for any real automorph $\mathbf{S}$ of f. Let $\mathbf{U} = \mathbf{U}(\mathbf{S})$ be an integral unimodular matrix such that $g(\mathbf{SUx})$ is Minkowski-reduced. Then

$$f(\mathbf{SUx}) = f(\mathbf{Ux})$$

is Hermite-reduced and integrally equivalent to f. By Theorem 2.1, Corollary 2, $f(\mathbf{Ux})$ is one of a finite set of forms, say

$$f_j(\mathbf{x}) \qquad (1 \leqslant j \leqslant J).$$

For each j $(1 \leqslant j \leqslant J)$ pick once and for all a real automorph $\mathbf{S}_j$ of f and an integral unimodular matrix $\mathbf{U}_j$ such that

$$g_j(\mathbf{x}) = g(\mathbf{S}_j\mathbf{U}_j\mathbf{x})$$

is Minkowski-reduced and such that

$$f(\mathbf{S}_j\mathbf{U}_j\mathbf{x}) = f(\mathbf{U}_j\mathbf{x}) = f_j(\mathbf{x}).$$

Now let $\mathbf{S}$ be again any real automorph of f and let $\mathbf{U} = \mathbf{U}(\mathbf{S})$ be as at the beginning of the proof. Then $f(\mathbf{Ux})$ must be $f_j(\mathbf{x})$ for some j: that is

$$f(\mathbf{Ux}) = f_j(\mathbf{x}) = f(\mathbf{U}_j\mathbf{x}).$$

Hence

$$\mathbf{W} = \mathbf{U}\mathbf{U}_j^{-1}$$

is an automorph of f.

The form

$$h(\mathbf{x}) = g(\mathbf{SUx})$$

is Minkowski-reduced and so by Theorem 2.1, Corollary 1, the coefficients of h are bounded by a constant K depending only on n and the determinant of f. On the other hand,

$$h(\mathbf{x}) = g_j(\mathbf{Vx})$$

where

$$\mathbf{V} = (\mathbf{S}_j\mathbf{U}_j)^{-1}\,\mathbf{SU}.$$

By Lemma 5.2 of Chapter 12 applied to $\mathbf{x} = \mathbf{Ve}_l$ ($\mathbf{e}_1, \ldots, \mathbf{e}_n$ being the unit vectors) the elements of the real matrix $\mathbf{V}$ are bounded by a constant depending only on K and g_j. But there are only a finite number of j and so the elements of $\mathbf{V}$ are bounded by a constant independent of $\mathbf{S}$.

Finally,

$$\begin{aligned}\mathbf{SW} &= \mathbf{SUU}_j^{-1}\\ &= \mathbf{S}_j\mathbf{U}_j\mathbf{V}\mathbf{U}_j^{-1}\end{aligned}$$

has elements bounded by a constant independent of $\mathbf{S}$. On replacing $\mathbf{S}$ by $\mathbf{S}^{-1}$ we have the conclusion of the Theorem.

COROLLARY. *There is a constant* $C = C(f)$ *depending only on* f *with the following property: let* $\mathbf{b} \in \mathbf{R}^n$ *have* $f(\mathbf{b}) \neq 0$. *Then there is an integral automorph* $\mathbf{T}$ *of* f *such that*

$$\|\mathbf{Tb}\| \leqslant C|f(\mathbf{b})|^{\frac{1}{2}}.$$

Proof. Here we are using the notation $\|\mathbf{x}\| = \max|x_j|$. Without loss of generality suppose that $f(\mathbf{b}) > 0$ and let $\mathbf{a}$ be any fixed vector with $f(\mathbf{a}) > 0$. By Theorem 4.1 of Chapter 2, there is a real automorph $\mathbf{S}$ of f such that $\mathbf{Sb} = \lambda\mathbf{a}$, where

$$\lambda > 0,\ \lambda^2 = f(\mathbf{b})/f(\mathbf{a}).$$

Let $\mathbf{T}$ be the integral automorph given by the Theorem. Then

$$\begin{aligned}\mathbf{Tb} &= (\mathbf{TS}^{-1})(\mathbf{Sb})\\ &= \lambda(\mathbf{TS}^{-1})\,\mathbf{a},\end{aligned}$$

where the elements of $(\mathbf{TS}^{-1})\,\mathbf{a}$ are bounded.

3. BINARY FORMS

In this section we study the automorphs of indefinite integral binary forms and integral representation by binary forms. We shall use the theory of Hermite reduction in the proofs. There is an older theory of reduction of binary forms which we could have used but it does not generalize naturally to $n > 2$: we discuss it briefly at the end of the section.

We must first consider real automorphs.

LEMMA 3.1. *The group* $O_{\mathbf{R}}^{+}(f)$ *of real proper automorphs of an indefinite binary form is isomorphic to the multiplicative group* $\mathbf{R}^*$ *of the non-zero reals. Improper real automorphs exist and they all have order* 2.

Proof. For f is real-equivalent to

$$f_0(x, y) = xy. \tag{3.1}$$

Clearly the proper automorphs are

$$\mathbf{S}\colon\ x \to sx,\ y \to s^{-1}y \qquad (s \in \mathbf{R}^*) \tag{3.2}$$

and the improper automorphs are

$$\mathbf{S}\colon x \to sy,\ y \to s^{-1}x \qquad (s \in \mathbf{R}^*). \tag{3.3}$$

The truth of the Lemma is now evident.

To avoid a proliferation of indices we shall put

$$\mathbf{x} = (x, y) \tag{3.4}$$

and write our form in the shape

$$f(x, y) = ax^2 + 2bxy + cy^2 \tag{3.5}$$

with the matrix

$$\mathbf{F} = \begin{pmatrix} a & b \\ b & c \end{pmatrix} \tag{3.6}$$

and determinant

$$d = d(f) = ac - b^2. \tag{3.7}$$

We shall suppose that f is classically integral, that is

$$a, b, c \in \mathbf{Z} \tag{3.8}$$

and *primitive:*

$$\text{g.c.d.}(a, b, c) = 1. \tag{3.9}$$

We shall have to distinguish between *properly primitive* forms, i.e. those with

$$\text{g.c.d.}(a, 2b, c) = 1 \tag{3.10}$$

and *improperly primitive* ones: i.e. those for which $\frac{1}{2}f(x, y) \in \mathbf{Z}[x, y]$ but $\frac{1}{2}f(x, y)$ is not classically integral.

We dispose first of the isotropic forms.

LEMMA 3.2. *Let f be an isotropic binary integral form.*

(i) *The only proper integral automorphs* $\mathbf{T}$ *of f are* $\pm\mathbf{I}$.

(ii) *There are only a finite number of integral representations $f(a, b) = e$ of a given integer $e \neq 0$.*

Proof. (i) We have

$$f(x, y) = kU(x, y)\, V(x, y) \tag{3.11}$$

where $k \in \mathbf{Z}$ and $U(x, y)$, $V(x, y)$ are primitive integral linear forms. Since $\mathbf{T}$ is proper,

$$\left.\begin{aligned} U(\mathbf{Tx}) &= tU(\mathbf{x}) \\ V(\mathbf{Tx}) &= t^{-1}\, V(\mathbf{x}). \end{aligned}\right\} \tag{3.12}$$

Here $t \in \mathbf{Z}$ since $\mathbf{T}$ is integral and U is primitive. Similarly $t^{-1} \in \mathbf{Z}$; and so $t = \pm 1$.

(ii) If,

$$f(a, b) = e$$

then

$$U(a, b) = u, \ V(a, b) = v$$

where $u, v \in \mathbf{Z}$ and $kuv = e$. There are only finitely many choices for the pair (u, v) and each such pair determines at most one pair (a, b) of integers.

We now consider anisotropic forms. Their proper automorphs are related to the solutions (t, u) of "*Pell's equation*"

$$t^2 + du^2 = 1. \tag{3.13}$$

THEOREM 3.1. *The group $O^+(f)$ of proper integral automorphs of a primitive indefinite anisotropic binary form consists of the $\mathbf{T} = \pm \mathbf{T}_0^v$ $(v \in \mathbf{Z})$ where $\mathbf{T}_0$ is of infinite order. The $\mathbf{T} \in O^+(f)$ are the matrices*

$$\mathbf{T} = \begin{pmatrix} t - bu & -cu \\ au & t + bu \end{pmatrix}, \tag{3.14}$$

where (3.13) *holds and*

$$t, u \in \mathbf{Z} \quad \textit{if } f \textit{ is properly primitive} \tag{3.15$_1$}$$

$$2t, 2u, t - u \in \mathbf{Z} \quad \textit{if } f \textit{ is improperly primitive}. \tag{3.15$_2$}$$

Proof. $O^+(f)$ contains $\pm\mathbf{I}$ and is a subgroup of $O^+_{\mathbf{R}}(f)$, which is isomorphic to $\mathbf{R}^*$ by Lemma 3.1. By Theorem 2.2, $O^+(f)$ cannot consist simply of $\pm\mathbf{I}$. On the other hand, the image of $O^+(f)$ in $O^+_{\mathbf{R}}(f) \simeq \mathbf{R}^*$ is clearly discrete. The only discrete subgroups of $\mathbf{R}^*$ containing but strictly greater than $\{\pm 1\}$ are $\{\pm \eta_0^v, v \in \mathbf{Z}\}$ for some $\eta_0 \neq \pm 1$. If $\mathbf{T}_0$ is the automorph corresponding to η_0, then $\mathbf{T}_0$ has the properties of the first sentence of the enunciation.

It remains to obtain (3.14). Let

$$\mathbf{T} = \begin{pmatrix} l & m \\ n & r \end{pmatrix} \tag{3.16}$$

be a proper integral automorph: that is

$$\mathbf{T}'\mathbf{F}\mathbf{T} = \mathbf{F} \tag{3.17}$$

and

$$\det \mathbf{T} = 1. \tag{3.18}$$

Then

$$\mathbf{F}\mathbf{T} = (\mathbf{T}')^{-1}\,\mathbf{F}, \tag{3.19}$$

where

$$(\mathbf{T}')^{-1} = \begin{pmatrix} r & -n \\ -m & l \end{pmatrix} \tag{3.20}$$

by (3.18). The equation (3.19) is linear in l, m, n, r. On equating elements on both sides, using (3.6), we have

$$a(r - l) = 2nb,$$
$$c(l - r) = 2mb,$$
$$am + cn = 0.$$

This implies that (3.14) holds for some t, u which, in the first place, need only be rational. The condition that $\mathbf{T}$ be integral then readily gives (3.15). Finally, (3.13) follows from (3.14) and (3.18). Conversely, (3.14) subject to (3.15) gives a $\mathbf{T} \in O^+(f)$. This concludes the proof.

There is another description of $O^+(f)$. Put

$$D = |d| = -d. \tag{3.21}$$

Then

$$\begin{aligned} af(x, y) &= (ax + by)^2 - Dy^2 \\ &= (ax + by + y\sqrt{D})(ax + by - y\sqrt{D}) \\ &= L(x, y)\, M(x, y) \text{ (say).} \end{aligned} \tag{3.22}$$

It is then easy to verify that

$$L(\mathbf{Tx}) = \eta L(\mathbf{x}),\ M(\mathbf{Tx}) = \eta^{-1} M(\mathbf{x}), \tag{3.23}$$

where

$$\eta = t + u\sqrt{D}. \tag{3.24}$$

If $\mathbf{T} = \pm \mathbf{T}_0^v$ then $\eta = \pm \eta_0^v$ in an obvious notation. On replacing η_0 by $\pm \eta_0^{\pm 1}$ we may suppose that $\eta_0 > 1$. Then the corresponding solution t_0, u_0 of Pell's equation is called the *fundamental solution*.

For some values of d there may be a solution of the equation

$$t^2 + du^2 = -1 \tag{3.25}$$

which is like Pell's equation (3.13) except for the -1 on the right-hand side. There is no simple criterion for deciding when (3.25) has a solution: the question will be discussed in Chapter 14. We have the

COROLLARY. *Suppose that* t, u *is a solution of* (3.25) *satisfying the integrality conditions* (3.15). *Then* $\mathbf{T}$ *given by* (3.14) *is an improper equivalence of* f *with* $-f$. *Suppose, further, that* $t_1 > 0, u_1 > 0$, *is the solution of* (3.25) *for which* $\eta_1 = t_1 + u_1\sqrt{D}$ *is smallest. Then* $\eta_0 = \eta_1^2$.

Proof. Clear.

Note. When it exists, η_1 rather than η_0 is sometimes called the fundamental solution.

We now have a more explicit version of Theorem 2.2, Corollary for $n = 2$.

LEMMA 3.3. *Let* $\mathbf{b} \neq \mathbf{0}$. *Then there is a* $\mathbf{T} \in O^+(f)$ *such that*

$$|L(\mathbf{Tb})| \leqslant |\eta_0 f(\mathbf{b})|^{\frac{1}{2}}$$

$$|M(\mathbf{Tb})| \leqslant |\eta_0 f(\mathbf{b})|^{\frac{1}{2}}$$

where η_0 *is given by* (3.24) *with a fundamental solution* (t_0, u_0) *and where* L, M *are defined by* (3.22).

Proof. We define $v \in \mathbf{Z}$ by

$$\eta_0^v |L(\mathbf{b})| \leqslant |\eta_0 f(\mathbf{b})|^{\frac{1}{2}} < \eta_0^{v+1} |L(\mathbf{b})|.$$

Then by (3.23)

$$\eta_0^{-\frac{1}{2}} |f(\mathbf{b})|^{\frac{1}{2}} < |L(\mathbf{T}_0^v \mathbf{b})| \leqslant |\eta_0 f(\mathbf{b})|^{\frac{1}{2}}.$$

But $L(\mathbf{T}_0^v \mathbf{b})\, M(\mathbf{T}_0^v \mathbf{b}) = f(\mathbf{b})$ and so

$$\eta_0^{-\frac{1}{2}} |f(\mathbf{b})|^{\frac{1}{2}} \leqslant |M(\mathbf{T}_0^v \mathbf{b})| < |\eta_0 f(\mathbf{b})|^{\frac{1}{2}}.$$

Now put $\mathbf{T} = \mathbf{T}_0^v$.

COROLLARY 1. *One can determine effectively representatives of all the orbits of representations of a given integer $b \neq 0$ by f.*

Proof. Clear.

COROLLARY 2. *One can determine effectively whether two integral binary forms are equivalent or not.*

Proof. We consider only when f, g are indefinite anisotropic as remaining cases are trivial. Let $f(\mathbf{x}), g(\mathbf{x})$ be classically integral forms. We have to find whether there is a basis $\mathbf{b}, \mathbf{c}$ of $\mathbf{Z}^2$ such that $f(x\mathbf{b} + y\mathbf{c}) = g(x, y)$. If $\mathbf{b}, \mathbf{c}$ have this property then so have $\mathbf{Tb}, \mathbf{Tc}$ for any $\mathbf{T} \in O(f)$. Hence without loss of generality $\mathbf{b}$ is one of the representatives given by the preceding Corollary. For each possible $\mathbf{b}$ it is straightforward to determine whether or not a suitable $\mathbf{c}$ exists.

Example. There are no integer solutions a, b of $a^2 - 82b^2 = 2$. This equation is soluble in $\mathbf{Z}_p$ for all p, so this shows that there is no "local to global" principle for integral representations. Here $f(x, y) = x^2 - 82y^2$ and there is the solution $t_1 = 9$, $u_1 = 1$ of $t_1^2 - 82u_1^2 = -1$. By Theorem 3.1., Corollary, we may use η_1 instead of η_0 in Lemma 3.3 provided that we look at the representations of $+2$ and -2 together. Thus if there is a solution of $f(a, b) = \pm 2$ at all, then there is one with

$$|a + b\sqrt{82}| \leqslant (2\eta_1)^{\frac{1}{2}}$$
$$|a - b\sqrt{82}| \leqslant (2\eta_1)^{\frac{1}{2}}$$

and so

$$|b|\sqrt{82} \leqslant (2\eta_1)^{\frac{1}{2}} = \{2(9 + \sqrt{82})\}^{\frac{1}{2}}.$$

This implies $b = 0$, a contradiction.

It follows that $x^2 - 82y^2$ and $2x^2 - 41y^2$ are not equivalent, although they are in the same genus.

The question whether a given binary form has an improper automorph will be of central importance in Chapter 14.

THEOREM 3.2. *A necessary and sufficient condition for a regular real binary form f to have an improper integral automorph is that it be integrally equivalent to a form*

$$f'(x, y) = a'x^2 + 2b'xy + c'y^2 \tag{3.26}$$

with either

$$b' = 0 \tag{3.27}$$

or

$$2b' = a'. \tag{3.28}$$

Note. The theorem applies both to definite and to indefinite forms, both to isotropic and anisotropic forms. In Chapter 14 we shall be interested only in integral forms.

We require

LEMMA 3.4. *Let* $\mathbf{T}$ *be an integral* 2×2 *matrix with*

$$\mathbf{T}^2 = \mathbf{I}, \qquad \det \mathbf{T} = -1. \tag{3.29}$$

Then there is an integral matrix $\mathbf{S}$ *with* $\det \mathbf{S} = +1$ *such that*

$$\mathbf{S}^{-1}\mathbf{TS} = \begin{pmatrix} 1 & w \\ 0 & -1 \end{pmatrix}, \tag{3.30}$$

where

$$w = 0 \quad or \quad 1. \tag{3.31}$$

Proof. By (3.29) the eigenvalues of $\mathbf{T}$ are ± 1. Hence there is an integral vector $\mathbf{a}$ with

$$\mathbf{Ta} = \mathbf{a} \tag{3.32}$$

and without loss of generality $\mathbf{a}$ is primitive. Let $\mathbf{b}$ be a vector such that $\mathbf{a}, \mathbf{b}$ is a basis of $\mathbf{Z}^2$ and $\det(\mathbf{a}, \mathbf{b}) = +1$. Then $\mathbf{a}, \mathbf{Tb}$ is also a basis of $\mathbf{Z}^2$ and since $\det \mathbf{T} = -1$ we have

$$\mathbf{Tb} = w\mathbf{a} - \mathbf{b} \tag{3.33}$$

for some $w \in \mathbf{Z}$. If we replace $\mathbf{b}$ by $v\mathbf{a} + \mathbf{b}$ ($v \in \mathbf{Z}$) we replace w by $w + 2v$. Hence we have (3.31) by suitable choice of v. If we now take for $\mathbf{S}$ the matrix transferring to the basis $\mathbf{a}$, $\mathbf{b}$, then we have what is required.

Proof of Theorem 3.2. Let $\mathbf{T}$ be an improper integral automorph of f and let $\mathbf{S}$ be given by Lemma 3.3. Then $f'(\mathbf{x}) = f(\mathbf{S}^{-1}\mathbf{x})$ has an automorph given by the right-hand side of (3.30). It is then readily verified that (3.27) or (3.28) holds according as $w = 0$ or $w = 1$. Conversely, (3.26) subject to (3.27) or (3.28) has an obvious improper integral automorph.

There is a classical theory of the reduction of indefinite binary forms which is different from Hermite's. It is closely related to the theory of continued fractions and does not generalize naturally to forms in $n > 2$ variables. We sketch the classical theory briefly here. The rest of this section will not be referred to again and the reader may prefer to omit it and go on directly to Section 4.

To avoid minor complications we consider real indefinite binary forms f which are anisotropic over $\mathbf{Q}$. Such a form is called *classically reduced* if it is of the shape

$$f(x, y) = a(x + \theta y)(x - \phi y) \tag{3.34}$$

where

$$0 < \phi < 1 < \theta. \tag{3.35}$$

Clearly (3.35) is equivalent to the conditions

$$\left.\begin{aligned} f(1,0)\,f(0,1) &< 0 \\ f(1,0)\,f(1,1) &> 0 \\ f(1,0)\,f(-1,1) &< 0 \end{aligned}\right\}. \tag{3.36}$$

LEMMA 3.5. *There are only finitely many classically integral classically reduced binary forms of given determinant* $d < 0$.

Proof. Let

$$f(x, y) = ax^2 + 2bxy + cy^2 \tag{3.37}$$

be such a form. Then

$$-4d = a^2(\theta + \phi)^2. \tag{3.38}$$

Since $\theta + \phi > 1$ by (3.25), this gives only finitely many possibilities for $a \in \mathbf{Z}$. But now, since θ and ϕ are positive by (3.35), we have

$$\begin{aligned} |2b| &= |a(\theta - \phi)| \\ &\leqslant |a(\theta + \phi)| \\ &= 2|d|^{\frac{1}{2}} \end{aligned} \tag{3.39}$$

by (3.38). Hence there are only finitely many possibilities for $b \in \mathbf{Z}$. Finally, c is uniquely determined by $ac - b^2 = d$.

LEMMA 3.6. *Every real indefinite* $\mathbf{Q}$*-anisotropic binary form* f *is integrally equivalent to a classically reduced form.*

Proof (sketch). We have

$$f(x, y) = U(x, y)\,V(x, y) \tag{3.40}$$

for some (not uniquely determined) real linear forms U, V. By hypothesis $U(a, b) = 0$ for $a, b \in \mathbf{Q}$ implies $a = b = 0$, and similarly for V. Further,

$$\begin{aligned} |\det(U, V)| &= \tfrac{1}{2}|d|^{\frac{1}{2}} \\ &= \delta \text{ (say)}. \end{aligned} \tag{3.41}$$

The set of points $\boldsymbol{\xi} = (\xi, \eta)$ with

$$\xi = U(a,b), \quad \eta = V(a,b) \qquad (a, b \in \mathbf{Z}) \tag{3.42}$$

form a lattice Λ in $\mathbf{R}^2$ in the sense of the Geometry of Numbers. Clearly we can find (in infinitely many ways) a $\boldsymbol{\xi}_0 = (\xi_0, \eta_0) \in \Lambda$ such that there are no points of Λ other than $\mathbf{0} = (0,0)$ in

$$|\xi| < |\xi_0|, \qquad |\eta| < |\eta_0|. \tag{3.43}$$

Amongst the $(\xi, \eta) \in \Lambda$ with $|\eta| < |\eta_0|$ there must be one with smallest ξ. Let it be $\boldsymbol{\xi}_1 = (\xi_1, \eta_1)$. Then

$$|\eta_1| < |\eta_0|, \qquad |\xi_0| < |\xi_1| \tag{3.44}$$

and there is no point of Λ other than $\mathbf{0}$ in

$$|\eta| < |\eta_0|, \qquad |\xi| < |\xi_1|. \tag{3.45}$$

On replacing $\boldsymbol{\xi}_j$ by $\pm\boldsymbol{\xi}_j$ we may suppose without loss of generality that

$$0 < \xi_0 < \xi_1. \tag{3.46}$$

Then

$$\eta_0 \eta_1 < 0, \tag{3.47}$$

since otherwise $\boldsymbol{\xi}_1 - \boldsymbol{\xi}_0$ would lie in (3.45).

By (3.41) the fact that the only $\boldsymbol{\xi} \in \Lambda$ in (3.45) is $\boldsymbol{\xi} = \mathbf{0}$ implies by Minkowski's Linear Forms Theorem (Theorem 2.4 of Chapter 5) that

$$\xi_1 |\eta_0| \leqslant \delta.$$

Hence

$$\begin{aligned} |\det(\boldsymbol{\xi}_0, \boldsymbol{\xi}_1)| &\equiv \xi_1|\eta_0| + \xi_0|\eta_1| \\ &< 2\xi_1|\eta_0| \\ &\leqslant 2\delta. \end{aligned}$$

By hypothesis

$$\boldsymbol{\xi}_j = (U(\mathbf{a}_j), V(\mathbf{a}_j)) \qquad (j = 0, 1)$$

and so

$$0 < |\det(\mathbf{a}_0, \mathbf{a}_1)| < 2;$$

that is

$$\det(\mathbf{a}_0, \mathbf{a}_1) = \pm 1.$$

Hence $\mathbf{a}_0, \mathbf{a}_1$ is a basis for $\mathbf{Z}^2$. It follows that f is integrally equivalent to the form

$$\begin{aligned} f_0(x,y) &= (x\xi_0 + y\xi_1)(x\eta_0 + y\eta_1) \\ &= a_0(x + \theta_0 y)(x - \phi_0 y) \end{aligned}$$

where

$$a_0 = \xi_0 \eta_0$$

and

$$0 < \phi_0 = -\eta_1/\eta_0 < 1 < \theta_0 = \xi_1/\xi_0.$$

This concludes the proof of the Lemma. We now pursue the argument a little further. On taking $\boldsymbol{\xi}_1$ instead of $\boldsymbol{\xi}_0$ in the construction, we find a $\boldsymbol{\xi}_2 \in \Lambda$ such that the pair $\boldsymbol{\xi}_1, \boldsymbol{\xi}_2$ enjoys similar properties to the pair $\boldsymbol{\xi}_0, \boldsymbol{\xi}_1$. An easy argument shows that there is a pair $\boldsymbol{\xi}_{-1}, \boldsymbol{\xi}_0$ which has similar properties. In this way we get an infinite sequence of points of Λ:

$$\boldsymbol{\xi}_m = (\xi_m, \eta_m) \qquad (-\infty < m < \infty)$$

such that

$$0 < \xi_m < \xi_{m+1} \tag{3.48}$$

and

$$\eta_m \eta_{m+1} < 0, \qquad |\eta_{m+1}| < |\eta_m|.$$

To each m there corresponds a classically reduced form

$$\begin{aligned} f_m(x, y) &= (x\xi_m + y\xi_{m+1})(x\eta_m + y\eta_{m+1}) \\ &= a_m(x + \theta_m y)(x - \phi_m y). \end{aligned} \tag{3.49}$$

Since $(\boldsymbol{\xi}_{m-1}, \boldsymbol{\xi}_m)$ and $(\boldsymbol{\xi}_m, \boldsymbol{\xi}_{m+1})$ are bases of Λ and

$$\det(\boldsymbol{\xi}_m, \boldsymbol{\xi}_{m+1}) = -\det(\boldsymbol{\xi}_m, \boldsymbol{\xi}_{m-1}) \tag{3.50}$$

(as is easy to see), there is a $w_m \in \mathbf{Z}$ such that

$$\boldsymbol{\xi}_{m+1} - \boldsymbol{\xi}_{m-1} = w_m \boldsymbol{\xi}_m.$$

Further,

$$w_m > 0$$

by (3.48). It follows that

$$\theta_m = \frac{\xi_{m+1}}{\xi_m} = w_m + \frac{1}{\theta_{m-1}}.$$

Recursion gives an expression for θ_m as a continued fraction involving the w_j $(j \leqslant m)$. Similarly ϕ_m can be expressed as a continued fraction in the w_j $(j > m)$. We do not follow this line of thought further here. [See e.g. Cassels (1957)].

COROLLARY. *Every classically reduced form which is properly equivalent to f occurs as one of the f_m.*

Proof (sketch). For such a form corresponds to a pair of points $\boldsymbol{\xi}_0^*, \boldsymbol{\xi}_1^*$ of Λ which satisfy the analogues of (3.44)–(3.47). A simple geometric argument then shows that $\boldsymbol{\xi}_0^* = \boldsymbol{\xi}_m, \boldsymbol{\xi}_1^* = \boldsymbol{\xi}_{m+1}$ for some integer m.

Now suppose that the initial form $f(x, y)$ is integral as well as anisotropic. Then the infinite sequence of forms $f_m(x, y)$ given by (3.49) can contain only finitely many distinct forms by Lemma 3.5. The existence of automorphs other than $\pm\mathbf{I}$ follows immediately. More precisely we have

LEMMA 3.7. *Let $f(x, y)$ be an indefinite anisotropic integral form. Then the sequence w_m constructed above is periodic. Let $s > 0$ be the least even period and let $\mathbf{T}_0$ be determined by*

$$\mathbf{T}_0\xi_0 = \xi_s, \qquad \mathbf{T}_0\xi_1 = \xi_{s+1}. \tag{3.51}$$

Then the group of proper integral automorphs is generated by $\mathbf{T}_0$ and $-\mathbf{I}$.

Further, f is improperly equivalent to $-f$ precisely when the sequence w_m has an odd period.

Proof (sketch). Suppose that

$$f_{m+r} = \pm f_m \tag{3.52}$$

for some integers m and $r > 0$. Then from the construction it is clear that $w_n = w_{n+r}$ for all integers n: that is, r is a period. Conversely, if $w_n = w_{n+r}$ for all integers n, then $\theta_m = \theta_{m+r}, \phi_m = \phi_{m+r}$ for all integers m and so by (3.49) we have (3.51) on comparing determinants. Since $f_m(1, 0)\, f_{m+1}(1, 0) < 0$, the sign in (3.52) must be $(-1)^r$. Since the forms f_{m+1}, f_m are improperly equivalent by (3.50), the final statement of the enunciation follows. Similarly, if r is even, the transformation taking ξ_0, ξ_1 into ξ_r, ξ_{r+1} is a proper automorph of f. The rest of the proof follows that of Lemma 3.6, Corollary, and is left to the reader.

COROLLARY. *Suppose that f is primitively integral and has determinant d. Then Pell's equation with -1 [display (3.25)] has a solution precisely when the sequence w_m has an odd period.*

Proof. See Theorem 3.1, Corollary.

4. CONSTRUCTION OF AUTOMORPHS

Before embarking on the systematic theory we discuss special devices for constructing integral automorphs of integral quadratic forms.

We have seen (Chapter 2, Section 4) that the group of rational automorphs of a form f is generated by the symmetries

$$\tau_{\mathbf{a}}: \mathbf{x} \to \mathbf{x} - \frac{2f(\mathbf{a}, \mathbf{x})}{f(\mathbf{a})}\mathbf{x}, \tag{4.1}$$

where $\mathbf{a}$ is a rational vector with $f(\mathbf{a}) \neq 0$. The symmetry $\tau_{\mathbf{a}}$ is integral provided that the form f is classically integral and $\mathbf{a}$ is integral, $f(\mathbf{a}) = \pm 1$ or ± 2. Cases of this construction are given in Chapter 9, examples 15–18. We also recall that the spinor group gives an approach to the construction of elements of $O(f)$ (Chapter 10, Section 4).

Another possibility is exemplified by

$$f_0(\mathbf{x}) = ax^2 - by^2 - cz^2, \tag{4.2}$$

where a, b, c are positive integers. If ab is not a perfect square, the binary form $ax^2 - by^2$ has an infinite group of automorphs and these extend to automorphs of f_0 by keeping z fixed. Similarly, if ac is not a perfect square, there is an infinite group of automorphs keeping y fixed. If neither ab nor ac is a perfect square, these two sets of automorphs generate a nonabelian subgroup of $O(f_0)$. If we are lucky, then we have the whole of $O(f_0)$ but in any case we have made substantial progress towards finding it (cf example at end of Section 6).

The last example suggests a study of automorphs $\mathbf{T}$ of a form f which fix a given integral vector $\mathbf{b}$. Without loss of generality $\mathbf{b}$ is primitive and so it is sufficient to consider the case when

$$\mathbf{T}\mathbf{e}_1 = \mathbf{e}_1 \tag{4.3}$$

where, as usual, $\mathbf{e}_1, \ldots, \mathbf{e}_n$ are the unit vectors.† The cases $f(\mathbf{e}_1) \neq 0$ and $f(\mathbf{e}_1) = 0$ require separate treatment. The first case has, in effect, been considered in Chapter 9, Section 6, but we repeat the simple argument.

Suppose, first, that $f_{11} = f(\mathbf{e}_1) \neq 0$. Then

$$f_{11} f(\mathbf{x}) = (f_{11}x_1 + f_{12}x_2 + \ldots + f_{1n}x_n)^2 + g(x_2, \ldots, x_n) \tag{4.4}$$

for some integral form g. The condition (4.3) is equivalent to saying that the transformation $\mathbf{y} = \mathbf{T}\mathbf{x}$ has the shape

$$\left.\begin{aligned} y_1 &= x_1 + \sum_{2 \leqslant j \leqslant n} t_j x_j \\ (y_2, \ldots, y_n) &= \mathbf{S}(x_2, \ldots, x_n), \end{aligned}\right\} \tag{4.5}$$

where $\mathbf{S}$ is an automorph of g. Conversely an automorph $\mathbf{S}$ of g can be extended to an automorph (4.5) of f provided that $t_2, \ldots, t_n \in \mathbf{Z}$ can be found appropriately: and this can certainly be done whenever $\mathbf{S}$ is congruent to the unit matrix modulo f_{11}. We have thus proved

THEOREM 4.1. *Every* $\mathbf{T} \in O(f)$ *with* (4.3) *induces an* $\mathbf{S} \in O(g)$. *The set of such* $\mathbf{S}$ *is a subgroup of* $O(g)$ *of finite index, which includes all the* $\mathbf{S} \equiv \mathbf{I}$ (mod f_{11}).

When (4.3) holds and $f(\mathbf{e}_1) = 0$ it is convenient to simplify the shape of f by further unimodular transformations. We have

$$f(\mathbf{x}) = 2x_1(f_{12}x_2 + \ldots + f_{1n}x_n) + \ldots.$$

† When n is odd and $\mathbf{T}$ is proper there is always a $\mathbf{b}$ such that $\mathbf{T}\mathbf{b} = \mathbf{b}$ by Chapter 2, Example 9, and so we have (4.3) by suitable choice of basis.

By a suitable unimodular transformation on $x_2, \ldots, x_n$ we may suppose that $f_{13} = f_{14} = \ldots = f_{1n} = 0$ and then

$$f(\mathbf{x}) = 2x_2(f_{12}x_1 + f_{23}x_3 + \ldots + f_{2n}x_n) + \text{terms in } x_3, \ldots, x_n. \tag{4.6}$$

One could now proceed to "complete the square" with respect to the pair of variables x_1, x_2 and obtain an analogue of Theorem 4.1 but with a form g in $n - 2$ variables. There is however the further possibility:

THEOREM 4.2. *Let f be given by* (4.6) *and let $k_3, \ldots, k_n$ be any integers with $k_j \equiv 0 \pmod{2f_{12}}$. Then there are integers $l_2, \ldots, l_n$ such that the transformation*

$$\left.\begin{aligned} x_1 &\to x_1 + l_2x_2 + l_3x_3 + \ldots + l_nx_n \\ x_2 &\to x_2 \\ x_j &\to x_j + k_jx_2 \qquad (j > 2) \end{aligned}\right\} \tag{4.7}$$

is an automorph of f.

Proof. Clear.

Example. $2x_1x_2 + x_3^2$ has the automorphs

$$x_1 \to x_1 - 2k^2x_2 - 2kx_3,\ x_2 \to x_2,\ x_3 \to x_3 + kx_2.$$

It also has the automorphs obtained by interchanging the roles of $x_1 . x_2$: and together they generate a non-abelian group.

We note that the automorph (4.7) is "unipotent", that is all its eigenvalues are 1. In the general theory of algebraic groups the unipotent elements play a very special role (see e.g. Borel (1969)). The automorphism group of a quadratic form contains unipotent elements other than $\mathbf{I}$ only if the form is isotropic in >2 variables.

Finally, we note that there is a close relation between the integral automorphism groups of two quadratic forms which are rationally equivalent.

LEMMA 4.1. *Let $f(\mathbf{x})$ and $g(\mathbf{x}) = f(\mathbf{Sx})$ be quadratic forms in n variables where $\mathbf{S}$ is a non-singular $n \times n$ matrix with rational elements. If $\mathbf{T} \in O(f)$ and if $\mathbf{S}^{-1}\mathbf{TS} = \mathbf{U}$ (say) is integral then $\mathbf{U} \in O(g)$. The set of such $\mathbf{T}$ form a subgroup of finite index in $O(f)$ and the set of $\mathbf{U}$ is a subgroup of finite index in $O(g)$.*

Proof. $\mathbf{U}$ is certainly integral if $\mathbf{T} \equiv \mathbf{I} \bmod s_1s_2$ where s_1, s_2 are non-zero integers such that $s_1\mathbf{S}^{-1}$ and $s_2\mathbf{S}$ are integral matrices. This implies the statement that the set of $\mathbf{T}$ is of finite index. The rest is even more trivial.

5. ISOTROPIC TERNARY FORMS

We examine first the special case

$$f_0(\mathbf{x}) = x_1x_3 - x_2^2 \tag{5.1}$$

to which we shall later reduce the study of general isotropic ternary forms.

For (5.1) we invoke the theory of binary quadratic forms because $f_0(\mathbf{x})$ is just the determinant of the form

$$\phi_{\mathbf{x}}(\mathbf{u}) = x_1u_1^2 + 2x_2u_1u_2 + x_3u_2^2. \tag{5.2}$$

Denote by GL the group of matrices

$$\mathbf{S} = \begin{pmatrix} \alpha & \beta \\ \gamma & \delta \end{pmatrix} \tag{5.3}$$

where

$$\alpha, \beta, \gamma, \delta \in \mathbf{Q} \tag{5.4}$$

and

$$\lambda = \lambda(\mathbf{S}) = \alpha\delta - \beta\gamma \neq 0. \tag{5.5}$$

The determinant of $\phi_{\mathbf{x}}(\mathbf{Su})$ is λ^2 times the determinant of $\phi_{\mathbf{x}}(\mathbf{u})$. This implies that

$$\mathbf{T}(\mathbf{S}) = \lambda^{-1}\begin{pmatrix} \alpha^2 & 2\alpha\gamma & \gamma^2 \\ \alpha\beta & \alpha\delta + \beta\gamma & \gamma\delta \\ \beta^2 & 2\beta\delta & \delta^2 \end{pmatrix} \tag{5.6}$$

is an automorph of f_0. It can be verified directly that

$$\det(\mathbf{T}(\mathbf{S})) = +1. \tag{5.7}$$

LEMMA 5.1. *The map* $\mathbf{S} \to \mathbf{T}(\mathbf{S})$ *is a group homomorphism of* GL *onto* $O_{\mathbf{Q}}^+(f_0)$. *The kernel is the set of scalar multiples of* $\mathbf{I}$.

Note. This is the spin representation of $O_{\mathbf{Q}}^+$ under a light disguise, but we are proceeding independently of Chapters 10 and 11.

Proof. The fact that $\mathbf{S} \to \mathbf{T}(\mathbf{S})$ is a group homomorphism is clear from the way that $\mathbf{T}(\mathbf{S})$ was introduced: and the statement about the kernel is readily verified. It remains to show that every $\mathbf{T} \in O_{\mathbf{Q}}^+(f_0)$ can be put in the shape $\mathbf{T} = \mathbf{T}(\mathbf{S})$.

Let $\mathbf{T} = (t_{ij})$ be any element of $O_{\mathbf{Q}}^+(f_0)$. On comparing the coefficients of x_1^2 in the identity $f_0(\mathbf{Tx}) = f_0(\mathbf{x})$ we have

$$t_{21}^2 - t_{31}t_{11} = 0 \tag{5.8}$$

and so

$$t_{11} = \mu\alpha_1^2, \qquad t_{21} = \mu\alpha_1\beta_1, \qquad t_{31} = \mu\beta_1^2 \tag{5.9}$$

for some

$$\mu, \alpha_1, \beta_1 \in \mathbf{Q}. \tag{5.10}$$

Choose $\gamma_1, \delta_1 \in \mathbf{Q}$ such that $\alpha_1\delta_1 - \beta_1\gamma_1 \neq 0$ and let

$$\mathbf{S}_1 = \begin{pmatrix} \alpha_1 & \beta_1 \\ \gamma_1 & \delta_1 \end{pmatrix}. \tag{5.11}$$

Then the first columns of $\mathbf{T}(\mathbf{S}_1)$ and of $\mathbf{T}$ are proportional, and so on considering $\{\mathbf{T}(\mathbf{S}_1)\}^{-1}\mathbf{T}$ instead of $\mathbf{T}$ we may suppose without loss of generality that

$$t_{21} = t_{31} = 0. \tag{5.12}$$

On comparing the coefficients of x_1x_2 in $f_0(\mathbf{Tx}) = f_0(\mathbf{x})$ we now have

$$t_{32} = 0. \tag{5.13}$$

Hence

$$1 = \det \mathbf{T} = t_{11}t_{22}t_{33}, \tag{5.14}$$

so $t_{33} \neq 0$. On comparing coefficients of x_3^2 in $f_0(\mathbf{Tx}) = f_0(\mathbf{x})$ there is a $\gamma_2 \in \mathbf{Q}$ such that

$$t_{13} = t_{33}\gamma_2^2, \qquad t_{23} = t_{33}\gamma_2. \tag{5.15}$$

On considering $\{\mathbf{T}(\mathbf{S}_2)\}^{-1}\mathbf{T}$ instead of $\mathbf{T}$, where

$$\mathbf{S}_2 = \begin{pmatrix} 1 & 0 \\ \gamma_2 & 1 \end{pmatrix}, \tag{5.16}$$

we are reduced to considering the case when $\mathbf{T}$ is diagonal. And that is easy.

We now obtain the group $O^+(f_0)$ of integral automorphs. Denote by $SL^{\pm}$ the subgroup of GL consisting of the $\mathbf{S}$ given by (5.3) where

$$\alpha, \beta, \gamma, \delta \in \mathbf{Z} \tag{5.17}$$

and

$$\lambda(\mathbf{S}) = \alpha\delta - \beta\gamma = \pm 1. \tag{5.18}$$

LEMMA 5.2. *The map* $\mathbf{S} \to \mathbf{T}(\mathbf{S})$ *is a homomorphism of* $SL^{\pm}$ *onto* $O^+(f_0)$ *with kernel* $\pm\mathbf{I}$.

Proof. The proof follows that of the previous Lemma closely so we note only the additional details. In (5.9) we may suppose that α_1, β_1 are integers without common factor, so we can choose $\gamma_1, \delta_1 \in \mathbf{Z}$ such that $\alpha_1\delta_1 - \beta_1\gamma_1 = 1$. Then $\mathbf{S}_1 \in SL^{\pm}$ and we have the reduction to (5.12). Now (5.14) implies that $t_{33} = \pm 1$ and so γ_2 given by (5.15) is now in $\mathbf{Z}$. Hence $\mathbf{S}_2 \in SL^{\pm}$ and again we are reduced to the diagonal case: which is trivial.

Denote by SL^+ the subgroup of $SL^\pm$ consisting of the $\mathbf{S}$ with $\det \mathbf{S} = 1$ and let Θ be the image of SL^+ in $O^+(f_0)$. Then Θ is a subgroup of $O^+(f_0)$ of index 2. [It is the set of elements of $O^+(f_0)$ of spinor norm $+1$.]

LEMMA 5.3. *Let $m \in \mathbf{Z}, m \neq 0$. The orbits of the $\mathbf{b}$ with $f_0(\mathbf{b}) = m$ under the action of Θ correspond bijectively to the proper equivalence classes of classically integral binary forms of determinant m.*

Proof. Clear.

COROLLARY. *The orbits under Θ can be effectively determined.*

Proof. We saw in Section 3 that the proper equivalence classes of binary forms of given determinant can be uniquely determined.

LEMMA 5.4. *The orbits under $O^+(f_0)$ of the integral $\mathbf{b}$ with $f_0(\mathbf{b}) = m$, where m is given, can be effectively determined.*

Proof. Follows from Lemma 5.2 and Lemma 5.3, Corollary.

We now consider the general integral isotropic form $f(\mathbf{x})$. There is an integer $m \neq 0$ and a nonsingular 3×3 integral matrix $\mathbf{M}$ such that

$$mf(\mathbf{x}) = f_0(\mathbf{M}\mathbf{x}). \tag{5.19}$$

Hence any integral representation $\mathbf{b}$ of a given integer b by f gives rise to the integral representation $\mathbf{Mb}$ of mb by f_0. Conversely, all the representations of mb by f_0 are of the type $\mathbf{T(S)c}$ where $\mathbf{c}$ runs through a finite set and $\mathbf{S}$ runs through $SL^\pm$. Further, $\mathbf{T(S)c}$ comes from an integral representation of b by $f(\mathbf{x})$ whenever

$$\mathbf{M}^{-1}\mathbf{T(S)c} \in \mathbf{Z}^3. \tag{5.20}$$

The condition (5.20) for fixed $\mathbf{c}$ is a collection of congruences modulo $\det \mathbf{M}$ on the elements $\alpha, \beta, \gamma, \delta$ of $\mathbf{S}$. Thus we may determine whether b is represented by f and, if so, have a collection of formulae for all the representations.

By (5.19) and Lemma 4.1 we can determine the automorphism group of $f(\mathbf{x})$. One can thus determine all the orbits for $O^+(f)$ of the integral representations of a given integer b. The numerical details could be tedious in any given case.

6. REPRESENTATION BY ANISOTROPIC TERNARIES

In this section we show how to determine all the orbits of representations of a given integer by an indefinite anisotropic integral ternary form f. The argument will depend on results about the action of the proper orthogonal group $O^+(f)$ which will be proved in Section 8 but must first be described here. We do not at first use the hypothesis that f is anisotropic.

An indefinite ternary form $f(x_1, x_2, x_3) \in \mathbf{Q}[x_1, x_2, x_3]$ can be written in the shape

$$f(\mathbf{x}) = a_1 L_1(\mathbf{x})^2 + a_2 L_2(\mathbf{x})^2 + a_3 L_3(\mathbf{x})^2 \tag{6.1}$$

where $L_j(\mathbf{x}) \in \mathbf{Q}[\mathbf{x}]$ are linear forms and $a_j \in \mathbf{Q}^*$ $(1 \leqslant j \leqslant 3)$. On taking $-f$ instead of f if need be, we may suppose that

$$a_1 > 0, \qquad a_2 > 0, \qquad a_3 < 0. \tag{6.2}$$

Then

$$f(\mathbf{x}) = X_1^2 + X_2^2 - X_3^2 \tag{6.3}$$

where

$$X_j = |a_j|^{\frac{1}{2}} L_j \in \mathbf{R}[\mathbf{x}]. \tag{6.4}$$

Put

$$\xi_j = X_j / X_3 \qquad (j = 1, 2). \tag{6.5}$$

We write

$$\boldsymbol{\xi} = (\xi_1, \xi_2) = \pi(\mathbf{x}). \tag{6.6}$$

The map

$$\pi\colon \mathbf{x} \to \boldsymbol{\xi} \tag{6.7}$$

is well-defined on $\mathbf{R}^3$ except for the $\mathbf{x}$ on the plane $X_3(\mathbf{x}) = 0$: in particular $\pi(\mathbf{x})$ is always well-defined whenever $f(\mathbf{x}) < 0$.

We denote by $\mathscr{D}$ the open unit disc

$$\mathscr{D}\colon \xi_1^2 + \xi_2^2 < 1. \tag{6.8}$$

Clearly (i) $\pi(\mathbf{x}) \in \mathscr{D}$ if $f(\mathbf{x}) < 0$, (ii) $\pi(\mathbf{x})$ is on the boundary of $\mathscr{D}$ if $f(\mathbf{x}) = 0$, while (iii) if $f(\mathbf{x}) > 0$, then either $\pi(\mathbf{x})$ is outside the closure of $\mathscr{D}$ or is not defined.

A linear transformation $\mathbf{x} \to \mathbf{T}\mathbf{x}$ of $\mathbf{R}^3$ onto itself induces via π a fractional linear transformation of the plane which we shall denote by

$$\boldsymbol{\xi} \to \mathbf{T}(\boldsymbol{\xi}). \tag{6.9}$$

In particular, if

$$\mathbf{T} \in O_{\mathbf{R}}(f) \tag{6.10}$$

then (6.9) is a fractional-linear transformation of $\mathscr{D}$ onto itself.

The action of the group

$$O^+(f) = O_{\mathbf{Z}}^+(f) \subset O_{\mathbf{R}}^+(f) \tag{6.11}$$

on $\mathscr{D}$ has a *fundamental domain* with pleasant properties.

THEOREM 6.1. *Suppose that f is anisotropic. Then there is a convex polygon $\mathscr{C}$ with vertices in $\mathscr{D}$ and with the following properties:*

(i) *If $\boldsymbol{\alpha} \in \mathscr{D}$, there is a $\mathbf{T} \in O^+(f)$ such that $\mathbf{T}(\boldsymbol{\alpha}) \in \mathscr{C}$.*

(ii) *If $\boldsymbol{\alpha}$ is in the interior of $\mathscr{C}$ and $\mathbf{T}(\boldsymbol{\alpha}) \in \mathscr{C}$, $\mathbf{T} \in O^+(f)$, then $\mathbf{T} = \mathbf{I}$.*

COROLLARY. *$\mathscr{C}$ lies in a disc*

$$\xi_1^2 + \xi_2^2 \leqslant r^2 \tag{6.12}$$

where

$$0 < r < 1. \tag{6.13}$$

The proof of Theorem 6.1 is reserved for Section 8: at the end of this section we shall discuss one particular f in some detail. We note in passing that for isotropic forms the conclusions of Theorem 6.1 do not hold, but there is a weaker form in which the vertices of $\mathscr{C}$ are allowed to be on the boundary of $\mathscr{D}$. For $x_1x_3 - x_2^2$ this follows from Section 5 and the reduction theory of definite binary forms: and the general isotropic case can readily be reduced to this. It follows that the Corollary of Theorem 6.1 is definitely false for all isotropic forms.

We now use Theorem 6.1 to treat the integral representations of an integer m by f and must consider the cases $m < 0$ and $m > 0$ separately.

We say that an integral vector $\mathbf{b}$ with $f(\mathbf{b}) = m < 0$ is a *reduced representation* (with respect to the given fundamental domain $\mathscr{C}$) if

$$\pi(\mathbf{b}) \in \mathscr{C}.$$

THEOREM 6.2. *Let f be an integral anisotropic form which is $\mathbf{R}$-equivalent to $X_1^2 + X_2^2 - X_3^2$ and let $m < 0$ be an integer. Then*

(i) *every integral representation $\mathbf{c}$ of m by f is equivalent to a reduced representation in the sense that $\mathbf{c} = \mathbf{Tb}$ where $\mathbf{T} \in O^+(f)$ and $\mathbf{b}$ is reduced.*

(ii) *the reduced integral representations $\mathbf{b}$ can be found effectively.*

Proof. The statement (i) is an immediate consequence of the definition of a fundamental domain [i.e. of Theorem 6.1(i)], so it remains to verify (ii). Put

$$\mathbf{B} = \mathbf{X}(\mathbf{b}), \qquad \boldsymbol{\beta} = \pi(\mathbf{b}) \tag{6.14}$$

(where $\mathbf{X}$ is defined by (6.4)). Since $\boldsymbol{\beta} \in \mathscr{C}$ by hypothesis, Theorem 6.1, Corollary, gives

$$\beta_1^2 + \beta_2^2 \leqslant r^2,$$

that is

$$B_1^2 + B_2^2 \leqslant r^2 B_3^2.$$

On the other hand,

$$B_1^2 + B_2^2 - B_3^2 = f(\mathbf{b}) = m = -|m|.$$

It follows that

$$B_3^2 \leqslant |m|/(1 - r^2).$$

We thus have bounds for $B_j = X_j(\mathbf{b})$ $(j = 1, 2, 3)$ and so for the integers b_1, b_2, b_3. This concludes the proof.

Before discussing representations of $m > 0$ we require a further definition. Two points $\boldsymbol{\alpha}$, $\boldsymbol{\beta}$ of $\mathbf{R}^2$ will be said to be *copolar* if

$$\alpha_1\beta_1 + \alpha_2\beta_2 = 1. \tag{6.15}$$

The set of $\boldsymbol{\alpha}$ copolar to a given $\boldsymbol{\beta}$ thus lie on a line (the *polar line* of $\boldsymbol{\beta}$). This is a familiar concept from elementary geometry. It is readily verified that if $\mathbf{T} \in O^+(f)$ then $\mathbf{T}(\boldsymbol{\alpha})$, $\mathbf{T}(\boldsymbol{\beta})$ are copolar whenever $\boldsymbol{\alpha}$, $\boldsymbol{\beta}$ are [for $\mathbf{T}$ preserves the quadratic form $X_1^2 + X_2^2 - X_3^2$ and so also the corresponding bilinear form].

We say that an integral vector $\mathbf{b}$ with $f(\mathbf{b}) = m > 0$ is a *reduced representation* (with respect to the given fundamental domain $\mathscr{C}$) if either (i) $\pi(\mathbf{b})$ is not defined [i.e. $X_3(\mathbf{b}) = 0$] or (ii) there is an $\boldsymbol{\alpha} \in \mathscr{C}$ such that $\boldsymbol{\alpha}$ and $\boldsymbol{\beta} = \pi(\mathbf{b})$ are copolar. [If the origin is in $\mathscr{C}$, which will usually be the case, then (i) can be regarded as a special case of (ii) in the sense of projective geometry. For in case (i) the point $\pi(\mathbf{b})$ is "at infinity" and the "line at infinity" is the polar line of the origin.]

THEOREM 6.3. *Let f be an integral anisotropic form which is $\mathbf{R}$-equivalent to $X_1^2 + X_2^2 - X_3^2$ and let $m > 0$ be an integer. Then*

(i) *every integral representation $\mathbf{c}$ of m by f is equivalent to a reduced representation.*

(ii) *the reduced integral representations $\mathbf{b}$ are finite in number and can be found effectively.*

Proof. (i) If $\boldsymbol{\gamma} = \pi(\mathbf{c})$ is not defined then $\mathbf{c}$ itself is reduced, by definition. Otherwise $\boldsymbol{\gamma}$ is outside the closure of $\mathscr{D}$ and so the polar line of $\boldsymbol{\gamma}$ meets $\mathscr{D}$ [it is, in fact, the line joining the points of contact of the tangents from $\boldsymbol{\gamma}$ to the circle $\xi_1^2 + \xi_2^2 = 1$]. Let $\boldsymbol{\delta}$ be any point of $\mathscr{D}$ which is copolar to $\boldsymbol{\gamma}$. There is a $\mathbf{T} \in O^+(f)$ such that $\boldsymbol{\alpha} = \mathbf{T}(\boldsymbol{\delta}) \in \mathscr{C}$. Then $\mathbf{b} = \mathbf{Tc}$ is reduced. [Note that there is a wide variety in the possible choice of $\boldsymbol{\delta}$, and so of $\mathbf{T}$, in contrast to the case $m < 0$. This is because for $\mathbf{b} \in \mathbf{Z}^3$ the set of $\mathbf{T} \in O^+(f)$ with $\mathbf{Tb} = \mathbf{b}$ is finite if $f(\mathbf{b}) < 0$ but infinite if $f(\mathbf{b}) > 0$: the last statement being left as an exercise for the reader.]

(ii) Suppose that $\mathbf{b}$ is reduced, $f(\mathbf{b}) = m > 0$. We must distinguish the two cases in the definition of a reduced representation. If $B_3 = X_3(\mathbf{b}) = 0$, then $B_1^2 + B_2^2 = m$ and so there are bounds for $\mathbf{B}$ and hence for $\mathbf{b}$. If, however, $B_3 \neq 0$, so $\boldsymbol{\beta} = \pi(\mathbf{b})$ is defined, let $\boldsymbol{\alpha} \in \mathscr{C}$ be copolar with $\boldsymbol{\beta}$. Then

$$(\alpha_1^2 + \alpha_2^2)(\beta_1^2 + \beta_2^2) \geqslant (\alpha_1\beta_1 + \alpha_2\beta_2)^2 = 1,$$

so

$$\beta_1^2 + \beta_2^2 \geqslant r^{-2}$$

by Theorem 6.1, Corollary. Hence

$$B_3^2 \leqslant r^2(B_1^2 + B_2^2), \qquad B_1^2 + B_2^2 - B_3^2 = m > 0$$

and so

$$B_1^2 + B_2^2 \leqslant m/(1 - r^2).$$

We thus have bounds for **B**, and hence for **b**. This concludes the proof.

We conclude this section by considering $O^+(f)$ for the special form

$$f(\mathbf{x}) = x_1^2 + x_2^2 - 3x_3^2$$

as an illustrative example. The justification for some of the statements made will have to wait until the next section.

We can take

$$X_1 = x_1, \qquad X_2 = x_2, \qquad X_3 = 3^{\frac{1}{2}}x_3,$$

so

$$\xi_j = x_j/3^{\frac{1}{2}}x_3 \qquad (j = 1, 2).$$

To avoid the irrationality $3^{\frac{1}{2}}$ it is, however, convenient to write

$$\eta_j = x_j/x_3 = 3^{\frac{1}{2}}\xi_j \qquad (j = 1, 2)$$

so that $\mathscr{D}$ is given by

$$\mathscr{D}: \eta_1^2 + \eta_2^2 < 3.$$

There is an automorph $\mathbf{T}_1$ given by

$$\mathbf{T}_1 x_1 = 2x_1 + 3x_3, \qquad \mathbf{T}_1 x_2 = x_2, \qquad \mathbf{T}_1 x_3 = x_1 + 2x_3.$$

Its action on the $\boldsymbol{\eta}$-plane is given by

$$\mathbf{T}_1(\boldsymbol{\eta}) = \boldsymbol{\eta}'$$

where

$$\eta_1' = (2\eta_1 + 3)/(\eta_1 + 2), \qquad \eta_2' = \eta_2/(\eta_1 + 2).$$

In particular $\mathbf{T}_1$ takes the line $\eta_1 = -1$ into the line $\eta_1 = +1$. On interchanging x_1 and x_2 we have the automorph $\mathbf{T}_2$ with

$$\mathbf{T}_2 x_1 = x_1, \qquad \mathbf{T}_2 x_2 = 2x_2 + 3x_3, \qquad \mathbf{T}_2 x_3 = x_2 + 2x_3.$$

Denote by $\mathscr{E} \subset \mathscr{D}$ the square

$$\mathscr{E}: |\eta_1| \leqslant 1 \qquad |\eta_2| \leqslant 1.$$

Let $\Gamma \subset O^+(f)$ be the group generated by $\mathbf{T}_1, \mathbf{T}_2$. It is easy to convince oneself that the transforms $\mathbf{T}(\mathscr{E})$ $(\mathbf{T} \in \Gamma)$ cover the whole of $\mathscr{D}$ without overlapping [i.e. if $\mathbf{T}_3(\mathscr{E})$ and $\mathbf{T}_4(\mathscr{E})$ have inner points in common then they coincide, $\mathbf{T}_3, \mathbf{T}_4 \in \Gamma$], see Fig. 13.1: for example there are 6 transforms $\mathbf{T}(\mathscr{E})$, $\mathbf{T} \in \Gamma$ which surround the point (1. 1). That the $\mathbf{T}(\mathscr{E})$ actually do cover $\mathscr{D}$ without overlapping will follow from the results of Section 7.

FIGURE 13.1

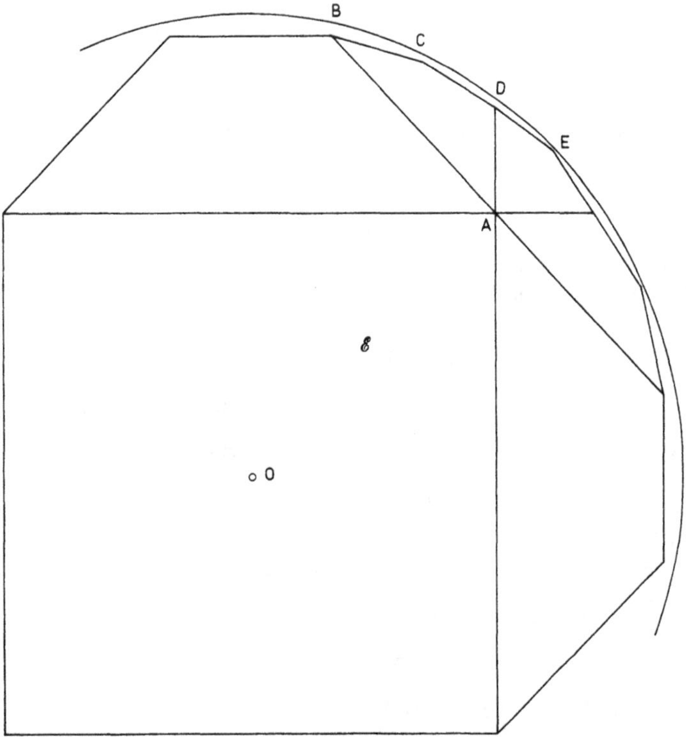

$A = (1, 1)$ $B = (1/3, 5/3)$ $C = (5/7, 11/7)$
$D = (1, 7/5)$ $E = (11/9, 11/9)$ $O = (0, 0)$

The circle has radius $\sqrt{3}$. The points B, C, D, E lie strictly inside it, though this is not obvious because of the finite thickness of the lines. The Figure illustrates the first stage in the construction of a fundamental domain for the group of integral automorphs of the form $x_1^2 + x_2^2 - 3x_3^2$ and shows $\mathscr{E}$ together with some of the $\mathbf{T}(\mathscr{E})$. The $\mathbf{T}(\mathscr{E})$ cover the interior $\mathscr{D}$ of the circle without overlapping: they have no accumulation point in $\mathscr{D}$ but accumulate rapidly towards every point of the boundary.

We next consider the set Δ of $\mathbf{S} \in O^+(f)$ such that $\mathbf{S}(\mathbf{0}) \in \mathscr{E}$ where $\mathbf{0} = (0, 0)$ is the centre of $\mathscr{D}$. For such an $\mathbf{S}$ the vector $\mathbf{b} = \mathbf{S}(0, 0, 1) \in \mathbf{Z}^3$ satisfies $f(\mathbf{b}) = -3$ and $\pi(\mathbf{b}) \in \mathscr{E}$. By using the same argument as in the proof of Theorem 6.2 one deduces readily that $\mathbf{b} = (0, 0, 1)$. Hence Δ is a group and consists of the transformations

$$x_j \to \varepsilon_j x_j \qquad \varepsilon_j^2 = 1 \quad (j = 1, 2, 3) \qquad \varepsilon_1 \varepsilon_2 \varepsilon_3 = +1$$

and

$$x_1 \to \varepsilon_1 x_2, \qquad x_2 \to \varepsilon_2 x_1, \qquad x_3 \to \varepsilon_3 x_3, \qquad \varepsilon_j^2 = 1, \qquad \varepsilon_1 \varepsilon_2 \varepsilon_3 = -1.$$

FIGURE 13.2

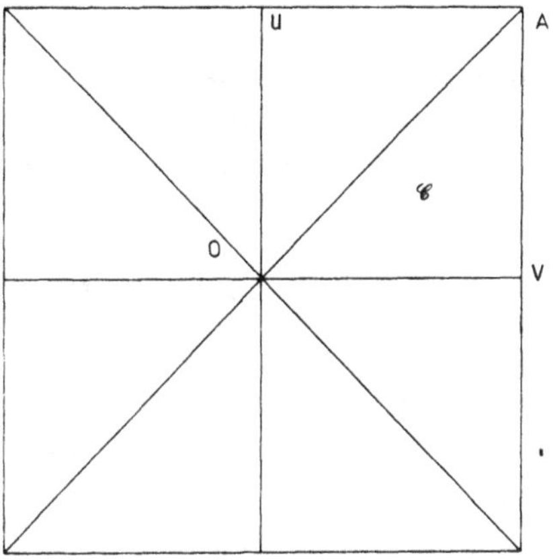

$A = (1, 1) \qquad U = (0, 1) \qquad V = (1, 0) \qquad O = (0, 0)$

This Figure illustrates the final stage in the construction of a fundamental domain $\mathscr{C}$ for the integral automorphs of $x_1^2 + x_2^2 - 3x_3^2$. The large square is the boundary of the set $\mathscr{E}$ constructed in the first stage (cf. Fig. 13.1).

Let $\mathscr{C}$ be the closed triangle with vertices at $\boldsymbol{\eta} = (0, 0), (1, 0), (1, 1)$. Then the $\mathbf{S}(\mathscr{C}), \mathbf{S} \in \Delta$ cover $\mathscr{E}$ without overlapping.

Now let $\mathbf{T}$ be any element of $O^+(f)$. Then $\mathbf{T}(\mathbf{0}) \in \mathbf{T}^*(\mathscr{E})$ for some $\mathbf{T}^* \in \Gamma$ and hence $\mathbf{U}(\mathbf{0}) \in \mathscr{E}$ *for* $\mathbf{U} = (\mathbf{T}^*)^{-1}\mathbf{T}$. It follows that $\mathbf{U} \in \Delta$. This has two consequences. In the first place $\mathscr{C}$ is a fundamental domain for $O^+(f)$. Secondly, every $\mathbf{T} \in O^+(f)$ is of the shape $\mathbf{T} = \mathbf{T}^*\mathbf{U}$, $\mathbf{T}^* \in \Gamma$, $\mathbf{U} \in \Delta$: in particular $O^+(f)$ is generated by $\mathbf{T}_1, \mathbf{T}_2$ and the $\mathbf{U} \in \Delta$.

7. THE NON-EUCLIDEAN PLANE

In this section we first investigate the group M (say) of real fractional-linear transformations which map the disc

$$\mathscr{D}\colon\ \xi_1^2 + \xi_2^2 < 1 \tag{7.1}$$

onto itself. The study of the properties of $\mathscr{D}$ invariant under M is essentially the study of the Beltrami–Klein model of non-euclidean geometry. We shall use this in Section 8 to give a proof and interpretation of Theorem 6.1.

LEMMA 7.1. (i) *The group M acts transitively on $\mathscr{D}$.*

(ii) *The group $ST_{\mathbf{0}}$ of elements of M which leave the origin* $\mathbf{0}$ *invariant* (*the* stabilizer *of* $\mathbf{0}$) *consists of the rotations* (*in the usual euclidean sense*) *around* $\mathbf{0}$ *and the reflections about lines passing through* $\mathbf{0}$.

COROLLARY. *There is a 1–1 correspondence between the points of* $\mathscr{D}$ *and the cosets* $M/ST_{\mathbf{0}}$.

The Corollary is, of course, an immediate consequence of the Lemma. The Lemma itself is an immediate consequence of the next Lemma, which gives a parametric representation of the group M. We denote by $SL_{\mathbf{R}}^{\pm}$ the group of matrices

$$\mathbf{S} = \begin{pmatrix} \alpha & \beta \\ \gamma & \delta \end{pmatrix} \qquad (\alpha, \beta, \gamma, \delta \in \mathbf{R}) \tag{7.2}$$

with

$$\lambda = \lambda(\mathbf{S}) = \alpha\delta - \beta\gamma = \pm 1. \tag{7.3}$$

LEMMA 7.2. (i) *the group* M *is isomorphic to* $SL_{\mathbf{R}}^{\pm}/\{\pm\mathbf{I}\}$.

(ii) *the above isomorphism can be chosen in such a way that the stabilizer* $ST_{\mathbf{0}}$ *of* $\mathbf{0}$ *is* $K^{\pm}/\{\pm\mathbf{I}\}$, *where* $K^{\pm} \subset SL_{\mathbf{R}}^{\pm}$ *is the group consisting of the*

$$\begin{pmatrix} \alpha & \beta \\ -\beta & \alpha \end{pmatrix} \qquad (\lambda = \alpha^2 + \beta^2 = 1) \tag{7.4}$$

and the

$$\begin{pmatrix} \alpha & \beta \\ \beta & -\alpha \end{pmatrix} \qquad (\lambda = -(\alpha^2 + \beta^2) = -1). \tag{7.5}$$

Proof. We use the machinery introduced at the beginning of Section 6 with

$$f_0(\mathbf{x}) = -x_1x_3 + x_2^2, \tag{7.6}$$

$$X_1 = \tfrac{1}{2}(x_3 - x_1), \qquad X_2 = x_2, \qquad X_3 = \tfrac{1}{2}(x_3 + x_1) \tag{7.7}$$

and

$$\xi_1 = X_1/X_3, \qquad \xi_2 - X_2/X_3. \tag{7.8}$$

As in Section 6, these substitutions induce a homomorphism $O_{\mathbf{R}}^{+}(f_0) \to M$ and it is readily verified that this is in fact an isomorphism.

On the other hand the arguments of Section 5 are readily modified to show that every element of $O^{+}(f_0)$ is of the shape (5.6), where by homogeneity we may suppose that $\lambda = \alpha\delta - \beta\gamma = \pm 1$. [We have changed the sign of f_0 from that used in Section 5 but that is immaterial.]

Finally, an element $\mathbf{S}$ of $SL_{\mathbf{R}}^{\pm}$ maps into an element of the stabilizer of $\mathbf{0}$ precisely when the corresponding element of $O^{+}(f)$ given by (5.6) takes $\mathbf{x} = (1, 0, 1)$ into a scalar multiple of itself. It is then readily verified that $\mathbf{S}$ has the shape (7.4) or (7.5). This completes the proof of Lemma 7.2. As already remarked, Lemma 7.1 follows readily from it.

COROLLARY. *Under the above action of* $SL_{\mathbf{R}}^{\pm}$ *the points of* $\mathscr{D}$ *are identified with the cosets* $SL_{\mathbf{R}}^{\pm}/K_{\mathbf{R}}^{\pm}$ *and also with* $SL_{\mathbf{R}}^{+}/K_{\mathbf{R}}^{+}$ *(where* $K^{+} = K^{\pm} \cap SL^{+}$ *is the set of* (7.4)).

LEMMA 7.3. *There is a metric on $\mathscr{D}$ which is invariant under the action of M. It is uniquely defined up to a scalar multiple.*

Proof. We can actually write an explicit formula for the metric namely

$$d(\boldsymbol{\xi}, \boldsymbol{\eta}) = \arg\cosh\left\{\frac{1 - \xi_1\eta_1 - \xi_2\eta_2}{(1 - \xi_1^2 - \xi_2^2)^{\frac{1}{2}}(1 - \eta_1^2 - \eta_2^2)^{\frac{1}{2}}}\right\} \tag{7.9}$$

for the distance between $\boldsymbol{\xi}, \boldsymbol{\eta}$; but we shall not need this.

We use the methods of differential geometry and are cavalier about the details. Let $\boldsymbol{\xi}$ and $\boldsymbol{\xi} + d\boldsymbol{\xi}$ be two infinitesimally near points of $\mathscr{D}$. Then there is a fractional linear transformation $\mu_0 \in M$ such that $\mu_0\boldsymbol{\xi} = \mathbf{0}$. The transformation μ_0 is not unique but any other μ with the same property is of the shape $\mu = \mu_1\mu_0$ where μ_1 is a rotation about the origin or a reflection, by Lemma (7.1). Hence the infinitesimal euclidean distance from $\mathbf{0} = \mu(\boldsymbol{\xi})$ to $\mu(\boldsymbol{\xi} + d\boldsymbol{\xi})$ is independent of μ. We define it to be the infinitesimal non-euclidean distance ds from $\boldsymbol{\xi}$ to $\boldsymbol{\xi} + d\boldsymbol{\xi}$. Clearly this is the only possible definition of an invariant metric which coincides with the euclidean metric in the infinitesimal neighbourhood of the origin. On taking polar coordinates about $\mathbf{0}$:

$$\xi_1 = r\cos\theta, \qquad \xi_2 = r\sin\theta \tag{7.10}$$

it can be verified that

$$(ds)^2 = (1 - r^2)^{-2}(dr)^2 + (1 - r^2)^{-1}(rd\theta)^2. \tag{7.11}$$

COROLLARY 1. *The geodesics are line segments.*

Proof. The masochist can verify this by the standard methods of differential geometry but it is easier to use a trick. Let $\boldsymbol{\xi}, \boldsymbol{\eta}$ be any two points of $\mathscr{D}$. There is a fractional linear transformation $\mu \in M$ such that $\mu(\boldsymbol{\xi}) = \mathbf{0}$ and $\mu(\boldsymbol{\eta}) = (R, 0)$ with $0 < R < 1$. It is clear from (7.11) that the geodesic from $\mathbf{0}$ to $(R, 0)$ is the line segment joining them. But fractional linear transformations take line segments into line segments.

COROLLARY 2. *The set of points equidistant* (*in the non-euclidean metric*) *from two given points $\boldsymbol{\xi}, \boldsymbol{\eta}$ is the set of points in $\mathscr{D}$ on a certain line.*

Note. We shall speak of this as the "perpendicular bisector". Here both "perpendicular" and "bisector" must be understood in the non-euclidean sense.

Proof. There is a $\mu \in M$ such that $\mu(\boldsymbol{\xi}) = (-R, 0), \mu(\boldsymbol{\eta}) = (R, 0)$ for some $R > 0$, as is easily verified. In this case the required locus is the set of $(0, t)$, $-1 < t < 1$ by symmetry. The general result now follows.

Corollary 3. *The perpendicular bisector of* $\mathbf{0}$ *and* $(r, 0)$, *where* $r > 0$, *is the line segment* (s, ξ_2) *where*

$$\left(\frac{1-s}{1+s}\right)^2 = \frac{1-r}{1+r} \tag{7.12}$$

and $s^2 + \xi_2^2 < 1$.

Proof. After Corollary 1 it is enough to verify that if $(s, 0)$ is equidistant from $(0, 0)$ and $(r, 0)$ then s is given by (7.12). This can be checked as in Corollary 2. Alternatively one can note that fractional linear transformations preserve cross-ratios and so the cross-ratio of $-1, 0, s, +1$ must be equal to that of $-1, s, r, +1$.

Finally we note that there is an invariant measure on $\mathscr{D}$ with a very simple interpretation.

Lemma 7.4. *Let* $\mathscr{S} \subset \mathscr{D}$ *and let* $\pi^{-1}(\mathscr{S})$ *be the set of points* $\mathbf{X} = (X_1, X_2, X_3)$ *with*

$$(X_1/X_3, X_2/X_3) \in \mathscr{S}$$

and

$$0 \geqslant X_1^2 + X_2^2 - X_3^2 \geqslant -1.$$

Define $m(\mathscr{S})$ *to be the ordinary Lebesgue measure of* $\pi^{-1}(\mathscr{S})$. *Then* $m(\mathscr{S})$ *for subsets* $\mathscr{S}$ *of* $\mathscr{D}$ *is an invariant measure under* M.

Proof. We have seen that the $\mu \in M$ are all induced by elements $\mathbf{T}$ of $O^+(f_1)$, where $f_1(\mathbf{X}) = X_1^2 + X_2^2 - X_3^2$. In an obvious notation

$$\mathbf{T}(\pi^{-1}(\mathscr{S})) = \pi^{-1}(\mathbf{T}(\mathscr{S})).$$

Since $\det \mathbf{T} = 1$, this gives the result.

As an appendix we describe the relation between the Beltrami–Klein model of the non-euclidean plane and other models. Let H be the "upper half plane" of complex numbers z with strictly positive imaginary parts. This defines a positive definite quadratic form

$$\begin{aligned} g_z(\mathbf{u}) &= (u_1 + zu_2)(u_1 + \bar{z}u_2) \\ &= x_1u_1^2 + 2x_2u_1u_2 + x_3u_2^2 \end{aligned}$$

and so a point

$$\xi = \xi(z) \in \mathscr{D}$$

according to the prescription (7.7) and (7.8). This is clearly a bijection between $\mathscr{D}$ and H. If

$$\mathbf{S} = \begin{pmatrix} \alpha & \beta \\ \gamma & \delta \end{pmatrix} \in SL_{\mathbb{R}}^{\pm}$$

has determinant $\lambda(\mathbf{S}) = +1$ then the corresponding action on H is the conformal one:

$$z \to \frac{\beta + \delta z}{\alpha + \gamma z},$$

but if $\lambda(\mathbf{S}) = -1$ it is the anti-conformal

$$z \to \frac{\beta + \delta \bar{z}}{\alpha + \gamma \bar{z}}.$$

When $\alpha, \beta, \gamma, \delta \in \mathbf{Z}$ this gives the familiar representation of the unimodular group.

The transformation

$$w = \frac{z - i}{z + i}$$

takes H into the unit disc

$$|w| < 1.$$

Again, the elements of $SL_{\mathbf{R}}^{+}$ act conformally and those of $SL_{\mathbf{R}}^{-}$ act anti-conformally. It can be verified that the point $w = u + iv$ corresponding to $\boldsymbol{\xi} = (\xi_1, \xi_2)$ is given by

$$u = 2\xi_1/(1 + \xi_1^2 + \xi_2^2), \qquad v = 2\xi_2/(1 + \xi_1^2 + \xi_2^2).$$

8. PROOF OF THEOREM 6.1

In this section we use the language of non-euclidean geometry to discuss further the representation of the group $O^+(f)$ discussed at the beginning of Section 6 and, in particular, to prove Theorem 6.1. The form f is integral and indefinite. We retain the notation of Section 6. In particular, if $\mathbf{T} \in O^+(f)$ then the result of its action on $\boldsymbol{\xi} \in \mathscr{D}$ is denoted by $\mathbf{T}(\boldsymbol{\xi})$.

Lemma 8.1. *Let $\boldsymbol{\eta} \in \mathscr{D}$ and $r, 0 < r < 1$ be given. Then there are only finitely many $\mathbf{T} \in O^+(f)$ such that $\mathbf{T}(\boldsymbol{\eta}) = \boldsymbol{\xi}$ lies in $\xi_1^2 + \xi_2^2 \leqslant r^2$.*

Proof. Note that we do not require f to be anisotropic. Suppose, first, that $\boldsymbol{\eta} = \pi(\mathbf{a})$ where $\mathbf{a} \in \mathbf{Z}^3$ and put $m = f(\mathbf{a}) < 0$. Then $\mathbf{T}(\boldsymbol{\eta}) = \pi(\mathbf{Ta})$ and $f(\mathbf{Ta}) = m$. The finiteness of the set of $\mathbf{T}$ follows as in the proof of Theorem 6.2.

For general $\boldsymbol{\eta}$ we use the invariant metric constructed in Section 7. The condition $\xi_1^2 + \xi_2^2 \leqslant r^2$ is the same as $d(\mathbf{0}, \boldsymbol{\xi}) \leqslant d_0$ for some d_0. Let $\boldsymbol{\eta}_1$ be any element of $\mathscr{D}$ of the shape $\boldsymbol{\eta}_1 = \pi(\mathbf{a})$, $\mathbf{a} \in \mathbf{Z}^3$. Then

$$\begin{aligned} d(\mathbf{0}, \mathbf{T}\boldsymbol{\eta}_1) &\leqslant d(\mathbf{0}, \mathbf{T}\boldsymbol{\eta}) + d(\mathbf{T}\boldsymbol{\eta}, \mathbf{T}\boldsymbol{\eta}_1) \\ &= d(\mathbf{0}, \mathbf{T}\boldsymbol{\eta}) + d(\boldsymbol{\eta}, \boldsymbol{\eta}_1) \end{aligned}$$

by the invariance of the metric. Hence $d(\mathbf{0}, \mathbf{T}\boldsymbol{\eta}) \leqslant d_0$ implies $d(\mathbf{0}, \mathbf{T}\boldsymbol{\eta}_1) \leqslant d_1$ with $d_1 = d_0 + d(\boldsymbol{\eta}, \boldsymbol{\eta}_1)$, and we are reduced to the previous case.

LEMMA 8.2. *Suppose that f is anisotropic. Then there is a d_2 depending only on f with the following property: to every $\boldsymbol{\xi} \in \mathscr{D}$ there is a $\mathbf{T} \in O^+(f)$ such that*

$$d(\boldsymbol{\xi}, \mathbf{T}(\mathbf{0})) \leqslant d_2. \tag{8.1}$$

Proof. By Lemma 2.3 (or, alternatively, by Lemma 7.1(i)) there is an $\mathbf{S} \in O^+_{\mathbf{R}}(f)$ such that $\mathbf{S}(\mathbf{0}) = \boldsymbol{\xi}$. By Theorem 2.2 (with $\mathbf{S}^{-1}, \mathbf{T}^{-1}$ instead of $\mathbf{S}, \mathbf{T}$) there is a $\mathbf{T} \in O^+(f)$ such that the elements of $\mathbf{U}$ (say) $= \mathbf{T}^{-1}\mathbf{S}$ are bounded by a constant k depending only on f. By the invariance of the metric,

$$\begin{aligned} d((\boldsymbol{\xi}, \mathbf{T}(\mathbf{0})) &= d(\mathbf{T}^{-1}(\boldsymbol{\xi}), \mathbf{0}) \\ &= d(\mathbf{U}(\mathbf{0}), \mathbf{0}). \end{aligned}$$

The truth of Lemma 8.2 now follows from the following very general Lemma.

LEMMA 8.3. *Let $k > 0$ be given. Then there is a constant d_2 depending only on k and f with the following property: suppose that the elements of the matrix $\mathbf{U} \in O^+_{\mathbf{R}}(f)$ are at most k in absolute value. Then*

$$d(\mathbf{0}, \mathbf{U}(\mathbf{0})) \leqslant d_2.$$

Proof. Let $X_j(\mathbf{x})$ $(1 \leqslant j \leqslant 3)$ be the linear forms such that

$$f(\mathbf{x}) = X_1^2 + X_2^2 - X_3^2$$

and

$$\xi_j = X_j/X_3.$$

Let $\mathbf{a} \in \mathbf{R}^3$ be such that

$$X_1(\mathbf{a}) = X_2(\mathbf{a}) = 0, \qquad X_3(\mathbf{a}) = 1.$$

It is easy to see that

$$B_j = X_j(\mathbf{Ua})$$

are bounded by a constant depending only on k and f: and

$$B_1^2 + B_2^2 - B_3^2 = 1$$

because $\mathbf{U}$ is an automorph. Then

$$\mathbf{U}(\mathbf{0}) = \boldsymbol{\eta} = (\eta_1, \eta_2)$$

where

$$\eta_j = B_j/B_3;$$

so

$$\begin{aligned} \eta_1^2 + \eta_2^2 &= (B_1^2 + B_2^2)/B_3^2 \\ &= (B_1^2 + B_2^2)/(1 + B_1^2 + B_2^2) \\ &\leqslant r_0^2 \end{aligned}$$

where $r_0 < 1$ depends only k and f. This completes the proof of Lemma 8.3 and so also of Lemma 8.2.

The next stage in the proof of Theorem 6.1 is the introduction of the notion of a Dirichlet domain. By Lemma 8.1, for any ξ the infimum

$$\inf_{\mathbf{T} \in O^+(f)} d(\xi, \mathbf{T}(\mathbf{0})) \tag{8.2}$$

is attained. The set of ξ for which the infimum is attained at $\mathbf{T} = \mathbf{I}$ (and, possibly, elsewhere) is called the *Dirichlet domain* $\mathscr{E}$ (with respect to $\mathbf{0}$). By the invariance of the metric we clearly have

$$\mathscr{D} = \bigcup_{\mathbf{T} \in O^+(f)} \mathbf{T}(\mathscr{E}). \tag{8.3}$$

The set of ξ in $\mathscr{E}$ is defined by the infinite collection of inequalities

$$d(\xi, \mathbf{0}) \leqslant d(\xi, \mathbf{T}(\mathbf{0})) \qquad (\mathbf{T} \in O^+(f)). \tag{8.4}$$

By Lemma 7.3, Corollary 2, each of the inequalities (8.4) states that ξ lies in a certain closed half-plane. Hence $\mathscr{E}$ is convex and relatively closed in $\mathscr{D}$.

So far we have not made use of the hypothesis of Theorem 6.1 that f is anisotropic.

LEMMA 8.4. *Suppose that f is anisotropic and let d_2 be given by Lemma* 8.2. *Then $\mathscr{E}$ is defined by the inequalities* (8.4) *taken over the* $\mathbf{T}$ *for which*

$$d(\mathbf{0}, \mathbf{T}(\mathbf{0})) \leqslant 2d_2. \tag{8.5}$$

Proof. Lemma 8.2. implies that (8.2) is bounded above by d_2. In particular $\mathscr{E}$ is contained in the disc

$$d(\xi, \mathbf{0}) \leqslant d_2. \tag{8.6}$$

Let $\varepsilon > 0$ be arbitrarily small and suppose that $\xi \in \mathscr{D}$ is any point with

$$d(\xi, \mathbf{0}) < d_2 + \varepsilon. \tag{8.7}$$

By Lemma 8.2. there is a $\mathbf{T}$ (depending on ξ) such that

$$d(\xi, \mathbf{T}(\mathbf{0})) \leqslant d_2. \tag{8.8}$$

By (8.7) and (8.8) we have

$$d(\mathbf{0}, \mathbf{T}(\mathbf{0})) < 2d_2 + \varepsilon. \tag{8.9}$$

Hence a ξ satisfying (8.7) is in $\mathscr{E}$ provided that it satisfies (8.4) for all the $\mathbf{T}$ with (8.9). But $\mathscr{E}$ is actually contained in (8.6) and so it easy to see that $\mathscr{E}$ is defined by (8.4) and (8.9) alone. Finally on letting $\varepsilon \to 0$ and using Lemma 8.1 we may replace (8.9) by (8.5), as required.

COROLLARY. *$\mathscr{E}$ is a closed polygon whose vertices all lie in $\mathscr{D}$.*

Proof. It is a polygon because there are only finitely many $\mathbf{T}$ with (8.5) by Lemma 8.1. The vertices are in $\mathscr{D}$ by (8.6).

Now let $ST \subset O^+(f)$ be the stabilizer of $\mathbf{0}$. It is a finite group of rotations or reflections about $\mathbf{0}$ and clearly $\mathbf{S}(\mathscr{E}) = \mathscr{E}$ for $\mathbf{S} \in ST$. Let m be the order of ST. Then clearly the part of $\mathscr{E}$ in an appropriate sector of angle $2\pi/m$ at $\mathbf{0}$ is a fundamental domain $\mathscr{C}$. This concludes the proof of Theorem 6.1.

We have already treated a numerical example of Theorem 6.1 at the end of Section 6. There are others in Fricke–Klein (1897) which are treated in terms of the action of $O^+(f)$ on the upper half-plane described at the end of Section 7.

We conclude this section by showing that $O^+(f)$ is finitely generated and obtaining a set of generators.

THEOREM 8.1. *Suppose that f is anisotropic. Let $\mathscr{P}$ be any open set containing $\mathscr{E}$ (the Dirichlet domain of $\mathbf{0}$) and let Σ be the set of $\mathbf{T} \in O^+(f)$ such that $\mathbf{T}(\mathscr{E})$ meets $\mathscr{P}$. Then $O^+(f)$ is generated by Σ and the stabilizer of $\mathbf{0}$.*

Proof. We have

$$\mathscr{E} \subset \mathscr{P} \subset \bigcup_{\mathbf{T} \in \Sigma} \mathbf{T}(\mathscr{E}). \tag{8.10}$$

Let Γ be the group generated by the $\mathbf{T} \in \Sigma$. Then

$$\bigcup_{\mathbf{T} \in \Gamma} \mathbf{T}(\mathscr{E}) = \bigcup_{\mathbf{T} \in \Gamma} \mathbf{T}(\mathscr{P})$$

$$= \mathscr{R} \text{ (say).} \tag{8.11}$$

Hence $\mathscr{R}$ is open because $\mathscr{P}$ is open. On the other hand, the intersection of $\mathscr{R}$ with any closed disc

$$d(\mathbf{0}, \boldsymbol{\xi}) \leqslant d_1 \tag{8.12}$$

is closed because (8.12) meets only finitely many $\mathbf{T}(\mathscr{E})$. Hence $\mathscr{R}$ contains (8.12). Since d_1 is arbitrary, we must have

$$\mathscr{R} = \mathscr{D}. \tag{8.13}$$

Now let $\mathbf{T}$ be any element of $O^+(f)$. By (8.11) and (8.13) we have $\mathbf{T}(\mathbf{0}) \in \mathbf{S}(\mathscr{E})$ for some $\mathbf{S} \in \Gamma$. Then $\mathbf{S}^{-1}\mathbf{T}(\mathbf{0}) \in \mathscr{E}$ and so $\mathbf{S}^{-1}\mathbf{T}$ is in the stabilizer of $\mathbf{0}$. This concludes the proof.

We have already treated a numerical example of Theorem 8.1 at the end of Section 6.

9. QUATERNARY FORMS

There are descriptions of the real orthogonal groups of indefinite quaternary forms in terms of non-euclidean geometry which are somewhat analogous to that for ternary forms investigated in Section 7. These descriptions can similarly be used to treat the integral automorphism groups. In this section we shall only consider the real orthogonal groups. Up to sign there are two real equivalence classes of indefinite quaternaries and these must be treated differently.

Consider first

$$f(\mathbf{x}) = f_{3,1}(\mathbf{x}) = x_1^2 + x_2^2 + x_3^2 - x_4^2. \tag{9.1}$$

We introduce

$$\xi_j = x_j/x_4 \qquad (1 \leqslant j \leqslant 3) \tag{9.2}$$

and consider the unit ball

$$\mathscr{B}\colon\ \xi_1^2 + \xi_2^2 + \xi_3^2 < 1. \tag{9.3}$$

The $\mathbf{T} \in O_{\mathbf{R}}^{+}(f)$ act as fractional-linear transformations of $\mathscr{B}$ onto itself. We have, therefore, to deal with non-euclidean 3-space.

We can interpret $-f(\mathbf{x})$ as the determinant of the Hermitian form

$$\begin{aligned} h_{\mathbf{x}}(u_1, u_2) = (x_3 + x_4)u_1\bar{u}_1 + (x_1 - ix_2)u_1\bar{u}_2 \\ + (x_1 + ix_2)\bar{u}_1 u_2 + (x_3 - x_4)u_2\bar{u}_2. \end{aligned} \tag{9.4}$$

Thus any 2×2 complex-valued matrix

$$\mathbf{A} = \begin{pmatrix} a_{11} & a_{12} \\ a_{21} & a_{22} \end{pmatrix} \qquad (a_{ij} \in \mathbf{C}) \tag{9.5}$$

with

$$|\det \mathbf{A}| = 1 \tag{9.6}$$

induces an automorphism $\mathbf{T}(\mathbf{A}) \in O_{\mathbf{R}}^{+}(f)$ by means of the substitution

$$\mathbf{u} \to \mathbf{A}\mathbf{u}. \tag{9.7}$$

In other words, $\mathbf{T}(\mathbf{A})$ is given by

$$\mathbf{X} \to \bar{\mathbf{A}}'\mathbf{X}\mathbf{A}, \tag{9.8}$$

where

$$\mathbf{X} = \begin{pmatrix} x_3 + x_4 & x_1 + ix_2 \\ x_1 - ix_2 & -x_3 + x_4 \end{pmatrix}. \tag{9.9}$$

It may be checked by direct calculation that $\det(\mathbf{T}(\mathbf{A})) = +1$. It is however simpler to note that $\det(\mathbf{T}(\mathbf{A})) = \pm 1$ is a continuous function of $\mathbf{A}$ and that the set of $\mathbf{A}$ is connected.

The set of $\mathbf{T}(\mathbf{A})$ form a subgroup Θ of $O_{\mathbf{R}}^{+}(f)$ of finite index and Θ acts transitively on $\mathscr{B}$. The $\mathbf{A}$ stabilizing $\mathbf{x} = (0, 0, 0, 1)$ are those with

$$\bar{\mathbf{A}}'\mathbf{A} = \mathbf{I};$$

that is, those in the unitary group K. We have thus identified $\mathscr{B}$ with the set of cosets G/K where G is the set of $\mathbf{A}$ with (9.5) and (9.6) and where K, the unitary group, is a maximal compact subgroup.

The use of the above representation of automorphs of $f_{3,1}(\mathbf{x})$ was, apparently, introduced by Bianchi to study the integral automorphs of forms real-equivalent to $f_{3,1}$. There is a discussion in Fricke–Klein (1879), pp. 577–584.

We now pass to the consideration of

$$f(\mathbf{x}) = f_{2,2}(\mathbf{x}) = x_1^2 + x_2^2 - x_3^2 - x_4^2. \tag{9.10}$$

This is the determinant of

$$\mathbf{X} = \begin{pmatrix} x_1 + x_3 & x_4 + x_2 \\ x_4 - x_2 & x_1 - x_3 \end{pmatrix}. \tag{9.11}$$

Let $\mathbf{A}, \mathbf{B}$ be 2 real 2×2 matrices with

$$\det \mathbf{A} = \det \mathbf{B} = 1. \tag{9.12}$$

The automorph $\mathbf{T}(\mathbf{A}, \mathbf{B}) \in O_{\mathbf{R}}^{+}(f)$ is given by

$$\mathbf{X} \to \mathbf{AXB}. \tag{9.13}$$

The set of $\mathbf{T}(\mathbf{A}, \mathbf{B})$ is a subgroup Θ of finite index in $O_{\mathbf{R}}^{+}(f)$.

Now denote by K the subgroup of Θ which preserves

$$\operatorname{Trace} \mathbf{X}'\mathbf{X} = 2(x_1^2 + x_2^2 + x_3^2 + x_4^2). \tag{9.14}$$

Then K is a subgroup of the orthogonal group of $x_1^2 + x_2^2 + x_3^2 + x_4^2$ and so compact. We have an explicit description of K. In fact $\mathbf{T}(\mathbf{A}, \mathbf{B}) \in K$ means that identically

$$\begin{aligned} \operatorname{Trace} \mathbf{X}'\mathbf{X} &= \operatorname{Trace}(\mathbf{B}'\mathbf{X}'\mathbf{A}'\mathbf{AXB}) \\ &= \operatorname{Trace}(\mathbf{X}'\mathbf{A}'\mathbf{AXBB}'). \end{aligned} \tag{9.15}$$

[Recall that $\operatorname{Trace} \mathbf{UV} = \operatorname{Trace} \mathbf{VU}$ for any two $n \times n$ matrices.] A sufficient condition for this is

$$\mathbf{A}'\mathbf{A} = \mathbf{B}'\mathbf{B} = \mathbf{I}. \tag{9.16}$$

We now show that the condition (9.16) is also necessary. For let $\mathbf{C}, \mathbf{D}$ be any real matrices with

$$\mathbf{C}'\mathbf{C} = \mathbf{D}'\mathbf{D} = \mathbf{I}, \qquad \det \mathbf{C} = \det \mathbf{D} = 1. \tag{9.17}$$

On replacing $\mathbf{X}$ by $\mathbf{CXD}$ we leave the left-hand side of (9.15) unchanged but replace $\mathbf{A'A}$ and $\mathbf{B'B}$ on the right-hand side respectively by

$$\mathbf{C'A'AC}, \qquad \mathbf{DBB'D'}. \tag{9.18}$$

We may choose $\mathbf{C}, \mathbf{D}$ so that the matrices (9.18) are diagonal: and the verification that (9.16) is necessary for the truth of (9.15) when $\mathbf{A'A}, \mathbf{BB'}$ are diagonal is quite straightforward.

We have thus identified Θ/K with the product of two copies of SL^+/K_2 where SL^+ is the group of real matrices $\mathbf{A}$ with $\det \mathbf{A} = 1$ and where K_2 consists of the $\mathbf{A} \in SL^+$ with $\mathbf{A'A} = \mathbf{I}$. Comparison with Lemma 7.2, Corollary, shows that Θ/K is the direct product of two copies of the non-euclidean plane.

In conclusion, we note an alternative approach to $f_{2,2}$. It is equivalent to the form

$$f'(\mathbf{y}) = y_1y_4 - y_2y_3.$$

The equation

$$f'(\mathbf{y}) = 0$$

defines a conic in 3-dimensional projective space. This conic contains the two families of lines

$$\frac{y_2}{y_1} = \frac{y_3}{y_4} = \alpha$$

and

$$\frac{y_3}{y_1} = \frac{y_2}{y_4} = \beta,$$

where α, β are parameters. An element $\mathbf{T}$ of $O^+_{\mathbf{R}}(f')$ either preserves each of these families as a whole or interchanges them: and there is a subgroup of $O^+_{\mathbf{R}}(f')$ of index 2 which preserves each family. It then follows by general geometric considerations that $\mathbf{T}$ must induce fractional linear transformations on α and β:

$$\alpha \to \frac{a + b\alpha}{c + d\alpha}, \qquad \beta \to \frac{r + s\beta}{t + u\beta};$$

where $a, b, c, d, r, s, t, u \in \mathbf{R}$ depend only on $\mathbf{T}$: and this shows that $\mathbf{T}$ is of the shape (9.13).

One can do a similar analysis of (9.1) but working in the complex field $\mathbf{C}$. There are two pencils of lines which are taken into each other by complex conjugation. The actions on the two pencils of lines must therefore also be conjugate over $\mathbf{R}$.

10. REAL AUTOMORPHS: GENERAL CASE

In this section we generalize the results for indefinite ternaries and quaternaries by obtaining a representation† $\mathscr{E}$ of $O_{\mathbf{R}}(f)/K$ where f is a general indefinite form and K is a maximal compact subgroup. The action of $O_{\mathbf{R}}(f)$ on $\mathscr{E}$ has a shape reminiscent of the ordinary modular group.

Let

$$f(\mathbf{x}) = f_{m,l}(\mathbf{x}) = x_1^2 + \ldots + x_m^2 - x_{m+1}^2 - \ldots - x_{m+l}^2 \tag{10.1}$$

where

$$m > 0, \qquad l > 0, \qquad m + l = n. \tag{10.2}$$

The corresponding matrix is

$$\mathbf{F} = \begin{pmatrix} \mathbf{I}_m & \mathbf{0} \\ \mathbf{0} & -\mathbf{I}_l \end{pmatrix} \tag{10.3}$$

where $\mathbf{I}_m, \mathbf{I}_l$ are the unit $m \times m$ and $l \times l$ matrices.

A matrix

$$\mathbf{W} = \begin{pmatrix} \mathbf{U} & \mathbf{0} \\ \mathbf{0} & \mathbf{V} \end{pmatrix} \tag{10.4}$$

is an automorph when

$$\mathbf{U}'\mathbf{U} = \mathbf{I}_m; \qquad \mathbf{V}'\mathbf{V} = \mathbf{I}_l. \tag{10.5}$$

The group K of such $\mathbf{W}$ is compact.

A matrix

$$\mathbf{M} = \begin{pmatrix} \mathbf{A} & \mathbf{B} \\ \mathbf{C} & \mathbf{D} \end{pmatrix} \tag{10.6}$$

is in $O_{\mathbf{R}}(f)$ provided that

$$\mathbf{A}'\mathbf{A} = \mathbf{I}_m + \mathbf{C}'\mathbf{C}, \tag{10.7}$$

$$\mathbf{D}'\mathbf{D} = \mathbf{I}_l + \mathbf{B}'\mathbf{B}, \tag{10.8}$$

$$\mathbf{A}'\mathbf{B} = \mathbf{C}'\mathbf{D}. \tag{10.9}$$

It follows from (10.7) and (10.8) that the quadratic forms given by the matrices $\mathbf{A}'\mathbf{A}$ and $\mathbf{D}'\mathbf{D}$ are strictly positive definite. In particular, $\mathbf{A}$ and $\mathbf{D}$ are non-singular matrices and so we can write

$$\mathbf{E} \text{ (say)} = (\mathbf{A}')^{-1}\mathbf{C}' = \mathbf{B}\mathbf{D}^{-1}. \tag{10.10}$$

On substituting for $\mathbf{C}, \mathbf{B}$ in (10.7) and (10.8) we get

$$(\mathbf{A}\mathbf{A}')^{-1} = \mathbf{I}_l - \mathbf{E}\mathbf{E}' \tag{10.11}$$

$$(\mathbf{D}\mathbf{D}')^{-1} = \mathbf{I}_m - \mathbf{E}'\mathbf{E}. \tag{10.12}$$

† This use of $\mathscr{E}$ must not be confused with the notation of Section 8.

LEMMA 10.1. *Let $\mathscr{E}$ denote the set of all $l \times m$ real matrices* $\mathbf{E}$ *such that*

$$\sup_{\mathbf{u},\mathbf{v}} \mathbf{u}'\mathbf{E}\mathbf{v} < 1, \tag{10.13}$$

where the supremum is over all $\mathbf{u} \in \mathbf{R}^l$, $\mathbf{v} \in \mathbf{R}^m$ *such that*

$$\mathbf{u}'\mathbf{u} = \mathbf{v}'\mathbf{v} = 1. \tag{10.14}$$

Then (10.10) *sets up a bijection between* $\mathscr{E}$ *and the set of cosets* $O_{\mathbf{R}}(f)/K$.

Proof. The three statements (i) the right-hand side of (10.11) is the matrix of a positive definite form (ii) the right-hand side of (10.12) is positive definite and (iii) (10.13) holds, are equivalent, as may be easily verified. [Note that if $\mathbf{w} \in \mathbf{R}^n$ then $\mathbf{u}'\mathbf{w}$ with $\mathbf{u}'\mathbf{u} = 1$ attains its maximum when $\mathbf{u} = |\mathbf{w}|^{-1}\mathbf{w}$. Put $\mathbf{w} = \mathbf{E}\mathbf{v}$.] We have already seen that $\mathbf{M}$ defines $\mathbf{E}$ uniquely. Further, $\mathbf{M}$ and $\mathbf{MW}$ determine the same $\mathbf{E}$ by (10.5). Conversely if $\mathbf{E}$ is given, there are square matrices $\mathbf{A}, \mathbf{D}$ satisfying (10.11) and (10.12) and then $\mathbf{B}, \mathbf{C}$ are determined by (10.10). The matrices $\mathbf{A}, \mathbf{D}$ are not uniquely determined by (10.11) and (10.12): but if $\mathbf{A}, \mathbf{D}$ are solutions then the general solutions are $\mathbf{AU}, \mathbf{DV}$ where (10.5) holds.

LEMMA 10.2. *Let* $\mathbf{M}, \mathbf{M}_1, \mathbf{M}_2 \in O_{\mathbf{R}}(f)$ *and suppose that*

$$\mathbf{M}_2 = \mathbf{M}\mathbf{M}_1. \tag{10.15}$$

Then

$$\mathbf{E}_2 = (\mathbf{A}\mathbf{E}_1 + \mathbf{B})(\mathbf{C}\mathbf{E}_1 + \mathbf{D})^{-1} \tag{10.16}$$

in the notation (10.6), *where* $\mathbf{E}_1, \mathbf{E}_2 \in \mathscr{E}$ *correspond to* $\mathbf{M}_1, \mathbf{M}_2$.

Proof. This is a straightforward verification. We have in an obvious notation

$$\begin{aligned}\mathbf{B}_2 &= \mathbf{A}\mathbf{B}_1 + \mathbf{B}\mathbf{D}_1\\ &= (\mathbf{A}\mathbf{E}_1 + \mathbf{B})\,\mathbf{D}_1\end{aligned} \tag{10.17}$$

and

$$\begin{aligned}\mathbf{D}_2 &= \mathbf{C}\mathbf{B}_1 + \mathbf{D}\mathbf{D}_1\\ &= (\mathbf{C}\mathbf{E}_1 + \mathbf{D})\,\mathbf{D}_1.\end{aligned} \tag{10.18}$$

Then (10.16) follows, since $\mathbf{E}_2 = \mathbf{B}_2\mathbf{D}_2^{-1}$.

11. HERMITE REDUCTION: ISOTROPIC FORMS

In this section we discuss the extension of the results of Section 2 to isotropic forms. We show first that the conclusion of Theorem 2.1 holds also for isotropic forms although the proof is much more delicate. As we have already noted, the conclusion of Theorem 2.2 ceases to hold for any isotropic form. We conclude this section by discussing the rather weaker theorem which is true for isotropic forms, but we do not give a proof.

THEOREM 11.1. *There are only finitely many classically integral Hermite reduced forms of given determinant* $d \neq 0$.

We enunciate the critical steps in the proof as Lemmas and take the general properties of Hermite reduction from Section 2.

LEMMA 11.1. *Let* $\mathbf{F}, \mathbf{G}$ *be the matrices of an indefinite form* f *and of a Hermite majorant* g. *Then*

$$(\mathbf{F}\mathbf{G}^{-1})^2 = \mathbf{I}. \tag{11.1}$$

Proof. By hypothesis there is an $n \times n$ matrix $\mathbf{L}$, where n as usual is the dimension of f, such that

$$\mathbf{F} = \mathbf{L}'\begin{pmatrix} \mathbf{I}_m & \mathbf{0} \\ \mathbf{0} & -\mathbf{I}_l \end{pmatrix}\mathbf{L} \tag{11.2}$$

and

$$\mathbf{G} = \mathbf{L}'\mathbf{L} \tag{11.3}$$

for some m, l with $m + l = n$. The verification of (11.1) is now immediate.

Following Siegel we write (11.1) in the shape

$$\mathbf{F}'\mathbf{G}^{-1}\mathbf{F} = \mathbf{G} \tag{11.4}$$

($\mathbf{F}' = \mathbf{F}$ by definition). Thus $\mathbf{F}$ can be considered as an integral (but not in general unimodular) transformation giving a rational equivalence between the quadratic forms with matrices $\mathbf{G}^{-1}$ and $\mathbf{G}$.

We now require the theory of Siegel domains and use the notation of Chapter 12. Denote by $\mathbf{J}$ the $n \times n$ matrix

$$\mathbf{J} = (j_{ik})$$

with

$$j_{ik} = \begin{cases} 1 & \text{if } i + k = n + 1 \\ 0 & \text{otherwise.} \end{cases}$$

Then

$$\mathbf{x} \to \mathbf{J}\mathbf{x}$$

inverts the order of the variables $x_1, \ldots, x_n$. Note that

$$\mathbf{J}^{-1} = \mathbf{J}' = \mathbf{J}. \tag{11.5}$$

LEMMA 11.2. *There is an* $\eta_1 > 0$ *depending only on* $\eta > 0$ *and* n *with the following property. Suppose that* $\mathbf{G}$ *is the matrix of a quadratic form in the Siegel domain* $\mathscr{S}_n(\delta, \eta)$. *Then* $\mathbf{J}\mathbf{G}^{-1}\mathbf{J}$ *is the matrix of a form in* $\mathscr{S}_n(\delta, \eta_1)$.

Proof. We have

$$\mathbf{G} = \mathbf{C'HC}$$

for matrices

$$\mathbf{H} = \begin{pmatrix} h_1 & 0 & \dots & 0 & 0 \\ 0 & h_2 & \dots & 0 & 0 \\ & & \dots & & \\ 0 & 0 & \dots & 0 & h_n \end{pmatrix}$$

and

$$\mathbf{C} = \begin{pmatrix} 1 & c_{12} & c_{13} & \dots & c_{1n} \\ 0 & 1 & c_{23} & \dots & c_{2n} \\ & \dots & & \dots & \\ 0 & 0 & 0 & \dots & 1 \end{pmatrix}$$

where, by the definition of a Siegel domain,

$$0 < h_j \leqslant \delta h_{j+1} \qquad (1 \leqslant j < n)$$

and

$$|c_{ij}| \leqslant \eta \qquad (1 \leqslant i < j \leqslant n).$$

Hence

$$\mathbf{J}\mathbf{G}^{-1}\mathbf{J} = \mathbf{C}_1'\mathbf{H}_1\mathbf{C}_1,$$

where

$$\mathbf{H}_1 = \mathbf{J}\mathbf{H}^{-1}\mathbf{J}$$

and

$$\mathbf{C}_1 = \mathbf{J}(\mathbf{C}')^{-1}\mathbf{J}.$$

The rest of the proof is easy.

LEMMA 11.3. *Let* $\mathbf{G}$ *be the matrix of a quadratic form in a Siegel domain* $\mathscr{S}_n(\delta, \eta)$ *and let* $\mathbf{T}$ *be a non-singular integral matrix. Let* $\mathbf{m}_1, \dots, \mathbf{m}_n$ *be linearly independent integral vectors giving the successive minima of the form whose matrix is*

$$\mathbf{T'GT}.$$

Then the elements of the vectors $\mathbf{Tm}_j$ *are bounded by a constant depending only on* n, δ, η *and* $\det \mathbf{T}$ *(but not otherwise on* $\mathbf{G}$ *and* $\mathbf{T}$*).*

Proof. This is a generalization of Lemma 4.2, Corollary, of Chapter 12. The reader should have no difficulty in making the appropriate modification of the proof. [Essentially, the set of $\mathbf{Tb}, \mathbf{b} \in \mathbf{Z}^n$ form a sublattice Λ of $\mathbf{Z}^n$. One looks at the values taken by the form $\mathbf{G}$ on Λ. Cf. Siegel (1940)].

Proof of Theorem 11.1 Let f be a classically integral form of given determinant d and let g be a Minkowski-reduced Hermite majorant of f. Let $\mathbf{m}_1, \dots, \mathbf{m}_n$ be linearly independent vectors giving the successive minima of g. By Lemma 4.2, Corollary, of Chapter 12 the elements of the $\mathbf{m}_j$ are bounded by a constant depending only on n.

By Lemma 1.2 of Chapter 12, the form g is in a Siegel domain $\mathscr{S}_n(\delta,\eta)$ where δ,η depend only on n. By Lemma 11.2 the form given by $\mathbf{JG}^{-1}\mathbf{J}$ is in $\mathscr{S}_n(\delta,\eta_1)$, where η_1 depends only on n.

By Lemma 11.1 we have

$$\begin{aligned}\mathbf{G} &= \mathbf{FG}^{-1}\mathbf{F}\\ &= \mathbf{T}'(\mathbf{JG}^{-1}\mathbf{J})\mathbf{T}\end{aligned}$$

where

$$\mathbf{T} = \mathbf{JF}$$

and we have used (11.5) and $\mathbf{F}' = \mathbf{F}$. Lemma 11.3 (with $\mathbf{JG}^{-1}\mathbf{J}$ instead of $\mathbf{G}$ and η_1 instead of η) the elements of $\mathbf{Tm}_j$ are bounded by a constant depending only on η,δ,η_1 and $\det \mathbf{T}$, that is only on n and $d = \det \mathbf{F}$.

Since the $\mathbf{m}_j$ and the $\mathbf{Tm}_j$ are bounded, so also is the matrix $\mathbf{T} = \mathbf{JF}$. This is what was required.

We now discuss what becomes of Theorem 2.2 for isotropic forms. For any regular real form f the group $O_{\mathbf{R}}^+(f)$ has a natural topology. When f is integral, we can define $\mathbf{S}_1,\mathbf{S}_2 \in O_{\mathbf{R}}^+(f)$ to be equivalent if $\mathbf{S}_2 = \mathbf{TS}_1$ for some $\mathbf{T} \in O^+(f)$ (i.e. $\mathbf{T}$ is an integral automorph). Theorem 2.2 then says that if f is anisotropic, then there is a compact subset of $O_{\mathbf{R}}^+(f)$ which contains representatives of all equivalence classes. In fact the group $O_{\mathbf{R}}^+(f)$ has not merely a topology but an invariant (Haar) measure which is unique up to a multiplicative constant. [cf. Lemma 7.4]. We now have

THEOREM 11.2 [not proved here]. *Suppose that f is not an isotropic binary form. Then there is a subset of $O_{\mathbf{R}}^+(f)$ of finite invariant measure which contains representatives of all equivalence classes modulo $O_{\mathbf{R}}^+(f)$.*

Here we have supposed $O^+(f)$ to act on $O_{\mathbf{R}}^+(f)$ on the left but we could equally have made it act on the right. The isotropic binaries are a genuine exception to the Theorem.

We must refer to Siegel (1940) for the proof, which uses the representation of $O^+(f)$ constructed in Section 10. In fact there is an explicit description of the invariant measure, and the induced measure of the quotient space $O^+(f)\backslash O_{\mathbf{R}}^+(f)$ plays an important part in the analytic theory: see Appendix B.

12. EFFECTIVENESS

It was pointed out at the end of Section 1 of Chapter 9 that it is not immediately obvious that we can always solve the following problems:

Problem 1. Does a given integral quadratic form f represent integrally a given integer $b \neq 0$?

Problem 2. Are two given integral quadratic forms f_1, f_2 integrally equivalent? If so, can we give explicitly an integral transformation which takes f_1 into f_2?

The solution of Problem 1 can be reduced to that of Problem 2 by the type of argument used in the beginning of Section 6 of Chapter 9: the details are left to the reader. In this section we sketch an algorithm due to Siegel (1972) for solving Problem 2. We have in fact in various places already produced algorithms which together solve Problem 2. We shall recall them at the end of this section and give some further comments.

Siegel's algorithm follows on from the ideas of the previous section combining them with the results of Chapter 12, Section 7. As in Chapter 12, we regard forms of dimension n as points in the space $\mathbf{R}^N$, $N = \frac{1}{2}n(n+1)$. The set of all positive definite forms is denoted as before by $\mathscr{P}^\circ$. We introduce the new notation $\Phi = \Phi(f)$ for the set of all Hermite majorants of the regular form f.

LEMMA 12.1. *$\Phi(f)$ is a closed subset of*† $\mathscr{P}^\circ$.

Proof. Follows immediately from Lemma 11.1.

LEMMA 12.2. *$\Phi(f)$ consists of the $g(\mathbf{Sx})$ where $\mathbf{S}$ runs through $O^+_{\mathbf{R}}(f)$ and g is any element of $\Phi(f)$.*

Proof. In (11.2) we may suppose that

$$\det \mathbf{L} > 0 \tag{12.1}$$

since otherwise we can premultiply $\mathbf{L}$ by the diagonal matrix whose first element is -1 and whose remaining diagonal elements are $+1$. If $\mathbf{L}_1, \mathbf{L}_2$ are two solutions of (11.2) and (12.1), we must have $\mathbf{L}_2 = \mathbf{L}_1\mathbf{S}$ where $\mathbf{S} \in O^+_{\mathbf{R}}(f)$. The corresponding majorants by (11.3) are $\mathbf{G}_1 = \mathbf{L}_1'\mathbf{L}_1$ and $\mathbf{G}_2 = \mathbf{L}_2'\mathbf{L}_2 = \mathbf{S}'\mathbf{G}_1\mathbf{S}$.

COROLLARY. *$\Phi(f)$ is a connected set.*

Proof. For $\Phi(f)$ is the continuous image of the connected set $O^+_{\mathbf{R}}(f)$ [cf. Chapter 2, example 14].

LEMMA 12.3. *Let integers $D > 0$ and $n > 0$ be given. Then the coefficients of an Hermite-reduced integral form f of dimension n with $|d(f)| = D$ are at most equal to B where $B = B(n, D)$ can be effectively calculated.*

Proof. It is left to the reader to check that the proof of Theorem 11.1 can be reworked so as to give effectively computable bounds at every stage.

† Indeed $\Phi(f)$ is closed considered as a subset of $\mathbf{R}^N$, since $d(g) = |d(f)|$ is constant for $g \in \Phi(f)$: compare the *Note* after Theorem 5.1 of Chapter 12.

COROLLARY. *There are at most A distinct Hermite-reduced integral forms of dimension n with $|d(f)| = D$, where $A = A(n, D)$ can be effectively calculated.*

Proof. Clear.

Now let f be any integral form of dimension n and put $D = |d(f)|$. Any $g \in \Phi(f)$ belongs to at least one set $\mathbf{T}\mathscr{R}$ where $\mathbf{T} \in SL$ and $\mathscr{R}$ is the set of Minkowski-reduced forms. Then the form† $\mathbf{T}^{-1}g$ is Minkowski-reduced and so $\mathbf{T}^{-1}f$ is Hermite-reduced: further, every Hermite-reduced form h equivalent to f arises as a $\mathbf{T}^{-1}f$ in this way. Denote by $\Phi(f;h)$ the set of corresponding $g \in \Phi(f)$: that is

$$\Phi(f;h) = \cup\{\Phi(f) \cap \mathbf{T}\mathscr{R}\}, \tag{12.2}$$

the union being over the $\mathbf{T}$ for which

$$\mathbf{T}^{-1}f = h. \tag{12.3}$$

LEMMA 12.4. *$\Phi(f;h)$ is closed in $\mathscr{P}^{\circ}$.*

Proof. Each intersection $\Phi(f) \cap \mathbf{T}\mathscr{R}$ is closed by Lemma 5.3(ii) of Chapter 12 and Lemma 12.1. The truth of the Lemma now follows readily from the local finiteness of the cover of $\mathscr{P}^{\circ}$ by the $\mathbf{T}\mathscr{R}$ [Theorem 7.6 of Chapter 12].

LEMMA 12.5. *There is an E depending only on n with the following property:*

Suppose that $\Phi(f;h_1)$ and $\Phi(f;h_2)$ have a non-empty intersection. Then $h_2 = \mathbf{S}h_1$ for some $\mathbf{S} \in SL$ all of whose elements are at most E in absolute value.

Proof. Let $g \in \Phi(f;h_1) \cap \Phi(f;h_2)$. Then there are $\mathbf{T}_1, \mathbf{T}_2$ such that

$$h_j = \mathbf{T}_j^{-1}f \qquad (j = 1,2)$$

and that the forms

$$\mathbf{T}_j^{-1}g \qquad (j = 1,2)$$

are both Minkowski-reduced. On applying Theorem 1.2 of Chapter 12 to $\mathbf{T}_1^{-1}g$ we deduce that the elements of $\mathbf{T}_2^{-1}\mathbf{T}_1 = \mathbf{S}$ (say) are bounded by the constants there denoted by C_1. Since $h_2 = \mathbf{S}h_1$ the result follows.

THEOREM 12.1. *Let f_1, f_2 be integral Hermite-reduced forms and suppose that they are equivalent. Then there is a $\mathbf{T} \in SL$ such that*

$$f_2 = \mathbf{T}f_1$$

and such that the elements of $\mathbf{T}$ do not exceed a constant K depending only on the dimension n and the determinant of the f_j. Further K can be specified effectively.

† Recall that $\mathbf{T}^{-1}g$ is the form g^* (say) given by $g^*(\mathbf{x}) = g(\mathbf{T}^{-1}\mathbf{x})$.

Proof. We apply the preceding analysis where f is any form equivalent to f_1 and f_2 [e.g. $f = f_1$]. The sets $\Phi(f;h)$, where h is Hermite-reduced, are a finite cover of $\Phi(f)$. By Lemma 12.1 and Lemma 12.2, Corollary, there is a sequence h_j $(1 \leqslant j \leqslant J)$ of distinct Hermite-reduced forms such that

$$h_1 = f_1, \qquad h_J = f_2$$

and

$$\Phi(f;h_j) \cap \Phi(f;h_{j+1})$$

is non-empty for $1 \leqslant j < J$. By Lemma 12.5 we have

$$h_{j+1} = \mathbf{S}_j h_j \qquad (1 \leqslant j < J)$$

where the elements of $\mathbf{S}_j$ are bounded by E. Then $f_2 = \mathbf{T} f_1$ where

$$\mathbf{T} = \mathbf{S}_{J-1}\mathbf{S}_{J-2}\ldots\mathbf{S}_1.$$

Since $J \leqslant A$ by Lemma 12.3, Corollary, the elements of $\mathbf{T}$ are bounded by a constant K of the kind required.

We note that Theorem 12.1 does in fact give an effective solution of Problem 2. Clearly it is enough to solve the problem when f_1, f_2 are reduced. They are certainly not equivalent unless they have the same dimension and determinant. When this condition is satisfied then we have "only" to consider all the forms $\mathbf{T}f_1$ where $\mathbf{T}$ runs through all the matrices in SL whose elements are bounded by the K of the theorem. If f_2 is not equal to any of these $\mathbf{T}f_1$ then f_1, f_2 cannot be equivalent. Of course, the constant K obtained by the argument above is so large that the method could hardly be applied in practice.

In conclusion we enumerate other ways of tackling Problem 2 in special cases which together cover all possibilities. It is clearly enough to consider only f_1, f_2 in the same genus.

(i) f_1 and f_2 are definite forms. Then the solution of Problem 2 is trivial, as already remarked in Chapter 9, Section 1.

(ii) f_1 and f_2 are indefinite binary. Then Section 3 of this Chapter gives a decision procedure.

(iii) f_1 and f_2 are indefinite and $n \geqslant 3$. Then they are equivalent precisely when they are in the same spinor genus [Theorem 1.4 of Chapter 11] and Chapter 11 gives an effective procedure for deciding this. This answers the first question in Problem 2. One can answer the second question by analysing the arguments of Chapter 11 but that is not really necessary. From a purely logical point of view the second question is a mere rhetorical flourish once we have a decision procedure for the first question. Suppose we know that f_1 and f_2 are equivalent and let $\mathbf{T}_1, \mathbf{T}_2, \ldots$ be an enumeration of SL. Then consider the $\mathbf{T}_j f_1$ $(j = 1, 2, \ldots)$ in order. One of them must be equal to f_2 and this is a perfectly respectable way of finding a $\mathbf{T}$ with $\mathbf{T}f_1 = f_2$.

The procedures (i), (ii) and (iii) together cover all cases. We should however also mention

(iv) f_1 and f_2 both anisotropic. Let $b = f_2(1, 0, \ldots, 0)$. Then Theorem 2.2. Corollary gives us effectively a set of vectors $\mathbf{b}_j$ such that $f_1(\mathbf{b}_j) = b$ and such that the set contains at least one representative of each orbit of primitive representations of b by f_1. By "completing the square" with respect to each of the $\mathbf{b}_j$ as in Chapter 9, Section 6, we can use induction on the dimension n. The details are left to the reader.

NOTES

For generalizations to algebraic number fields, see Humbert (1949) and Ramanathan (1951, 1952). There is an analogue of Hermite reduction for fields with a discrete valuation in Springer (1956).

There is a table of classes of indefinite ternaries of low discriminant in Dickson (1930). See also Venkov (1945), where a problem in the geometry of numbers produces a number of anisotropic ternaries.

Section 2. Watson (1957) gives a weak generalization of Theorem 2.2. to all indefinite forms f with $n \geqslant 4$. He shows that if f represents an integer $b \neq 0$ then there is a bounded representation, but does not show that every orbit contains a bounded representation.

Section 7. For accounts of non-euclidean geometry see Bonola (1955), Coxeter (1942), Greenberg (1972) or Klein (1928).

Section 11. Theorem 11.1 was asserted by Hermite and proved by him for ternaries under the tacit assumption that they were anisotropic. According to Siegel (1940) "Die Hermitesche Behauptung is tatsächlich für jeden Fall ausnahmlos richtig. Hierfür wurde 1902 von Stouff ein Beweis gegeben; aber dieser Beweis ist so umständlich, dass er noch nicht einmal in dem sonst so ausführlichen Werke von Bachmann vollständig dargestellt worden ist."

EXAMPLES

[*Note.* There are examples on integral automorphs in some earlier chapters, especially Chapter 9, examples 14–19.]

1. Let $ax^2 + 2bxy + cy^2$ be a positive definite primitive classically integral form. Show that its only proper integral automorphs are $\pm\mathbf{I}$ except when $d = ac - b^2$ is 1 or 3. Determine all the proper automorphs in the exceptional cases.

2. Find a complete set of generators of the group of integral automorphs of $3x^2 + 2xy - 27y^2$.

3. (i) Show that 5 is not represented integrally by $x^2 - 79y^2$.

(ii) Deduce that the forms $x^2 - 79y^2$ and $5x^2 + 4xy - 15y^2$ are not equivalent, though they are in the same genus.

(iii) Find all the proper equivalence classes of classically integral forms $ax^2 + 2bxy + cy^2$ with $ac - b^2 = -79$.

4. (i) Show that 4 is not represented primitively by $x^2 + xy - 36y^2$.

(ii) Deduce that $x^2 + xy - 36y^2$ and $4x^2 + xy - 9y^2$ are not equivalent.

(iii) Determine all the classes of integer-valued quadratic forms $Ax^2 + Bxy + Cy^2$ with $B^2 - 4AC = 145$.

5. Let k be a positive integer. Show that $kx^2 + kxy - y^2$ does not represent any integer m in $0 < m < k$.

6. Let $f(x, y) = ax^2 + 2bxy + cy^2$ be a classically integral form with $a > 0 > c$. Let t, u be a solution of Pell's equation $t^2 + (ac - b^2)u^2 = 1$ with $t > 0, u > 0$. Show that every integral solution of $f(x, y) = e$ where $e > 0$ is in the same orbit (under the group of integral automorphs of f) as a solution with

$$x > 0, \quad y > 0, \qquad -aux + (t - bu)y \leqslant 0.$$

7. Show from first principles that there are only finitely many Hermite-reduced integral forms $ax^2 + 2bxy + cy^2$ with given determinant d.

[*Note*: the difficult case is $a = 0$.]

8. Let $f(x, y, z)$. be a classically integral form which is real equivalent to $X^2 + Y^2 - Z^2$ Let $\mathbf{b} \in \mathbf{Z}^3$ and let G consist of integral automorphs $\mathbf{T}$ of f such that $\mathbf{Tb} = \mathbf{b}$.

(i) if $f(\mathbf{b}) < 0$ show that G is finite.

(ii) if f is anisotropic and $f(\mathbf{b}) > 0$ show that G is infinite.

(iii) if f is isotropic and $f(\mathbf{b}) > 0$ show that G is usually infinite and discuss the cases when it is finite.

9. Find the group of integral automorphs of $f(x, y, z) = 3x^2 - 2y^2 - z^2$.
[*Hint*: f is isotropic and so $f = r(X_1X_3 - X_2^2)$ for some $r \in \mathbf{Q}$ and some linear forms X_1, X_2, X_3.]

10. Find a set of generators for the automorphism group of $f(x, y, z) = 3x^2 + 2y^2 - z^2$.

11. (i) Find a set of generators for the automorphism group of $f(x, y, z) = x^2 - 3y^2 - 2yz - 23z^2$.

(ii) Show that there are precisely two orbits of integral representations of 1 by f.

[*Note*. cf. end of Chapter 9, Section 1 where this form is referred to. Although this example could be done by hand it is perhaps best to use a computer to search for integral representations of 1 with small x, y, z.]

12. Find the group of integral automorphs of $x^2 + xy + xz + y^2 + z^2$.

13. Show that the order of the group of proper automorphs of a positive definite form f of dimension n must divide

$$g(n) = \prod_{q \text{ prime}} q^{\gamma(q)},$$

where

$$\gamma(2) = n - 1 + [n/2] + [n/2^2] + \dots$$

$$\gamma(q) = [n/(q-1)] + [n/q(q-1)] + [n/q^2(q-1)] + \dots \qquad (q > 2).$$

[*Hint*. Chapter 6, example 7.]

CHAPTER 14

Composition of Binary Quadratic Forms

1. INTRODUCTION

In this chapter it is convenient to take integral quadratic forms with the *non-classical* definition:

$$f(x, y) = ax^2 + bxy + cy^2 \tag{1.1}$$

where

$$a, b, c \in \mathbf{Z}. \tag{1.2}$$

Such a form is *primitive* if

$$\text{g.c.d.}(a, b, c) = 1. \tag{1.3}$$

Further, we shall use the *discriminant*

$$D = D(f) = b^2 - 4ac. \tag{1.4}$$

This is related to the determinant $d(f)$ by

$$D(f) = -4d(f). \tag{1.5}$$

We note that

$$D \equiv 0 \quad \text{or} \quad 1 \pmod 4 \tag{1.6}$$

and that

$$b \equiv D \pmod 2. \tag{1.7}$$

Further, by *equivalence class* we shall always in this chapter mean class for *proper equivalence*. It was shown by Gauss (who, however, worked with the classical definition of integrality) that the equivalence classes of primitive forms with given discriminant D have a natural structure as an abelian group $\mathscr{G}$. This group is closely related to the group of ideal classes in an appropriate quadratic ring, and this was essentially Gauss' approach, but we shall follow Dirichlet. The details will be given in Section 2 but the essential idea is that if

$$f_j(x, y) = a_j x^2 + bxy + c_j y^2 \qquad (j = 1, 2, 3) \tag{1.8}$$

are three primitive forms of discriminant D with the same central coefficient b and if

$$a_3 = a_1 a_2, \tag{1.9}$$

then the corresponding classes $\mathscr{C}_j$ satisfy

$$\mathscr{C}_3 = \mathscr{C}_1 \mathscr{C}_2. \tag{1.10}$$

It turns out that this definition is independent of the choice of the $f_j \in \mathscr{C}_j$. The unit element $\mathscr{E}$ of the group is the class of

$$f_0 = \begin{cases} x^2 - \frac{1}{4}Dy^2 & \text{if } D \text{ is even} \\ x^2 + xy + \frac{1}{4}(1 - D)y^2 & \text{if } D \text{ is odd.} \end{cases} \tag{1.11}$$

Further, the reciprocal $\mathscr{C}^{-1}$ of the class $\mathscr{C}$ consists of the forms which are improperly equivalent to those of $\mathscr{C}$.

Gauss considered the map

$$\mathscr{C} \to \mathscr{C}^2 \tag{1.12}$$

of $\mathscr{G}$ into itself. He showed that a class is in the image $\mathscr{G}^2$ precisely when it is in the *principal genus*, which is, by definition, the genus of the class $\mathscr{C}_0$ to which f_0 (given by (1.11)) belongs. More generally, the cosets of $\mathscr{G}$ modulo $\mathscr{G}^2$ are the genera. It follows that each genus contains the same number of classes, a fact which is special to binary forms.

Gauss observed that these facts gave a way to determine the number of genera. Since $\mathscr{G}$ is a finite group, the number of genera, that is the order of $\mathscr{G}/\mathscr{G}^2$, is equal to the order of the kernel $\mathscr{A}$ (say) of (1.12). A class $\mathscr{C}$ is in $\mathscr{A}$ by definition if $\mathscr{C}^2 = \mathscr{E}$, that is

$$\mathscr{C} = \mathscr{C}^{-1}. \tag{1.13}$$

Hence $\mathscr{C} \in \mathscr{A}$ if and only if the forms in it are improperly equivalent to themselves. Such forms and the classes to which they belong are called *ambiguous*.

The forms of the shape

$$ax^2 + cy^2 \tag{1.14}$$

and

$$ax^2 + axy + cy^2 \tag{1.15}$$

are clearly ambiguous. Conversely, it can be shown that, excluding some special cases, an ambiguous class contains precisely two forms (1.14) or (1.15) if $D < 0$ and precisely four such forms if $D > 0$. This enables us to determine the order of $\mathscr{A}$ and so of the genus group $\mathscr{G}/\mathscr{G}^2$.

The results just discussed give a proof of the existence of genera of binary forms with prescribed local properties subject to the obvious necessary conditions (Theorem 1.2 of Chapter 9). We proved this before by invoking Dirichlet's theorem on primes in arithmetic progression†. Gauss' proof, which we are now presenting, is the original one and antedates Dirichlet's theorem. In fact the "elementary" proof of Dirichlet's theorem [Selberg (1949)] makes essential use of the existence of genera.

There are two other places in the text where we have made essential use of Dirichlet's theorem. The first is the proof of the existence of rational forms with prescribed local properties (Theorem 1.3 of Chapter 6). This easily follows from the theorem on the existence of genera and so requires no special discussion. The third place where Dirichlet's theorem was used for the case $n = 4$ of the Strong Hasse Principle (Section 5 of Chapter 6). We shall show in Section 7 that this, too, can be disposed of without recourse to Dirichlet's theorem.

The fact that $\mathscr{G}/\mathscr{G}^2$ is the group of genera gives us some information about the 2-component of the class group $\mathscr{G}$. The methods of this chapter can be used to give further information about the 2-component; for example to decide when it contains elements of order 4. A curious by-product is information about when the "*negative Pell's equation*"

$$t^2 - Du^2 = -4 \tag{1.16}$$

has an integral solution t, u. We have seen that the ordinary Pell's equation with $+4$ as the right-hand side is always soluble provided that $D > 0$ and D is not a perfect square; and the solutions give the automorphs of forms f of discriminant D (Section 3 of Chapter 13). The equation (1.16) may or may not be soluble. When it is, every f of discriminant D is improperly equivalent to $-f$. [Theorem 3.1, Corollary of Chapter 13].

2. COMPOSITION OF BINARY FORMS

In this section we establish the existence of a group law on the (proper) equivalence classes of primitive binary forms of given discriminant D and develop its main properties. The group law is traditionally called "*composition*". It is convenient to introduce the notation

$$f = [a, b, c] \tag{2.1}$$

† via Theorem 1.3 of Chapter 6.

for the form

$$f(x, y) = ax^2 + bxy + cy^2. \tag{2.2}$$

We shall use "$\sim$" to denote (proper) equivalence.

LEMMA 2.1. *Let $f = [a, b, c]$ be a primitive form and let M be any integer. Then f represents an integer prime to M.*

Note. This is, of course, a special case of a theorem about forms in any number of variables.

Proof. Let p be a prime dividing M. We consider three cases (not mutually exclusive).

(i) $p \nmid a$. If $p \nmid x$ and $p | y$ then $f(x, y)$ is prime to p.
(ii) $p \nmid c$. Similar.
(iii) $p | a$, $p | c$, so $p \nmid b$. Then $p \nmid x$, $p \nmid y$ ensures that $f(x, y)$ is prime to p.

LEMMA 2.2. *Suppose that*

$$[a_1, b, c_1] \sim [a_2, b, c_2], \tag{2.3}$$

where the two forms are primitive and have the same middle coefficient. Let l be an integer,

$$l | c_1 \qquad l | c_2 \tag{2.4}$$

and suppose that

$$\text{g.c.d.}(a_1, a_2, l) = 1. \tag{2.5}$$

Then

$$[la_1, b, l^{-1}c_1] \sim [la_2, b, l^{-1}c_2]. \tag{2.6}$$

Proof. By hypothesis there is an integral matrix

$$\mathbf{T} = \begin{pmatrix} r & s \\ t & u \end{pmatrix} \qquad ru - ts = 1$$

such that

$$\mathbf{T}'\begin{pmatrix} 2a_1 & b \\ b & 2c_1 \end{pmatrix}\mathbf{T} = \begin{pmatrix} 2a_2 & b \\ b & 2c_2 \end{pmatrix},$$

where $\mathbf{T}'$ is the transpose. Then

$$(\mathbf{T}')^{-1} = \begin{pmatrix} u & -t \\ -s & r \end{pmatrix},$$

and so

$$\begin{pmatrix} 2a_1 & b \\ b & 2c_1 \end{pmatrix}\begin{pmatrix} r & s \\ t & u \end{pmatrix} = \begin{pmatrix} u & -t \\ -s & r \end{pmatrix}\begin{pmatrix} 2a_2 & b \\ b & 2c_2 \end{pmatrix}.$$

On equating the matrix elements on both sides and eliminating r, u, we obtain

$$a_1 s + c_2 t = 0$$
$$a_2 s + c_1 t = 0.$$

But now $l | s$ by (2.4), (2.5). Then the matrix

$$\mathbf{T}^* = \begin{pmatrix} r & l^{-1}s \\ lt & u \end{pmatrix}$$

is integral and gives the required equivalence between the forms (2.6).

In what follows we shall not often be concerned with the value of the third coefficient c of a form $[a, b, c]$. When $a \neq 0$, which will always be the case, the value of c is determined by a, b and the discriminant D. We therefore adopt the convention of using an asterisk (*) to denote a coefficient whose value is immaterial to the argument.

We shall say that two primitive forms

$$f_j = [a_j, b_j, c_j] \qquad (j = 1, 2) \tag{2.7}$$

of discriminant D are *concordant* if (i) $a_1 a_2 \neq 0$, (ii) the two middle coefficients are the same, say

$$b_1 = b_2 = b \tag{2.8}$$

and (iii) the form

$$f_3 = [a_1 a_2, b, *] \tag{2.9}$$

of discriminant D is integral. As already remarked in Section 1, f_3 is then necessarily primitive. It will be called the *composition* of f_1 and f_2.

Note. When g.c.d.$(a_1, a_2) = 1$, condition (iii) follows automatically from (i) and (ii). For then the integrality of (2.7) implies that $D - b^2$ is divisible by $4a_1$ and $4a_2$: hence it is divisible by $4a_1 a_2$, and so (2.9) is integral.

LEMMA 2.3. *Let $\mathscr{C}_1, \mathscr{C}_2$ be two classes of primitive forms of discriminant $D \neq 0$. Then there are concordant forms $f_j = [a_j, b, *] \in \mathscr{C}_j$. Further, they may be chosen so that a_1, a_2 are prime to one another and to any integer M given in advance.*

Proof. By Lemma 2.1 the class $\mathscr{C}_1$ represents some integer a_1 prime to M and $\mathscr{C}_2$ represents some a_2 prime to $a_1 M$. Hence there are forms

$$[a_j, b_j, *] \in \mathscr{C}_j \qquad (j = 1, 2).$$

A substitution

$$x \to x + l_j y, \qquad y \to y$$

replaces b_j by

$$b_j^* = b_j + 2a_j l_j.$$

We have $b_1 \equiv b_2 \pmod 2$ by (1.7) and so, since a_1, a_2 are coprime, we may choose l_1, l_2 so that $b_1^* = b_2^* = b$ (say), as required.

LEMMA 2.4. *Let $\mathscr{C}_1, \mathscr{C}_2$ be two classes of primitive forms of discriminant $D \neq 0$. Then there is a class $\mathscr{C}$ such that the composition of $f_j \in \mathscr{C}_j$ $(j = 1, 2)$ always lies in $\mathscr{C}$.*

Proof. Let

$$[a_j', b', *] \in \mathscr{C}_j \qquad (j = 1, 2) \tag{2.10}$$

and

$$[a_j'', b'', *] \in \mathscr{C}_j \qquad (j = 1, 2) \tag{2.11}$$

be two pairs of concordant forms. We have to show that

$$[a_1'a_2', b', *] \sim [a_1''a_2'', b'', *]. \tag{2.12}$$

By Lemma 2.3 with $M = a_1'a_2'a_1''a_2''$ there is a concordant pair

$$[a_j, b, *] \in \mathscr{C}_j \qquad (j = 1, 2) \tag{2.13}$$

such that

$$\text{g.c.d.}(a_1, a_2) = \text{g.c.d.}(a_1 a_2, a_1'a_2'a_1''a_2'') = 1. \tag{2.14}$$

By symmetry it will be enough to show that

$$[a_1'a_2', b', *] \sim [a_1 a_2, b, *]. \tag{2.15}$$

Since $a_1 a_2$ is prime to $a_1'a_2'$ by (2.14), we can, as in the proof of Lemma 2.3, find an integer B such that simultaneously

$$B \equiv b \pmod{2a_1 a_2}$$

$$B \equiv b' \pmod{2a_1'a_2'}.$$

Then we have

$$[a_j, b, *] \sim [a_j, B, *] \qquad (j = 1, 2) \tag{2.16}$$

$$[a_1 a_2, b, *] \sim [a_1 a_2, B, *] \tag{2.17}$$

and

$$[a_j', b', *] \sim [a_j', B, *] \qquad (j = 1, 2) \tag{2.18}$$

$$[a_1'a_2', b', *] \sim [a_1'a_2', B, *]. \tag{2.19}$$

Next we have

$$[a_1, B, *] \sim [a_1', B, *] \tag{2.20}$$

by (2.10), (2.13), (2.16) and (2.18). The conditions of Lemma 2.2 are satisfied† by $l = a_2'$, and so

$$[a_1 a_2', B, *] \sim [a_1' a_2', B, *]. \tag{2.21}$$

Similarly by Lemma 2.2 with $l = a_1$ we have

$$[a_1 a_2, B, *] \sim [a_1 a_2', B, *]. \tag{2.22}$$

Finally, the required equivalence (2.15) follows from (2.17), (2.19), (2.21) and (2.22). This completes the proof of the Lemma.

It is easy to see that any two forms of the same discriminant D which represent 1 are properly equivalent. We shall denote this equivalence class by $\mathscr{E}$. Then $f_0 \in \mathscr{E}$ where f_0 is given by (1.11).

THEOREM 2.1. *If $\mathscr{C}_1, \mathscr{C}_2$ are primitive classes of discriminant D, write*

$$\mathscr{C} = \mathscr{C}_1 \mathscr{C}_2,$$

where $\mathscr{C}$ is the class given by Lemma 2.4. This rule of composition gives the set $\mathscr{G}$ of primitive classes of discriminant D the structure of a finite abelian group. The neutral element ("one") of the group is the class $\mathscr{E}$ just defined. Further, the inverse $\mathscr{C}^{-1}$ of the class $\mathscr{C}$ is the class containing the forms improperly equivalent to those of $\mathscr{C}$.

Proof. The finiteness of $\mathscr{G}$ is a special case of Theorem 1.1 of Chapter 9 and

$$\mathscr{C}_1 \mathscr{C}_2 = \mathscr{C}_2 \mathscr{C}_1$$

follows at once from the definition.

Next for associativity. Let $\mathscr{C}_1, \mathscr{C}_2, \mathscr{C}_3$ be any three classes. Then we can choose successively a_1, a_2, a_3 represented by these classes so that a_2 is prime to a_1 and a_3 to $a_1 a_2$. By an argument already invoked on several occasions, there is a b such that

$$[a_j, b, *] \in \mathscr{C}_j \qquad (j = 1, 2, 3).$$

Then it is easily seen that $(\mathscr{C}_1 \mathscr{C}_2)\mathscr{C}_3$ and $\mathscr{C}_1(\mathscr{C}_2 \mathscr{C}_3)$ are both the class containing the integral form

$$[a_1 a_2 a_3, b, *].$$

† That condition (2.4) is satisfied follows from the type of argument used in the *Note* before Lemma 2.3. More precisely, $D - B^2$ is divisible by $4a_1'a_2'$ because (2.10) are concordant and also by $4a_1$, where a_1 is prime to $a_1'a_2'$ by construction.

Hence

$$(\mathscr{C}_1\mathscr{C}_2)\mathscr{C}_3 = \mathscr{C}_1(\mathscr{C}_2\mathscr{C}_3).$$

Clearly

$$\mathscr{E}\mathscr{C} = \mathscr{C}\mathscr{E} = \mathscr{C}$$

for any class $\mathscr{C}$. Finally, $[c, b, a]$ is improperly equivalent to $[a, b, c]$ and their composition is

$$[ac, b, 1],$$

which represents 1 and so is in $\mathscr{E}$. This confirms the statement about $\mathscr{C}^{-1}$ and so completes the proof.

The following simple consequence of Theorem 2.1 will be useful later:

LEMMA 2.5. *Let $n > 1$ be an integer. A necessary and sufficient condition that a class $\mathscr{C}$ be of the form $\mathscr{C} = \mathscr{C}_1^n$ is that $\mathscr{C}$ represent primitively an integer of the shape w^n, where w is prime to $2D$.*

Proof. Suppose, first, that $\mathscr{C}$ represents w^n. Then it contains a form

$$f = [w^n, b, c]$$

and some b, c. Since $D = b^2 - 4w^n c$, the middle term b is prime to w. Hence

$$f_1 = [w, b, w^{n-1}c]$$

is primitive. Clearly the class $\mathscr{C}_1$ of f_1 does what is required.

Now suppose that $\mathscr{C} = \mathscr{C}_1^n$. By Lemma 2.1 the class $\mathscr{C}_1$ represents an integer w prime to $2D$ and so contains a form

$$f_1 = [w, b_1, c_1].$$

As before, b_1 is prime to w. The substitution $x \to x + ly$, $y \to y$ replaces c_1 by

$$c_2 = l^2 w + b_1 l + c_1.$$

We can find an integer l such that c_2 is divisible by w^{n-1} (say by induction on n or by the use of Hensel's Lemma and the Chinese Remainder Theorem). Hence $\mathscr{C}_1$ contains a form

$$f_1 = [w, b, w^{n-1}c].$$

Then $\mathscr{C} = \mathscr{C}_1^n$ contains $[w^n, b, c]$, as required.

3. DUPLICATION AND GENERA

By the *principal genus* we mean the genus containing the principal class $\mathscr{E}$: that is, the set of forms equivalent to (1.11) over every $\mathbf{Z}_p$ (including $p = \infty$).

THEOREM 3.1. (Gauss). *A necessary and sufficient condition that a class $\mathscr{C}$ be in the principal genus is that $\mathscr{C} = \mathscr{C}_1^2$ for some class $\mathscr{C}_1$.*

Proof. (i) Suppose, first, that f is a form in the principal genus. Then by Lemma 9.1, Corollary, of Chapter 6 with $a = 1$ (or, alternatively, by Theorem 7.1 of Chapter 9), the form f represents properly a square w^2 prime to any given integer, and in particular, prime to $2D$. Hence $\mathscr{C} = \mathscr{C}_1^2$ for some class $\mathscr{C}_1$ by Lemma 2.5.

(ii) Now suppose that $\mathscr{C} = \mathscr{C}_1^2$. Then by Lemma 2.5 the class $\mathscr{C}$ represents some w^2 prime to $2D$. Hence $\mathscr{C}$ represents 1 over $\mathbf{Z}_p$ for $p = \infty$ and all $p \mid 2D$, and so is in the principal genus.

In order to discuss genera other than the principal genus we require some trivia about $\mathbf{Z}_p$-equivalence. We denote by $\mathscr{G}_p$ the set of equivalence classes of primitive $\mathbf{Z}_p$-integral forms of discriminant D for $\mathbf{Z}_p$-equivalent. Then by mimicking the argument of Section 2 it is easy to see that $\mathscr{G}_p$ has a natural structure as an abelian group. There is a natural group homomorphism

$$\mathscr{G} \to \mathscr{G}_p \tag{3.1}$$

which maps $\mathscr{C} \in \mathscr{G}$ into the $\mathbf{Z}_p$-class which contains it.

The following result is now immediate:

COROLLARY (to Theorem 3.1). *The genera are precisely the cosets of $\mathscr{G}$ modulo $\mathscr{G}^2$. In particular, any two genera contain the same number of classes.*

We now investigate the $\mathscr{G}_p$ more fully. Since every p-adic form is improperly $\mathbf{Z}_p$-equivalent to itself (Lemma 3.2, Corollary of Chapter 8), any element of $\mathscr{G}_p$ other than the identity is of order 2. This is confirmed in the enumeration which follows, which will be required later.

LEMMA 3.1. *Let p be an odd prime. Then*

(i) *if $p \nmid D$ the group $\mathscr{G}_p$ is trivial.*

(ii) *if $p \mid D$ there are precisely two classes, with representatives $x^2 - \frac{1}{4}Dy^2$ and $rx^2 - \frac{1}{4}r^{-1}Dy^2$, where r is any unit which is not in $(\mathbf{Q}_p^*)^2$ (i.e. a quadratic non-residue).*

LEMMA 3.2 ($p = \infty$).

(i) *if $D > 0$ the group $\mathscr{G}_\infty$ is trivial.*

(ii) *if* $D < 0$, *then* $\mathcal{G}_\infty$ *has two elements represented respectively by* $x^2 - \frac{1}{4}Dy^2$ *and* $-x^2 + \frac{1}{4}Dy^2$.

LEMMA 3.3 ($p = 2$).

(i) *if* $2 \nmid D$, *then* $\mathcal{G}_2$ *is trivial.*

(ii) *if* $2 | D$, *then*

$$D = 4d \qquad (d \in \mathbf{Z}_2) \tag{3.2}$$

and there are the following cases:

(iiα) $d \equiv 1 \pmod 4$. *Then* $\mathcal{G}_2$ *is trivial.*

(iiβ) $d \equiv -1 \pmod 4$. *Then* $\mathcal{G}_2$ *is of order* 2.

(iiγ) $2 | d$ *but* $2^3 \nmid d$. *Then* $\mathcal{G}_2$ *is of order* 2.

(iiδ) $2^3 | d$. *Then* $\mathcal{G}_2$ *is non-cyclic of order* 4.

Proofs. Only case (ii) of Lemma 3.3 requires discussion, the rest of the three lemmas being trivial. If (3.2) holds, every class contains a form

$$f_u = ux^2 - u^{-1}dy^2$$

where

$$u = 1, 3, 5 \text{ or } 7.$$

Further,

$$f_u \sim f_v$$

if and only if f_u represents v modulo 8. With this hint, the proof is easily completed.

Since $\mathcal{G}_p$ is trivial for all but finitely many p, the product group

$$\prod_{p,\ \text{inc}\ \infty} \mathcal{G}_p \tag{3.3}$$

is finite. In Section 5 we shall be concerned to find the image of $\mathcal{G}$ in (3.3) under the homomorphism (3.1). For later reference we enunciate the following consequence of Lemmas 3.1, 3.2 and 3.3.

LEMMA 3.4. *Let* λ *be the number of distinct odd primes dividing* D *and put*

$$\mu = \begin{cases} 1 & \text{if } D < 0 \\ 0 & \text{if } D > 0 \end{cases}$$

$$\nu = \begin{cases} 0 & \text{if } 2 \nmid D \text{ or } D = 4d,\ d \equiv 1 \pmod 4 \\ 2 & \text{if } 2^5 | D \\ 1 & \text{otherwise.} \end{cases}$$

Then the order of the group (3.3) *is*

$$2^{\lambda+\mu+\nu}.$$

4. AMBIGUOUS FORMS AND CLASSES

In this section we consider the group $\mathscr{A}$ of classes $\mathscr{C}$ such that $\mathscr{C}^2 = \mathscr{E}$ or, what is the same thing,

$$\mathscr{C}^{-1} = \mathscr{C}. \tag{4.1}$$

Such classes are called *ambiguous classes.* By Theorem 2.1 they are the classes which are improperly equivalent to themselves.

In Section 3 of Chapter 13 it was shown that an improper automorph $\mathbf{S}$ of a binary form f satisfies

$$\mathbf{S}^2 = \mathbf{I}. \tag{4.2}$$

There is a primitive vector $\mathbf{b}$ such that

$$\mathbf{Sb} = -\mathbf{b} \tag{4.3}$$

and this can be extended to a basis $\mathbf{a}, \mathbf{b}$ of $\mathbf{Z}^2$ with $\det(\mathbf{a}, \mathbf{b}) = 1$. Further,

$$\mathbf{Sa} = \mathbf{a} + w\mathbf{b} \tag{4.4}$$

for some $w \in \mathbf{Z}$. On replacing $\mathbf{a}$ by $\mathbf{a} + u\mathbf{b}$ with integral u, we replace w by $w - 2u$. Hence we may suppose without loss of generality that

$$w = 0 \quad \text{or } 1. \tag{4.5}$$

Hence f is equivalent to a form of one of the two types

$$f_a = [a, 0, c] \qquad D = -4ac \tag{4.6}$$

$$g_a = [a, a, c] \qquad D = a(a - 4c). \tag{4.7}$$

We shall refer to these as *ambiguous forms* of the first and second kind respectively. We have just proved

LEMMA 4.1. *Every ambiguous class contains at least one ambiguous form.*

We note that $a \mid D$ in both (4.6) and (4.7). It is a trivial matter to enumerate the primitive forms of the shape (4.6) and (4.7). The reader will have no difficulty in filling in the details, or he may refer forward to the proof of Lemma 6.1 which gives fuller information. For later reference we give the results as

LEMMA 4.2. *Let λ be the number of distinct odd prime divisors of D. Then the number of (primitive) ambiguous forms is given by the following table:*

D	*First kind*	*Second kind*	*Total*
D odd		$2^{\lambda+1}$	$2^{\lambda+1}$
$D = 4d$, $d \equiv 1(4)$	$2^{\lambda+1}$		$2^{\lambda+1}$
$D = 4d$, $d \equiv -1(4)$	$2^{\lambda+1}$	$2^{\lambda+1}$	$2^{\lambda+2}$
$2^3 \mid D, 2^5 \nmid D$	$2^{\lambda+2}$		$2^{\lambda+2}$
$2^5 \mid D$	$2^{\lambda+2}$	$2^{\lambda+2}$	$2^{\lambda+3}$

We shall not, however, be so much concerned with the number of ambiguous forms as with the number of ambiguous classes. To find this, we must determine the number of distinct ambiguous forms in an ambiguous class: and for this we must study the argument at the beginning of this section more closely. We note first that if $\mathbf{T}$ is a proper and $\mathbf{S}$ an improper automorph, then $\mathbf{TS}$ is improper and the application of (4.2) to it gives:

$$\mathbf{TSTS} = \mathbf{I}. \tag{4.8}$$

Clearly the improper automorph $\mathbf{S}$ determines the primitive vector $\mathbf{b}$ in (4.3) up to sign and so determines the resulting form (4.6) or (4.7) uniquely. Now let $\mathbf{S}_1$ be another improper automorph which leads to the *same* form (4.6) or (4.7) as $\mathbf{S}$ does and let $\mathbf{b}_1$ be the corresponding eigenvector. Then we must have

$$\mathbf{b}_1 = \mathbf{Tb}$$

for some proper automorph $\mathbf{T}$. But then $\mathbf{b}_1$ is clearly an eigenvector of $\mathbf{TST}^{-1}$, so

$$\begin{aligned}\mathbf{S}_1 &= \mathbf{TST}^{-1}\\ &= \mathbf{T}^2\mathbf{S}\end{aligned}$$

by (4.8).

On the other hand, the general improper automorph $\mathbf{S}_2$ is of the shape

$$\mathbf{S}_2 = \mathbf{TS}$$

where $\mathbf{T}$ runs through the group O^+ of proper automorphs. Hence the number of ambiguous forms in a given ambiguous class is the order of the quotient group $O^+/(O^+)^2$. This makes sense since we know that the proper orthogonal group O^+ is abelian and indeed is (i) of order 2 if $D < -4$ (ii) cyclic of order 4 or 6 if $D = -4$ or -3 (iii) cyclic of order 2 if D is a square and (iv) the product of a cycle of order 2 and an infinite cycle if $D > 0$ is not a perfect square. [Chapter 13, Section 3].

On computing $O^+/(O^+)^2$ for these groups, we have

LEMMA 4.3. *The number of ambiguous forms in an ambiguous class is* 4 *if D is positive but not a perfect square: otherwise the number is* 2.

COROLLARY. *The number of ambiguous classes is obtained by dividing the number in the last column of the table in Lemma* 4.2 *by* 4 *or* 2 *respectively.*

5. EXISTENCE THEOREM

Let $\mathscr{G}$ as always be the group of classes of primitive integral forms of discriminant D and let $\mathscr{G}_p$ be the corresponding group over $\mathbf{Z}_p$. In Section 3 we saw that there are natural homomorphisms

$$\mathscr{G} \to \mathscr{G}_p \tag{5.1}$$

and so

$$\mathscr{G} \to \prod_{p,\ \mathrm{inc}\ \infty} \mathscr{G}_p. \tag{5.2}$$

THEOREM 5.1. *If D is a perfect square, the image of* (5.2) *is the entire group. Otherwise, the image is a subgroup of index* 2.

Proof. We have already sketched the argument in Section 1. By Theorem 3.1 the kernel of (5.2) is precisely $\mathscr{G}^2$ and so the image of (5.2) is isomorphic to $\mathscr{G}/\mathscr{G}^2$.

By a general property of finite abelian groups, the order of $\mathscr{G}/\mathscr{G}^2$ is equal to the order of the kernel of the map $\mathscr{C} \to \mathscr{C}^2$ of $\mathscr{G}$ into itself. The kernel is the group $\mathscr{A}$ of ambiguous classes whose order is given by Lemma 4.3 Corollary. Thus we know the order of the image of (5.2).

The order of the whole group

$$\prod_{p,\ \mathrm{inc}\ \infty} \mathscr{G}_p \tag{5.3}$$

is given by Lemma 3.4. A minor miracle ensures that the order of (5.3) is equal to the order of $\mathscr{A}$ when D is a square, otherwise twice the order of $\mathscr{A}$. This concludes the proof.

COROLLARY. *For all p (including* ∞*) let* f_p *be a primitive integral p-adic form of discriminant D. Suppose that*

$$\prod_{p,\ \mathrm{inc}\ \infty} c_p(f_p) = 1, \tag{5.4}$$

where c_p is the Hasse–Minkowski invariant. Then there is a global integral form f which is $\mathbf{Z}_p$-equivalent to f_p for each p.

Proof. We know that (5.4) is necessary. The condition (5.4) determines a subgroup of (5.3) which is of index 2 if D is not a square, otherwise the whole group. This subgroup contains the image of (5.2) and so must coincide with it.

We note that the proof of Theorem 5.1 did not use the product formula for the Norm Residue Symbol, or, what is equivalent, the law of quadratic reciprocity. Indeed Theorem 5.1 can be used to furnish a proof of quadratic reciprocity ["Gauss' second proof", Gauss (1801), Sections 257–262; or see Dirichlet-Dedekind (1863), Supplement X.] We give just one case and will need

$$(-1/p) = \begin{cases} +1 & \text{if } p \equiv 1\,(4) \\ -1 & \text{if } p \equiv -1\,(4). \end{cases} \tag{5.5}$$

This can be proved by quadratic forms arguments, but it is in any case an almost immediate consequence of the definition, and we shall take it as known. We shall now prove the equation

$$(r/s)(s/r) = -1 \tag{5.6}$$

when the primes r, s satisfy

$$r \equiv s \equiv -1 \pmod 4). \tag{5.7}$$

We consider the forms of discriminant $4rs$. Since $rs \equiv 1\,(4)$, the local group $\mathscr{G}_p$ is trivial except for $p = r$ and $p = s$, and then it is of order 2. For $p = r, s$ we denote by χ_p the non-trivial character on $\mathscr{G}_p$. The $\mathbf{Z}_r$-equivalence class of f is given by

$$\chi_r(f) = (a/r),$$

where a is any r-adic unit represented by f over $\mathbf{Z}_r$, and χ_s is similarly defined. The form

$$f_0 = [1, 0, -rs]$$

has

$$\chi_r(f_0) = \chi_s(f_0) = +1.$$

The form $-f_0$ has

$$\chi_r(-f_0) = \chi_s(-f_0) = -1$$

by (5.5) and (5.7). Theorem 5.1 now implies that

$$\chi_r(f) = \chi_s(f) \tag{5.8}$$

for any form f with $D = 4rs$, in particular for

$$f = [r, 0, -s].$$

Here

$$\chi_r(f) = (-s/r) = -(s/r) \tag{5.9}$$

and

$$\chi_s(f) = (r/s). \tag{5.10}$$

Then (5.6) follows from (5.8), (5.9) and (5.10).

Because of its importance we note explicitly the application of the results of this section for rational equivalence:

LEMMA 5.1. *Let $D \in \mathbf{Q}^*$. For all p including ∞ let $g_p(x, y)$ be a $\mathbf{Q}_p$-form of discriminant D. Suppose that the Hasse–Minkowski symbol $c_p(g_p)$ is $+1$ for all except finitely many p and that*

$$\prod_{p,\,\mathrm{inc}\,\infty} c_p(g_p) = 1.$$

Then there is a $\mathbf{Q}$-form of discriminant D which is equivalent to g_p over $\mathbf{Q}_p$ for each p.

Proof. For each p we may replace g_p by a form of discriminant D which is $\mathbf{Q}_p$-equivalent to it. If g_p is equivalent to

$$g_0 = x^2 - \tfrac{1}{4}Dy^2$$

then we replace g_p by g_0. We can certainly do this if $p \neq \infty$, $p \nmid 2D$ and $c_p(g_p) = +1$. In particular there are only finitely many p for which $g_p \neq g_0$. We can now find an integer $m \neq 0$ such that

$$f_p(x, y) = mg_p(x, y)$$

is $\mathbf{Z}_p$-integral and primitive for every p. By Theorem 5.1, Corollary, there is a global form f which is $\mathbf{Z}_p$-equivalent to f_p for every p. Then the form $m^{-1}f = g$ does what is required.

6. THE 2-COMPONENT OF THE CLASS-GROUP AND PELL'S EQUATION

The results already obtained give a procedure for investigating the 2-component of the class group $\mathscr{G}$. We have the two bits of information:

(i) the group $\mathscr{A}$ of ambiguous classes consists precisely of the elements of order 2 in $\mathscr{G}$ together with the principal class $\mathscr{E}$. (see Section 4).

(ii) a class $\mathscr{C}$ is of the shape $\mathscr{C} = \mathscr{C}_1^2$ precisely when $\mathscr{C}$ is in the principal genus (Theorem 3.1).

The simplest case is when $\mathscr{E}$ is the only class of $\mathscr{A}$ in the principal genus. Then $\mathscr{G}$ cannot contain any element of order 4 and $\mathscr{A}$ must be the complete 2-component of $\mathscr{G}$.

If this does not happen, then there is some $\mathscr{C} \neq \mathscr{E}$ in $\mathscr{A}$ which is in the principal genus. Then

$$\mathscr{C} = \mathscr{C}_1^2 \tag{6.1}$$

for some $\mathscr{C}_1$. Hence $\mathscr{C}_1$ is of order 4. If $\mathscr{C}_2$ is one solution of (6.1) then the most general solution is

$$\mathscr{C}_1 = \mathscr{C}_2\mathscr{C}_3, \tag{6.2}$$

where $\mathscr{C}_3$ runs through the elements of $\mathscr{G}$ of order 2, i.e. through $\mathscr{A}$. If none of (6.2) is in the principal genus then $\mathscr{C}$ cannot be put in the shape $\mathscr{C} = \mathscr{C}_4^4$. If, however, one of (6.2) is in the principal genus, then such a $\mathscr{C}_4$ exists: it has order 8 and we can continue the process.

In fact, however, the arguments of the preceding sections do not so much give us the group $\mathscr{A}$ of ambiguous classes as the set $\mathscr{B}$ (say) of ambiguous forms. It turns out that if we attempt to work with $\mathscr{B}$ rather than $\mathscr{A}$, we obtain information not merely about the 2-component of $\mathscr{A}$ but also about Pell's equation

$$t^2 - Du^2 = 4.$$

This is not surprising because Pell's equation determines the automorphs of forms of discriminant D and these played an important rôle in the arguments of Section 4.

The key is the following

LEMMA 6.1. *The set $\mathscr{B}$ of ambiguous forms can be given a structure of abelian group of exponent 2 in such a way that the map*

$$\mathscr{B} \to \mathscr{A}, \tag{6.4}$$

which takes a form into its equivalence class, is a group homomorphism.

Proof.† For typographical convenience we modify our notation for the ambiguous forms and write

$$f(a) = [a, 0, c] \qquad D = -4ac \tag{6.5}$$

$$g(a) = [a, a, c] \qquad D = a(a - 4c). \tag{6.6}$$

We denote the principal form by e, so

$$e = \begin{cases} [1, 0, -\tfrac{1}{4}D] = f(1) & (D \text{ even}) \\ [1, 1, \tfrac{1}{4}(1 - D)] = g(1) & (D \text{ odd}). \end{cases} \tag{6.7}$$

† The somewhat artificial proof which follows is motivated by the multiplication of the corresponding ideals (cf. *Notes*).

This will be the unit in the group law on $\mathscr{B}$. We denote the operation on $\mathscr{B}$ by an asterisk ($*$).

Since we want $\mathscr{B}$ to have exponent 2 we must have

$$f(a) * f(a) = e \tag{6.8}$$

$$g(a) * g(a) = e \tag{6.9}$$

for all a such that $f(a)$ or $g(a)$ is defined. Further, we put

$$f(a_1) * f(a_2) = f(a_3), \tag{6.10}$$

where

$$a_1 a_2 = a_3\{\gcd(a_1, a_2)\}^2. \tag{6.11}$$

The rest of the prescription for $*$ differs according to the 2-adic nature of D and we split cases:

(i) D odd, so $D \equiv 1 \pmod 4$. Here there are no ambiguous forms of the type $f(a)$. We have

$$g(1) = e \tag{6.12}$$

and we prescribe

$$g(a_1) * g(a_2) = g(a_3) \tag{6.13}$$

where a_1, a_2, a_3 are related by (6.11).

(ii) $D = 4d$, $d \equiv 1$ (4). Then every element of $\mathscr{B}$ is of the type $f(a)$, so (6.10) is enough to define ($*$).

(iii) $D = 4d, d \equiv -1$ (4). The elements of $\mathscr{B}$ are of the type $f(\alpha)$ and $g(2\alpha)$ where α is odd. We define $*$ by the following table

	$f(\alpha_1)$	$g(2\alpha_1)$
$f(\alpha_2)$	$f(\alpha_3)$	$g(2\alpha_3)$
$g(2\alpha_2)$	$g(2\alpha_3)$	$f(\alpha_3)$

where

$$\alpha_1 \alpha_2 = \alpha_3\{\gcd(\alpha_1, \alpha_2)\}^2. \tag{6.14}$$

(iv) $2^3 | D$ but $2^5 \nmid D$. Here every element of $\mathscr{B}$ is of the type $f(a)$ and so (6.10) suffices.

(v) $2^5 | D$, say $D = 2^{2+\delta} D_1$ where $2 \nmid D_1$ and $\delta \geqslant 3$. The ambiguous forms are of the shape $f(\alpha)$, $f(2^\delta \alpha)$, $g(2^2\alpha)$ and $g(2^\delta \alpha)$, where α is odd. We define $*$ by the table:

	$f(\alpha_1)$	$f(2^\delta\alpha_1)$	$g(2^2\alpha_1)$	$g(2^\delta\alpha_1)$
$f(\alpha_2)$	$f(\alpha_3)$	$f(2^\delta\alpha_3)$	$g(2^2\alpha_3)$	$g(2^\delta\alpha_3)$
$f(2^\delta\alpha_2)$	$f(2^\delta\alpha_3)$	$f(\alpha_3)$	$g(2^\delta\alpha_3)$	$g(2^2\alpha_3)$
$g(2^2\alpha_2)$	$g(2^2\alpha_3)$	$g(2^\delta\alpha_3)$	$f(\alpha_3)$	$f(2^\delta\alpha_3)$
$g(2^\delta\alpha_2)$	$g(2^\delta\alpha_3)$	$g(2^2\alpha_3)$	$f(2^\delta\alpha_3)$	$f(\alpha_3)$

where $\alpha_1, \alpha_2, \alpha_3$ are related by (6.14).

It is readily verified that $*$ so defined is a group law and that (6.4) is a homomorphism. This completes the proof of Lemma 6.1.

We shall now investigate the kernel of (6.4). A form $h \in \mathscr{B}$ is in the kernel if it is equivalent to the principal form e, that is if h represents 1. There is one obvious element of the kernel, namely

$$e' = \begin{cases} g(-D) = [-D, -D, \tfrac{1}{4}(1-D)] & \text{if } 2 \nmid D \\ f(-d) = [-d, 0, 1] & \text{if } D = 4d, \end{cases} \tag{6.15}$$

since $g(-D)(1, -2) = 1$ and $f(-d)(0, 1) = 1$. We note that

$$e' * f(a) = f(a'), \qquad 4aa' = -D \tag{6.16}$$

and

$$e' * g(a) = g(a''), \qquad aa'' = -D; \tag{6.17}$$

and it is easy to verify directly that $f(a) \sim f(a')$, $g(a) \sim g(a'')$.

When $D < 0$, (i.e. the forms are definite) Lemma 4.3 tells us that the kernel of the map $\mathscr{B} \to \mathscr{A}$ has order precisely 2 and indeed when $D \neq -4$ it is not difficult to see from the theory of reduction of binary forms that e and e' are the only elements of $\mathscr{B}$ which can represent $+1$. The remaining case $D = -4$ is anomalous: then $e = e'$ but nevertheless the kernel has order 2.

When $D > 0$ but is not a perfect square, Lemma 4.3 tells us that there are two further elements in the kernel besides e and e'. We can obtain them from a knowledge of the solution of Pell's equation

$$t^2 - Du^2 = 4. \tag{6.18}$$

By the theory of automorphs of binary forms we know that this has a solution, and the solutions with, say, $t > 0$ are all in an appropriate sense "powers" of a fundamental solution.

Suppose then that t, u is an integral solution of (6.8). Then

$$Du^2 = (t+2)(t-2) \tag{6.19}$$

where $t + 2$, $t - 2$ can have only 1, 2 and 4 as their greatest common divisor. We consider their factorization in the light of (6.19) and distinguish three cases.

(i) t is odd, so D and u are odd. Then

$$t + 2 = D_1 u_1^2, \qquad t - 2 = D_2 u_2^2$$

where

$$D = D_1 D_2, \qquad u = u_1 u_2.$$

On eliminating t we have

$$D_1 u_1^2 - D_2 u_2^2 = 4.$$

Since u_1, u_2 are odd, we can put

$$u_1 = u_2 + 2v$$

and then

$$1 = g(D_1)(v, u_2).$$

(ii) $t \equiv 2 \pmod 4$. Then

$$t + 2 = 4X, \qquad t - 2 = 4Y,$$

where

$$X - Y = 1$$

and

$$16XY = Du^2.$$

Suppose, first, that D is even, so $D = 4d$. Then $X = d_1 u_1^2$, $X_2 = d_2 u_2^2$ for integers d_1, d_2, u_1, u_2 with $d_1 d_2 = d$. Then

$$1 = f(d_1)(u_1, u_2).$$

If, however, D is odd then $X = D_1 u_1^2$, $Y = D_2 u_2^2$ with $D_1 D_2 = D$ and then

$$\begin{aligned} 1 &= D_1 u_1^2 - D_2 u_2^2 \\ &= g(D_1)(u_1 - u_2, 2u_2). \end{aligned}$$

(iii) $t \equiv 0\,(4)$. Then

$$t + 2 = 2X, \qquad t - 2 = 2Y$$

where X, Y are odd and

$$X - Y = 2, \qquad 4XY = Du^2.$$

If $D = 4d$ is even, we have

$$d_1 u_1^2 - d_2 u_2^2 = 2$$

so $d = d_1 d_2 \equiv 3\,(4)$. Put $u_1 = u_2 + 2v$. Then

$$1 = g(2d_1)(v, u_2).$$

If, however, D is odd we have $X = D_1u_1^2$, $Y = D_2u_2^2$ with $D = D_1D_2$ and then

$$D_1u_1^2 - D_2u_2^2 = 2.$$

This is a contradiction since an odd discriminant satisfies $D \equiv 1\ (4)$.

To sum up the argument so far: every solution of Pell's equation (6.18) leads to a representation of 1 by some form $h \in \mathscr{B}$ which is, consequently, in the kernel of $\mathscr{B} \to \mathscr{A}$. The arguments also run in the opposite direction but more simply, the details being left to the reader: a representation of 1 by an $h \in \mathscr{B}$ gives rise to a solution (t, u) of Pell's equation. Further, it is readily verified that if $h = e$ or $h = e'$, then (t, u) is the "square" of another solution of Pell's equation. Hence if (t, u) is not a "square", for example if it is a fundamental solution, then h is an element of the kernel of $\mathscr{B} \to \mathscr{A}$ other than e and e'. We have, in fact, now virtually reproved Lemma 4.3 as well as obtaining new information.

We can now revert to the algorithm described at the beginning of this section. If we do not know a solution of Pell's equation, or do not wish to use it, then we might work with $\mathscr{B}$ instead of $\mathscr{A}$. If $h \in \mathscr{B}$ we can find successively if there are classes $\mathscr{C}_m$ such that

$$h \in \mathscr{C}_m^M \qquad M = 2^m.$$

If m can be arbitrarily large, then h is in the kernel of $\mathscr{B} \to \mathscr{A}$. If not, then we obtain elements of the 2-component of $\mathscr{G}$ and a tedious but simple examination shows that we can compute the precise structure of the 2-component.

Rather than discuss this process in the abstract we discuss a concrete example. It is in fact more convenient to work with $\mathscr{B}_1$, the quotient of $\mathscr{B}$ by the subgroup $\{e, e'\}$ of order 2. If $D > 0$ is not a perfect square, then the kernel of $\mathscr{B}_1 \to \mathscr{A}$ is of order 2.

Our example is

$$D = 4rs$$

where r, s are primes and

$$r \equiv s \equiv 1\ (4).$$

The only local groups $\mathscr{G}_p$ which are not trivial are those with $p = r$ and $p = s$. Hence there are two genera, which we denote by $\varepsilon = +1$ and $\varepsilon = -1$. If a is prime to r and represented by $\mathscr{C}$, then $\mathscr{C}$ is in the genus ε given by

$$(a/r) = \varepsilon:$$

and similarly for s instead of r [cf. (5.8).].

There are eight forms in $\mathscr{B}$ so there are four cosets modulo $\{e, e'\}$. They can be represented by:

$$f(1) = [1, 0, -rs] \tag{6.20}$$

$$f(-1) = [-1, 0, rs] \tag{6.21}$$

$$f(r) = [r, 0, -s] \tag{6.22}$$

$$f(s) = [s, 0 - r]. \tag{6.23}$$

It follows at once that the group $\mathscr{A}$ is of order 2, so $\mathscr{G}$ contains only one element of (precise) order 2: hence the 2-component of $\mathscr{G}$ is cyclic.

Since

$$(-1/r) = (-1/s) = +1,$$

the form $f(-1)$ is always in the principal genus. We note further that, by the law of quadratic reciprocity,

$$(r/s) = (s/r) \qquad (=\eta,\ \text{say}).$$

Suppose, first, that $\eta = -1$. Then $f(r)$ and $f(s)$ are both in the genus $\varepsilon = -1$. Hence they cannot be in the kernel. Since the kernel is of order 2, the form $f(-1)$ must be in the kernel. In particular there must be an integral solution of $x^2 - rsy^2 = -1$. Further, the 2-component of the class group $\mathscr{G}$ has order precisely 2.

Now suppose that $\eta = +1$. Then all the forms of $\mathscr{B}$ are in the principal genus. In particular, there is an integer w prime to $2rs$ and integers x, y such that

$$f(r)(x, y) = rx^2 - sy^2 = w^2. \tag{6.24}$$

Hence $f(r)$ is in the square of the class $\mathscr{C}_r$ of

$$[w, b, c]$$

for some b, c. The class $\mathscr{C}_r$ is not determined uniquely but its genus is: for any other candidate must be of the shape $\mathscr{C}_r\mathscr{C}_1$ where $\mathscr{C}_1$ is ambiguous and so in the principal genus. The character ε of $\mathscr{C}_r$ is given by

$$\varepsilon = (w/r) = (w/s).$$

We now show that ε can be described in terms of r and s alone. For this we must introduce the so-called "conditional biquadratic symbol". If a is a quadratic residue of the prime $p \equiv 1\ (4)$ we write

$$[a/r]_4 = 1$$

if there is a solution b of $a \equiv b^4 \pmod{r}$: otherwise

$$[a/r]_4 = -1.$$

Clearly

$$[a_1a_2/r]_4 = [a_1/r]_4\,[a_2/r]_4$$

in the sense that if two of the symbols are defined then so is the third and the equation holds.

With this notation we have

$$(w/r) = [w^2/r]_4 = [-sy^2/r]_4 = [s/r]_4\,[-1/r]_4\,(y/r).$$

We now evaluate (y/r). Let q be an odd prime dividing y. Then

$$(r/q) = 1$$

by (6.24), and so

$$(q/r) = 1$$

by quadratic reciprocity. Hence

$$(y/r) = (2/r)^t$$

where 2^t is the precise power of 2 dividing y.

We must now divide cases. Suppose, first, that $r \equiv 1\ (8)$. Then

$$[-1/r]_4 = (2/r) = +1$$

and so

$$(w/r) = [s/r]_4. \tag{6.25}$$

Next, suppose that $r \equiv 5\ (8)$. Then

$$[-1/r]_4 = (2/r) = -1.$$

On the other hand, if $r \equiv 5\ (8)$, then $t = 1$, as is seen by regarding (6.14) as a congruence modulo 8. Hence again (6.25) holds.

To sum up the argument so far: the form $f(r)$ is in a class $\mathscr{C}_r^2$ for some $\mathscr{C}_r$. The genus of $\mathscr{C}_r$ is independent of the choice of $\mathscr{C}_r$ and has character

$$\varepsilon_r = [s/r]_4.$$

Similarly $f(s) \in \mathscr{C}_s^2$ for some class $\mathscr{C}_s$ with character

$$\varepsilon_s = [r/s]_4.$$

We now consider the various possible values of ε_r, ε_s.

Case (*i*) $\varepsilon_r = \varepsilon_s = -1$. Then neither $f(r)$ nor $f(s)$ is in a class $\mathscr{C}^4$: in particular they cannot be in the kernel. Hence $f(-1)$ must be in the kernel and $x^2 - sry^2 = -1$ is soluble. The 2-component of $\mathscr{G}$ is cyclic of order 4.

Case (*ii*) $\varepsilon_r\varepsilon_s = -1$, say $\varepsilon_r = +1$, $\varepsilon_s = -1$. Then $f(r) \in \mathscr{C}^4$ for some $\mathscr{C}$ but $f(s)$ is not in the fourth power of a class. But $f(1), f(-1)$, $f(r)$, $f(s)$ are equivalent in pairs, so $f(1), f(r)$ must be in the kernel of $\mathscr{B} \to \mathscr{A}$, whereas $f(-1), f(s)$ are not. In particular, $rx^2 - sy^2 = 1$ is soluble but $x^2 - rsy^2 = -1$ is not soluble. The 2-component of $\mathscr{G}$ is cyclic of order 4.

Case (iii) $\varepsilon_r = \varepsilon_s = +1$. Here both $f(r)$ and $f(s)$ are in classes $\mathscr{C}^4$ for some $\mathscr{C}$. We have no information about the kernel of $\mathscr{B} \to \mathscr{A}$. The 2-component of $\mathscr{G}$ is cyclic of order at least 8.

7. ELIMINATION OF DIRICHLET'S THEOREM

As has already been remarked, there is some methodological interest in developing the theory of quadratic forms without use of Dirichlet's theorem about the existence of primes in arithmetic progressions.

This theorem has been used in two places.

(i) Theorem 1.3 of Chapter 6. Here the Dirichlet theorem is used only for binaries and Lemma 5.1 is precisely the assertion of that theorem for this case.

(ii) The proof of the case $n = 4$ of the Strong Hasse Principle (Section 5 of Chapter 6). We shall show that this case can be deduced from an application of the Weak Hasse Principle. Since the latter can be proved without Dirichlet's Theorem (Section 11 of Chapter 6), this will do what we want.

Let then $f(x_1, \dots, x_4)$ be a regular quadratic form with coefficients in $\mathbf{Q}$ which is isotropic over every $\mathbf{Q}_p$ (including $p = \infty$). We have to show that f is isotropic over $\mathbf{Q}$. By hypothesis, for each p there is a binary form g_p such that f is equivalent over $\mathbf{Q}_p$ to

$$g_p + H,$$

where H is a hyperbolic plane. Then

$$d(g_p) = -d(f)$$

and

$$\prod_{p,\,\text{inc}\,\infty} c_p(g_p) = 1,$$

since

$$c_p(f) = c_p(g_p)c_p(H)\left(\frac{-d(f), -1}{p}\right).$$

Hence by Lemma 5.1 there is a global form g which is equivalent to g_p over each $\mathbf{Q}_p$.

Consider the form in six variables

$$h(x_1, \dots, x_6) = f(x_1, \dots, x_4) - g(x_5, x_6).$$

Over each $\mathbf{Q}_p$ the form h is equivalent to the sum of three hyperbolic planes.

By the Weak Hasse Principle (in the formulation that an element of the global Witt group which is everywhere locally trivial must be trivial) it follows that h is globally the sum of three hyperbolic planes. This means that there are six linear forms $t_1, \dots, t_6$ in $x_1, \dots, x_6$ with rational coefficients such that

$$h(x_1, \dots, x_6) = t_1 t_4 + t_2 t_5 + t_3 t_6.$$

We can now solve the three homogeneous equations

$$t_j(a_1, a_2, a_3, a_4, 0, 0) = 0 \qquad (j = 1, 2, 3)$$

for rationals $a_1, \dots, a_4$ not all 0. Then

$$\begin{aligned} 0 &= h(a_1, \dots, a_4, 0, 0) \\ &= f(a_1, \dots, a_4), \end{aligned}$$

as required. [For a variant of this proof invoking the Geometry of Numbers instead of the Weak Hasse Principle, see Cassels (1959a).]

NOTES

In Sections 2–5 we have followed Gauss (1801) except that instead of his definition of composition we have used the equivalent one of Dirichlet (1851) [or cf. Dirichlet–Dedekind (1863), Pall (1948)]. Gauss defines the class $\mathscr{C}$ of a form $f(x, y)$ to be the composition of the classes $\mathscr{C}_j$ containing the forms f_j $(j = 1, 2)$ if there are integral bilinear forms

$$X(x_1, y_1; x_2, y_2), \qquad Y(x_1, y_1; x_2, y_2)$$

such that identically

$$f(X, Y) = f_1(x_1, y_1)\, f_2(x_2, y_2).$$

The theory of composition over $\mathbf{Z}$ is equivalent to the theory of ideals in rings of algebraic integers of dimension 2 over $\mathbf{Z}$. We sketch the connection briefly assuming that the reader is familiar with algebraic number theory. For simplicity we consider only fundamental discriminants.† [For a fuller account, see Jones (1949) or Jones (1950). For a generalization, see Kaplansky (1968).]

Let $\Delta \neq 1$ be a square-free rational integer and put

$$\delta^2 = \Delta.$$

The ring I of integers of $\mathbf{Q}(\delta)$ has $\mathbf{Z}$-basis $1, \omega$ where

$$\omega = \begin{cases} \frac{1}{2}(1 + \delta) & \text{if } \Delta \equiv +1 \pmod 4 \\ \delta & \text{otherwise.} \end{cases}$$

† For definition, see Example 1.

Then

$$D = \begin{cases} \Delta \\ 4\Delta \end{cases}$$

is a fundamental discriminant.

Let J be an ideal in I, say with **Z**-basis α, β. Then

$$N(x\alpha + y\beta) = jf(x, y)$$

where N denotes the norm for $\mathbf{Q}(\delta)/\mathbf{Q}$, the form $f(x, y)$ has discriminant D and $j > 0$ is the norm of the ideal J. If we replace α, β by another **Z**-basis of J, then f is replaced by a form properly or improperly equivalent to it.

If $\lambda \in I$, the ideal λJ has basis $\lambda\alpha$, $\lambda\beta$ and so gives rise to the form $\pm f$. We say that ideals J_1, J_2 are equivalent in the wide sense if there are $\lambda_1, \lambda_2 \in I$ (not zero) such that

$$\lambda_1 J_1 = \lambda_2 J_2. \tag{§}$$

The argument above shows that then J_1, J_2 give rise to the same wide classes of quadratic forms $\pm f$. Conversely, suppose that the bases α_1, β_1 and α_2, β_2 of the ideals J_1, J_2 give the same quadratic form f; that is

$$N(x\alpha_1 + y\beta_1) = j_1 f(x, y)$$
$$N(x\alpha_2 + y\beta_2) = j_2 f(x, y)$$

where j_1, j_2 are the norms of J_1, J_2. Then $\theta_1 = \beta_1/\alpha_1$ and $\theta_2 = \beta_2/\alpha_2$ are both roots of the equation $f(\theta, -1) = 0$. There are thus two possibilities. The first is that $\theta_1 = \theta_2$; in which case J_1, J_2 are equivalent in the wide sense defined above. The other possibility is that θ_2 is the conjugate of θ_1; and then J_2 is widely equivalent to the conjugate of J_1.

An ideal J of I has a basis of the special shape

$$\left.\begin{aligned} \alpha &= l \\ \beta &= m + n\omega \end{aligned}\right\},$$

where $l, m, n \in \mathbf{Z}$ and $l > 0, n > 0$ [here $n | l$, $n | m$ because $l\omega \in J$, $\omega\beta \in J$, but we do not need this]. Hence J determines not merely a wide equivalence class but a narrow (=proper) equivalence class of forms. Further, it can be shown that the narrow equivalence classes determined by J_1, J_2 in (§) are the same provided that

$$N(\lambda_1)\, N(\lambda_2) > 0. \tag{§§}$$

We say that the ideals J_1, J_2 are equivalent (in the narrow sense) if (§), (§§) hold. The argument above has set up a 1–1 correspondence between (proper) classes of quadratic forms and narrow classes of ideals. We define the product $J = J_1 J_2$ of ideals J_1, J_2 to be the ideal generated by the $\gamma_1\gamma_2$ with $\gamma_j \in J_j$ ($j = 1, 2$). Then the form f belonging to J is the composition of the forms f_1, f_2 belonging to J_1, J_2.

Brandt has shown that the Gaussian definition of composition extends to certain classes of quaternary forms with square determinant. The situation is related to ideals in rings of generalized quaternions but is much more complicated than the binary case: for example the classes do not form a group but only a "groupoid" [Brandt (1924, 1925, 1925a, 1928) and the survey Brandt (1943). For a generalization see Kaplansky (1969)].

The absence of any further generalization of composition to general n is related to a theorem of Hurwitz (1898, 1923) [see also Eckman (1943)]. To explain it, we recall that for $n = 2, 4, 8$ there are bilinear forms $X_j(\mathbf{y}, \mathbf{z})$ $(1 \leqslant j \leqslant n)$ in the variables $\mathbf{y} = (y_1, \ldots, y_n)$, $\mathbf{z} = (z_1, \ldots, z_n)$ such that

$$\Sigma X_j^2 = (\Sigma y_j^2)(\Sigma z_j^2). \tag{£}$$

The formulae for $n = 2, 4$ are given by the multiplication rules in $\mathbf{Z}[i]$ and in the quaternions and that for $n = 8$ by that in a non-associative system known as the Octonions or Cayley numbers. The theorem of Hurwitz asserts that for any other values of $n > 1$ there can be no bilinear forms X_j with real coefficients such that (£) holds. Indeed a topological theorem of J. F. Adams about maps of Hopf invariant 1 tells us that (£) cannot hold for $n \neq 2, 4, 8$ even if we allow the X_j to be any continuous functions of the real vectors $\mathbf{y}$, $\mathbf{z}$.

It was therefore the more surprising when Pfister (1956) invented the theory of multiplicative forms which can be regarded as a partial generalization of composition and which has important consequences for the theory of quadratic forms over general fields [Pfister (1966), see also Lam (1973), Lorenz (1970), or Scharlau (1969)]. For example it is one very special consequence of Pfister's theory that for every $n = 2^m$ and for every field k the set of non-zero values taken by $\sum_1^n x_j^2$ is a group under multiplication.

There is a very substantial body of theory and numerical data about the class-numbers and class-groups of binary forms, much of it expressed in the language of quadratic number-fields. We give here only a few remarks and references and discuss definite forms first. [For class-number formulae, see Appendix B, Section 2.]

Gauss constructed extensive tables of class-numbers of classically integral definite forms of given determinant and these are reproduced in the second volume of his *Werke* [Gauss (1870)]. For tables giving the group-structure see Wada (1970). A computer program has been devised by Shanks (1969) for computing the class-groups for very large positive discriminants: he uses it both as an efficient tool for factorizing large integers and to find class-groups of special types. See also Buell (1977).

Siegel (1935a) showed that the class number $h(D)$ for discriminant D tends to ∞ as $D \to -\infty$ [a quantitative generalization for algebraic number-fields is known as the Siegel–Brauer theorem, see Stark (1975)]. The only $D < 0$ for which $h(D) = 1$ are

$$-3, \ -4, \ -7, \ -8, \ -11, \ -19, \ -43, \ -67, \ -163$$
$$-12, \ -16, \ -28, \ -27$$

(those in the first row being fundamental): see Stark (1967), and Baker (1975) Chapter 5. There are 101 known values of D for which there is one class in each genus. It is conjectured that there are no more [cf. Briggs and Chowla (1954), Grosswald (1963), Baker and Schinzel (1971)]. There are 65 of these D which are of the form $D = 4e$, that is, which correspond to classically integral forms. The values of e are the idoneal numbers (*numeri idonei*) of Euler, who used them to factorize large integers [cf. Mathews (1892), Chapter 9].

There is extensive information about indefinite binary forms in the tables of Ince (1934). He not only gives the class group and fundamental solution of Pell's equation but all the reduced forms and their equivalences. There are many primes p for which the class-number is 1; indeed it is consistent with the numerical evidence that there are infinitely many [see Shanks (1969) and Hendy (1975)]. There are a number of tables in the literature of varying extent and accuracy giving the continued fractions for $\sqrt{n}$ for integer n and sometimes also for $\frac{1}{2}(\sqrt{n} + 1)$ with $n \equiv 1\ (4)$. In some cases fundamental solutions of Pell's equation are listed. The parity of the length of the period determines whether there is a solution of Pell's equation with -1 [cf. Chapter 13, Section 3]. The most extensive tables appear to be Patz (1955) and Kortum and McNiel (1968): both give the continued fraction for $\sqrt{n}$ for $n \leqslant 10000$ and the latter gives the solutions of Pell's equation. With a modern computer such tables can be constructed with ease.

Given any integer n there are both positive and negative discriminants for which the class-number is divisible by n [Nagell (1922), Yamamoto (1970) and Weinberger (1973)]. A discriminant D is said to be irregular if the p-component of the class-group is non-cyclic for some $p \neq 2$. Such discriminants exist, both positive and negative [cf. Shanks (1969) and Neild and Shanks (1974)].

Section 6. The treatment here follows Rédei (1953a), where it is discussed in relation to the 2-class group of a quadratic number-field, cf. also Rédei (1953). For the restricted biquadratic residue symbol, see Fröhlich (1959). For recent work on the 2-component of the class group see Shanks (1971a), Bauer (1972) and Hasse (1975).

EXAMPLES

1. Define the integer $D \neq 0$ to be a *discriminant* if $D \equiv 0$ or $1 \pmod 4$. A discriminant D is a *fundamental discriminant* if it is not of the shape $D = D_0 b^2$ where D_0 is a discriminant and $b^2 > 1$.

   (i) Show that every discriminant D can be written uniquely in the shape $D = D_0 b^2$ where D_0 is a fundamental discriminant.

   (ii) Show that $\pm 8, -4, (-1)^{(p-1)/2} p$ (p an odd prime) are fundamental discriminants and that every fundamental discriminant is uniquely expressible as a produce of distinct members of this set.

2. Let D be a discriminant and define the *Kronecker symbol* $\left(\frac{D}{\overline{\overline{b}}}\right)$ as follows.

   (α) If p is odd and $p \nmid D$ then

$$\left(\frac{D}{\overline{\overline{p}}}\right) = \left(\frac{D}{p}\right)$$

where the right-hand side is the ordinary quadratic residue symbol [$= +1$ if $D \in (\mathbf{Q}_p^*)^2$; $= -1$ otherwise].

   (β) If $2 \nmid D$, then

$$\left(\frac{D}{\overline{\overline{2}}}\right) = \begin{cases} +1 & \text{if } D \equiv 1 \pmod 8 \\ -1 & \text{if } D \equiv 5 \pmod 8). \end{cases}$$

   (γ) If $b = \prod p^{\beta(p)}$ is prime to D, then

$$\left(\frac{D}{\overline{\overline{b}}}\right) = \prod \left(\frac{D}{\overline{\overline{p}}}\right)^{\beta(p)}.$$

   (i) Show that if $p \nmid D$ then p is represented by some integral-valued form $f(x, y)$ of discriminant D if and only if $\left(\frac{D}{\overline{\overline{p}}}\right) = +1$.

   (ii) If D is odd, show that

$$\left(\frac{D}{\overline{\overline{b}}}\right) = \left(\frac{b}{D}\right)$$

where the right-hand side is the ordinary Jacobi symbol

$$\left[= \prod \left(\frac{b}{p}\right)^{\delta(p)} \quad \text{if } D = \pm \prod p^{\delta(p)}\right].$$

(iii) If $b_1 \equiv b_2 \pmod D$ and b_1, b_2 are prime to D show that

$$\left(\frac{D}{b_1}\right) = \left(\frac{D}{b_2}\right).$$

(iv) Extend the definition of $\left(\frac{D}{b}\right)$ to $b < 0$ by periodicity (mod D). Show that

$$\left(\frac{D}{-1}\right) = \begin{cases} +1 & \text{if } D > 0 \\ -1 & \text{if } D < 0. \end{cases}$$

[*Hint.* Use quadratic reciprocity. Alternatively, use the product formula for the Hilbert Norm Residue Symbol to show that

$$\left(\frac{D}{b}\right) = \prod_{p|D} \left(\frac{D, b}{p}\right).]$$

3. Define $\left(\frac{D}{p}\right)$ as in the preceding question for $p \nmid D$ and put $\left(\frac{D}{p}\right) = 0$ if $p|D$. Let $f(x, y)$ be a primitive integral-valued form of discriminant D. Show that the number N of pairs of integers $x, y \pmod p$ such that $f(x, y)$ is prime to p is given by

$$N = (p - 1)\left(p - \left(\frac{D}{p}\right)\right).$$

[*Note.* $p = 2$ is not excluded.]

4. Let $D < 0$ be a discriminant and q a prime such that $\left(\frac{D}{q}\right) = +1$. Show that the number $h(D)$ of classes of forms of determinant D satisfies

$$1 + h(D) \geqslant \log(\tfrac{1}{4}(|D| - 1))/\log q.$$

[*Hint.* Let $g(x, y)$ be a form of discriminant D representing q. If g has order M in the class group then q^M is representable by the form f_0 given by (1.11) and so $\geqslant \frac{1}{4}(|D| - 1)$.]

5. (i) Let $N > 1$ be arbitrary. Show that there is a discriminant $D < 0$ whose class group contains an element of order N.

(ii) Show that D can be taken to be a fundamental discriminant.

[*Hint.* Choose integer $c > 1$ and an integer $M > 0$ with $N|M$. Choose D such that c^M is represented by the $f_0(x, y)$ in (1.11), say $f_0(e, 1) = c^M$ for some e. Then c is representable by some form g of discriminant D. Show that c, M, e can be chosen so that the order of g in the class group is divisible by N. For (ii) take M to be even.]

[*Note.* The results of examples 4 and 5 are due to Nagell (1922). That of example 5 has been rediscovered several times with variants of the proof. The corresponding result to example 5 for $D > 0$ is also true, see Yamamoto (1970) or Weinberger (1973).]

6. Let $f(x, y) = ax^2 + bxy + cy^2$ be a primitive integer-valued form of determinant $D = b^2 - 4ac \neq 0$. Show that a necessary and sufficient condition for f to be in the principal genus is that it should be primitively represented by the form

$$g(x, y, z) = y^2 - xz.$$

This last condition means that there is a basis $\mathbf{r}, \mathbf{s}, \mathbf{t}$ of $\mathbf{Z}^3$ such that

$$g(x\mathbf{r} + y\mathbf{s}) = f(x, y). \qquad (\pounds)$$

[*Note.* This is a special case of the generalization of Chapter 9, Section 6, to the representation of forms by forms.]

[*Hint.* If (£) holds show that g represents 1 over $\mathbf{Z}_p$ for every p. Conversely, if f is in the principal genus, show that there are $l, m, n \in \mathbf{Z}$ such that the determinant $d(G)$ of the form

$$G(x, y, z) = ax^2 + bxy + cy^2 + lxz + myz + nz^2$$

satisfies

$$4d(G) = -f(m, -l) + nD = -1.$$

Hence show that G is integrally equivalent to g.]

7. With the notation of the previous question let

$$\mathbf{r} = (r_1, r_2, r_3), \mathbf{s} = (s_1, s_2, s_3) \text{ and put}$$

$$e_1 = r_2 s_3 - r_3 s_2, \qquad e_2 = r_3 s_1 - r_1 s_3, \qquad e_3 = r_1 s_2 - r_2 s_1$$

and

$$h(x, y) = e_1 x^2 - e_2 xy + e_3 y^2.$$

(i) Show that $h(x, y)$ is a primitive integer-valued form of discriminant D.

(ii) Let $\mathbf{T}$ be an integral automorph of g given by (5.6) of Chapter 13 with $\alpha, \beta, \gamma, \delta \in \mathbf{Z}$ and $\lambda = \alpha\delta - \beta\gamma = +1$. Put

$$\mathbf{r}^* = \mathbf{T}\mathbf{r}, \qquad \mathbf{s}^* = \mathbf{T}\mathbf{s}.$$

Show that

$$g(x\mathbf{r}^* + y\mathbf{s}^*) = f(x, y).$$

[Why?] Clearly if Λ is not indecomposable then it can be expressed as the orthogonal sum of two or more indecomposable sublattices. In the indefinite case this decomposition need not be unique. For example let Λ be the binary lattice with basis $\mathbf{b}_1$, $\mathbf{b}_2$ and let

$$\phi(x_1\mathbf{b}_1 + x_2\mathbf{b}_2) = x_1^2 - 2x_2^2.$$

Then Λ is the orthogonal sum of the 1-dimensional sublattices Γ_1, Γ_2 with basis $\mathbf{b}_1, \mathbf{b}_2$ respectively: but it is also the orthogonal sum of $\mathbf{T}\Gamma_1, \mathbf{T}\Gamma_2$, where $\mathbf{T}$ is any automorph of Λ. A similar argument shows that the orthogonal decomposition of a decomposable lattice Λ can never be unique when ϕ is indefinite, except, possibly, in the very special case when $n = 2$ and ϕ is isotropic.

For definite ϕ we have, however, the following theorem. It is due to Eichler but the elegant proof is that of Kneser (1954).

THEOREM 2.1. *Let Λ be a lattice in the positive definite* **Q**-*quadratic space V, ϕ. Then the orthogonal decomposition of Λ into indecomposable sublattices is unique.*

Proof. After multiplying ϕ by a suitable integer we may suppose without loss of generality that ϕ takes integer values on Λ.

We discuss first the decomposition of vectors of Λ. If

$$\mathbf{b} = \mathbf{b}_1 + \mathbf{b}_2, \tag{2.1}$$

where

$$\mathbf{b}_1, \mathbf{b}_2 \neq \mathbf{0}, \quad \mathbf{b}_1, \mathbf{b}_2 \in \Lambda, \qquad \phi(\mathbf{b}_1, \mathbf{b}_2) = 0, \tag{2.2}$$

we say that $\mathbf{b} \in \Lambda$ is decomposable. Otherwise $\mathbf{b}$ is *indecomposable*.

If a decomposition (2.1), (2.2) exists, then

$$\phi(\mathbf{b}) = \phi(\mathbf{b}_1) + \phi(\mathbf{b}_2)$$

and so

$$\phi(\mathbf{b}_j) < \phi(\mathbf{b}) \qquad (j = 1, 2).$$

An easy induction shows that every $\mathbf{b} \in \Lambda$ can be expressed as the sum of a collection of indecomposable vectors $\mathbf{b}_j$. The argument does not show that the $\mathbf{b}_j$ are uniquely determined by $\mathbf{b}$ nor that they are mutually orthogonal.

Suppose that Λ is the orthogonal sum of sublattices Γ_j $(1 \leqslant j \leqslant J)$. Any $\mathbf{b} \in \Lambda$ is of the form $\mathbf{b} = \Sigma\mathbf{c}_j$ $(\mathbf{c}_j \in \Gamma_j)$ and so, if $\mathbf{b}$ is irreducible, it must belong to precisely one of the Γ_j. Further, if $\mathbf{b}_1, \mathbf{b}_2$ are indecomposable and $\phi(\mathbf{b}_1, \mathbf{b}_2) \neq 0$, then $\mathbf{b}_1, \mathbf{b}_2$ must belong to the same Γ_j. This remark motivates the following construction.

Denote by I the set of all indecomposable vectors of Λ. We say that $\mathbf{b}_1, \mathbf{b}_2 \in I$ are equivalent if there is a sequence $\mathbf{c}_k \in I$ $(1 \leqslant k \leqslant K)$ such that

$$\mathbf{b}_1 = \mathbf{c}_1, \qquad \mathbf{b}_2 = \mathbf{c}_K, \qquad \phi(\mathbf{c}_k, \mathbf{c}_{k+1}) \neq 0 \qquad (1 \leqslant k < K).$$

This is an equivalence relation. Let I_r $(1 \leqslant r \leqslant R)$ be the equivalence classes and let Δ_r be the sublattice of Λ generated by the elements of I_r $(1 \leqslant r \leqslant R)$. Clearly Λ is the orthogonal sum of the Δ_r and each Δ_r is indecomposable. Further, the remarks above show that if Λ is the orthogonal sum of sublattices Γ_j, then each Δ_r must be contained in one of the Γ_j. If the Γ_j are also indecomposable it follows that they must coincide with the Δ_r up to order.

3. CLASS-NUMBERS OF GENERA AND SPINOR-GENERA

In contrast to indefinite forms [cf. Chapter 11] it is rare for a genus (or a spinor-genus) of definite forms to have only one class. Watson has shown that every genus of definite forms in $n \geqslant 11$ variables has at least 2 classes and he has much information about the one-class genera in $n \leqslant 10$ variables. Further, the number of classes in a genus (or, indeed, in a spinor-genus) tends to infinity as the number n of variables tends to infinity. [For all this cf. Watson (1975), Gersten (1972) and the references given there.]

The precise results for low values of n require a substantial amount of detailed case-by-case discussion. The estimates of the class-number for large n which have been obtained by elementary arguments are much weaker than those which can be obtained by analytic means [Magnus (1937), cf. Appendix B, Section 3]. Hence we shall content ourselves with sketching a proof of

THEOREM 3.1. *Let $c(n)$ be the minimum number of classes in a genus of positive definite integer-valued forms in n variables. Then $c(n) \to \infty$ as $n \to \infty$.*

Note. We shall indicate at the end how the word "genus" may be replaced by "spinor-genus".

Proof. It is convenient to modify our notations and conventions. We shall be concerned with pairs Λ, ϕ consisting of a $\mathbf{Z}$-lattice Λ and a regular positive definite quadratic form ϕ on Λ. Normally we shall be concerned with only one quadratic form on each lattice and will omit mention of the forms. Thus a statement "the lattices Λ, Γ are isomorphic" is shorthand for "the lattices Λ, Γ together with their quadratic forms are isomorphic". The lattices under consideration will no longer be supposed to lie in the same vector space. We shall write $\Lambda = \Gamma + \Delta$ to indicate that Λ is the orthogonal sum of lattices isomorphic to Γ and Δ: and similarly $\Lambda = r\Gamma$ will mean that Λ is the orthogonal sum of r lattices each isomorphic to Γ.

We shall need to know the existence and properties of two special lattices. The first is Γ_8, the 8-dimensional lattice with an integer-valued quadratic form ϕ_8 (say) of determinant 1. [Here "determinant" is in a sense appropriate to integer-valued forms. Classically, $2\phi_8$ is an improperly primitive integral form of determinant 1.] We shall denote by Γ_{16} the 16-dimensional lattice with an integer-valued quadratic form ϕ_{16} of determinant 1, which is not isomorphic to $2\Gamma_8$. We need to know that Γ_{16} and $2\Gamma_8$ are in the same genus. [For all this cf. Chapter 9, example 10 and the examples to this Appendix.]

Suppose now that a given genus $\mathscr{G}$ of lattices contains a lattice

$$r\Gamma_8 + \Delta \tag{3.1}$$

for some $r > 1$ and some lattice Δ. Then $\mathscr{G}$ also contains the lattice

$$j\Gamma_{16} + (r - 2j)\Gamma_8 + \Delta \tag{3.2}$$

for all j, $0 \leqslant j \leqslant \frac{1}{2}r$. These lattices are not isomorphic by Theorem 2.1. Hence the class number of $\mathscr{G}$ is at least† $[\frac{1}{2}r]$.

Not every genus $\mathscr{G}$ contains a lattice of the type (3.1) however. We now, following Watson, introduce an operation on a genera which (i) does not increase the number of classes in the genus and (ii) by repeated application leads to a genus which does contain a lattice of the type (3.1), where r is large when n is large. The operation is as follows. Let ϕ be a positive definite integer-valued quadratic form on a lattice Λ and let p be a prime. Denote by Λ_1 the set of $\mathbf{c} \in \Lambda$ such that

$$\phi(\mathbf{b} + \mathbf{c}) \equiv \phi(\mathbf{b}) \pmod p \qquad (\text{all } \mathbf{b} \in \Lambda).$$

Then [cf. Chapter 8, example 8] Λ_1 is a lattice,

$$\Lambda \supset \Lambda_1 \supset p\Lambda_1$$

and

$$\phi(\mathbf{c}) \equiv 0 \pmod p \qquad (\text{all } \mathbf{c} \in \Lambda_1).$$

Hence the quadratic form

$$\phi_1(\mathbf{x}) = p^{-1}\phi(\mathbf{x})$$

is integer-valued on Λ_1. Clearly, isomorphic lattices Λ, ϕ produce isomorphic lattices Λ_1, ϕ_1. Further, if Λ'_1, ϕ'_1 is any lattice in the same genus as Λ_1, ϕ_1 then it is easy to construct§ a lattice $\Lambda' \supset \Lambda'_1$ such that Λ', ϕ', with $\phi' = p\phi'_1$, is in the same genus as Λ, ϕ and such that Λ', ϕ' gives rise to Λ'_1, ϕ'_1 by the process just described. Hence the genus $\mathscr{G}_1$ of Λ_1, ϕ_1 contains at most as many classes as the genus $\mathscr{G}$ of Λ, ϕ.

We now repeat the process, possibly with a different prime p.‡ A local

† This is very weak. In fact the genus of $3\Gamma_8$ contains 22 indecomposable isomorphism classes in addition to the decomposable $3\Gamma_8$ and $\Gamma_8 + \Gamma_{16}$ [Niemeier (1973)]. Analytic methods [cf. Appendix B] show that the number of classes in the genus of $r\Gamma_8$ increases very fast when r increases.

‡ $\mathscr{G}_1 = \mathscr{G}$ unless $p = 2$ or p divides the determinant.

§ e.g. as indicated below.

study [Chapter 8, example 8] shows that starting from a given genus $\mathscr{G}$ one obtains only a finite set of genera altogether. Let $\mathscr{G}_0$ be the genus which can be obtained from $\mathscr{G}$ by repeated operations and for which the determinant is least. The local study then shows that $\mathscr{G}_0$ contains a lattice of the type (3.1) for which $r \geqslant (n-6)/16$. Since $\mathscr{G}_0$ has class number $\geqslant [r/2]$ and the class number of $\mathscr{G}$ is at least that of $\mathscr{G}_0$, this completes the proof of Theorem 3.1.

It remains only to fulfil the promise to show that one can replace "genus" by "spinor-genus" in the enunciation. There are two points which require attention. First we must show that if Λ_1, ϕ_1 and Λ'_1, ϕ'_1 are in the same spinor genus then we can choose Λ', ϕ' in the same spinor genus as Λ, ϕ. We can suppose (i) that Λ_1 and Λ'_1 are in the same quadratic space V, ψ, (ii) that ψ induces ϕ_1, ϕ'_1 on Λ_1, Λ'_1 respectively, and (iii) that for every prime q there is an $\alpha_q \in \Theta(V_q)$ such that $\alpha_q(\Lambda_1)_q = (\Lambda'_1)_q$ [here the suffix q, as usual, denotes the localization at q]. Then we can take Λ' to be the lattice such that $(\Lambda')_q = \alpha_q \Lambda_q$ for all q. Secondly we must show that a spinor genus contains a lattice of the type (3.1) if the genus containing it does, at least† when Δ has dimension $\geqslant 3$: and this follows as in the proof of Theorem 7.1 of Chapter 11.

4. REPRESENTATIONS OF INTEGERS BY DEFINITE FORMS

Since there is usually more than one class in a genus, general questions on representability tend to be much more difficult for definite than for indefinite forms. There is a long history, particularly relating to representation for ternaries. Some general theorems are proved in Chapter 11, Section 8, and we refer to this and the notes at the end of that Chapter.

NOTES

There is a brief survey of definite forms (also over algebraic fields) together with copious references in O'Meara (1976) Section 9.

EXAMPLES

1. Let $n = 8l$ and let Λ be the lattice with basis $\mathbf{e}_1, \ldots, \mathbf{e}_n$ in the quadratic space V, ϕ for which

$$\phi(\Sigma x_j \mathbf{e}_j) = \Sigma x_j^2.$$

† If dim $\Delta \leqslant 2$, replace r by $r-1$.

Let Δ be the sublattice of Λ consisting of the points

$$\Sigma u_j \mathbf{e}_j, \qquad u_j \in \mathbf{Z}, \qquad \Sigma u_j \equiv 0 \pmod 2$$

and let $\Gamma = \Gamma_n$ be the lattice generated by Δ and

$$\mathbf{b} = \tfrac{1}{2}\Sigma \mathbf{e}_j.$$

(i) Show that the quadratic form given by ϕ on Γ has determinant 1.

(ii) Show that the values taken by ϕ on Γ are all in $2\mathbf{Z}$.

(iii) Show that Γ is indecomposable.

(iv) Deduce that Γ_{16} and $2\Gamma_8$ are in the same genus but are not equivalent.

APPENDIX B

Analytic Methods

1. INTRODUCTION

In this book no use has been made of the more sophisticated theory of functions of real variables or of that of complex variables. There are results which can only, or can best, be obtained from these theories: in return, the consideration of quadratic forms has had a profound effect on complex function theory and, more recently, on general harmonic analysis. To deal adequately with this area would require both another book and another author. The object of this Appendix is merely to survey briefly two somewhat related lines of enquiry.

One line of enquiry goes back to Dirichlet's discovery of a formula for the number of classes of integral binary forms of given discriminant. Although the formula itself is expressed in elementary terms, the proof uses the methods of analysis and no really elementary proof has since been discovered. [See Notes.] The quite simple analytic proof will be sketched in Section 2. By stages Dirichlet's formula has been generalized and in 1935 Siegel gave very suggestive and powerful formulae for the weight of a genus of positive definite forms and for the weight of the representation of an integer (or, more generally, of a form) by such a genus: subsequently he extended the formulae to indefinite forms. In Section 3 we shall explain but not prove Siegel's formulae and we shall illustrate their application. Quite recently it has been found that Siegel's formulae can be very pregnantly expressed in terms of the measure theory of adeles on the orthogonal group. This will be explained in Section 4.

The other line of enquiry also goes back to the beginning of the 19th century. Put

$$\Theta(z) = 1 + 2z + 2z^4 + 2z^9 + \dots = \sum_{n=-\infty}^{\infty} z^{n^2}, \tag{1.1}$$

the series converging for all complex numbers z in $|z| < 1$. If we put

$$\theta(s) = \Theta(z), \qquad z = e^{\pi i s}, \qquad (\mathrm{Im}\, s > 0),$$

the function $\theta(s)$ behaves in a simple manner under the action of the *modular group*

$$s \to (as + b)/(cs + d)$$

where

$$a, b, c, d \in \mathbf{Z}, \qquad ad - bc = +1.$$

Clearly

$$\theta(s + 2) = \theta(s).$$

Less trivially,

$$\theta(-1/s) = \frac{\sqrt{s}}{\sqrt{i}}\,\theta(s) \tag{1.1 bis}$$

where $\sqrt{s}, \sqrt{i}$ are normalized to have both their real and their imaginary parts positive. Jacobi used the theory of such functions to prove the identities

$$\{\Theta(z)\}^2 = 1 + 4\sum_{m \geqslant 0} \frac{(-1)^m z^{2m+1}}{1 - z^{2m+1}} \tag{1.2}$$

$$\{\Theta(z)\}^4 = 1 + 8\sum_{\substack{m \geqslant 1 \\ 4 \nmid m}} \frac{m z^m}{1 - z^m} \tag{1.3}$$

$$\{\Theta(z)\}^6 = 1 + 16\sum_{m > 0} \frac{m^2 z^m}{1 - z^{2m}} - 4\sum_{m \geqslant 0} (-1)^m \frac{(2m+1)^2 z^{2m+1}}{1 - z^{2m+1}} \tag{1.4}$$

and

$$\{\Theta(z)\}^8 = 1 + 16\Sigma \frac{m^3 z^m}{1 - (-1)^m z^m}. \tag{1.5}$$

These power series identities are, of course, equivalent to formulae for the number $a_f(m)$ of integral representations of m by the form f where

$$f(\mathbf{x}) = x_1^2 + \ldots + x_n^2$$

and $n = 2, 4, 6$ and 8 (the formulae for $n = 2$ and 4 had already been obtained by elementary means).

Glaisher and Ramanujan found identities for $\{\Theta(z)\}^n$ for all even values of $n \leqslant 24$ but now the formulae may involve terms defined only as the coefficients of certain modular forms. One of the simpler examples is Ramanujan's identity

$$691\{\Theta(z)\}^{24} = E(z) + C(z) \tag{1.6}$$

with

$$E(z) = 16\sum_{m\geq 1} \frac{m^{11}z^m}{1-(-1)^m z^m} \tag{1.7}$$

and

$$C(z) = -33152G(-z) - 65536G(z^2) \tag{1.8}$$

where

$$G(z) = z\Pi(1 - z^m)^{24}. \tag{1.9}$$

Here $G(z)$ is the important "discriminant" function from modular form theory: its coefficients are the notorious Ramanujan function $\tau(m)$.

The treatment of $\{\Theta(z)\}^n$ for odd n is much more difficult than for even n. The cases $n = 5$ and 7 were elucidated by Mordell and Hardy but the formulae for the number of representations as a sum of 5 or 7 squares had already been obtained by Minkowski and H. J. S. Smith by other means. There is a good discussion of representation as sums of squares from the present point of view in Hardy (1940) and a briefer one in Hardy and Wright (1938), both giving historical references.

Largely by the work of Hecke, the theory of modular forms was extended to cover all positive definite forms in an even number of variables. The theory is an extremely powerful one: for example the "Hecke operators" give the sort of information which for binary forms is given by the theory of composition. It will be discussed in Section 5. There are extensions to definite forms in an odd number of variables and to indefinite forms but they are less clear-cut and will only be alluded to. Siegel has developed a theory of modular forms of several variables ("the Siegel modular group") to deal with quadratic forms but this we shall not discuss.

2. BINARY FORMS

The discriminant

$$D = b^2 - 4ac \tag{2.1}$$

of the integer-valued binary form

$$f(x, y) = ax^2 + bxy + cy^2 \tag{2.2}$$

satisfies

$$D \equiv 0 \text{ or } 1 \pmod 4). \tag{2.3}$$

By $h(D)$ we denote (i) for $D > 0$ the number of classes (for proper integral equivalence) of primitive forms with discriminant D and (ii) for $D < 0$ the number of such classes of positive definite forms.

If

$$D = D_0 m^2 \qquad (m > 1), \tag{2.4}$$

and D_0 satisfies (2.3), then the determination of $h(D)$ can be made elementarily to depend on that of $h(D_0)$ [cf. Chapter 9, example 12]. Although it is not really necessary, we shall therefore suppose in what follows that D is not of the form (2.4) or, as we shall say, that D is a *fundamental discriminant*. By

$$(D/a) \tag{2.5}$$

we shall mean the Kronecker symbol [cf. Chapter 14, example 2].

THEOREM 2.1. *Let D be a fundamental discriminant. Then*

(i)
$$2|D|h(D) = w\sum_{0<r<|D|} (D/r)r, \tag{2.6}$$

for $D < 0$, *where*

$$w = \begin{cases} 6 & \textit{if } D = -3 \\ 4 & \textit{if } D = -4 \\ 2 & \textit{otherwise.} \end{cases} \tag{2.7}$$

(ii)
$$(\log \eta)h(D) = -\sum_{0<r<D} (D/r)\log(\sin(\pi r/d)) \tag{2.8}$$

for $D > 0$, *where* $\eta > 1$ *is a fundamental unit.* [*For the definition of a fundamental unit, see Chapter* 13, *Section* 3.]

We recall that two representations of a given integer m by a form $f(x, y)$ are in the same orbit if they can be taken into one another by a proper automorph of f. The number of orbits of such representations (primitive or not) will be denoted by $a^*(m) = a^*_f(m)$. Write†

$$A^*(M) = A^*_f(M) = \sum_{\substack{(m,D)=1 \\ 0 \leq m \leq M}} a^*_f(m) \tag{2.9}$$

LEMMA 2.1

$$\lim_{M\to\infty} M^{-1}A^*(M) = \begin{cases} (2\pi/w\sqrt{|D|})\prod_{p|D}(1-p^{-1}) & (D<0) \\ (\log\eta/\sqrt{D})\prod_{p|D}(1-p^{-1}) & (D>0). \end{cases} \tag{2.10}$$

Proof. Suppose first that $D < 0$. Then w is the number of proper automorphs of f and the total number of representations of m by f is $wa^*(m)$. Hence $wA^*(M)$ is the number of integer solutions (x, y) of

$$f(x, y) \leq M \tag{2.11}$$

subject to the condition

$$(f(x, y), D) = 1. \tag{2.12}$$

For $p|D$ the congruence $f(x, y) \equiv 0 \pmod p$ is equivalent to a linear congruence in x and y. Hence the number of solutions of (2.11) and (2.12) is

† We write (m, D) for g.c.d.(m, D).

approximately

$$\prod_{p|D}(1 - p^{-1})$$

times the area of (2.11), which is

$$2\pi M/\sqrt{|D|}.$$

This gives (2.10) for $D < 0$. The proof for $D > 0$ is similar except that now there is precisely one representative of each orbit satisfying Lemma 3.3 of Chapter 13; and so one has to calculate the area of a region bounded by two radii vectores and a segment of a hyperbola.

It would be possible to work with (2.10) directly [e.g. Landau (1927)] but it is technically easier to take a weighted average:

COROLLARY.

$$\lim_{s\to 1+} (s - 1) \sum_{\substack{m>0\\(m,D)=1}} a^*(m)m^{-s}$$

$$= \begin{cases} (2\pi/w\sqrt{|D|}) \prod_{p|D}(1 - p^{-1}) & (D < 0) \\ (\log \eta/\sqrt{D}) \prod_{p|D}(1 - p^{-1}) & (D > 0). \end{cases} \qquad (2.12\ bis)$$

Proof. Follows from (2.10) by partial summation, since

$$\lim_{s\to 1+} (s - 1)\sum m^{-s} = 1.$$

LEMMA 2.2. *An integer $m > 0$ prime to D is represented primitively by some primitive form f of determinant D only if every prime divisor p of m satisfies $(D/p) = +1$. When this condition is satisfied, the total number of orbits of primitive representations of m by all the classes of forms is 2^μ, where μ is the number of distinct prime factors of m.*

Proof. For the classes of representations correspond bijectively to the set of forms

$$mx^2 + uxy + vy^2,$$

where

$$u^2 - 4mv = D$$

and

$$-m < u \leqslant m.$$

It is not difficult to see that there are 2^μ such forms.

COROLLARY

$$\sum_f{}^* \sum_{\substack{m\geqslant 0\\(m,D)=1}} \bar{a}_f(m)m^{-s} = \prod_{(D/p)=+1} \left(\frac{1 + p^{-s}}{1 - p^{-s}}\right), \qquad (2.13)$$

where $s > 1$, $\sum_f^*$ is a sum over all the classes of forms of determinant D, and $\bar{a}_f(m)$ is the number of orbits of primitive representations of m by f.

Proof. The lemma asserts that the left-hand side of (2.13) is

$$\prod_{(D/p)=1} (1 + 2p^{-s} + 2p^{-2s} + 2p^{-3s} + \ldots),$$

which agrees with (2.13).

For comparison with (2.11) we need to replace $\bar{a}_f(m)$ by $a_f^*(m)$ on the left-hand side of (2.13). Clearly

$$a_f^*(m) = \sum_{u^2|m} \bar{a}_f(m/u^2). \tag{2.14}$$

Hence

$$\begin{aligned}\sum_f^* \sum_{\substack{m \geq 0 \\ (m,D)=1}} a_f^*(m) &= \prod_{p \nmid D} \frac{1}{(1 - p^{-2s})} \prod_{(D/p)=+1} \left(\frac{1 + p^{-s}}{1 - p^{-s}}\right) \\ &= \prod_{p \nmid D} \frac{1}{(1 - p^{-s})(1 - (D/p)p^{-s})} \cdot \end{aligned} \tag{2.15}$$

We now let $s \to 1+$ and compare with (2.12 *bis*). In the first place,

$$\begin{aligned}\lim_{s \to 1+} \prod \frac{1}{(1 - (D/p)p^{-s})} &= \lim_{s \to 1+} \sum (D/m) m^{-s} \\ &= \sum_{m \geq 1} (D/m) m^{-1}. \end{aligned} \tag{2.16}$$

Here the series is readily shown to converge, and we may disregard the condition $p \nmid D$ in (2.15) because $(D/p) = 0$ by definition for $p|D$. On the other hand

$$\lim_{s \to 1+} (s - 1) \prod_p \frac{1}{(1 - p^{-s})} = \lim (s - 1) \Sigma m^{-s} = 1. \tag{2.17}$$

Bearing in mind the condition $p \nmid D$ in (2.15) it follows from (2.11) (2.15) (2.16) and (2.17) that

$$h(D) = \begin{cases} \dfrac{w\sqrt{|D|}}{2\pi} \Sigma(D/m)\, m^{-1} & \text{for } D < 0 \\[2ex] \dfrac{\sqrt{|D|}}{\log \eta} \Sigma(D/m)\, m^{-1} & \text{for } D > 0. \end{cases} \tag{2.18}$$

The proof of Theorem 2.1 will thus be complete when we have proved

LEMMA 2.3. *Let D be a fundamental discriminant. Then*

$$\Sigma(D/m)\,m^{-1} = \begin{cases} -\pi|D|^{-\frac{3}{2}} \sum\limits_{0<r<|D|} (D/r)\,r & (D<0) \\ -D^{-\frac{1}{2}} \sum\limits_{0<r<D} \log\sin(\pi r/D) & (D>0). \end{cases} \tag{2.19}$$

Proof. We require Gauss' sum

$$\sum_{r \bmod D} (D/r)\,e^{2\pi i r m/|D|} = \begin{cases} (D/m)\sqrt{D} & (D>0) \\ (D/m)\,i\sqrt{|D|} & (D<0) \end{cases} \tag{2.20}$$

for a fundamental discriminant† D; which we assume as known.‡ We also need

$$\Sigma \frac{e^{im\phi}}{m} = -\log(2\sin\tfrac{1}{2}\phi) + i(\tfrac{1}{2}\pi - \tfrac{1}{2}\phi) \tag{2.21}$$

for

$$0 < \phi < 2\pi, \tag{2.22}$$

which is readily obtained from the power series expansion for $-\log(1-z)$ on letting $z \to e^{2\pi i\phi}$ since the series on the left of (2.21) is readily seen to converge.

The required result now follows on putting $\phi = 2\pi r/|D|$ in (2.21), multiplying by (D/r), summing over $0 < r < |D|$ and recalling that

$$(D/-1) = \begin{cases} 1 & \text{if } D>0 \\ -1 & \text{if } D<0 \end{cases}$$

(Chapter 14, example 2). The details are left to the reader.

3. SIEGEL'S FORMULAE

In generalizing Dirichlet's class-number formula to forms of higher dimension, there are additional complications: two genera of the same determinant need not contain the same number of classes and two classes of definite forms in the same genus need not have the same number of automorphs. Eisenstein (1847a) observed that the appropriate generalization to definite forms should give the *weight* of a genus (in the sense of Chapter 9, Section 6) and gave without proof formulae for the weight of positive definite ternary forms subject to some restrictions. In a paper published just before his death [Eisenstein (1852)] he showed that in some very special cases the

† This is the place where we use that D is a fundamental discriminant. If $D = D_0m^2$, where D_0 is a fundamental discriminant, then Gauss' sum for D is obtained from that for D_0 by multiplication by factors $p - (D_0/p)$ for $p|m$. Compare Chapter 9, example 12(ii).

‡ See example 5 to this Appendix.

weight of a genus can be obtained in a purely elementary manner [the argument is reproduced in Lemma 6.6 of Chapter 9 and also in Bachmann (1898)] but it was left to H. J. S. Smith (1867) to give a proof of Eisenstein's formulae.

In his inaugural dissertation Minkowski (1885) gave a formula for the weight of any genus of positive definite forms. His proof is by induction on the dimension n using Lemma 6.2 of our Chapter 9. The formula for $n > 2$ is a rather simple expression multiplied by a convergent infinite product, there being one factor for each prime p. Minkowski gives heuristic reasons why there should be a formula for the weight of this type. It should be observed that the formula (2.18) can be regarded as falling within this general pattern since the infinite sum there is equal to

$$\prod_p (1 - (D/p)\,p^{-1})^{-1}. \tag{3.1}$$

The convergence of (3.1) is, however, rather deep [cf. Landau (1909) Section 109] while that of the sum is straightforward.

Siegel (1935) observed that Minkowski's formula for the weight contained in general an erroneous power of 2. He gave a correct formula in a somewhat different shape and also a more general formula for the weight of the representations of a genus of forms of lower dimension by a positive definite genus: a particular case being the weight of the representations of a given integer. Siegel's proof is analytic. In a series of subsequent papers [all reproduced in Siegel (1966)] he generalized this formula to indefinite forms and also to forms over algebraic number-fields.

More recently, Siegel's formulae, both for definite and indefinite genera, have received a new interpretation and new proofs in terms of "Tamagawa numbers": this will be discussed in Section 4. In the rest of this section we enunciate Siegel's formulae for definite genera and consider some consequences. We do not give any proofs.

[In comparing the formulae given here with those of Siegel (1935) one should note that Siegel works with the equivalence of quadratic forms in the wide sense whereas we define a class of forms as a class for proper equivalence. Further, Siegel defines the weight of a genus to be $\Sigma 1/o(f)$ where $o(f)$ is the order of $O(f)$ and f runs through representatives of the classes in the wide sense. We have taken the weight to be $\Sigma 1/o^+(f)$, the sum being over representatives of proper equivalence classes. Consequently the weight of a genus in our sense is twice that in Siegel's sense.]

Siegel's formula for the weight $W(\mathscr{F})$ of a genus $\mathscr{F}$ of positive definite integral forms is

$$W(\mathscr{F}) = 2\beta_\infty \prod_p \beta_p \tag{3.2}$$

where the β_p and β_∞ are defined as follows.

Let $\mathbf{F}$ be the square matrix representing a form f in the genus $\mathscr{F}$, and let $t > 0$ be an integer. Consider the number N_t of $n \times n$ integral matrices $\mathbf{S} \bmod p^t$ such that

$$\mathbf{S}'\mathbf{F}\mathbf{S} \equiv \mathbf{F} \pmod{p^t}. \tag{3.3}$$

It can be shown that $p^{-\frac{1}{2}n(n-1)}N_t$ is independent of t when t is large enough (where n is the dimension and "large enough" may depend on p and $\mathscr{F}$). We define this value to be $2\beta_p^{-1}$. The factor 2 may be regarded as an expression of the fact that (3.3) implies that det $\mathbf{S}$ is p-adically near to ± 1 and we are interested only in the $\mathbf{S}$ for which it is near $+1$. When $p \neq 2$ and p does not divide the determinant d of the forms of $\mathscr{F}$, then $t = 1$ is already "large enough" and the value of β_p is given by Chapter 2, example 13(v), namely

$$\beta_p^{-1} = \begin{cases} (1 - p^{-2})(1 - p^{-4})\dots(1 - p^{-2m}) & (n = 2m + 1) \\ (1 - p^{-2})(1 - p^{-4})\dots(1 - p^{-2m+2})(1 - \varepsilon p^{-m}) & (n = 2m) \end{cases} \tag{3.4}$$

where

$$\varepsilon = \left(\frac{(-1)^m d}{p}\right), \qquad p \neq 2, \quad p \nmid d. \tag{3.5}$$

For $p = 2$ or $p \mid d$ the value of β_p requires a special consideration. In any case β_p depends only on f as a form over $\mathbf{Z}_p$ and so only on the genus $\mathscr{F}$.

The interpretation of β_∞ is analogous. We consider real matrices $\mathbf{G}$ in the neighbourhood of $\mathbf{F}$ and real square matrices $\mathbf{S}$ such that

$$\mathbf{S}'\mathbf{F}\mathbf{S} = \mathbf{G}. \tag{3.6}$$

Let $\mathscr{G}$ be a neighbourhood of $\mathbf{F}$ and let $\mathscr{S}$ be the set of matrices $\mathbf{S}$ such that $\mathbf{G} \in \mathscr{G}$. Let $v(\mathscr{G})$, $v(\mathscr{S})$ be the volumes of $\mathscr{G}$, $\mathscr{S}$ considered as subsets of $\frac{1}{2}n(n+1)$ - and n^2-dimensional space respectively. Then β_∞ is the limit of $2v(\mathscr{G})/v(\mathscr{S})$ as $\mathscr{G}$ shrinks to $\mathbf{F}$. [Note that det $\mathbf{S}$ is near to ± 1 and the factor 2 expresses that we are considering the $\mathbf{S}$ with det $\mathbf{S}$ near 1.] The numerical value is†

$$\beta_\infty = 2\pi^{-n(n+1)/4}\left\{\prod_{j=1}^{n} \Gamma(j/2)\right\} d^{(n+1)/2}, \tag{3.7}$$

where d is the determinant of $\mathscr{F}$.

We note that (3.4) implies that $\Pi\beta_p$ converges absolutely whenever $n > 2$. In fact one obtains a finite product multiplied by a finite product of terms $\zeta(2m)$ together with the value of an L-function if n is even. The resulting formulae are, however, less transparent than (3.2).

Let us now consider the case when $\mathscr{F} = \mathscr{F}_0$ is the genus of

$$f_0 = x_1^2 + \dots + x_n^2. \tag{3.8}$$

† Here Γ is the "gamma function" from analysis.

Then [Magnus (1937)] Siegel's formula gives

$$W(\mathscr{F}_0) = \frac{1}{2^{n-1}n!} \quad (n \leqslant 8)$$

and

$$W(\mathscr{F}_0) > \frac{1}{2^{n-1}\,n!} \qquad (n > 8),$$

particular values being

$$\frac{1}{2^8 9!}\cdot\frac{3^2 17}{137} \quad (n = 9), \qquad \frac{1}{2^9 10!}\cdot\frac{3^2 5^2}{137} \quad (n = 10).$$

Since f_0 itself has $2^{n-1}n!$ proper integral automorphs, it follows that $\mathscr{F}_0$ consists only of the class of f_0 for $n \leqslant 8$ but that it always contains further classes when $n > 8$. In fact

$$\log W(\mathscr{F}_0) = (n^2/4)(\log n - \tfrac{3}{2} - \log 2\pi) + O(n \log n) \tag{3.9}$$

as $n \to \infty$, so the number of classes in $\mathscr{F}_0$ increases very rapidly when n increases.

In the same paper Magnus (1937) uses Siegel's formula to show that there are only finitely many genera of primitive integral forms in $n \geqslant 3$ variables which contain only one class and that in particular every genus with $n \geqslant 35$ contains at least two classes. [For later results in this direction cf Appendix A.]

We turn now to Siegel's other formula. Let f, g be regular integral forms in n and $l < n$ variables. We say that an $l \times n$ integral matrix $\mathbf{X}$ gives an integral representation of g by f if

$$\mathbf{X}'\mathbf{F}\mathbf{X} = \mathbf{G} \tag{3.10}$$

where $\mathbf{F}$, $\mathbf{G}$ are the matrices of f, g. If $l = 1$ this reduces to the notation of an integral representation of an integer by f. When f is positive definite, which we are supposing, the number of integral representations of $\mathbf{G}$ by $\mathbf{F}$ is finite and will be denoted by $A(f, g)$.

The sum

$$S(\mathscr{F}, \mathscr{G}) = \sum_f A(f, g)/o^+(f), \tag{3.11}$$

where f runs through representatives of the (proper) equivalence classes in a genus $\mathscr{F}$, depends only on the genus $\mathscr{G}$ of g, as is implicit in the formula which follows.† The ratio

$$\mathscr{A}(\mathscr{F}, \mathscr{G}) = S(\mathscr{F}, \mathscr{G})/W(\mathscr{F}) \tag{3.12}$$

is a weighted average of $A(f, g)$ over the classes in the genus of $\mathscr{F}$, since

† Compare Lemma 6.2 of Chapter 9.

$W(\mathscr{F})$ is $\Sigma 1/o^{+}(f)$. Siegel's formula is

$$\mathscr{A}(\mathscr{F}, \mathscr{G}) = \prod_{p \text{ inc } \infty} \alpha_p(\mathscr{F}, \mathscr{G}) \qquad (l < n-1)$$

$$= \tfrac{1}{2} \prod_{p \text{ inc } \infty} \alpha_p(\mathscr{F}, \mathscr{G}) \qquad (l = n-1) \tag{3.13}$$

where $\alpha_p(\mathscr{F}, \mathscr{G})$ depends only on the p-adic behaviour of $\mathscr{F}$ and $\mathscr{G}$.

If we were to allow $l = n$ and so could take g to be in the genus $\mathscr{F}$, the expression (3.12) would clearly just reduce to $\{W(\mathscr{F})\}^{-1}$ and we could obtain (3.2) as a special case of (3.12). In fact the definitions of the $\alpha_p(\mathscr{F}, \mathscr{G})$ are the natural generalizations of the definitions of the β_p^{-1} except for the factor 2. More precisely, $\alpha_p(\mathscr{F}, \mathscr{G})$ is such that the number of solutions $\mathbf{X}$ modulo p^t of the congruence

$$\mathbf{X}'\mathbf{F}\mathbf{X} \equiv \mathbf{G} \pmod{p^t} \tag{3.14}$$

is

$$(p^t)^{ln - l(l+1)/2}\, \alpha_p(\mathscr{F}, \mathscr{G}) \tag{3.15}$$

for all sufficiently large t. The definition of $\alpha_\infty(\mathscr{F}, \mathscr{G})$ is the obvious modification of that of $(\frac{1}{2}\beta_\infty)^{-1}$ and gives the value

$$\alpha_\infty(\mathscr{F}, \mathscr{G}) = d^{-l/2} D^{(n-l-1)/2} \gamma_{nl} \tag{3.16}$$

where d, D are the determinants of $\mathscr{F}$, $\mathscr{G}$ respectively and where

$$\gamma_{nl} = \pi^{l(2n-l+1)/4} \prod_{n-l<j\leqslant n} \{\Gamma(j/2)\}^{-1}. \tag{3.17}$$

As Siegel remarks, the formula (3.13) has a very appealing interpretation. The $\alpha_p(\mathscr{F}, \mathscr{G})$ are a measure of the number of integral p-adic representations of a $g \in \mathscr{G}$ by an $f \in \mathscr{F}$ and $\mathscr{A}(\mathscr{F}, \mathscr{G})$ is an appropriately weighted average of the number of $\mathbf{Z}$-representations as g, f run through $\mathscr{G}, \mathscr{F}$ respectively.

The formula (3.13) is particularly useful when $\mathscr{F}$ is a genus of 1 class since then $\mathscr{A}(\mathscr{F}, \mathscr{G}) = A(f, g)$ is just the number of representations of g by f. In particular, when $\mathscr{F}_0$ is the genus of (3.8) and $n \leqslant 8, l = 1$, one obtains by manipulation of (3.13) the already known expressions for the number of representations of an integer by f_0 [Siegel (1935) Section 10]. When $n > 8$ this no longer happens. In fact what one gets had already been obtained† previously as the "singular series" of the Hardy–Littlewood "circle method" and it was already known that this gives the correct number of solutions for $n \leqslant 8$ but not for $n > 8$.

† There is a further discussion in Section 5.

4. TAMAGAWA NUMBERS

There is a very succinct formulation of Siegel's theorem which places it in the general context of linear algebraic groups. It makes use of the notion of *adele*, which generalizes that of idele in algebraic number theory [for which see, e.g., Cassels (1967)]. The ideas are closely related to those of Chapters 10, 11.

We must first recall our notation. Let V, ϕ be a quadratic space of dimension n over $\mathbf{Q}$. As usual, the localizations will be denoted by V_p where p is a prime or $p = \infty$. Similarly for a $\mathbf{Z}$-lattice $\Lambda \subset V$ the localizations will be denoted by Λ_p. They are $\mathbf{Z}_p$-lattices in V_p. We shall not require to define Λ_∞. The corresponding proper orthogonal groups are denoted by $O^+(V)$, $O^+(\Lambda), O^+(V_p)$ and $O^+(\Lambda_p)$, the two latter being regarded as topological groups with their p-adic topology.

An *adele* $\boldsymbol{\alpha} = \{\alpha_p\}$ is by definition a set of $\alpha_p \in O^+(V_p)$ given for all p including $p = \infty$ with the further property that $\alpha_p \in O^+(\Lambda_p)$ for almost all p. Here Λ is any $\mathbf{Z}$-lattice and the definition does not depend on the choice of Λ by Theorem 1.1 of Chapter 11. The difference between adeles and the objects considered in Chapter 11 is that we now have a component for $p = \infty$. The adeles form a group O_{A}^+ under componentwise multiplication:

$$\{\alpha_p\}\{\beta_p\} = \{\alpha_p\beta_p\}.$$

The group O_{A}^+ inherits a topology from the $O^+(V_p)$ in an obvious way. [The "restricted product topology", see, e.g., Cassels (1967).] The group $O^+(V)$ is embedded in O_{A}^+ by mapping $\alpha \in O^+(V)$ into the adele all of whose components are α. It can be shown that $O^+(V)$ is then a discrete subgroup of O_{A}^+.

The groups $O^+(V_p)$, being locally compact, have an invariant (Haar) measure μ_p which is unique up to multiplication by a constant. Since $O^+(V_p)$ is unimodular, the measure μ_p is invariant both on the left and on the right. If the arbitrary constants in the μ_p are chosen so that

$$\prod_{p \neq \infty} \mu_p(O^+(\Lambda_p)) \tag{4.1}$$

is convergent, the μ_p define a product measure μ on O_{A}^+. This measure is also invariant both on left and right.

The fact that O^+ is an algebraic group allows us to normalize the measure μ uniquely. Let N be the dimension of O^+ [so in fact $N = n(n-1)/2$]. There is an invariant differential ω on O^+ of dimension N which is defined over $\mathbf{Q}$, and ω is unique up to multiplication by a factor from $\mathbf{Q}^*$. The differential ω gives a means of constructing measures μ_p on the $O^+(V_p)$ as follows. Let ξ be a general point on O^+. We can write

$$\omega = h(\xi)\, dt_1 \ldots dt_N, \tag{4.2}$$

where $t_1, \ldots, t_N$ are local parameters at ξ and $h(\xi)$ is a function, everything being defined over $\mathbf{Q}$. For any set $\mathscr{S} \subset O^+(V_\infty)$ we put

$$\mu_\infty(\mathscr{S}) = \int_{\mathscr{S}} |h(\xi)|_\infty \, dt_1 \ldots dt_N. \tag{4.3}$$

It can be shown that this does in fact give a Haar measure on $O^+(V_\infty)$ depending only on the choice of ω. For $p \neq \infty$ the measures μ_p are defined by the p-adic analogue of (4.3). Further, the condition about the convergence of (4.1) is satisfied (with certain caveats if $n = 2$) and so ω determines a product measure μ on $O_{\mathbb{A}}^+$. If we replace ω by $c\omega$, $c \in \mathbf{Q}^*$, the measure μ_p is multiplied by $|c|_p$ and so μ is multiplied by

$$\prod_{p,\ \text{inc}\ \infty} |c|_p = 1.$$

This uniquely defined measure μ is the *Tamagawa measure* and will be denoted by τ.

Since $O^+(V)$ is a discrete subgroup of $O_{\mathbb{A}}^+$, the set $O_{\mathbb{A}}^+/O^+(V)$ inherits a measure, which we shall also denote by τ. We have now

THEOREM 4.1.

$$\tau(O_{\mathbb{A}}^+/O^+(V)) = 2. \tag{4.4}$$

There are proofs in Weil (1961) and Kneser (1974). As has already been noted, the notion of Tamagawa numbers makes sense for other algebraic groups and Weil's proof uses an induction which runs through other groups than orthogonal ones. Kneser's treatment is entirely within the framework of quadratic forms. For a discussion of the application of Theorem 4.1 see Kneser (1961) or Weil (1962), and for the background, including other applications of the theory of algebraic groups to quadratic forms, see Kneser (1967).

We shall content ourselves with sketching the deduction of one of Siegel's formulae from Theorem 4.1. The group $O_{\mathbb{A}}^+$ acts on the set of $\mathbf{Z}$-lattices in V. Indeed it was shown in Chapter 11 that if $\boldsymbol{\alpha} = \{\alpha_p\} \in O_{\mathbb{A}}^+$ and if Λ is a $\mathbf{Z}$-lattice, then there is a unique $\mathbf{Z}$-lattice Γ such that

$$\alpha_p \Lambda_p = \Gamma_p \quad (\text{all } p \neq \infty). \tag{4.5}$$

We shall write

$$\Gamma = \boldsymbol{\alpha}\Lambda. \tag{4.6}$$

In the language of Chapter 11 the lattices Λ, Γ are in the same genus. They are equivalent if $\boldsymbol{\alpha}$ can be chosen to be in $O^+(V)$.

Now let $\Lambda = \Lambda_1$ be fixed, let Λ_j $(1 \leqslant j \leqslant J)$ be representatives of the equivalence classes of lattices in the genus of Λ and let $\boldsymbol{\beta}_j \in O_{\mathbb{A}}^+$ be chosen so that

$$\boldsymbol{\beta}_j \Lambda = \Lambda_j. \tag{4.7}$$

Clearly $O_{\mathbb{A}}^{+}$ is the disjoint union of the sets (double cosets)

$$O^{+}(V)\,\boldsymbol{\beta}_j S(\Lambda), \tag{4.8}$$

where $S(\Lambda)$ is the stabilizer of Λ, that is

$$S(\Lambda) = \{\boldsymbol{\alpha} \in O_{\mathbb{A}}^{+} : \boldsymbol{\alpha}\Lambda = \Lambda\}. \tag{4.9}$$

Clearly

$$S(\Lambda_j) = \boldsymbol{\beta}_j S(\Lambda)\,\boldsymbol{\beta}_j^{-1}. \tag{4.10}$$

On taking cosets with respect to $O^{+}(V)$ and using the invariance of the measure τ under the $\boldsymbol{\beta}_j^{-1}$, we obtain

$$\begin{aligned} 2 &= \tau(O^{+}(V)\backslash O_{\mathbb{A}}^{+}) \\ &= \Sigma\tau(X_j), \end{aligned} \tag{4.11}$$

where

$$X_j = O^{+}(V)\backslash O^{+}(V)S(\Lambda_j). \tag{4.12}$$

[We have cosets on the left here while Theorem 4.1 is enunciated for right cosets. As the measure τ is invariant on both sides, this is harmless.]

Clearly

$$O^{+}(V) \cap S(\Lambda_j) = O^{+}(\Lambda_j), \tag{4.13}$$

and so (4.12) implies

$$X_j = O^{+}(\Lambda_j)\backslash S(\Lambda_j). \tag{4.12 bis}$$

By definition $S(\Lambda_j)$ is the product of $O^{+}(V_\infty)$ and the

$$O^{+}((\Lambda_j)_p) \qquad (p \neq \infty). \tag{4.12 ter}$$

The diagonal map $O^{+}(V) \to O_{\mathbb{A}}^{+}(V)$ injects $O^{+}(V)$ in to every local component, in particular $O^{+}(V_\infty)$. It follows that X_j is product of the (4.12 ter) and of

$$O^{+}(\Lambda_j)\backslash O^{+}(V_\infty) \tag{4.12 quater}$$

Hence

$$\tau(X_j) = \tau_\infty(O^{+}(\Lambda_j)\backslash O^{+}(V_\infty)) \prod_{p \neq \infty} \tau_p(O^{+}((\Lambda_j)_p)), \tag{4.14}$$

where τ_∞ and τ_p are the local components of the Tamagawa measure τ with respect to some differential ω. By (4.10) the measures

$$\tau_p(O^{+}((\Lambda_j)_p)) = \lambda_p \text{ (say)} \tag{4.15}$$

are independent of j, and depend only on the genus. On the other hand,

$$\sigma_j \text{ (say)} = \tau_\infty(O^{+}(\Lambda_j)\backslash O^{+}(V_\infty)) \tag{4.16}$$

is the measure of a fundamental domain of the group of real automorphs modulo the integral automorphs and does depend in general on j. The equation

(4.11) thus becomes

$$\sum_J \sigma_J = 2 \prod_{p \neq \infty} \lambda_p^{-1}. \tag{4.17}$$

Now suppose that we are in the positive definite case. Then $O^+(V_\infty)$ is compact and has finite measure λ_∞ (say). Clearly

$$\sigma_J = \lambda_\infty / o^+(\Lambda_J), \tag{4.18}$$

and (4.17) becomes

$$W(\mathscr{F}) = \Sigma 1/o^+(\Lambda_J) = 2\lambda_\infty \prod_{p \neq \infty} \lambda_p, \tag{4.19}$$

where $\mathscr{F}$ is the genus of the Λ_J. This can be identified with Siegel's formula (3.2) by suitable choice of the differential ω used to define the measures τ_p, τ_∞. Further, (4.17) is the generalization of (3.2) to indefinite forms.

Siegel's formula (3.13) and its generalization to indefinite forms may be obtained similarly by considering Tamagawa measure on the subgroup of $O^+(V)$ which leaves invariant a fixed subspace U of V.

5. MODULAR FORMS

Let $f(\mathbf{x})$ be an integer-valued positive definite quadratic form and put

$$\Theta(f, z) = \sum_{m \geqslant 0} a_f(m)\, z^m \tag{5.1}$$

where $a_f(m)$ is the number of integral representations of m (primitive or not). This is a modular form in a sense which will be described below, and the theory of modular forms gives the most natural framework to discuss its properties even though, as we shall see, some of them can, at least in special cases, be derived in a more elementary manner.

We shall require the notion of the *Mellin transform* $\mathscr{M}(g)$ of a power series

$$g(z) = g_0 + g_1 z + \ldots + g_m z^m + \ldots. \tag{5.2}$$

By definition $\mathscr{M}(g)$ is the Dirichlet series

$$\mathscr{M}(g)(s) = g_1 1^{-s} + g_2 2^{-s} + \ldots + g_m m^{-s} + \ldots. \tag{5.3}$$

We shall regard this as a purely formal operation, though it can be treated analytically. Note that the definition of $\mathscr{M}(g)$ does not involve the constant term g_0.

When the number n of variables is 2, the theory of modular forms adds nothing to what can be derived from the theory of composition of quadratic forms developed in Chapter 14 and some aspects of the theory of modular forms can best be regarded as a generalization of composition to $n > 2$. We shall therefore first look at (5.1) for binary forms using composition. Let $D < 0$ be a fundamental discriminant and let χ be any character of the group of proper equivalence classes of integer-valued forms $f(x, y)$ of discriminant D. Put

$$\Theta_\chi(z) = \sum_f \chi(f)\,\Theta(f, z), \tag{5.4}$$

the sum being over representatives f of all the proper equivalence classes of forms of discriminant D. It is significant that the constant term of Θ_χ is 0 except when χ is the principal character ($=1$ identically). The Mellin transforms $\mathcal{M}(\Theta_\chi)$ can be written as products

$$\mathcal{M}(\Theta_\chi) = \prod_{\varepsilon(p)=0} (1 - \beta_p p^{-s})^{-1} \prod_{\varepsilon(p)=+1} (1 - \beta_p p^{-s})^{-1}(1 - \gamma_p p^{-s})^{-1}$$

$$\prod_{\varepsilon(p)=-1} (1 - p^{-2s})^{-1}, \tag{5.5}$$

where (i) $\varepsilon(p) = (D/p)$, (ii) β_p, γ_p for $\varepsilon(p) = +1$ are the characters of the two classes of forms $px^2 \pm bxy + cy^2$ of discriminant D and (iii) β_p for $\varepsilon(p) = 0$ is the character of the unique class $px^2 + bxy + cy^2$ of discriminant D. This is easy to verify from the theory of composition. We can re-write (5.5) as

$$\mathcal{M}(\Theta_\chi) = \prod_p (1 - \lambda(p)\,p^{-s} + \varepsilon(p)\,p^{-2s})^{-1} \tag{5.6}$$

where

$$\lambda(p) = \lambda(\chi, p) = \begin{cases} \beta_p & \text{if } \varepsilon(p) = 0 \\ \beta_p + \gamma_p & \text{if } \varepsilon(p) = +1 \\ 0 & \text{if } \varepsilon(p) = -1. \end{cases} \tag{5.7}$$

We must now introduce briefly the notion of a modular form. For the proofs of statements made, the reader must consult the appropriate texts. There are good introductions with some discussion of the relevance to quadratic forms in Gunning (1962), Ogg (1969), Serre (1970) and Rankin (1977): there is a very fine exposition from a rather more advanced standpoint in Hecke (1940), which remains the standard reference.

Let $\Gamma(= SL^+)$ denote the group of matrices

$$\sigma = \begin{pmatrix} a & b \\ c & d \end{pmatrix} \qquad a, b, c, d \in \mathbf{Z} \qquad ad - bc = 1. \tag{5.8}$$

It acts on the upper half-plane H of complex numbers t with strictly positive imaginary part by putting

$$\sigma(t) = \frac{at + b}{ct + d}. \tag{5.9}$$

Let k be an integer. We define an action of Γ on a function $\phi(t)$ $(t \in \mathbf{C})$ by writing

$$\phi|_k\sigma = \psi \tag{5.10$_1$}$$

where

$$\psi(t) = (ct + d)^{-k}\,\phi(\sigma(t)). \tag{5.10$_2$}$$

It is readily verified that

$$(\phi|_k\sigma)|_k\tau = \phi|_k\tau\sigma$$

for $\tau, \sigma \in \Gamma$.

Let now Δ be a subgroup of Γ of finite index and let $\varepsilon\colon \Delta \to \mathbf{C}^*$ be a character of Δ of finite order. An *unrestricted modular form* of weight k belonging to Δ and ε is a regular† function $\phi(t)$ defined for $t \in H$ and such that

$$\phi|_k\sigma = \varepsilon(\sigma)\,\phi \tag{5.11}$$

for all $\sigma \in \Delta$. A *modular form* is an unrestricted modular form which satisfies certain finiteness conditions which must now be described.

The standard fundamental region $\mathscr{F}_\Gamma$ for Γ has a single cusp at infinity (customarily denoted by $i\infty$) in the sense that it approaches the boundary of H only at this one point. [It is perhaps easier to make a conformal transformation of H into the open unit disc to understand this point, cf. the discussion at the end of Chapter 13, Section 7.] A fundamental domain $\mathscr{F}_\Delta$ for Δ can be taken to be the union of a finite number of translations of $\mathscr{F}_\Gamma$ and so has several cusps in general. We say than an unrestricted modular form ϕ for Δ has a finite value v at $i\infty$ if $\phi(t)$ tends to v as t tends to $i\infty$ inside $\mathscr{F}_\Delta$. Similarly one defines the boundedness at other cusps by transporting the cusps to $i\infty$ with an element of Γ. A *modular form* is an unrestricted modular form ϕ which takes finite values at all the cusps. If ϕ takes the value 0 at all the cusps, then it is a *cusp form.* [German: Spitzenform.]

For given group Δ weight k and character ε, the set $M(k, \Delta, \varepsilon)$ of modular forms is clearly a vector space over $\mathbf{C}$. It is a fundamental theorem that this space is finite-dimensional. The set $S(k, \Delta, \varepsilon)$ of cusp-forms is a linear subspace of $M(k, \Delta, \varepsilon)$.

A special type of modular forms are the *Eisenstein series*. We give the definition only for $\Delta = \Gamma$ and $\varepsilon = 1$, that for general Δ, ε being analogous. Put

$$E_k(t) = \Sigma(ct + d)^{-k}, \tag{5.12}$$

the sum being over all coprime pairs of integers c, d. The series converges absolutely for $k \geqslant 4$ and is identically 0 for odd k. Clearly $E_k(t)$ is a modular form for Γ of weight k: in particular, $E_k(t)$ has a Fourier series, which can be shown to be

$$E_k(t) = 2\zeta(k) + \frac{2(2\pi i)^{k/2}}{(k-1)!}\sum_{r=1}^{\infty} s_{k-1}(r)\,e^{2\pi irt} \qquad (2|k) \tag{5.13}$$

† In the sense of complex variable theory (= analytic).

where

$$s_{k-1}(n) = \sum_{d|n} d^{k-1},$$

and

$$\zeta(k) = \sum_{r} r^{-k}$$

is the Riemann zeta function.

For general Δ, ε there are Eisenstein series corresponding to all the cusps. An Eisenstein series is non-zero at precisely one equivalence class of cusps under the action of Δ. A modular form is uniquely the sum of a linear combination of Eisenstein series and a cusp form.

We shall be concerned only with groups Δ and characters ε of a special type. Let $N \geqslant 1$ be an integer. The set of matrices (5.8) with

$$c \equiv 0 \pmod{N} \tag{5.13 bis}$$

form a subgroup $\Gamma_0(N)$ (say) of Γ. Let ε be any character of the multiplicative group of integers modulo N. Then $\varepsilon(\sigma) = \varepsilon(a)$ gives a character of $\Gamma_0(N)$. We shall write $M(k, N, \varepsilon)$ instead of $M(k, \Gamma_0(N), \varepsilon)$ for the space of modular forms, and similarly for the space $S(k, N, \varepsilon)$ of cusp forms. We note that $\sigma(t) = t$ for all t when $\sigma = -\mathbf{I}$ and so

$$\varepsilon(-1) = (-1)^k \tag{5.14}$$

by (5.11). Further, it follows from (5.11) with $c = 0, a = d = 1$, so $\varepsilon(\sigma) = 1$, that any $\phi \in M(k, N, \varepsilon)$ is periodic modulo 1 and hence has a Fourier series†

$$\phi(t) = \sum_{m \geqslant 0} \phi_m e^{2\pi i m t}. \tag{5.15}$$

Let $f(\mathbf{x})$ be a primitive integer-valued positive definite quadratic form in an even number

$$n = 2k \geqslant 4 \tag{5.16}$$

of variables. We write f in the shape

$$f(\mathbf{x}) = \tfrac{1}{2}\mathbf{x}'\mathbf{F}\mathbf{x}, \tag{5.17}$$

so that the elements of the matrix $\mathbf{F}$ are integers and those on the principal diagonal are even. There is a uniquely defined integer $N > 0$ such that

$$N\mathbf{F}^{-1} \tag{5.18}$$

corresponds similarly to a primitive integer-valued form. We call N the Stufe of f. It may easily be verified that N and $\det \mathbf{F}$ have the same prime factors, in general with different multiplicities. We define

$$\varepsilon(a) = \left(\frac{(-1)^k \det \mathbf{F}}{a}\right) \tag{5.19}$$

† Boundedness as $t \to i\infty$ implies $\phi_m = 0$ for $m < 0$.

where the symbol on the right-hand side is a Kronecker symbol†. It may be shown that $\varepsilon(a)$ is a character mod N and that

$$\theta(f, t) \in M(k, N, \varepsilon), \tag{5.20}$$

where

$$\theta(f, t) = \sum_{\mathbf{x}} e^{2\pi i f(\mathbf{x}) t}, \tag{5.21}$$

the sum being over all integral vectors $\mathbf{x}$. [The proof uses a Fourier transform argument generalizing one of the standard proofs that the Jacobi θ-functions of a single variable are modular functions; for the details see Schoeneberg (1939).] In the notation of (5.1) we have‡

$$\theta(f, t) = \Theta(f, e^{2\pi i t}). \tag{5.22}$$

Note that both the Stufe N and the character ε depend only on the genus of the form f.

The fact that $\theta(f, t)$ is a modular function has many more or less immediate consequences. In the first place

$$\theta(f, t) = \theta_e(f, t) + \theta_s(f, t) \tag{5.23}$$

where $\theta_s(f, t)$ is a cusp form and $\theta_e(f, t)$ is a linear combination of Eisenstein series. If there are no non-trivial cusp forms, then $\theta_s(f, t) = 0$: hence we have an explicit expression for $\theta(f, t)$ and so for the numbers $a_f(m)$ of representations. This is the case, for example, when f is the sum of n squares and $n = 2k \leqslant 8$. [Cf. formulae (1.2)–(1.5).] Here there is only one form in the genus. Rather more remarkable is the genus of improperly primitive integral forms of determinant 1 in 16 variables. Here $N = 1$ so $\varepsilon = 1$. There are two classes in the genus but no non-trivial cusp forms. It follows that

$$\theta(f_1, t) = \theta(f_2, t)$$

identically for forms f_1, f_2 in the two classes: that is, f_1 and f_2 represent every integer m the same number of times [Witt (1941)].

There are estimates for the coefficients of cusp forms which show that they are asymptotically of a smaller order of magnitude than those of Eisenstein series. It follows that the coefficients of $\theta_e(f, t)$ are asymptotic estimates for the numbers $a_f(m)$ of representations: in fact $\theta_e(f, t)$ is the "singular series" obtained by the application of the Hardy–Littlewood "circle method". There is a further interpretation. Let g run through the classes of a genus $\mathscr{G}$. Then it can be shown that

$$\sum_g \{o^+(g)\}^{-1}\theta(g, t) = W(\mathscr{G})\theta_e(f, t) \tag{5.24}$$

† cf. Chapter 14, example 2. Note that $(-)^k \det \mathbf{F} \equiv 1 \pmod 4$ if $\det \mathbf{F}$ is odd, as follows readily from the canonical form for 2-adic integral forms.

‡ For formal reasons it is convenient to use the substitution $z = e^{2\pi i t}$ here and $z = e^{\pi i s}$ in Section 1.

for every form f of $\mathscr{G}$, where $W(\mathscr{G})$ is the weight. The coefficient of $e^{2\pi im}$ in the Fourier series on the left-hand side is just $W(\mathscr{G})$ times the average number of representations of m given by Siegel's formula (3.12).

If a basis of the space $M(k, N, \varepsilon)$ is known from other considerations and f is a quadratic form such that $\theta(f, t) \in M(k, N, \varepsilon)$, then one can express $\theta(f, t)$ in terms of the basis: it is necessary only to consider the first few coefficients. For example when $k = 12$ and $N = 1$ (so $\varepsilon = 1$) all the cusp forms are multiples of the "discriminant function"

$$G(z) = z \prod_m (1 - z^m)^{24} \qquad (z = e^{2\pi it}), \tag{5.25}$$

whose Fourier coefficients are the Ramanujan numbers $\tau(m)$. It follows that the numbers $a_f(m)$ of representations of m by an improperly primitive form f of determinant 1 in 24 variables can be expressed in terms of the Fourier coefficients of the Eisenstein series E_{12} [cf. (5.13)] and of the Ramanujan numbers $\tau(m)$ [cf. Serre (1970)]. A case where there is more than one cusp form is the sum of 24 squares: here $G(-z)$ and $G(z^2)$ span the appropriate space of cusp forms and the Eisenstein series is given by (1.7). Equating the coefficients of the first few powers of z determines the coefficients in (1.6)–(1.8).

Again, if $f_1, \ldots, f_J$ are forms such that

$$\theta(f_j, t) \in M(k, N, \varepsilon) \qquad (1 \leqslant j \leqslant J)$$

for some k, N, ε and if J is greater than the dimension of $M(k, n, \varepsilon)$, then there must be a relation

$$\sum_j b_j \theta(f_j, t) = 0 \tag{5.26}$$

where the $b_j \in \mathbf{Z}$ are not all 0. Such relations can sometimes be proved by elementary means [e.g. Kneser (1967a), cf. Chapter 9, examples 20–22]. However the theory of modular forms not merely predicts when such relations ought to exist but gives a powerful method of proving them. A relation (5.26) necessarily holds when the first m_0 Fourier coefficients of the left-hand side vanish, where m_0 depends only on k, N and ε.

Hecke showed that (5.6) could also be generalized to forms in $n = 2k > 2$ variables and indeed to modular forms $\phi \in M(k, N, \varepsilon)$. We say that ϕ has a canonical Euler product if the Mellin transform of the corresponding power series in $z = e^{2\pi it}$ has the shape

$$\prod_p (1 - \lambda(p)p^{-s} + \varepsilon(p)p^{k-1-2s})^{-1}. \tag{5.27}$$

Hecke showed that the ϕ of this kind span $M(k, N, \varepsilon)$. The proof follows from the consideration of the action of a remarkable family of linear operators T_p on $M(k, N, \varepsilon)$, the so-called Hecke operators. For this we must refer to the literature. Eichler (1952) has shown that in some cases the action of the Hecke operators can be described in terms of integral matrices $\mathbf{A}$ such that $f_1(\mathbf{Ax}) = mf_2(\mathbf{x})$ where $m \in \mathbf{Z}$ and f_1, f_2 are quadratic forms of the same genus. However the elementary approach appears to be neither so powerful nor so perspicuous as the analytic one.

When f is positive definite but the number n of variables is odd, one requires the theory of modular forms "of half-integral weight". This appears to be neither so elegant nor so well-developed as that for forms of integral weight. For indefinite forms f, information can be derived from analytic functions but one is also led naturally to modular forms which are not analytic functions of the variable t: and here again the theory is deeper and less satisfactory than that discussed in this appendix.

NOTES

Section 2. The class number formulae for binary forms are special cases, at least for fundamental discriminants, of formulae for any number-field with abelian galois group.

When $m > 0$ is representable as the sum of 3 squares, Venkov (1928, 1931, 1970) verified the formulae for the class-number $h(-m)$ using quaternions. The proof, while technically elementary, is quite complicated: in any case it does not apply to $m \equiv 7 \pmod 8$. See also Davies (1976).

Section 3. The result of Magnus (1937) about class-numbers of definite quadratic forms has been generalized to definite forms over totally real algebraic number-fields. [Pfeuffer (1971).]

In a letter to Gauss which has only recently been published, Eisenstein [(1975), Vol. 2, pp. 860–865] gives formulae for the number of classes of definite ternary forms of given determinant. No details of the proof are given, but it appears [ibid., pp. 874–5] that he uses knowledge of the weight of the genus and of the number of forms of the given determinant with proper automorphs other than the identity [cf. Chapter 9, Example 15].

Section 5. There is a discussion of many aspects and applications of modular functions of a single variable in the Proceedings of the Summer School which is listed as Antwerp (1972).

We have followed the convention at Antwerp and defined the weight k of a modular form by (5.10_2), (5.11). There is much terminological confusion in the literature. Hecke uses the term dimension, which he defines to be $-k$. Gunning uses the term weight to mean $\frac{1}{2}k$.

There are examples of inequivalent forms which possess the same θ-series in Witt (1941), Kneser (1967a) and Kitaoka (1977).

For the application of modular forms in several variables to quadratic forms, see the relevant papers in Siegel's *Gesammelte Abhandlungen* [Siegel (1966)] and also Siegel (1934, 1967). Maass (1971).

EXAMPLES

1. In the notation of Theorem 2.1 show for a fundamental discriminant $D < 0$ that

$$2(2 - (D/2))\,h(D) = w\sum_{r=1}^{[|D|/2]} (D/r).$$

2. Denote by $\chi(r)$ a character on the multiplicative group of integers prime to n modulo n.

(i) If $n = lm$, where l, m are coprime, show that

$$\chi(r) = \psi(r)\,\phi(r),$$

where ψ, ϕ are characters modulo l and modulo m respectively.

(ii) Write

$$\tau(\chi, n) = \Sigma\chi(r)\,e(r/n)$$

where the sum is over the integers prime to n modulo n and where

$$e(x) = e^{2\pi i x}.$$

Show that

$$\psi(m)\,\phi(l)\,\tau(\chi, n) = \tau(\psi, l)\,\tau(\phi, m).$$

3. Show that apart from the trivial character ($=1$ identically) there is precisely one real-valued character modulo p (a given prime). Show that

$$\{\tau(\chi, p)\}^2 = \chi(-1)p.$$

[*Hint*. The left-hand side is

$$\sum_{r, s \bmod p} \chi(rs)\,e((r + s)/p).$$

Change the variables of summation to r, t where $s \equiv rt \bmod p$. The sum with respect to r is now 0 except when $t \equiv -1 \pmod p$.]

4. Let D be a fundamental discriminant. Show that

$$\sum_{r \bmod |D|} (D/r)\,e(r/|D|) = \begin{cases} \pm\sqrt{|D|} & (D > 0) \\ \pm i\sqrt{|D|} & (D < 0). \end{cases}$$

[*Hint*. Use example 2 to reduce to the cases $D = \pm p \equiv 1 \pmod 4$ where p is a prime, $D = -4$ and $D = \pm 8$. Then use example 3.]

Note. This is (2.20) except for the occurrence of the $\pm$ sign. For a selection of proofs that the sign is always + using widely differing techniques see Landau (1927, Bd. 1, pp. 153–171). One of these is given below.]

5. Let $n > 1$ be an integer and let $\sqrt{n}$ denote the positive square root. For

$$\mathbf{a} = (a_0, \ldots, a_{n-1})$$

let

$$\mathbf{Fa} = \mathbf{b} = (b_0, \ldots, b_{n-1})$$

be the Fourier Transform given by

$$b_j = \frac{1}{\sqrt{n}} \Sigma a_k \, e(jk/n).$$

Regard $\mathbf{F}$ as a linear map $\mathbf{C}^n \to \mathbf{C}^n$.

(i) Show that $\mathbf{F}^4 = \mathbf{I}$. Deduce that the eigenvalues of $\mathbf{F}$ are i^u ($u = 0, 1, 2, 3$) with multiplicities m_u, where

$$m_0 + m_1 + m_2 + m_3 = n.$$

(ii) Show that $\mathbf{F}^2$ is the transformation

$$a_0 \to a_0, a_j \to a_{n-j} \qquad (1 \leqslant j < n).$$

Hence determine $m_0 + m_2$ and $m_1 + m_3$.

(iii) If $n = p$ show that the trace of $\mathbf{F}$ is

$$(\sqrt{p})^{-1} \tau(\chi, p)$$

where χ is the character of order 2 on the multiplicative group of integers prime to p mod p. Deduce that

$$(m_1 - m_2)^2 + (m_3 - m_4)^2 = 1.$$

(iv) Evaluate $\det(\mathbf{F})$. Deduce the value of $\sum_u u m_u$ (mod 4) and hence determine m_0, m_1, m_2, m_3. [*Hint.* The determinant is an alternant (van der Monde determinant) and so expressible as a product of a root of unity and sines of rational angles.]

(v) Hence show that

$$\tau(\chi, p) = \begin{cases} \sqrt{p} & p \equiv +1 \pmod 4 \\ i\sqrt{p} & p \equiv -1 \pmod 4. \end{cases}$$

(vi) Generalize the above to determine the value of Gauss' sum

$$\sum_{j \bmod n} e(j^2/n)$$

distinguishing the cases n odd, $n \equiv 2 \pmod 4$ and $n \equiv 0 \pmod 4$. [See Schur (1921).]

References

Works are cited by the name of the author(s) and year of publication, e.g. Gauss (1801). Where there is more than one work by the same author in the same year, those after the first are distinguished by appending a, b, c, ... to the year, e.g. Shanks (1971), Shanks (1971a). Names of journals are as in *Mathematical Reviews*. Titles of papers in Russian are transliterated using the *Mathematical Reviews* rules.

There has been no attempt at an encyclopedic coverage of the immense literature on quadratic forms. The reader should be able to find further references from the bibliographies of the papers cited. For this reason there has been a tendency to give here the latest papers on a given topic even though they may be less important than earlier ones which have not been given. For earlier references see also Bachmann (1898), Dickson (1919) and Lehmer (1941).

Antwerp (1972). *Modular functions in one variable.* I, II, III, IV. Proceedings International Summer School, Antwerp, 1972; V, VI. Proceedings International Conference, Bonn, 1976. *Lecture Notes in Mathematics* (Springer) **320** (195 pp.), **349** (598 pp.), **350** (350 pp), **476** (151 pp.), **601** (294 pp.), **627** (339 pp.).

E. Artin (1957). *Geometric algebra.* Interscience, London and New York.

L. Aubry (1912). Solution de quelques questions d'analyse indéterminée. *Sphinx-Œdipe* **7**, 81–84.

G. Bachman (1964). *Introduction to* p*-adic numbers and valuation theory.* Academic Press, New York, London.

P. Bachmann (1898). *Die Arithmetik der quadratischen Formen.* Erste Abtheilung (1898). Zweite Abteilung (1923). Teubner, Leipzig.

P. Bachmann (1910). *Niedere Zahlentheorie* (2 vols.), Teubner, Leipzig. Reprint: Chelsea Pub. Co., New York (No date).

A. Baker (1975). *Transcendental number theory.* Cambridge University Press.

A. Baker and A. Schinzel (1971). On the least integers represented by genera of binary quadratic forms. *Acta Arith.* **18**, 137–144.

E. S. Barnes (1957). The complete enumeration of extreme senary forms. *Philos. Trans. Royal Soc. London* A **249**, 461–506.

H. Bauer (1972). Die 2-Klassenzahlen spezieller quadratischer Zahlkörper. *J. reine angew. Math.* **252**, 79–81.

B. J. Birch and H. Davenport (1958). Quadratic equations in several variables. *Proc. Cambridge Philos. Soc.* **54**, 135–138 (= Davenport, *Coll. Works* **3**, 1120–1123.)

H. Blaney (1948). Indefinite quadratic forms in n variables. *J. London Math. Soc.* **23**, 153–160.

R. Bonola (1912). *Non-euclidean geometry*. Open Court Publishing Co. Reprinted by Dover, 1955.
A. Borel (1969). *Introduction aux groupes arithmétiques. Actualités Sci. Ind.*, **1341**. Hermann, Paris.
Z. I. Borevič & I. R. Šafarevič (1966). *Number theory*. Translated from the Russian by Newcomb Greenleaf. Academic Press, New York, London.
H. Brandt (1924). Der Kompositionsbegriff bei den quaternären quadratischen Formen. *Math. Ann.* **91**, 300–315.
H. Brandt (1925). Die Hauptklassen in der Kompositionstheorie der quaternären quadratischen Formen. *Math. Ann.* **94**, 166–175.
H. Brandt (1925a). Über die Komponierbarkeit quaternärer quadratischer Formen. *Math. Ann.* **94**, 179–197.
H. Brandt (1928). Idealtheorie in Quaternionenalgebren *Math. Ann.* **99**, 1–29.
H. Brandt (1943). Zur Zahlentheorie der Quaternionen. *Jber. Deutsch. Math.-Verein* **53**, 23–57.
H. Brandt and O. Intrau (1948). Tabellen reduzierter positiver ternärer quadratischer Formen. *Abh. Sächs. Akad. Wiss. Math.-Nat. Kl.* **45**, No. 4, 261 pp.
W. E. Briggs and S. Chowla (1954). On discriminants of binary quadratic forms with a single class in each genus. *Canadian J. Math.* **6**, 463–470.
R. H. Bruck and H. J. Ryser (1949). The non-existence of certain finite projective planes. *Canadian J. Math.* **2**, 93–99.
A. Brumer (1978). Remarques sur les couples de formes quadratiques. *C.R. Acad. Sci. Paris Sér. A* **286**, 679–681.
D. A. Buell (1977). Small class-numbers and extreme values of *L*-functions of quadratic fields. *Math. Comp.* **31**, 786–796.
J. W. S. Cassels (1955). Bounds for the least solutions of homogeneous quadratic equations. *Proc. Cambridge Philos. Soc.* **51**, 262–264 and **52** (1956), 604.
J. W. S. Cassels (1957). *An introduction to diophantine approximation. Cambridge Tracts in Mathematics and Mathematical Physics* No. 45. Cambridge University Press.
J. W. S. Cassels (1959). *An introduction to the geometry of numbers. Die Grundlehren der Math. Wiss.* No. 99. Springer Verlag, Berlin, Göttingen, Heidelberg.
J. W. S. Cassels (1959a). Note on quadratic forms over the rational field. *Proc. Cambridge Philos. Soc.* **55**, 267–270.
J. W. S. Cassels (1962). Über die Äquivalenz 2-adischer quadratischer Formen. *Comment. Math. Helv.* **37**, 61–64.
J. W. S. Cassels (1964). On the representations of rational functions as sums of squares. *Acta Arith.* **9**, 79–82.
J. W. S. Cassels (1967). Global fields. Chapter 2 of *Algebraic number theory* (eds. J. W. S. Cassels and A. Fröhlich). Academic Press, London and New York.
J. W. S. Cassels (1976). An embedding theorem for fields. *Bull. Austral. Math. Soc.* **14**, 193–198 and 479–480.
J. W. S. Cassels, W. J. Ellison and A. Pfister (1971). On sums of squares and on elliptic curves over function fields. *J. Number Theory* **3**, 125–149.
J. W. S. Cassels and A. Fröhlich (1967). (Eds.) *Algebraic Number Theory* ("The Brighton Book"). Academic Press, London and New York
S. Chowla. See also W. E. Briggs.
J. H. Conway (1973). Invariants for quadratic forms. *J. Number Theory* **5**, 390–404.
H. S. M. Coxeter (1942). *Non-euclidean geometry. Mathematical Expositions* No. 2, University of Toronto Press.

H. Davenport (1949). On indefinite ternary quadratic forms. *Proc. London Math. Soc.* (2) **51**, 145–160. (= *Coll. Works*, **1**, 261–276.)

H Davenport (1971). Homogeneous quadratic equations. *Mathematika* **18**, 1–4. (= *Coll. Works* **3**, 1125–1128).

H. Davenport (1977). *Collected Works* (4 vols.), Academic Press, London.

H. Davenport. See also B. J. Birch.

R. W. Davies (1976). Class number formulae for imaginary quadratic fields. *J. reine angew. Math.* **286/287**, 369–379.

R. Dedekind. See also P. G. L. Dirichlet.

M. Deuring (1935). *Algebren. Ergebnisse der Math.* $\mathbf{4}_1$. Springer, Berlin. Reprint, Chelsea Pub. Co., New York, 1948.

L. E. Dickson (1919). *History of the theory of numbers* (3 vols). *Carnegie Institution of Washington Publications* No. 256. Reprint: Stechert, New York, 1934.

L. E. Dickson (1930). *Studies in the theory of numbers*. Univ. of Chicago Press.

J. Dieudonné (1955). *La géométrie des groupes classiques. Ergebnisse der Math. No. 5.* Springer-Verlag, Berlin, Göttingen, Heidelberg. [2nd edn. 1963.]

P. G. L. Dirichlet (1851). *De formarum binarium secundi gradus compositione.* Berolini, Typis Academicis. Reprinted in somewhat altered form in *J. reine angew. Math.* **47** (1854), 155–160, and in *Werke* II, 105–114 (Reimer, Berlin. 1897).

P. G. L. Dirichlet and R. Dedekind (1863). *Vorlesungen über Zahlentheorie.* von P. G. Lejeune Dirichlet: herausgegeben und mit Zusätzen versehen von R. Dedekind. Vieweg und Sohn, Braunschweig. [Subsequent editions such as the third (1879) contain substantially more material by Dedekind than does the first.]

A. H. Durfee (1977). Bilinear and quadratic forms on torsion modules. *Advances in Math.* **25**, 133–164.

A. G. Earnest and J. S. Hsia (1975). Spinor norms of local integral rotations, II. *Pacific J. Math.* **61**, 71–86.

B. Eckmann (1943). Gruppentheoretischer Beweis des Satzes von Hurwitz-Radon über die Komposition quadratischer Formen. *Comment. Math. Helv.* **15**, 358–366.

M. Eichler (1952). Die Ähnlichkeitsklassen indefiniter Gitter. *Math. Z.* **55**, 216–252.

M. Eichler (1952a). *Quadratische Formen und orthogonale Gruppen. Gundlehren d. Math. Wiss.* **63**, Springer, Berlin, Göttingen, Heidelberg.

G. Eisenstein (1847). *Mathematische Abhandlungen.* G. Reimer, Berlin. Reprint: G. Olms, 1967, Hildesheim. [This was published in his lifetime and does not contain many important papers.]

G. Eisenstein (1847a). Neue Theoreme der höheren Arithmetik. *J. reine angew. Math.* **35**, 117–136. (= *Math. Abh.*, 177–196; = *Math. Werke* **1**, 483–502.)

G. Eisenstein (1851). Tabelle der reducirten positiven quadratischen Formen, nebst den Resultaten neuerer Forschungen. *J. reine angew. Math.* **41**, 140–190. Anhang· 227–242 (= *Math. Werke* **2**, 637–702.)

G. Eisenstein (1852). Über die Vergleichung von solchen ternären quadratischen Formen, welche verschiedene Determinanten haben. *Sitzungsberichte der Preuss. Akad. Wiss. zu Berlin* **1852**, 350–389 (= *Math. Werke* **2**, 722–761.)

G. Eisenstein (1975). *Mathematische Werke* (2 vols). Chelsea Pub. Corp., New York.

W. J. Ellison. See also J. W. S. Cassels.

R. Fricke and F. Klein (1897). *Vorlesungen über die Theorie der automorphen Functionen* (2 vols.). Teubner, Leipzig.

R. Fricke. See also F. Klein.

R. Fricker (1971). Eine Beziehung zwischen der hyperbolischen Geometrie und der Zahlentheorie. *Math. Ann.* **191**, 293–312.

A. Fröhlich (1959). The restricted biquadratic symbol. *Proc. London Math. Soc.* (3) **9**, 189–207.
A. Fröhlich (1967). Quadratic forms à la local theory. *Proc. Cambridge Philos. Soc.* **63**, 579–586.
A. Fröhlich. See also J. W. S. Cassels.
C. F. Gauss (1801). *Disquisitiones arithmeticae.* Fleischer, Lipsiae (=Leipzig). [= *Werke* Bd 1. There are the following translations. Into English with the Latin title: Yale University Press, 1966. Into French: *Recherches arithmétiques*, Paris, 1807: reprints Hermann, 1910, Blanchard, 1953. Into German: *Untersuchungen über höhere Arithmetik.* Springer, Berlin, 1889: reprint Chelsea, 1965.]
C. F. Gauss (1870). *Werke.* K. Gesell. Wiss., Göttingen.
K. Germann (1963). Tabellen reduzierter, positiver quaternärer quadratischer Formen. *Comment. Math. Helv.* **38**, 56–83.
L. J. Gersten (972). The growth of class numbers of quadratic forms. *Amer. J. Math.* **94**, 221–236.
M. J. Greenberg (1972). *Euclidean and non-euclidean geometries.* W. H. Freeman. San Francisco.
E. Grosswald (1963). Negative discriminants of binary quadratic forms with one class in each genus. *Acta Arith.* **8**, 295–306.
R. C. Gunning (1962). *Lectures on modular forms. Annals of Mathematics Studies,* **48**.
Marshall Hall Jr. (1954). *Projective planes and related topics.* California Institute of Technology.
G. H. Hardy (1940). *Ramanujan.* Cambridge University Press.
G. H. Hardy and E. M. Wright (1938). *An introduction to the theory of numbers.* Oxford University Press.
H. Hasse (1923). Über die Darstellbarkeit von Zahlen durch quadratische Formen im Körper der rationalen Zahlen. *J. reine angew. Math.* **152**, 129–148. (= *Math. Abh.* Bd. 1, 3–22.)
H. Hasse (1923a). Über die Äquivalenz quadratischer Formen im Körper der rationalen Zahlen. *J. reine angew. Math.* **152**, 205–224. (=*Math. Abh.* Bd. 1, 23–42.)
H. Hasse (1924). Symmetrische Matrizen im Körper der rationalen Zahlen. *J. reine angew. Math.* **153**, 12–43. (=*Math. Abh.* Bd. 1, 43–74.)
H. Hasse (1924a). Darstellbarkeit von Zahlen durch quadratische Formen in einem beliebigen algebraischen Zahlkörper. *J. reine angew. Math.* **153**, 113–130. (=*Math. Abh.* Bd. 1, 75–92.)
H. Hasse (1924b). Äquivalenz quadratischer Formen in einem beliebigen algebraischen Zahlkörper. *J. reine angew. Math.* **153**, 158–162. (=*Math. Abh.* Bd. 1, 93–97.)
H. Hasse (1926). Bericht über neuere Untersuchungen und Probleme aus der Theorie der algebraischen Zahlkörper. *Jber. Deutsch. Math.-Verein* **35**, 1–55: **36** (1927), 233–311: *Ergänzungsband* **6** (1930), 1–204. Reprints: Teubner, Leipzig (1930) and Physica-Verlag, Würzburg, Wien (1965).
H. Hasse (1931). Beweis eines Satzes und Widerlegung einer Vermutung über das allgemeine Normenrestsymbol. *Nachr. Gesell. Wiss. Göttingen, Math.-Phys. Kl.* **1931**, 64–69. (=*Math. Abh.* Bd. 1, 155–160.)
H. Hasse (1975). An algorithm for determining the 2-Sylow subgroup of the divisor class group of a quadratic number field. *Symposia Mathematica, Instituto Naz. di alta Mat.,* **15**, 341–352. [Not in *Math. Abh.*]
H. Hasse (1975a). *Mathematische Abhandlungen* (3 vols.), de Gruyter, Berlin, New York.
M. A. Heaslet. See also J. V. Uspensky.

E. Hecke (1940). Analytische Arithmetik der positiven quadratischen Formen. *Danske Vid. Selsk. Math.-Fys. Medd.* **17**, No. 12 (134 pp.). (= *Math. Werke*, 789–918.)

E. Hecke (1959). *Mathematische Werke.* Vandenhoeck und Ruprecht, Göttingen.

M. D. Hendy (1975). The distribution of ideal class numbers of real quadratic fields. *Math. Comp.* **29**, 1129–1134 and **30** (1976), 679.

J. S. Hsia (1973). On the Hasse principle for quadratic forms. *Proc. Amer. Math. Soc.* **39**, 468–470.

J. S. Hsia. See also A. G. Earnest.

J. S. Hsia, Y. Kitaoka and M. Kneser (1978). Representations of positive definite quadratic forms. To appear in *J. reine angew. Math.*

P. Humbert (1940). Théorie de la réduction des formes quadratiques définies dans un corps algébrique K fini. *Comment. Math. Helv.* **12**, 263–306.

P. Humbert (1949). Réduction des formes quadratiques dans un corps algébrique fini. *Comment. Math. Helv.* **23**, 50–63.

A. Hurwitz (1896). Über die Zahlentheorie der Quaternionen. *Nachr. k. Gesell. Wiss. Göttingen, Math.-Phys. Kl.* **1896**, 313–340. (= *Math. Werke* II, 303–330.)

A. Hurwitz (1898). Über die Komposition der quadratischen Formen von beliebig vielen Variabeln. *Nachr. k. Gesell. Wiss. Göttingen, Math.-Phys. Kl.* **1898**, 309–316. (= *Math. Werke* II, 565–571.)

A. Hurwitz (1919). *Vorlesungen über die Zahlentheorie der Quaternionen.* Springer, Berlin.

A. Hurwitz (1923). Über die Komposition der quadratischen Formen. *Math. Ann.* **88**, 1–25. (= *Math. Werke* II, 641–666).

A. Hurwitz (1962). *Mathematische Werke* (2 vols.) Birkhäuser, Basel and Stuttgart.

D. Husemoller. See also J. Milnor.

E. L. Ince (1934). *Cycles of reduced ideals in quadratic fields. Mathematical Tables*, vol. IV. British Association.

O. Intrau. See also H. Brandt.

B. W. Jones (1935). A table of Eisenstein-reduced positive ternary quadratic forms of determinant $\leqslant 200$. *Bull. Nat. Res. Council, U.S.A.* **97**, 1–51.

B. W. Jones (1944). A canonical quadratic form for the ring of 2-adic integers. *Duke. Math. J.* **11**, 715–727.

B. W. Jones (1949). The composition of quadratic binary forms. *Amer. Math. Monthly*, **56**, 380–391.

B. W. Jones (1950). *The arithmetic theory of quadratic forms. Carus Math. Monographs* No. 10. Wiley.

B. W. Jones (1977). Quasi-genera of quadratic forms. *J. Number Theory* **9**, 393–412.

B. W. Jones and G. Pall (1939). Regular and semi-regular positive ternary quadratic forms. *Acta Math.* **70**, 165–191.

B. W. Jones and G. L. Watson (1956). On indefinite ternary quadratic forms. *Canadian J. Math.* **8**, 592–608.

I. Kaplansky (1968). Composition of binary quadratic forms. *Studia Math.* **31**, 85–92.

I. Kaplansky (1969). Submodules of quaternion algebras. *Proc. London Math. Soc.* (3) **19**, 219–232.

J. Kirmse (1924). Zur Darstellung total positiver Zahlen als Summe von vier Quadraten. *Math. Z.* **21**, 195–202.

Y. Kitaoka (1977). Positive definite forms with the same representation numbers. *Arch. Math.* (*Basel*) **28**, 495–497.
Y. Kitaoka (1977a). Scalar extension of quadratic lattices. *Nagoya Math. J.* **66**, 139–149.
Y. Kitaoka. See also J. S. Hsia.
F. Klein (1928). *Vorlesungen über nicht-euklidische Geometrie. Grundlehren der Math. Wiss.* 26. Springer, Berlin.
F. Klein and R. Fricke (1890). *Theorie der elliptischen Modulfunctionen* (2 vols.). Teubner, Leipzig.
F. Klein. See also R. Fricke.
H. D. Kloosterman (1926). On the representation of numbers in the form $ax^2 + by^2 + cz^2 + dt^2$. *Acta Math.* **49**, 407–464.
M. Kneser (1954). Zur Theorie der Kristallgitter. *Math. Ann.* **127**, 105–106.
M. Kneser (1956). Klassenzahlen indefiniter quadratischer Formen. *Arch. Math.* (*Basel*). **7**, 323–332.
M. Kneser (1957). Klassenzahlen definiter quadratischer Formen. *Arch. Math.* (*Basel*) **8**, 241–250.
M. Kneser (1959). Kleine Lösungen der diophantischen Gleichung $ax^2 + by^2 = cz^2$. *Abh. Math. Sem. Univ. Hamburg.* **23**, 163–173.
M. Kneser (1961). Darstellungsmasse indefiniter quadratischer Formen. *Math. Z.* **77**, 188–194.
M. Kneser (1965). Starke Approximation in algebraischen Gruppen I. *J. reine angew. Math.* **218**, 190–203.
M. Kneser (1966). Strong approximation. *Algebraic groups and discontinuous subgroups.* (*Proc. Sympos. Pure Math., Boulder, Colo.* 1965) pp. 187–196. Amer. Math. Soc.
M. Kneser (1967). Semi-simple algebraic groups. *Algebraic number theory* (eds J. W. S. Cassels and A. Fröhlich), Academic Press, London and New York, pp. 250–265.
M. Kneser (1967a). Lineare Relationen zwischen Darstellungsanzahlen quadratischer Formen. *Math. Ann.* **168**, 31–39.
M. Kneser (1974). *Quadratische Formen.* Mathematisches Institut, Göttingen. (Duplicated lecture notes.)
M. Kneser. See also J. S. Hsia.
N. Koblitz (1977). p-*adic numbers,* p-*adic analysis, and zeta-functions. Graduate texts in mathematics,* **58**. Springer, New York, Heidelberg, Berlin.
L. A. Kogan (1971). *O predstavlenii celyh čisel položitel'no opredelennymi kvadratičnymi formani.* FAN, Taškent.
R. Kortum and G. McNiel (1968). *A table of periodic continued fractions.* Lockheed Missiles and Space Division, Sunnydale, California. [Corrigenda: *Math. Comp.* **23** (1969), 217, 219.]
T. Y. Lam (1973). *The algebraic theory of quadratic forms. Mathematics Lecture Notes Series.* W. A. Benjamin, Reading, Mass. U.S.A.
E. Landau (1909). *Handbuch der Lehre von der Verteilung der Primzahlen.* (2 vols.) B. G. Teubner, Leipzig and Berlin.
E. Landau (1927). *Vorlesungen über Zahlentheorie.* (3 vols.) S. Hirzel, Leipzig.
A. M. Legendre (1798). *Essai sur la théorie des nombres.* Paris. [The third edition (Firmin Didot Frères, Paris, 1830) has the title *Théorie des nombres* and was reprinted by Hermann, Paris, 1900.]

D. H. Lehmer (1941). *Guide to tables in the theory of numbers. Bulletin of the National Research Council* **105**. National Academy of Sciences, Washington, D.C.

G. G. Lekkerkerker (1969). *Geometry of numbers. Bibliotheca Mathematica,* Vol. 8. Wolters-Noordhoff, Groningen; and North-Holland, Amsterdam, London.

Ju. V. Linnik (1939). Odna obščaja teorema o predstavlenii čisel otdel'nymi ternarnymi formani. *Izv. Akad. Nauk SSSR (ser. mat.)* **3**, 87–108.

Ju. V. Linnik (1940). O predstavlenii bol'ših čisel položitel'nymi ternarnymi formami. *Izv. Akad. Nauk SSSR (ser. mat.)* **4**, 363–402.

Ju. V. Linnik (1949). Kvaternioni i čisla Kèli [= Cayley]: nekotorye priloženija arifmetiki kvaternionov. *Uspehi Mat. Nauk* **4**, 49–90.

Ju. V. Linnik (1956). Asimptotičeskaja geometrija gaussovyh rodov: analog èrgodičeskoĭ teoremy. *Dokl. Akad. Nauk SSSR* **108**, 1018–1021.

Ju. V. Linnik, A. V. Malyšev (1953). Priloženija arifmetiki kvaternionov k teorii ternarnyh kvadratičnyh form i k razloženiju čisel na kubi. *Uspehi Mat. Nauk* **8**, 3–71; **10**, 243–244 [*A.M.S. Translations* (2) **3**, 91–162 (1956).]

J. Liouville (1858–65). Sur quelques formules générales qui peuvent être utiles dans la théorie des nombres. *J. Math. pures appl.* (2) **3** (1858), 143–152, 193–200, 201–208, 241–250, 273–288, 325–336; **4** (1859), 1–8, 72–80, 111–120, 195–204, 281–304; **5** (1860), 1–8; **9** (1864), 249–256, 281–288, 321–336, 389–400; **10** (1865), 135–144, 169–176.

R. Lipschitz (1886). *Untersuchungen über die Summen von Quadraten.* Max Cohen und Sohn, Bonn.

R. Lipschitz (1959). Correspondence. *Ann. Math.* **69**, 247–251.

G. A. Lomadze (1978). Formuly dlja čisel predstavlenii čisel nekotorymi reguljarnymi i polureguljarnymi kvadratičnymi formami, prinadležaščimi dvyhklassnym rodam. *Acta Arith.* **34**, 131–162.

F. Lorenz (1970). *Quadratische Formen über Körpern. Lecture notes in mathematics* no. 130. Springer, Berlin, Heidelberg, New York.

H. Maass (1971). *Siegel's modular forms and Dirichlet series. Lecture Notes in Mathematics,* **216**, Springer, Berlin, New York.

W. Magnus (1937). Über die Anzahl der in einem Geschlecht enthaltenen Klassen von positiv-definiten quadratischen Formen. *Math. Ann.* **114**, 465–475 + **115**, 643–644.

K. Mahler (1973). *Introduction to p-adic numbers and their functions. Cambridge tracts in mathematics,* **64**. Cambridge University Press.

A. V. Malyšev (1962). O predstavlenii celyh čisel položitel'nymi kvadratičnymi formami. *Trudy Mat. Inst. Steklov* **65** (212 pp.).

A. V. Malyšev. See also Ju. V. Linnik.

G. B. Mathews (1892). *Theory of numbers.* Deighton Bell, Cambridge. Reprint: Chelsea Pub. Co., New York.

J. Mennicke (1967). On the group of units of ternary quadratic forms with rational coefficients. *Proc. Roy. Soc. Edinburgh* A **67**, 309–352.

A. Meyer (1891). Zur Theorie der indefiniten quadratischen Formen. *J. reine angew. Math.* **108**, 125–139.

J. Meyer (1977). Präsentation der Einheitengruppe der quadratischen Form $F(X) = -X_0^2 + X_1^2 + \ldots + X_n^2$. *Arch. Math.* (*Basel*). **29**, 261–266.

J. Milnor and D. Husemoller (1973). *Symmetric bilinear forms. Ergebnisse der Math.* No. 73. Springer, Berlin, Heidelberg, New York.

H. Minkowski (1885). Untersuchungen über quadratische Formen. Inauguraldissertation, Königsberg. *Acta Math.* **7**, 201–258. (= *Ges. Abh.* Bd. 1, 157–202.)

H. Minkowski (1886). Über positive quadratische Formen. *J. reine angew. Math.* **99**, 1–9. (= *Ges. Abh.* Bd. 1, 149–156.)

H. Minkowski (1887). Zur Theorie der positiven quadratischen Formen. *J. reine angew. Math.* **101**, 196–202. (= *Ges, Abh.* Bd. 1, 212–218).

H. Minkowski (1905). Diskontinuitätsbereich für arithmetische Äquivalenz. *J. reine angew. Math.* **129**, 220–274. (= *Ges. Abh.* Bd. 2, 53–100.)

H. Minkowski (1911). *Gesammelte Abhandlungen* (2 vols). Teubner, Leipzig und Berlin. [Contains many important papers relating to quadratic forms. Only those referred to individually are separately listed.]

L. J. Mordell (1969). *Diophantine equations.* Academic Press, London and New York.

T. Nagell (1922). Über die Klassenzahl imaginär-quadratischer Zahlkörper. *Abh. Math. Sem. Univ. Hamburg* **1**, 140–150

C. Neild and D. Shanks (1974). On the 3-rank of quadratic fields and the Euler product. *Math. Comp.* **28**, 279–291.

H. V. Niemeier (1973). Definite quadratische Formen der Diskriminante 1 und Dimension 24. *J. Number Theory* **5**, 142–178.

A. Ogg (1969). *Modular forms and Dirichlet series.* Benjamin, New York and Amsterdam.

O. T. O'Meara (1958). The integral representation of quadratic forms over local fields. *Amer. J. Math.* **80**, 843–878.

O. T. O'Meara (1963). *Introduction to quadratic forms. Grundlehren der Math. Wiss.* No. 117. Springer, Berlin, Göttingen and Heidelberg.

O. T. O'Meara (1976). Hilbert's eleventh problem: the arithmetic theory of quadratic forms. *Mathematical developments arising from Hilbert problems* (ed. F. E. Browder). *Proceedings of Symposia in Pure Mathematics* **28**, (American Math. Soc.), pp. 379–400.

G. Pall (1945). The arithmetical invariants of quadratic forms. *Bull. Amer. Math. Soc.* **51**, 185–197.

G. Pall (1946). The completion of a problem of Kloosterman. *Amer. J. Math.* **68**, 47–58.

G. Pall (1946a). On generalized quaternions. *Trans. Amer. Math. Soc.* **59**, 280–332.

G. Pall (1948). Composition of binary quadratic forms. *Bull. Amer. Math. Soc.* (2) **54**, 1171–1175.

G. Pall (1949). Representation by quadratic forms. *Canad. J. Math.* **1**, 344–364.

G. Pall. See also A. Ross.

S. J. Patterson (1975). A lattice point problem in hyperbolic space. *Mathematika* **22**, 81–88.

W. Patz (1955). *Tafel der regelmässigen Kettenbrüche und ihrer vollständigen Quotienten für die Quadratwurzeln aus den natürlichen Zahlen von* 1-10000. Akademie-Verlag, Berlin. [A table with the same range but less data was published under a similar title by Becker und Erler, Leipzig, 1941, and reprinted by Edwards Bros., Ann Arbor, Mich. U.S.A., 1946.]

M. Peters (1969). Ternäre und quaternäre quadratische Formen und Quaternionenalgebren. *Acta Arith.* **15**, 329–365.

M. Peters (1973). Quadratische Formen über Zahlringen. *Acta Arith.* **24**, 157–164.

M. Peters (1978). Darstellungen durch definite ternäre quadratische Formen. *Acta Arith.* **34**, 57–80.

H. Pfeuffer (1971). Einklassige Geschlechter totalpositiver quadratischer Formen in totalreellen algebraischen Zahlkörpern. *J. Number Theory* **3**, 371–411.

A. Pfister (1965). Multiplikative quadratische Formen. *Arch. Math. (Basel)*, **16**, 363–370.

A. Pfister (1966). Quadratische Formen in beliebigen Körpern. *Invent. Math.* **1**, 116–132.

A. Pfister (1967). Zur Darstellung definiter Funktionen als Summe von Quadraten. *Invent. Math.* **4**, 229–237.

A. Pfister. See also J. W. S. Cassels.

C. Pommerenke (1959). Über die Gleichverteilung von Gitterpunkten auf m-dimensionalen Ellipsoiden. *Acta Arith.* **5**, 227–257.

S. Raghavan (1975). Bounds for minimal solutions of diophantine equations. *Nachr. Akad. Wiss. Göttingen, II, Math.-Phys. Kl.* **1975** No. 9, 109–114.

K. G. Ramanathan (1951). Theory of units of quadratic and hermitean forms. *Amer. J. Math.* **73**, 233–255.

K. G. Ramanathan (1952). Units of quadratic forms. *Ann. of Math.* (2) **56**, 1–10.

S. Ramanujan (1917). On the expression of a number in the form $ax^2 + by^2 + cz^2 + du^2$. *Proc. Cambridge Philos. Soc.* **19**, 11–21 [= *Collected Papers* (Cambridge University Press, 1927), 169–178. See also notes there on pp. 341–343].

R. A. Rankin (1977). *Modular forms and functions.* Cambridge University Press.

L. Rédei (1953). Bedingtes Artinsches Symbol mit Anwendung in der Klassenkörpertheorie. *Acta Math. Acad. Sci. Hungar.* **4**, 1–29.

L. Rédei (1953a). Die 2-Ringklassengruppe des quadratischen Zahlkörpers und die Theorie der Pellschen Gleichung. *Acta Math. Acad. Sci. Hungar.* **4**, 30–87.

R. Remak (1938). Über die Minkowskische Reduktion der definiten quadratischen Formen. *Compositio Math.* **5**, 368–391.

A. E. Ross (1946). On a problem of Ramanujan. *Amer. J. Math.* **68**, 29–46.

A. E. Ross and G. Pall (1946). An extension of a problem of Kloosterman. *Amer. J. Math.* **68**, 59–65.

H. J. Ryser. See R. H. Bruck.

S. S. Ryškov (1971). K teorii privedenija položitel'nyh kvadratičnyh form. *Dokl. Akad. Nauk SSSR* **198**, 1028–1031.

S. S. Ryškov (1972). O privedenii položitel'nyh kvadratičnyh form ot n peremennyh po Èrmitu, po Minkovskomu i po Venkovy. *Dokl. Akad. Nauk SSSR* **207**, 1054–1056.

S. S. Ryškov (1973). K teorii privedenija položitel'nyh kvadratičnyh form po Èrmitu-Minkovskomu. *Zap. Naučn. Sem. Leningrad. Otdel. Mat. Inst. Steklov* **33**, 37–64.

S. S. Ryškov (1974), Geometrija položitel'nyh kvadratičnyh form. *Proceedings of the International Congress of Mathematicians, Vancouver, 1974*, **1**, 501–506.

I. R. Šafarevič. See also Z. I. Borevič.

W. Scharlau (1969). *Quadratic forms. Queen's papers on pure and applied maths.* No. 22. Queen's University, Kingston, Ontario.

A. Schinzel. See A. Baker.

B. Schoeneberg (1939). Das Verhalten von mehrfachen Thetareihen bei Modulsubstitutionen. *Math. Ann.* **116**, 511–523.

I. Schur (1921). Über die Gauszschen Summen. *Nachr. d.k. Gesell. Göttingen Math.-Phys. Kl.* **1921**, 147–153. (=*Ges. Abh.* (Springer, Berlin, 1973) Bd. 2, 327–333).

A. Selberg (1949). An elementary proof of Dirichlet's theorem about primes in an arithmetic progression. *Ann. of Math.* (2) **50**, 297–304.

J.-P. Serre (1962). *Corps locaux. Actualités Sci. Ind.* **1296**, Hermann, Paris.

J.-P. Serre (1964). Formes bilinéaires symétriques entières à discriminant ± 1. *Séminaire Henri Cartan,* 14ᵉ année, 1961/1962. Exp. 14–15 (16 pp.).

J.-P. Serre (1970). *Cours d'arithmétique. Collection SUP No. 2.* Presses Universitaires de France, Paris. [Translated as *A course in arithmetic. Graduate texts in pure mathematics* No. 7, (1973). Springer, New York, Heidelberg, Berlin.]

D. Shanks (1969). On Gauss's class number problems. *Math. Comp.* **23**, 151–163.

D. Shanks (1971). Class number, a theory of factorization and genera. *Proc. Symposia in Pure Math.* **20** (1969 Institute on Number Theory), 415–440. American Math. Soc.

D. Shanks (1971a). Gauss's ternary form reduction and the 2-Sylow subgroup. *Math. Comp.* **25**, 837–853.

D. Shanks. See also C. Neild.

C. L. Siegel (1934). *Lectures on the analytic theory of quadratic forms.* 3rd revised edition: Robert Peppermüller, Göttingen, 1963.

C. L. Siegel (1935). Über die analytische Theorie der quadratischen Formen. *Ann. of Math.* **36**, 527–606. (= *Ges. Abh.,* Bd. 1, 326–405.)

C. L. Siegel (1935a). Über die Classenzahl quadratischer Körper. *Acta Arith.* **1**, 83–86 (= *Ges. Abh,* Bd. 1, 406, 409).

C. L. Siegel (1937). Über die analytische Theorie der quadratischen Formen, III. *Ann. of Math.* **38**, 212–291 (= *Ges. Abh.,* I, 469–548).

C. L. Siegel (1940). Einheiten quadratischer Formen. *Abh. Math. Sem. Hansischen Univ.* (Hamburg) **13**, 209–239. (= *Ges. Abh.,* Bd. 2, 138–168.)

C. L. Siegel (1941). Equivalence of quadratic forms. *Amer. J. Math.* **63**, 658–680. (= *Ges. Abh.* Bd. 2, 217–239.)

C. L. Siegel (1951). Indefinite quadratische Formen und Funktionentheorie I. *Math. Ann.* **124**, 17–54. (= *Ges. Abh.* Bd. 3, 105–142.)

C. L. Siegel (1959). *Zur Reduktionstheorie quadratischer Formen. Publ. Math. Soc. Japan* **5**. ix + 69 pp. [= *Ges. Abh.* Bd. 3, 275–327, which, however, omits the "durchaus lesenswerte Vorwort".]

C. L. Siegel (1966). *Gesammelte Abhandlungen* (3 vols.). Springer, Berlin, Heidelberg, New York. [Papers which are referred to in the text also listed separately. But this contains many other important papers on quadratic forms.]

C. L. Siegel (1967). *Lectures on quadratic forms. Tata Institute Lectures in Mathematics,* 7, (192 pp.).

C. L. Siegel (1972). Zur Theorie der quadratischen Formen. *Nachr. Akad. Wiss. Göttingen* **1972**, 21–46 [not in the *Ges. Abh.*].

C. L. Siegel (1973). Normen algebraischer Zahlen. *Nachr. Akad. Wiss Göttingen, Math.-Phys. Kl.* **1973**, 197–215 [not in *Ges. Abh.*].

H. J. S. Smith (1859). Report on the theory of numbers. *Report of the British Association* **1859**, 228–267; **1860**, 120–169; **1861**, 292–340; **1862**, 503–526; **1863**, 768–786; **1865**, 322–375. [= *Coll. Math. Papers* Vol. 1, 38–364. This is an invaluable source on early work in the theory of numbers, especially for those who cannot consult the originals in German and other foreign tongues.]

H. J. S. Smith (1867). On the orders and genera of ternary quadratic forms. *Philos. Trans. Roy. Soc. London* **157**, 255–298 (= *Coll. Math. Papers* Vol. 1, 455–509).

H. J. S. Smith (1894). *Collected mathematical papers* (2 vols.). Cambridge University Press. Reprint: Chelsea Pub. Co., New York, 1965.

T. A. Springer (1955). Quadratic forms over fields with discrete valuation. I. Equivalence classes of definite forms. *Nederl. Akad. Wetensch. Proc. Ser.* A **58** (= *Indag. Math.* **17**), 352–362.

T. A. Springer (1956). Quadratic forms over fields with discrete valuation. II. Norms. *Nederl. Akad. Wetensch. Proc. Ser.* A **59** (= *Indag. Math.* **18**), 238–246.

T. A. Springer (1957). Note on quadratic forms over algebraic number fields. *Nederl. Akad. Wetensch. Proc. Ser.* A **60** (= *Indag. Math.* **19**), 39–40.

T. A. Springer (1959). On the equivalence of quadratic forms. *Nederl. Akad. Wetensch. Proc. Ser.* A **62** (= *Indag. Math.* **21**), 241–253.

K. C. Stacey (1975). The enumeration of perfect septenary forms. *J. London Math. Soc.* (2) **10**, 97–104.

K. C. Stacey (1976). The perfect septenary forms with $\Delta_4 = 2$. *J. Austral. Math. Soc.* **22** (Ser. A), 144–164.

H. M. Stark (1967). A complete determination of the complex quadratic fields of class number one. *Mich. Math. J.* **14**, 1–27.

H. M. Stark (1975). The analytic theory of algebraic numbers. *Bull. Amer. Math. Soc.* **81**, 961–972.

P. P. Tammela (1973). K teorii privedenija položitel'nyh kvadratičnyh form. *Dokl. Akad. Nauk SSSR* **209**, 1299–1302.

P. P. Tammela (1975). K teorii privedenija položitel'nyh kvadratičnyh form. *Zap. Naučn. Sem. Leningrad. Otdel. Math. Inst. Steklov* **50**, 6–96.

P. P. Tammela (1977). Oblast' privedenii Minkovskogo dlja položitel'nyh form ot semi peremennyh. *Zap. Naučn. Sem. Leningrad. Otdel. Mat. Inst. Steklov* **67**, 108–143.

V. Tartakovskiĭ [= W. Tartakowsky] (1929). Die Gesamtheit der Zahlen, die durch eine quadratische Form $F(x_1, x_2, \ldots, x_s)$ ($s \geqslant 4$) darstellbar sind. *Izv. Akad. Nauk SSSR* **1929**, 111–122, 165–196. (= *Bull. Acad. Sci. de l'URSS.*)

S. B. Townes (1940). Table of reduced positive quaternary forms. *Ann. of Math.* **41**, 57–58. [Corrigendum: *Math. Rev.* **5** (1944), 141–142 (Chao Ko and S. C. Wang).]

J. V. Uspensky and M. A. Heaslet (1939). *Elementary number theory*. McGraw-Hill, New York and London.

B. L. van der Waerden (1956). Die Reduktionstheorie der positiven quadratischen Formen. *Acta Math.* **96**, 265–309.

B. A. Venkov (1922). Ob arifmetike kvaternionov I, II. *Izv. Rossiĭskoĭ Akad. Nauk* (= *Bull. Acad. Sci. Russie*) (6) **16**, 205–220; 221–246.

B. A. Venkov (1928). O čisle klassov binarnyh kvadratičnyh form otricatel'nyh opredeliteleĭ. *Izv. Akad. Nauk SSSR* (*Otd. Fiz.-Mat.*) **1928**, 375–392; 455–480 (= *Bull. Acad. Sci. de l'URSS*). Translated as Venkov (1931).

B. A. Venkov (1929). Ob arifmetike kvaternionov III, IV, V. *Izv. Akad. Nauk SSSR.* (*Otd. Fiz.-Mat.*) (= *Bull. Acad. Sci. URSS*) **1929**, 489–504; 535–562; 607–622.

B. A. Venkov [= Wenkov] (1931). Über die Klassenzahl positiver binärer quadratischer Formen. *Math. Z.* **33**, 350–374. [A translation of Venkov (1928).]

B. A. Venkov (1940). O privedenii položitel'nyh kvadratičnyh form. *Izv. Akad. Nauk SSSR* (*ser. mat.*) **4**, 37–52.

B. A. Venkov (1945). Ob èkstremal'noĭ probleme Markova dlja neopredel'ennyh troĭničnyh form. *Izv. Akad. Nauk SSSR* (*ser. mat.*) **9**, 429–494.

B. A. Venkov (1970). *Elementary number theory*. Wolters–Noordhoff, Gronigen (Netherlands).

È. B. Vinberg (1972). O gruppah edinic nekotoryh kvadratičnyh form. *Mat. Sbor.* (N.S.) **87** (**129**), 18–36.

È. B. Vinberg (1972a). Ob unimoduljarnyh celočislennyh kvadratičnyh formah. *Funkcional. Anal. i Priloz.* **6**, 24–31.

G. Voronoï (1908). Sur quelques propriétés des formes quadratiques positives parfaites. *J. reine angew. Math.* **133**, 97–178.
H. Wada (1970). Table of ideal class groups in imaginary quadratic fields. *Proc. Japan Acad.* **46**, 401–403.
C. T. C. Wall (1964). On the orthogonal groups of unimodular quadratic forms II. *J. reine angew. Math.* **213**, 122–136.
W. C. Waterhouse (1976). Pairs of quadratic forms. *Invent. Math.* **37**, 157–164.
W. C. Waterhouse (1977). A nonsymmetric Hasse–Minkowski Theorem. *Amer. J. Math.* **99**, 755–759.
G. L. Watson (1954). The representation of integers by positive ternary quadratic forms. *Mathematika* **1**, 104–140.
G. L. Watson (1955). Representation of integers by indefinite quadratic forms. *Mathematika* **2**, 32–38.
G. L. Watson (1957). Bounded representations of integers by quadratic forms. *Mathematika* **4**, 17–24.
G. L. Watson (1960). *Integral quadratic forms. Cambridge tracts in mathematics and mathematical physics.* No. 51. Cambridge University Press.
G. L. Watson (1960a). Quadratic diophantine equations. *Philos. Trans. Roy. Soc. London* A **253**, 227–254.
G. L. Watson (1963). The class-number of a positive quadratic form. *Proc. London Math. Soc.* (3) **13**, 549–576.
G. L. Watson (1975). One-class genera of positive quadratic forms in at least 5 variables. *Acta. Arith.* **26**, 309–327.
G. L. Watson (1976). Regular positive ternary quadratic forms. *J. London Math. Soc.* (2), **13**, 97–102.
G. L. Watson (1976a). The 2-adic density of a quadratic form. *Mathematika* **23**, 94–106.
G. L. Watson. See also B. W. Jones.
A. Weil (1961). *Adeles and algebraic groups.* Lecture notes. Institute for Advanced Study, Princeton.
A. Weil (1962). Sur la théorie des formes quadratiques. *Colloque sur la théorie des groupes algébriques, Bruxelles, 1962,* pp. 9–20. CBRM Librairie Universitaire, Louvain et Gauthier-Villars, Paris.
A. Weil (1965). Sur la formule de Siegel dans la théorie des groupes classiques. *Acta. Math.* **113**, 1–87.
P. J. Weinberger (1973). Real quadratic fields with class number divisible by *n*. *J. Number Theory* **5**, 237–241.
B. A. Wenkov, see Venkov.
E. E. Whitford (1912). *The Pell equation.* Published by the Author, New York.
M. F. Willerding (1948). Determination of all classes of positive quaternary quadratic forms which represent all (positive) integers. *Bull. Amer. Math. Soc.* **54**, 334–337.
E. Witt (1936). Theorie der quadratischen Formen in beliebigen Körpern. *J. reine angew. Math.* **176**, 31–44.
E. Witt (1941). Eine Identität zwischen Modulformen zweiten Grades. *Abh. Math. Sem. Hansischen Univ.* (Hamburg) **14**, 323–337.
E. M. Wright. See also G. H. Hardy.
Y. Yamamoto (1970). On unramified extensions of quadratic number fields. *Osaka J. Math.* **7**, 57–76.

Note on Determinants

The term "determinant" and the notation $d(.)$ are used in the book in several different but related ways in different contexts. For convenience they are summarized below.

(i) Let $f(\mathbf{x}) = \Sigma f_{ij}x_ix_j\,(f_{ji} = f_{ij})$ be a quadratic form over a field k. Then the determinant $d(f)$ of f is given by

$$d(f) = \det(f_{ij}),$$

the right-hand side being the determinant of a square matrix (p. 5). Here $d(f)$ is an element of k.

(ii) The determinant of a k-equivalence class of regular quadratic forms is an element of $k^*/(k^*)^2$, i.e. the set of determinants of all the forms in the class. By "abuse of language" if a form f is being considered as a representative of its class, then its determinant $d(f)$ is taken to be in $k^*/(k^*)^2$ rather than in k: this is done in Chapter 4, for example.

(iii) More generally, let I be a ring in the field k and let U be its group of units. Then the determinant of an I-equivalence class of k-valued regular quadratic forms is an element of k^*/U^2 (p. 7). When a form f is considered as a representative of its I-equivalence class, there may be an "abuse of language" similar to that under (ii).

(iv) When $k = \mathbf{Q}$, $I = \mathbf{Z}$ in the situation considered under (iii), we have $U = \{\pm 1\}$. Hence the determinant of a $\mathbf{Z}$-equivalence class of regular $\mathbf{Q}$-valued forms can be taken to be in $\mathbf{Q}^*$.

(v) Two classes in the same genus have the same determinant in the sense of (iv) (p. 139). Hence the determinant $d(\mathscr{F})$ of a genus $\mathscr{F}$ is taken to be in $\mathbf{Q}^*$.

(vi) The determinant $d(\phi)$ of a quadratic space V, ϕ is defined on p. 13. It is either 0 or an element of $k^*/(k^*)^2$.

In addition to the above, there are the notations $d(\Lambda)$ for the determinant of a lattice in $\mathbf{R}^n$ (p. 72), and $d(\Gamma/\Lambda)$ for the relative determinant of two lattices in the same vector-space (p. 104). The determinant of a square matrix $\mathbf{T}$ and of a vector-space endomorphism σ are denoted respectively by $\det \mathbf{T}$ and $\det(\sigma)$.

Index of Terminology

Index of Notation

This list gives symbols which have a constant meaning in the whole book, or in a substantial portion of it. In some cases, the symbols are used with different meanings in other portions of the book, but it is hoped that this will not cause confusion. The symbols below are in the order (i) latin letters (all founts), (ii) greek letters (all founts), (iii) other symbols such as special uses of brackets.

Note that vectors are usually denoted by bold lower case letters with the convention exemplified by $\mathbf{x} = (x_1, x_2, \ldots, x_n)$. Although the components are written horizontally for convenience, they are treated as column vectors. Matrices are usually denoted by bold capitals with the convention exemplified by $\mathbf{T} = (t_{ij})$. In 2 and 3 dimensions (x, y) and (x, y, z) are regarded as interchangeable with (x_1, x_2) and (x_1, x_2, x_3) respectively.

Symbol	Meaning	Page
$c(f)$, $c_p(f)$	The Hasse–Minkowski invariant of the form f	55
$C(V)$	The Clifford algebra	172
$C_i(V)$ $(i = 0, 1)$	The even and odd Clifford algebras	174
$C(\Lambda)$		182
D	The determinant $d(f)$ of f (Chapter 12). The discriminant $D(f)$ of a binary form (Chapter 14)	
$\mathscr{D}$	The unit disc	304
$d(f)$, $d(\phi)$, $d(\mathscr{C})$	The determinants of the quadratic form f, the quadratic space V, ϕ and the genus $\mathscr{C}$ respectively (see note on p. 403)	
$d(\Lambda)$	The determinant of a lattice in $\mathbf{R}^n$	72
$d(\Lambda/\Gamma)$	The relative determinant	104
$d(\xi, \eta)$	Non-euclidean distance	311
$\det(\sigma)$ $\det(\mathbf{T})$	The determinant of the autometry (or vector space endomorphism) σ and of the square matrix $\mathbf{T}$ respectively	6
$\dim U$	The dimension of the vector space U	
$\mathbf{e}_j$	Usually $\mathbf{e}_1, \ldots, \mathbf{e}_n$ is the basis of a vector space or lattice. Sometimes $\mathbf{e}_j = (0, \ldots, 0, 1, 0, \ldots, 0)$, where the "1" is in the jth place	
$\mathrm{e}(J)$		173